ONCOGENE PROTEINS: STRUCTURE, FUNCTIONS AND ANALYSES

Oncogene Proteins: Structure, Functions and Analyses

Peter B. Murphy and Jason R. Clarke
Editors

Nova Biomedical Books
New York

For permission to use material from this book please contact us:
Telephone 631-231-7269; Fax 631-231-8175
Web Site: http://www.novapublishers.com

NOTICE TO THE READER

Library of Congress Cataloging-in-Publication Data
Oncogene proteins. Structure, functions, and analyses / Peter B. Murphy and Jason R. Clarke (editors).
p. ; cm.
Includes bibliographical references and index.
ISBN 978-1-60456-259-0 (hardcover)
1. Tumor proteins. I. Murphy, Peter B. II. Clarke, Jason R. III. Title: Structure, functions, and analyses.
[DNLM: 1. Oncogene Proteins. QZ 202 O5416 2008]
RC254.6.O53 2008
616.99'4071--dc22

2007050813

Published by Nova Science Publishers, Inc. ≃ New York

Contents

Preface

Oncogene proteins are proteins coded by oncogenes. They include proteins resulting from the fusion of an oncogene and another gene.

This new book presents the latest research in the field from around the world.

Short Communication A - Metastases are the main cause of death in human solid cancers. Large efforts have been made to identify the genetic and biochemical characteristics related to tumour metastatic capabilities. Gene expression profile analysis, using DNA microarrays, has identified various gene signatures indicative of tumour aggressiveness, metastatic potential and organ tropism. Cancer-initiating cells have also recently emerged as being determinants of metastasis. In this review, the authors will attempt to evaluate the progress made in the field and define future directions, with particular focus on a subset of pathways and embryonic genes associated with malignant conversion, cancer cell and stem cell migration. The potential role of these genes in the functions of cancer-initiating cells and thereby their contribution to the metastatic potential of tumours will be discussed.

Short Communication B - Prostate cancer is the most frequent diagnosed cancer among men. Knowledge of genetic changes underlying the initiation, development, and progression of prostate cancer is accumulating rapidly. Molecular fingerprinting may help to distinguish indolent from aggressive prostate cancer. The crucial role of molecules participating in complex signal transduction networks makes them appealing targets. Novel strategies are developing for the treatment of prostate cancer by targeting multiple tumorigenic signaling elements. Here, the authors accumulate the remarks of relevant reviews that where published this year.

Chapter I - Postmortem and functional imaging studies have reported morphological changes in the brains of untreated depressed patients, and evidence has associated this with the loss or atrophy of glial and neuronal cells. In animal studies, the antidepressant drugs can reverse the stress-induced cellular loss by increasing neurogenesis in hippocampus. Therefore, investigation of the biochemistry of these changes could provide insight for a better understanding and more effective medication in mood disorders. Studies of protein and phospholipid kinases, some of them konwn for their role in cell transformation, were undertaken. Postmortem brains of subjects deceased by suicide or not, with or without documented major depression, were used. Proteins levels were also measured in cell culture, after drug activation, along with cell proliferation and cell survival.

Enzymatic activities and protein levels were measured for the following proteins: Akt-1, an homolog to the viral oncogen (v-Akt), with enzyme activity which facilitates growth factor-mediated cell survival. Akt-1 also regulates the activity of GSK-3ß, a protein kinase involved in negative control of IGF and other growth factors. GSK-3ß is both directly and indirectly inhibited by lithium, a mood stabilizer drug. The other proteins consisted in PTEN and PI3K. The latter is a lipid kinase which phosphorylates the PtdIns-3,4,5-P and involved in cell proliferation. Opposed to its action, is the phosphatase PTEN, a tumor suppressor implicated in a variety of sporadic cancers, which acts by dephosphorylating the PtdIns-3,4,5-P. With respect to Akt and GSK-3ß, results indicated no change either in their protein levels or in lithium-inhibitable total GSK3□/ß activity in human prefrontal cortex (PFC). The activity of Akt-1 decreased (ANOVA: $P = 0.003$), while GSK-3ß activity increased (ANOVA $P = 0.002$) in depressed subjects. The study of PI3K and PTEN proteins indicated decrease in activity of PI3K and increase in PTEN levels in the same samples. Taken together, the authors data indicated that the signaling pathway through PI3K/GSK-3ß is blunted in depression. In cell culture experiments, the study showed activation of protein kinases by mood stabilizers (Lithium, Valproate), with a potentiation of cell proliferation and survival. These drugs induced an increase in Akt mRNA and protein levels to support potentiation of cell proliferation. In conclusion, these studies suggest a role of protein kinases and phosphatases, mostly known as oncogens, acting as downstream effectors of neurotransmitters and growth factors, in psychopathology and are supportive of the neuroplasticity hypothesis in affective disorders.

Chapter II - Immunotherapy against cancers is a strategy to eradicate residual tumor cells by means of the tumor-specific cytotoxic T lymphocytes (CTLs) or tumor-related antigens specific monoclonal antibodies (moAb), in tumor-bearing individuals. This new proceeding may be a promising alternative to chemo-radiotherapy which intrinsically embraces organ toxicities. An ideal target of cancer immunotherapy is a protein which is exclusively and highly produced by tumor cells, but not by normal cells. Furthermore, the protein, which is closely relevant to tumorigenesis or biological property of tumor itself, is more desirable. From these points of view, oncogene proteins potentially can be ideal candidates for cancer immunotherapy. For example, immunogenicities of WT1, BCR-ABL, those are well-known oncogene proteins of leukemias, and several phase I/II clinical trials of vaccination by using peptides derived from those proteins as anticancer vaccines have already been conducted. Although excellent clinical trials and elaborative basic researches have been accumulating the lines of evidence to show substantial therapeutic benefits, several limitations of cancer vaccine strategy itself have been simultaneously uncovered. Those drawbacks of cancer immunotherapy comprise the invalidation of tumor-specific CTLs by regulatory T-lymphocytes and myeloid derived suppressor cells, and the immune evasion of tumor cells by down-modulation of surface HLA class I molecules. In addition, deterioration of intracellular antigen-processing machinery, and epigenetic alteration of target antigen as immunoediting have been also demonstated. In this chapter, focusing on three representative oncogene proteins ,WT1, BCR-ABL, HER2/neu, the authors are reviewing the present state of cancer immunotherapy, especially the vaccine strategy to generate tumor-specific CTLs in tumor-bearing individuals, with describing some of their recent works.

Chapter III - Aberrant intracellular signaling resulting from activation of oncogenic proteins, inactivation of tumor suppressors, and gene amplification of critical proteins are a key feature of human cancer. Phosphoinositol-3-kinase (PI3K)/Akt/mammalian target of rapamycin (mTOR) pathway is frequently dysregulated in many human malignancies and impacts a numbers of cellular processes including cell growth, survival, angiogenesis, migration, and cellular metabolism by mediating extracellular (mitogenic growth factor and stress) and intracellular (altered tyrosine receptor kinases and ras) signaling. Great progress has been made in understanding the regulation and roles of PI3K/Akt/mTOR signaling over the past ten years, creating an opportunity to manipulate this pathway therapeutically for cancer treatment. In this chapter, the authors discuss PI3K/Akt/mTOR pathway components and functions in tumorigenesis and tumor progression, as well as its alteration in cancer and relationship with different therapeutic approaches. Moreover, they summarize the current status of a variety of pharmacological agents that target this pathway in preclinical and clinical studies.

Chapter IV – In the last decade the role of dietary fats in cancer-related diseases has attracted considerable interest within the community. First reports of the beneficial role of ω-3 long-chain polyunsaturated fatty acids (ω-3 PUFA) in human health occurred in the mid-1970s with observations that the decrease in prevalence of cardiovascular diseases and cancer correlated with dietary intake of ω-3 PUFA. Current research interest is focused on the protective effects of oily fish-derived ω-3 PUFA, mostly eicosapentaenoic (EPA) and docosahexaenoic acid (DHA) that counteract the pro-carcinogenic effects of animal meat-derived ω-6 PUFA, typified by arachidonic acid. Both epidemiologic and experimental findings indicate that ω-3 PUFA, almost absent from typical Western diets today, exert protective effects against cancer progression. Although beneficial effects of ω-3 PUFA have been widely documented, the current and future use of these PUFA, as preventive or complementary treatments in cancer and other diseases, is hampered by the lack of mechanistic information. The ω-3 PUFA have been shown to be less potent than ω-6 PUFA in activation of important oncogenes such as protein kinase C, transcription factors *ras* and AP1 and could differentially modulate apoptosis mediated by NFκB and Bcl-2. It is now believed that multiple mechanisms contribute to the observed biological effects of ω-3 PUFA. PUFA were able to modulate cancer regulation through their roles as structural phospholipids of the cell membrane, affecting membrane fluidity, cellular signaling and cellular interactions. PUFA also exhibit numerous functional properties, either directly or after conversion to their respective lipid mediators. Both ω-6 and ω-3 PUFA compete for the same bio-conversion enzymes: cyclooxygenase (COX), lipoxygenase (LOX) and epoxygenase (CYP) enzymes to form different classes of eicosanoids (prostaglandins, thromboxanes, leukotrienes and epoxides). Consequently, when the supply of ω-3 PUFA is increased relative to ω-6 PUFA, the synthesis of the corresponding eicosanoids is also increased. These ω-3 PUFA-derived eicosanoids have generally been shown to be less inflammatory and to be weaker cancer-promoters than their ω-6 counterparts. One of the mechanisms by which ω-3 PUFA could play a significant role in carcinogenesis is in the inhibition of new tumor blood vessel formation, or angiogenesis, which is an important process affecting cancer growth and dissemination. Angiogenesis involves complex interplay between tumour and endothelial cells that culminates in proliferation, migration, and differentiation of new endothelium. This review is focused on a comprehensive analysis of published research studies aimed at characterization of the potential mechanisms by which ω-3 PUFA influence angiogenesis.

Chapter V - The IAP family member Survivin, is overexpressed in most cancers, with low or no expression in most differentiated adult tissues, making it an attractive target for cancer treatment. Survivin expression is a poor prognostic marker in a number of cancers. Clinical studies are progressing evaluating targeted Survivin knockdown using anti-sense oligonucleotides, immunotherapy using Survivin primed dentritic cells and disruption of Survivin stability by antagonism of Survivin interaction with Hsp90. Strategies based upon ribozymes directed against Survivin mRNA, or dominant-negative cDNA to block Survivin function are in pre-clinical stages. Survivin, like many genes, is alternately spliced and a number of new splice variants have recently been identified. In some cancers, expression of Survivin splice variants correlates with steroid receptor loss and p53, suggesting that like the wild-type form, some Survivin splice variants may have prognostic relevance. This chapter will review the current understanding of the function of Survivin splice variants and their expression and sub-cellular localization in neoplastic tissues, as well as critically evaluate the potential effect of current Survivin directed therapies on alternatively spliced Survivin isoforms.

Chapter VI - In metazoans including human, the mitochondrial pathway of apoptosis depends on orchestrated interactions between Bcl-2 and a number of closely related proteins. The Bcl-2 family of proteins is composed of pro- and anti-apoptotic members that regulate mitochondrial outer membrane permeabilization, a pivotal step in programmed cell death. Death inducers (e.g. Bax, Bak, BH3-only proteins) promote cytochrome c release from mitochondria, leading to the activation of intracellular cysteine proteases termed caspases that mediate cell demise. Conversely, pro-survival members such as Bcl-2 or Bcl-xL decrease cell susceptibility to apoptosis by preventing the activation of Bax-like or BH3-only proteins. Overexpression of anti-apoptotic proteins commonly occur in human tumors, and can contribute not only to disease development and progression but also to resistance to chemotherapy. Therefore, pro-survival Bcl-2 family members represent attractive targets for the development of novel anticancer drugs. On the other hand, alternative strategies are also emerging that use cytoprotective agents targeting pro-apoptotic Bcl-2 family proteins to treat disorders associated with excess apoptosis (e.g. stroke or neurodegenerative diseases). To date, three-dimensional structures of about a dozen Bcl-2 family members have been determined, contributing to a better understanding of Bcl-2 family proteins and to the design of selective inhibitors. After a brief overview of the apoptosis pathways, the authors review the recent small-molecule drugs targeting Bcl-2 family proteins with an emphasis on those discovered through fragment-based approaches. The authors also examine the literature on potential therapeutic use of non-BH3, biologically active fragments derived from Bcl-2 family proteins, which has begun to emerge in addition to BH3 mimetics.

Chapter VII - AKT/PKB is a serine/threonine kinase which represents a central node of the PI3K signaling, one of critical pathways contributing to pathogenesis of diverse cancers and diseases. A variety of extracellular stimuli such as growth factors, hormones, and cytokines activate PI3K, resulting in phosphorylation and activation of AKT. AKT in turn phosphorylates downstream substrates, *e.g.* p21, p27, MDM2, GSK3, BAD, Caspase-9, FOXO, mTOR, and TSC1/2, thereby exerting various cellular processes including cell growth, proliferation, survival, angiogenesis, migration and metabolism. In a past decade, enormous number of improvements regarding this oncogenic pathway have been achieved including the identification of direct AKT substrates and their downstream components as well as the development of targeted therapeutic small molecules. Nonetheless, novel intriguing aspects of the AKT pathway have been unraveled more recently. Accumulative evidence now suggest that subcellular compartmentalization of AKT is involved in carcinogenesis through nuclear-cytoplasmic specific networks. In endocrine-related cancers, antitumor action by hormonal agents and acquired hormone resistance are both likely to be mediated by the AKT pathway. The crosstalk between Ca^{2+} signaling and the AKT pathway has been demonstrated in cardiomyocytes and other cell types. This review highlights and discusses these lines of emerging evidence regarding the AKT signaling and their therapeutic significance.

Chapter VIII - Fes (also known as Fps), together with Fer, form the nonreceptor protein tyrosine kinase subfamily. The expression of Fes in normal tissues has been shown to be limited to monocytic hematopoietic cells, endothelial cells, and some neuronal and epithelial cells. Studies using mice expressing activated mutant Fes or cultured endothelial cells indicated that Fes might be involved in pathological angiogenesis, suggesting it could be a potential target for anti-angiogenic therapy. Fes was initially identified as a protooncogene product. However, recent studies have indicated that Fes may act also as a tumor suppressor protein in some carcinomas, such as colon cancer, based on the identification of inactivating mutated Fes in colon cancer tissues and downregulated expression of Fes in some cancer cells. Thus, during treatment of these cancers, inhibition of Fes activity may stimulate progression of the pathological process. In this short communication, the authors will review the recent reports on the roles of Fes in tumor progression and discuss the problems that must be solved before targeting of this kinase in any anti-cancer therapy.

Chapter IX - The human epidermal growth factor receptor 2 (HER2, also known as c-neu or erbB-2) is an important member of the epidermal growth factor receptor (EGFR) family. As a preferred heterodimer partner for other members of the EGFR family, HER2 plays a critical role in signaling that generates a wide variety of cellular responses in both normal and malignant cells. HER2 is amplified/overexpressed in a number of cancers, including breast, ovarian, gastric, colon, and non-small cell lung carcinomas. HER2 amplification/overexpression is associated with poor prognosis largely due to aggressive tumor growth, early metastasis and resistance to chemotherapy. HER2 is not only a prognostic marker for patients with cancers that overexpress HER2, but is also a predictive marker for response to treatment. A humanized monoclonal antibody trastuzumab (Herceptin®) has been developed from the murine anti-HER2 monoclonal antibody 4D5 and has become the first-line therapy for patients with breast cancers that overexpress HER2. In this review, the authors have updated information regarding molecules and pathways that are influenced by HER2 signaling. Structure-function analysis indicates that HER2 is constitutively activated and prepared to heterodimerize with HER3 and EGFR. MUC4 levels can modulate binding of trastuzumab and susceptibility to treatment. Affymetrix microarray analysis has identified a number of new candidates for targets regulated by HER2. A number of HER2-regulated genes have also been implicated in G1 cell cycle arrest (p27Kip1, Jab1, Mirk, hKIS, AKT, CDK2, cyclin D1 and cyclin G2), non-receptor kinase signaling (BRK), estrogen responsiveness (ERRa), apoptosis (Bim), motility ($alpha_6beta_4$ integrin, MEMO, S100A4), invasion (CXCR4, RECK) and angiogenesis (VEGF, IL-8, TSP1). Independent activation of one or more of these targets may contribute to resistance to trastuzumab. Targeting both HER2 and one or more of these HER2 regulated molecules might provide greater anti-tumor activity in cancers that overexpress the receptor.

Chapter X - The human kinome constitutes one of the most promising targets for the development of novel chemotherapeutic drugs. Deregulation of many of the 518 members of the human kinome is associated with inflammation, neurodegenerative diseases and cancer. The several drugs that have reached clinical trials and other compounds, such as imatinib (Gleevec) and gefitinib (Iressa), that are already in clinical use as chemotherapeutic drugs, proof that the inhibition of malfunctioning tyrosine and serine-threonine kinases is a valid approach to the treatment of numerous diseases. Several small molecule inhibitors that selectively target the ATP binding site of Aurora kinases have been developed in recent years. Aurora A function is important for the proper progression through mitosis as it regulates mitotic checkpoint proteins. Aurora B plays an essential role in chromosome segregation and bipolar chromosome orientation and condensation while Aurora C cooperates with Aurora B to regulate mitotic chromosome dynamics in mammalian cells. This work describes the unique structural characteristics of Aurora kinases that have conducted to the development of novel antioncogenic agents. It also discusses the potentiality of dual-specificity inhibitors to interfere with the function of Aurora kinases through the disruption of specific protein-protein interactions.

Chapter XI - Oral cancer is the seventh most common malignancy, worldwide. It constitutes a major health problem because it is still associated with poor prognosis and about 50% five-year survival, despite advances in surgery, radiation and chemotherapy treatment over the past two decades. Oral squamous cell carcinoma (OSCC) accounts for more than 90% of oral cancer cases. In the development of OSCC have been implicated environmental factors such as tobacco smoking and heavy alcohol consumption, infectious agents such as some types of human papilloma virus, genetic alterations involving oncogenes and tumor suppressor genes, as well as epigenetic changes. Recently, some epidemiological studies have also incriminated diabetes mellitus as a risk factor for the development of oral premalignant lesions and OSCC.

In order to investigate the possible association of diabetes with increased risk for oral cancer, as well as the underlying molecular mechanism of this association an experimental model of OSCC induced in diabetic and normal Sprauge–Dawley rats was established. Diabetes type I was induced by a single intraperitoneal injection of streptozotocin in 20 rats and verified by measuring glucose levels in blood samples obtained by tail prick in a strip-operated sensor. In 14 of the diabetic rats, as well as in 12 rats without diabetes, chemical carcinogenesis was induced by a five-month application of the carcinogen 4-nitroquinoline N-oxide on the hard palate of the animals. Six diabetic rats as well as 6 non-diabetic rats were used as controls of normal oral tissue without carcinogenesis. The four groups of animals were used in order to study the putative effect of diabetes on signal transduction pathways involved in the sequential stages of oral carcinogenesis: normal mucosa, hyperplasia, dysplasia, early invasion, well differentiated OSCC and moderately differentiated OSCC.

The diabetic rat model was used to investigate whether the molecular basis for the putative association of diabetes with oral oncogenesis involves the insulin receptor and its substrate-1 (IR and IRS-1, respectively), as well as the influence of diabetes on signal transduction pathways involving oncogene proteins EGFR, erbB2, erbB3, FGFR-2, FGFR-3, FAK, c-myc, N-ras, ets-1, H-ras, c-fos and c-jun, tumor suppressor gene proteins p53 and p16, apoptosis-related markers Bax and Bcl-2, and cell proliferation-related marker Ki-67. The obtained findings suggest that the insulin and insulin-like growth factor-1/IR/IRS-1 signalling pathway is altered by diabetes, leading to reduced cell adhesion and possibly increased risk for oral cancer. In addition, diabetes seems to promote the activation of Ras/Raf/MAPK signal transduction pathway, mainly by induction of receptors erbB2 and erbB3, resulting in increased cell proliferation. Finally, diabetes does not seem to influence apoptotic mechanisms during oral carcinogenesis. An interesting extrapolation of these findings is that the type I diabetic patients who are not insulin-regulated may have a greater risk for developing oral cancer than the general population does. This hypothesis remains to be examined by large epidemiological studies.

Chapter XII - Apoptotic cell death plays a key role in the pathogenesis, aggressiveness and therapy response of cancer. Besides pathways that accelerate cell division, a disruption of the cell death pathway may contribute to selective expansion of malignant clones. The proteins clusterin and bcl-2 have been reported to present a powerful antiapoptotic activity in some systems. Their implications in carcinogenesis and the progression of some carcinomas make clusterin and bcl-2 interesting genes worthy of investigation. Resistance to anticancer chemotherapeutic drugs remains a major obstacle in cancer chemotherapy and novel therapeutic strategies that target the molecular basis of chemo-resistance are required. The response of cytotoxic drugs is modulated by pro and anti-apoptotic proteins, and defects in apoptosis pathways or the activation of antiapoptotic mechanisms may confer resistance to cytotoxic drug treatment. Combined treatment with clusterin or bcl-2 inhibitors and chemotherapeutic drugs has been shown to be a novel and attractive strategy to inhibit the progression of some human carcinomas.

Chapter XIII - Oncogene protein is an important group of protein. It strongly related to the oncogenesis mechanism. Application of proteomics techniques to oncogene protein research can help finding new knowledge. Basic biocheminformatics applications for oncogene protein research included genomics, proteomics, transcriptomics, immunomics, metabolomics, cytomics and systomics. Expression analysis of the oncogene protein can also be performed by the in vivo, in vitro and in silico techniques. In addition, the application of other new biotechnology including nanomedicine technology can help driven the difficult oncogene protein research to the new easier generation.

In: Oncogene Proteins: Structure, Functions and Analyses ISBN: 978-1-60456-259-0
Editors: P.B. Murphy and J.R. Clarke, pp. 1-9

Short Communication A

Early Determinants of Metastatic Potential

Stéphane Ansieau[1,2,*], Jérémy Bastid[1,2,3,4] and Alain Puisieux[1,2,3,4]

[1] Inserm U590, Lyon, F-69008, France
[2] Centre Léon Bérard, Lyon, F-69008, France
[3] Université de Lyon, Lyon, F-69003, France
[4] Université Lyon 1, ISPB, Lyon, F-69003, France

Abstract

Metastases are the main cause of death in human solid cancers. Large efforts have been made to identify the genetic and biochemical characteristics related to tumour metastatic capabilities. Gene expression profile analysis, using DNA microarrays, has identified various gene signatures indicative of tumour aggressiveness, metastatic potential and organ tropism. Cancer-initiating cells have also recently emerged as being determinants of metastasis. In this review, we will attempt to evaluate the progress made in the field and define future directions, with particular focus on a subset of pathways and embryonic genes associated with malignant conversion, cancer cell and stem cell migration. The potential role of these genes in the functions of cancer-initiating cells and thereby their contribution to the metastatic potential of tumours will be discussed.

Metastases are the main cause of death by cancer. Defining the genetic determinants of the metastatic potential of tumours is therefore one of the major challenges in cancer research. Tumour progression was once defined as a multistep Darwinian selection of cells harbouring genetic and epigenetic alterations that give them resistance to growth control, immune attacks and environmental restraints. In this linear definition of cancer progression, the emergence of metastatic cells was generally described as a late event. Nowadays, gene expression profile

[*] E-mail address: ansieau@lyon.fnclcc.fr. Tel: 0033-478785160, Fax : 0033-478782821. Corresponding author: Stéphane Ansieau, Inserm U590, Centre Léon Bérard, 28 rue Laennec, F-69008 Lyon, France.

analysis and the likely characterization of early premetastatic cells have suggested a completely different hypothesis. It is now believed that cancer cell dissemination starts early, at least in some tumours, from the premalignant lesion, and premetastatic cells remain dormant until secondary events initiate their proliferation in the novel environment. As a consequence, metastatic cells should genetically evolve independently of the primary tumour. The cellular heterogeneity of tumours and the plasticity of tumour cells also led to the hypothesis that cancer cells might originally derive from stem cells. Dormancy and growth in appropriate niches, in particular, are characteristics common to all hypotheses. In this review, we will attempt to match the two models, to describe recent advances in the understanding of genetic events associated with tumour aggressiveness and to evaluate their interest for further therapies.

Determination of Metastatic Potential

Over the past decades, huge efforts have been provided in an attempt to identify the genes implied in tumour aggressiveness and metastasis using gene expression profile comparisons. Up to now, seven independent gene signatures related to various tumour features have been described. They have been obtained either by comparing the gene expression profiles of aggressive and less aggressive tumours ("poor-prognosis gene signature")[1,2], of tumorigenic breast-cancer cells and normal breast epithelial cells ("invasiveness gene signature")[3], or altered pathways ("oncogenic pathway signature")[4], but also by analysing the gene expression pattern of serum-stimulated fibroblasts in culture mimicking fibroblasts in wounded tissues ("wound-response")[5], and hypoxia-induced genes ("hypoxia signatures")[6]. In addition, the comparison of breast tumour subtypes and the determination of gene profiles associated with recurrence of oestrogen receptor-positive breast cancers have identified "intrinsic subtype"[7] and "recurrence score"[8] signatures, respectively. An examination of the functions of these genes has led to the identification of proteins implicated in various stages of tumour dissemination; these proteins could be organized in three independent groups: proteins promoting cell motility, invasion and angiogenesis (metastatic initiation group), proteins performing limited functions in the primary tumour and other specific functions in metastatic spread (metastatic progression group), and proteins specifically required for secondary tumour growth (metastatic virulence group)[9]. Although of particular interest, these analyses barely highlight a complex network of proteins favouring tumour progression and dissemination rather than pin-point the key genetic determinants of metastasis. Combining the gene signatures does not further restrict the number of genes implicated in metastases; results of a statistical analysis suggest that they represent different views of the same process [10]. A recent study comparing the gene profiles of various single cell progenies derived from the pleural effusion of a breast cancer patient who relapsed years after removal of the primary tumour identified determinants of metastatic potential and tropism. Four of these genes were recently shown to be essential for tumour vascularisation and metastatic progression [11]. Nevertheless, the use of various shRNA combinations either specifically affects metastasis or additionally inhibits primary tumour growth. An associated growth advantage would not be compatible with the restricted population of metastatic competent cells in primary tumours. As several of these poor prognosis and premetastatic predisposition signatures are common to various cancers, they might represent genes expressed by stromal cells, including vasculature, connective tissue or immune cells, rather than by the tumour cells themselves. To distinguish between these two possibilities, an immunohistochemical expression analysis at the cellular level in tumour sections will be required. As metastases are clonal in origin, such studies will also help further determine whether concomitant expression of several, if not all, of these markers is required for metastatic spread.

As expected, loss of function is also associated with metastases. Several metastatic suppressors have been identified, with functions related to cell signalling (MKK4, NM23, RHOGD12), cell adhesion (KAI-1) and cell-to-cell communication (BRMS1). In most of the cases, the expression of these genes has no impact on the growth of primary tumour cells but specifically affects their metastatic ability [12].

There is accumulating evidence that host genetics also influence tumour metastasis. For example, polymorphisms affecting the Ras-GAP encoding gene Sipa1 strongly affect the metastatic dissemination of polyoma middleT (PyMT) transgene-induced mammary cells in mice [13]. Comparison of metastatic gene signatures with the quantitative trait loci suggests that differential metastatic gene expression is linked to polymorphisms [14], a hypothesis in agreement with mathematical models which exclude that cancer cell spread results from genetic alterations[15].

Emerging evidence also suggests that successful metastasis development implies an intimate cross-talk between the tumour and its environment. Recently, the laboratory of R.A. Weinberg elegantly demonstrated, by subcutaneous injection of breast cancer cell lines and mesenchymal stem cells (MCS) into *nude* mice, that the recruitment of MCS to the primary tumour enhances their metastatic potential by secretion of the CCL5 chemokine, favouring extravasation in the lung [16]. Accordingly, metastasis signatures include genes with stromal origin [17,18]. Besides, stromal cells have a direct impact on malignant potential; fully transformed breast cancer cells can be reversed to normal phenotype by contact with a non-permissive stroma [19]. Cancer cells and associated stroma-secreted chemokines are known to control the proliferation, the differentiation and the migration of the vascular cell network required for the establishment of a correct niche (named the premetastatic niche) capable to retain migrating cancer cells and to favour their propagation in their novel environment. Indeed, bone-marrow VEGFR1-expressing hematopoietic precursor cells and VEGFR2-expressing endothelial precursor cells are recruited to the premetastatic niche to remodel the matrix, favour tumour cell homing and locally promote neoangiogenesis. A unique array of tumour chemokines is suggested to specifically dictate the pattern of tumour spread (for review see [20]).

Taken together, these parameters suggest that the eradication of metastases can probably be achieved in different ways. Nevertheless, as metastases are clonal in origin and can, as shown in at least one patient, derive from single cells with independent behaviour and various tropisms [21], the only way to definitively eradicate the risk of metastasis dissemination might be to eliminate metastasis-initiating cells, provided that the existence of such cells is confirmed (see below).

Characterization of Metastasis-Initiating Cancer Cells

A large scale analysis has achieved further significant progress in the understanding of the requirements for tumour progression and dissemination and the consequences of gene expression deregulation. However, the question of whether dissemination occurs early or late is still debated. Determining whether metastatic cells develop independently of the primary tumour is of particular importance for developing future therapies. Cytokeratin-positive cells from the bone marrow of breast cancer patients which differed from normal cells only by subchromosomal DNA losses and, in some cases, by limited gene amplifications have been characterized. Results suggest that these cells might be representative of the earliest stages of metastasis and derive form premalignant tissues [21]. Lack of chromosomal instability, in particular, suggests that cell dissemination could even start before the telomere crisis. This model does not exclude that other metastatic cells in other cancer types might emerge later during tumour progression. In prostate cancers, for instance, cells with potential metastatic capabilities are detected at the invasive front of the primary tumour [22]. Further characterization of these cells is required to evaluate their interest for future therapies. Conversely, their detection is of limited interest for prognosis. Although disseminated cells in the bone marrow or the lymph nodes may be prognostic indicators of metastasis, they will not necessarily evolve in metastatic competent cells and their microenvironment and genetic background will not necessarily be compatible with metastasis dissemination. Accordingly, only 30-50% of patients with micrometastatic disease in their bone marrow will develop evident metastasis in the next five years [23].

Do Metastasis-Initiating Cells Display Stem Cell Properties?

The terms “cancer stem cell” or “cancer initiating cells” describe cancers cells that display proliferative capabilities but also the property to undergo differentiation to the phenotypically diverse cell populations that constitute the tumour. The origin of these cells remains a matter of debate. They either derive from adult stem cells or from more differentiated cells that have acquired dedifferentiation and self-renewing properties following oncogenic mutations. Now, several lines of evidence clearly establish that leukemias arise from mutations of normal haematopoietic progenitor cells [24-26]. Whether these haematopoietic progenitors are the only haematopoietic cells whose lifespan is compatible with the accumulation of mutations required for disease development, yet remains to be determined. Several solid cancers have now been proposed to derive from cancer stem cells. The most likely example is germinal teratocarcinoma, a tumour occurring mainly in young adults, at any site along the migration pathway of germinal cells from the brain to the gonads, during development [27]. It is also well accepted that neuroblastoma arises from the embryonic neural crest [28]. The pluripotency of neural crest cells is modified during migration to their final destination [29]. According to the differentiation stage at which the oncogenic mutations occur, neuroblastoma is either anaplastic and aggressive or more differentiated and classified as benign ganglioneuroma [28]. Cancer-initiating cells have been isolated from colon, pancreas, and breast cancers on the basis of the expression of markers whose specificity for stem cells largely remains to be demonstrated [22,30,31]. $CD133^+$ cells have been defined as a unique subpopulation of colon and pancreas cancer cells having tumorigenic properties in *nude* mice and differentiation capabilities *in vitro.* Unlike colon and pancreas cancers in which tumour progression is mostly dependent on a single pathway, no single dominant pathway has emerged in breast tumours. In addition, two types of luminal-restricted and one type of myoepithelial-restricted progenitors have been identified [32,33]. Nevertheless, all human mammary tumour-initiating cells yet isolated from patients display a luminal phenotype, raising the possibility that all breast tumour-initiating cells derive from the corresponding stem cells [31], a hypothesis largely supported by the large size of the progenitor cell pool in the luminal cell compartment[34]. Interestingly, a recent observation indicates that a subpopulation of pancreatic cancer-initiating cells displays metastatic capabilities. As indicated by the expression of the CXCR4 receptor, these cells are found specifically located at the invasive front of the primary tumour [22]. The plasticity of metastatic cells could explain both the clonality and heterogeneity of metastases. It also constitutes ground for a common requirement of specific niches for the homing and proliferation of both metastatic and stem cells. Whether dormant single cells also display stem cell properties remains to be elucidated.

Transdifferentiation, an Essential Event in Tumour Progression?

Cancers have been proposed to originate from alterations in stem cells. Another alternative is possible, namely that some oncogenic stresses can induce genomic reprogramming to promote cell dedifferentiation and acquisition of plasticity properties. Various examples clearly refute the dogma that differentiation is a unidirectional, irreversible mechanism. For instance, differentiated cells can be recruited to produce new cell types in regenerating amphibian eye and limb [35,36]. Recently, hypoxia was found to influence phenotypic heterogeneity by altering the expression of specific genes. Hypoxic neuroblastoma and breast cancer cells loose their differentiated gene expression pattern and develop stem cell-like phenotypes [37]. As low differentiation stage is associated with poor prognosis in neuroblastoma and breast cancers, hypoxia-induced dedifferentiation could contribute to tumour heterogeneity and aggressiveness. Hypoxia has been found to induce the expression of *Snail* and *Slug*, two well documented markers of NC precursor cells [38-40]. *Twist* is also a NC precursor marker. Interestingly, all three genes have been associated with cancer progression. In particular, Twist has been shown to counteract N-Myc anti-apoptotic properties and to contribute to neuroblastoma cancer progression [41]. We recently demonstrated that Twist could also inhibit premature senescence, suggesting that Twist expression plays a major role in failsafe program escape during the transition from the benign to the malignant stage. Interestingly, in epithelial cells, the Twist-associated failsafe program escape concomitantly induces epithelio-mesenchymal transition (EMT), a process promoting both motility and invasiveness (Ansieau et al., submitted). Although yet to be demonstrated, a similar link between early failsafe program escape and EMT might be mediated by Snail and Slug, in link with their anti-apoptotic properties. As Twist expression is associated with potential induction of stem cell markers (R.A. Weinberg, oral communication, CNIO 2007; A-P. Morel and A. Puisieux, unpublished data), one could assume that the selective pressure to bypass failsafe programs in premalignant tissues impacts both cell plasticity and invasiveness. Extending this hypothesis from neuroblastoma to other cancer cell types might also be possible, as a recent hypothesis suggests that neural crest cells might be at the origin of all adult stem cells [42,43]. Such a hypothesis might explain the pluripotency and the premalignant genotype of metastatic cells.

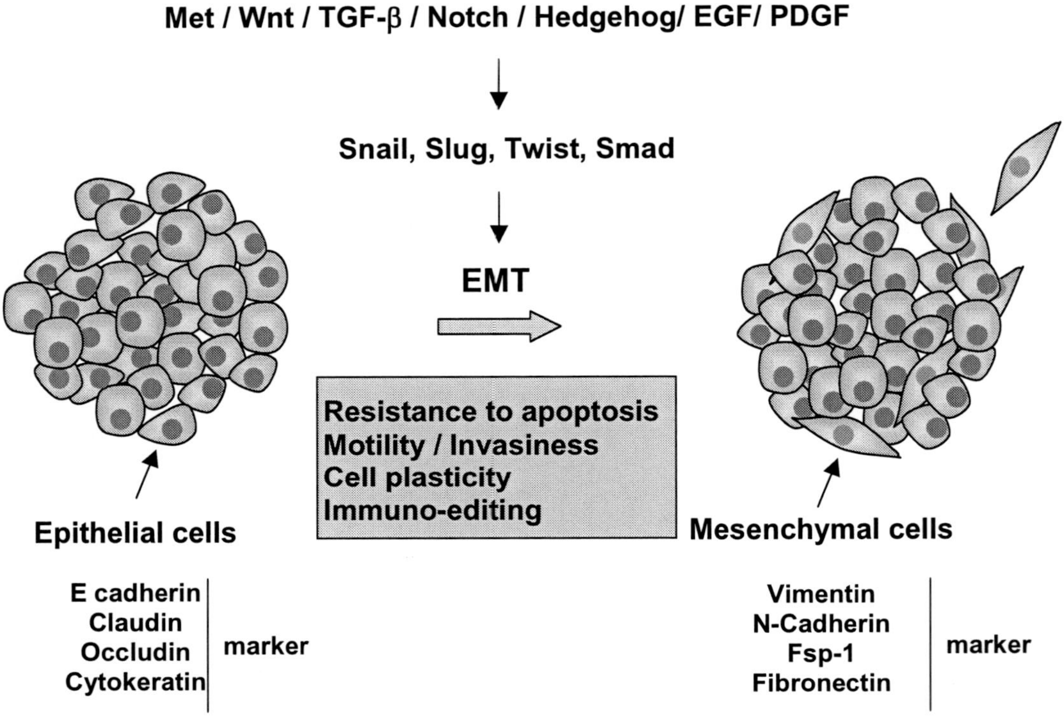

Various pathways are known to induce EMT either by inhibiting epithelial markers or by inducing mesenchymal markers, a process associated with resistance to apoptosis, cell migration, acquisition of stemness properties and immunoediting.

Conclusion

The extensive works in gene profiling gave rise to the identification of various gene signatures, all suggested to reflect tumour aggressiveness. Nevertheless, the clinical interest of these studies is limited as soon we consider that metastases derive from premalignant tissues or originate from a cell population potentially under-represented in the primary tumour mass. Further progress is imperatively required to identify the genes, specifically expressed either in cancer cells or in the surrounding stroma, that contribute to establish permissive conditions for cell dissemination. Targeting all these genes simultaneously should be extremely efficient although likely difficult to achieve, in particular if gene combinations vary with the tumour type and the metastatic tropism. Clinical trials with anti-angiogenic drugs have recently been developed. However, the potential benefit of the treatment might be lost if hypoxia promotes EMT and favours the migration of metastatic cells to less hostile environment. The emerging notion of "cancer-initiating cells" opens new hopes. Indeed, it is reasonable to assume that pathways implicated in self-renewing and stem cell migration, initiated by the Wnt, Notch or hedgehog proteins, by the SDF1-CXCR4 axis and the transcription factors implied in EMT and cell plasticity, such as Snail and Twist, constitute attractive targets for future therapeutic approaches.

Acknowledgments

The authors thank Dominique Reynaud for help in manuscript preparation.

References

[1] van, d., V *et al.* A gene-expression signature as a predictor of survival in breast cancer. *N. Engl. J. Med.* **347**, 1999-2009 (2002).

[2] Wang, Y. *et al.* Gene-expression profiles to predict distant metastasis of lymph-node-negative primary breast cancer. *Lancet* **365**, 671-679 (2005).

[3] Liu, R. *et al.* The prognostic role of a gene signature from tumorigenic breast-cancer cells. *N. Engl. J. Med.* **356**, 217-226 (2007).

[4] Bild, A. H. *et al.* Oncogenic pathway signatures in human cancers as a guide to targeted therapies. *Nature* **439**, 353-357 (2006).

[5] Chang, H. Y. *et al.* Robustness, scalability, and integration of a wound-response gene expression signature in predicting breast cancer survival. *Proc. Natl. Acad. Sci. U. S. A* **102**, 3738-3743 (2005).

[6] Chi, J. T. *et al.* Gene expression programs in response to hypoxia: cell type specificity and prognostic significance in human cancers. *PLoS. Med.* **3**, e47 (2006).

[7] Calza, S. *et al.* Intrinsic molecular signature of breast cancer in a population-based cohort of 412 patients. *Breast Cancer Res.* **8**, R34 (2006).

[8] Paik, S. Development and clinical utility of a 21-gene recurrence score prognostic assay in patients with early breast cancer treated with tamoxifen. *Oncologist.* **12**, 631-635 (2007).

[9] Nguyen, D. X. & Massague, J. Genetic determinants of cancer metastasis. *Nat. Rev. Genet.* **8**, 341-352 (2007).

[10] Fan, C. *et al.* Concordance among gene-expression-based predictors for breast cancer. *N. Engl. J. Med.* **355**, 560-569 (2006).

[11] Gupta, G. P. *et al.* Mediators of vascular remodelling co-opted for sequential steps in lung metastasis. *Nature* **446**, 765-770 (2007).

[12] Steeg, P. S. Metastasis suppressors alter the signal transduction of cancer cells. *Nat. Rev. Cancer* **3**, 55-63 (2003).

[13] Park, Y. G. *et al.* Sipa1 is a candidate for underlying the metastasis efficiency modifier locus Mtes1. *Nat. Genet.* **37**, 1055-1062 (2005).

[14] Hunter, K. Host genetics influence tumour metastasis. *Nat. Rev. Cancer* **6**, 141-146 (2006).

[15]Michaelson, J. S. *et al.* Spread of human cancer cells occurs with probabilities indicative of a nongenetic mechanism. *Br. J. Cancer* **93**, 1244-1249 (2005).
[16]Karnoub, A. E. *et al.* Mesenchymal stem cells within tumour stroma promote breast cancer metastasis. *Nature* **449**, 557-563 (2007).
[17]Allinen, M. *et al.* Molecular characterization of the tumor microenvironment in breast cancer. *Cancer Cell* **6**, 17-32 (2004).
[18]Kurose, K. *et al.* Genetic model of multi-step breast carcinogenesis involving the epithelium and stroma: clues to tumour-microenvironment interactions. *Hum. Mol. Genet.* **10**, 1907-1913 (2001).
[19]Bissell, M. J. & LaBarge, M. A. Context, tissue plasticity, and cancer: are tumor stem cells also regulated by the microenvironment? *Cancer Cell* **7**, 17-23 (2005).
[20]Kaplan, R. N., Rafii, S. & Lyden, D. Preparing the "soil": the premetastatic niche. *Cancer Res.* **66**, 11089-11093 (2006).
[21]Schardt, J. A. *et al.* Genomic analysis of single cytokeratin-positive cells from bone marrow reveals early mutational events in breast cancer. *Cancer Cell* **8**, 227-239 (2005).
[22]Hermann, P. C. *et al.* Distinct populations of cancer stem cells determine tumor growth and metastatic activity in human pancreatic cancer. *Cell Stem Cell* **1**, 313-323 (2007).
[23]Braun, S. *et al.* A pooled analysis of bone marrow micrometastasis in breast cancer. *N. Engl. J. Med.* **353**, 793-802 (2005).
[24]Rowley, J. D. The role of chromosome translocations in leukemogenesis. *Semin. Hematol.* **36**, 59-72 (1999).
[25]Schmidt, C. A. & Przybylski, G. K. What can we learn from leukemia as for the process of lineage commitment in hematopoiesis? *Int. Rev. Immunol.* **20**, 107-115 (2001).
[26]Pardal, R., Clarke, M. F. & Morrison, S. J. Applying the principles of stem-cell biology to cancer. *Nat. Rev. Cancer* **3**, 895-902 (2003).
[27]O'Hare, M. J. Teratomas, neoplasia and differentiation: a biological overview. I. The natural history of teratomas. *Invest Cell Pathol.* **1**, 39-63 (1978).
[28]Shimada, H. *et al.* Histopathologic prognostic factors in neuroblastic tumors: definition of subtypes of ganglioneuroblastoma and an age-linked classification of neuroblastomas. *J. Natl. Cancer Inst.* **73**, 405-416 (1984).
[29]Trentin, A., Glavieux-Pardanaud, C., Le Douarin, N. M. & Dupin, E. Self-renewal capacity is a widespread property of various types of neural crest precursor cells. *Proc. Natl. Acad. Sci. U. S. A* **101**, 4495-4500 (2004).
[30]O'Brien, C. A., Pollett, A., Gallinger, S. & Dick, J. E. A human colon cancer cell capable of initiating tumour growth in immunodeficient mice. *Nature* **445**, 106-110 (2007).
[31]Al-Hajj, M., Wicha, M. S., ito-Hernandez, A., Morrison, S. J. & Clarke, M. F. Prospective identification of tumorigenic breast cancer cells. *Proc. Natl. Acad. Sci. U. S. A* **100**, 3983-3988 (2003).
[32]Stingl, J., Eaves, C. J., Kuusk, U. & Emerman, J. T. Phenotypic and functional characterization in vitro of a multipotent epithelial cell present in the normal adult human breast. *Differentiation* **63**, 201-213 (1998).
[33]Stingl, J., Eaves, C. J., Zandieh, I. & Emerman, J. T. Characterization of bipotent mammary epithelial progenitor cells in normal adult human breast tissue. *Breast Cancer Res. Treat.* **67**, 93-109 (2001).
[34]Villadsen, R. *et al.* Evidence for a stem cell hierarchy in the adult human breast. *J. Cell Biol.* **177**, 87-101 (2007).

[35]Brockes, J. P. & Kumar, A. Plasticity and reprogramming of differentiated cells in amphibian regeneration. *Nat. Rev. Mol. Cell Biol.* **3**, 566-574 (2002).

[36]Gardiner, D. M., Endo, T. & Bryant, S. V. The molecular basis of amphibian limb regeneration: integrating the old with the new. *Semin. Cell Dev. Biol.* **13**, 345-352 (2002).

[37]Axelson, H., Fredlund, E., Ovenberger, M., Landberg, G. & Pahlman, S. Hypoxia-induced dedifferentiation of tumor cells--a mechanism behind heterogeneity and aggressiveness of solid tumors. *Semin. Cell Dev. Biol.* **16**, 554-563 (2005).

[38]Imai, T. *et al.* Hypoxia attenuates the expression of E-cadherin via up-regulation of SNAIL in ovarian carcinoma cells. *Am. J. Pathol.* **163**, 1437-1447 (2003).

[39]Evans, A. J. *et al.* VHL promotes E2 box-dependent E-cadherin transcription by HIF-mediated regulation of SIP1 and snail. *Mol. Cell Biol.* **27**, 157-169 (2007).

[40]Kurrey, N. K., K A & Bapat, S. A. Snail and Slug are major determinants of ovarian cancer invasiveness at the transcription level. *Gynecol. Oncol.* **97**, 155-165 (2005).

[41]Valsesia-Wittmann, S. *et al.* Oncogenic cooperation between H-Twist and N-Myc overrides failsafe programs in cancer cells. *Cancer Cell* **6**, 625-630 (2004).

[42]Pierret, C., Spears, K., Maruniak, J. A. & Kirk, M. D. Neural crest as the source of adult stem cells. *Stem Cells Dev.* **15**, 286-291 (2006).

[43]Motohashi, T., Aoki, H., Chiba, K., Yoshimura, N. & Kunisada, T. Multipotent cell fate of neural crest-like cells derived from embryonic stem cells. *Stem Cells* **25**, 402-410 (2007).

In: Oncogene Proteins: Structure, Functions and Analyses
Editors: P.B. Murphy and J.R. Clarke, pp. 11-15
ISBN: 978-1-60456-259-0

Short Communication B

Update in Prostate Cancer Pathogenesis

A. Papatsoris*[*], *S. Giannakopoulos, C. Kalaitzis and S. Touloupidis
Department of Urology, Democritus University of Thrace, Alexandroupolis, Greece

Abstract

Prostate cancer is the most frequent diagnosed cancer among men. Knowledge of genetic changes underlying the initiation, development, and progression of prostate cancer is accumulating rapidly. Molecular fingerprinting may help to distinguish indolent from aggressive prostate cancer. The crucial role of molecules participating in complex signal transduction networks makes them appealing targets. Novel strategies are developing for the treatment of prostate cancer by targeting multiple tumorigenic signaling elements. Here, we accumulate the remarks of relevant reviews that where published this year.

Introduction

Prostate cancer is the commonest solid-organ malignancy in males and it is estimated that one in six men will develop this cancer in their lifetime. Inevitably, all patients will develop androgen-independent disease, which remains the main obstacle to improve survival [1]. Unfortunately, existing first and second line hormonal treatments are condemned to extent survival for only a few years. These sobering data have forced researchers to explore in depth the molecular mechanisms implicated in the emergence of androgen independence and of prostate cancer invasion.

[*] E-mail address: agpapatsoris@yahoo.gr. Correspondence: Athanasios Papatsoris MD, MSc, PhD, FEBU, Elefteriou Venizelou 30, Agios Vasilios, Rio-Patra, 26500, Greece

Precursor Lesions

Several morphological lesions have been proposed that may act as potential precursor lesions of prostate cancer [2,3]. These are the morphologically distinct entities of focal atrophy or post-atrophic hyperplasia (PAH), atypical adenomatous hyperplasia (AAH) or adenosis, and prostatic intraepithelial neoplasia (PIN). The diagnostic criteria of low and high grade PIN (LGPIN and HGPIN, respectively), and of lesions suspicious for cancer (LSC) have been recently established [2].

Potential markers of prostate cancer precursor lesions include fatty acid synthetase, alpha-methylacyl coenzyme A racemase (AMARC), and A80 [2]. Furthermore, management strategies of precursor lesions may include repeat prostate biopsy, chemoprevention, radical prostatectomy and radiotherapy.

Diagnostic Approaches

Serum prostatic specific antigen (PSA) testing has been used for over 20 years in the diagnosis and management of prostate cancer. Although highly sensitive, it suffers from a lack of specificity, showing elevated serum levels in a variety of other clinical entities [4]. Hence, the use of PSA screening to diagnose and monitor prostate cancer is associated with several drawbacks. Ongoing prostate cancer trials will hopefully provide an opportunity for biological sample collection along with the potential to validate novel biomarkers [5]. The goal of current research is to apply genomic assays to noninvasive specimens such as blood and urine [4].

Epigenetic alterations, mainly promoter hypermethylation of prostate cancer-related genes might be used as cancer biomarkers [6-8]. From the numerous cancer-related genes hypermethylated in prostate cancer the following are strong candidates to become routine biomarkers: GSTP1, APC, RARbeta2, cyclin D2, MDR1, and PTGS2 [6]. Knowledge of genetic changes underlying the initiation, development, and progression of prostate cancer is accumulating rapidly. With increasing knowledge, it may be possible to distinguish indolent from aggressive prostate tumors by molecular fingerprinting.

Molecular Pathogenesis

The molecular pathology of prostate cancer is complex as not only multiple genes are involved, but additional environmental factors such as diet and inflammation are also involved [9]. Evidence is accumulating to implicate infection and inflammation as contributors to prostate cancer [9,10]. It has been demonstrated that proliferative inflammatory atrophy ties inflammatory responses to prostatic carcinogenesis [10].

Androgen receptor (AR) acquires a role in licensing for DNA replication in prostate cancer cells [11]. In its role as a licensing factor, AR must be degraded during mitosis in order to allow relicensing in the subsequent cell cycle [12]. Acute enhanced expression of AR in prostate cancer cells results in its incomplete degradation in mitosis. Lack of mitotic AR degradation inhibits subsequent cell proliferation due to the inability to relicense all origins of replication needed for the next round of cell division [12]. A new therapeutic approach has been suggested based upon stabilization of AR [12]. Furthermore, there are interactions between certain deregulated signaling pathways such as AR, estrogen receptor, epidermal growth factor receptor (EGFR) and Wnt/beta-catenin cascades in controlling the proliferation, survival and invasion of tumor prostate epithelial cells [13]. Hedgehog signaling has an important role in prostate development and it appears to be a characteristic feature of prostate cancer [14]. The development of hedgehog inhibitors based on the action of cyclopamine holds promise for novel treatments to slow or arrest tumor growth [14].

Activation of Ras-mitogen activated protein kinase (MAPK) signalling network is known to affect the AR activity [15]. This multi-partite network propagates chemical signals from the cell surface to the nucleus through sequential kinase signalling and intensive cross-talk. During prostate carcinogenesis, crucial components of this network are deregulated, thus affecting cellular proliferation, apoptosis and metastasis [15]. Various molecules of Ras-MAPK network represent appealing selective targets for prostate cancer therapeutics [15].

Autocrine and paracrine events regulated by nerve growth factor (NGF) and relevant low and high affinity receptors seem to play a significant role in prostate carcinogenesis [16]. Studies reveal that low affinity receptor p75(NTR) is both a tumor suppressor of growth and a metastasis suppressor of human prostate cancer cells [16]. Furthermore, p75(NTR) is progressively lost during prostate carcinogenesis. An imbalance between p75(NTR) and high affinity tropomyosin receptor kinase A (TrkA)-mediated signals may be involved in the progression of prostate cancer through increased proliferation and reduced apoptosis [16].

There seems to be an association of circulating levels of insulin-like growth factor (IGF)-1 and prostate cancer risk. The activation of type I IGF receptor (IGF-IR) by IGF-I and/or IGF-II, has mitogenic and antiapoptotic effects on normal and malignant prostate cells [17]. It has been demonstrated that IGF-1 is mainly involved in the early stages of prostate carcinogenesis. Also, it has been suggested that IGF-1 induces ligand-independent activation of the androgen receptor and enhances the expression of matrix metalloproteinase-2 and urokinase plasminogen activator. Additionally, progression to androgen independence has been linked to deregulation of the IGF-1/IGF-1R axis.

Future Perspectives

Molecular knowledge of prostate cancer pathogenesis is rapidly expanding, revealing novel potential therapeutic targets [18-20]. Together with the emerging technologies of "omics" (genomics, proteomics, pharmacogenomics and metallomics), the potential of appropriate design of selective, narrow spectrum-targeted therapies appear now realistic. Many novel therapeutic agents are under development and are gradually entering clinical testing in all phases of the prostate carcinogenesis process. As prostatic carcinogenesis represents a multi-step process, of paramount importance is the characterization of crucial candidate molecular target differences during various stages of prostate cancer progression. Better elucidation of the timing of these changes would establish whether novel biological agents are effective in either preventing or treating prostate cancer.

Novel therapeutic agents under development include growth factor receptor inhibitors, small molecules targeting signal transduction pathways, apoptosis and cell-cycle regulators, angiogenesis and metastasis inhibitors, differentiation agents, telomerase inactivators, and epigenetic therapeutics. Incorporation of these agents into existing treatment regimens will guide us in the development of a multidisciplinary treatment strategy of hormone refractory prostate cancer. In most preclinical studies, these novel agents reduce the growth rate of established tumours rather than cause tumour regression. This suggests that their use will be most beneficial when administered in combination with standard treatment modalities.

Epilogue

The time has come for concerted, large-scale multidisciplinary investigations that incorporate molecular genetics, pathobiochemistry, histopathology, and endocrinology into traditional clinical studies as such collaboration will lead to a deeper understanding of the etiologic pathways of prostate cancer, ultimately yielding better preventive, diagnostic, and therapeutic strategies.

References

[1] Schrijvers, D. Androgen-independent prostate cancer. *Recent Results Cancer Res*, 2007, 175, 239-249.
[2] Chrisofos, M; Papatsoris, AG; Lazaris, A; Deliveliotis, C. Precursor lesions of prostate cancer. *Crit Rev Clin Lab Sci*, 2007, 44, 1-28.
[3] Algaba, F; Trias, I; Arce, Y. Natural history of prostatic carcinoma: the pathologist's perspective. *Recent Results Cancer Res*, 2007, 175, 9-24.
[4] Reynolds, MA; Kastury, K; Groskopf, J; Schalken, JA; Rittenhouse, H. Molecular markers for prostate cancer. *Cancer Lett*, 2007, 249, 5-13.
[5] Flaig, TW; Nordeen, SK; Lucia, MS; Harrison, GS; Glodé, LM. Conference report and review: current status of biomarkers potentially associated with prostate cancer outcomes. *J Urol*, 2007, 177, 1229-1237.
[6] Costa, VL; Henrique, R; Jerónimo, C. Epigenetic markers for molecular detection of prostate cancer. *Dis Markers*, 2007, 23, 31-41.
[7] Dobosy, JR; Roberts, JL; Fu, VX; Jarrard, DF. The expanding role of epigenetics in the development, diagnosis and treatment of prostate cancer and benign prostatic hyperplasia. *J Urol*, 2007, 177, 822-831.
[8] Li, LC. Epigenetics of prostate cancer. *Front Biosci*, 2007, 112, 3377-3397.
[9] Nelson, WG. Prostate cancer prevention. *Curr Opin Urol*, 2007, 17, 157-167.
[10] De Marzo, AM; Platz, EA; Sutcliffe, S; Xu, J; Grönberg, H; Drake, CG; Nakai, Y; Isaacs, WB; Nelson, WG. Inflammation in prostate carcinogenesis. *Nat Rev Cancer,* 2007, 7, 256-269.
[11] Chmelar, R; Buchanan, G; Need, EF; Tilley, W; Greenberg, NM. Androgen receptor coregulators and their involvement in the development and progression of prostate cancer. *Int J Cancer,* 2007, 120, 719-733.
[12] Vander Griend, DJ; Litvinov, IV; Isaacs, JT. Stabilizing androgen receptor in mitosis inhibits prostate cancer proliferation. *Cell Cycle*, 2007, 15, 6, 647-651.
[13] Vecchione, A; Gottardo, F; Gomella, LG; Wildemore, B; Fassan, M; Bragantini, E; Pagano, F; Baffa, R. Molecular genetics of prostate cancer: clinical translational opportunities. *J Exp Clin Cancer Res*, 2007, 26, 25-37.

[14] Shaw, A; Bushman, W. Hedgehog signaling in the prostate. *J Urol*, 2007, 177, 832-838.
[15] Papatsoris, AG; Karamouzis, MV; Papavassiliou, AG. The Power and Promise of 'Re-wiring' the MAPK Network in Prostate Cancer Therapeutics. *Mol Cancer Ther*, 2007, 6, 811-819.
[16] Papatsoris, AG; Liolitsa, D; Deliveliotis, C. Manipulation of the nerve growth factor network in Prostate Cancer. *Exp Opin Invest Drugs*, 2007, 16, 303-309.
[17] Monti, S; Proietti-Pannunzi, L; Sciarra, A; Lolli, F; Falasca, P; Poggi, M; Celi, FS; Toscano, V. The IGF axis in prostate cancer. *Curr Pharm Des*, 2007, 13, 719-727.
[18] Langeberg, WJ; Isaacs, WB; Stanford, JL. Genetic etiology of hereditary prostate cancer. *Front Biosci*, 2007, 112, 4101-4110.
[19] Chokkalingam, AP; Stanczyk, FZ; Reichardt, JK; Hsing, AW. Molecular epidemiology of prostate cancer: hormone-related genetic loci. *Front Biosci*, 2007, 112, 3436-3460.
[20] Downes, MR; Byrne, JC; Pennington, SR; Dunn, MJ; Fitzpatrick, JM; Watson, RW. Urinary markers for prostate cancer. *BJU Int*, 2007, 99, 263-268.

In: Oncogene Proteins: Structure, Functions and Analyses ISBN: 978-1-60456-259-0
Editors: P.B. Murphy and J.R. Clarke, pp. 17-53

Chapter I

Molecular Determinants of Neuroplasticity in Affective Disorders. The study of protein kinases in human brain and in cell culture

Félicien Karege[1], Nader Perroud[1], Sandra Burkhardt[2], Michèle Schwald[1], Eladia Ballmann[1] Romano La Harpe[2] and Alain Malafosse[1]

Geneva University Hospitals. Department of Medical Genetics and Laboratory[1] and the Institute of Forensic Medecine[2], University of Geneva, Geneva, Switzerland

Abstract

Postmortem and functional imaging studies have reported morphological changes in the brains of untreated depressed patients, and evidence has associated this with the loss or atrophy of glial and neuronal cells. In animal studies, the antidepressant drugs can reverse the stress-induced cellular loss by increasing neurogenesis in hippocampus. Therefore, investigation of the biochemistry of these changes could provide insight for a better understanding and more effective medication in mood disorders. Studies of protein and phospholipid kinases, some of them konwn for their role in cell transformation, were undertaken. Postmortem brains of subjects deceased by suicide or not, with or without documented major depression, were used. Proteins levels were also measured in cell culture, after drug activation, along with cell proliferation and cell survival.

Enzymatic activities and protein levels were measured for the following proteins: Akt-1, an homolog to the viral oncogen (v-Akt), with enzyme activity which facilitates growth factor-mediated cell survival. Akt-1 also regulates the activity of GSK-3ß, a protein kinase involved in negative control of IGF and other growth factors. GSK-3ß is both directly and indirectly inhibited by lithium, a mood stabilizer drug. The other proteins consisted in PTEN and PI3K. The latter is a lipid kinase which phosphorylates the PtdIns-3,4,5-P and involved in cell proliferation. Opposed to its action, is the phosphatase PTEN, a tumor suppressor implicated in a variety of sporadic cancers, which acts by dephosphorylating the PtdIns-3,4,5-P. With respect to Akt and GSK-3ß, results indicated no change either in their protein levels or in lithium-inhibitable total GSK3α/ß activity in human prefrontal cortex (PFC). The activity of Akt-1 decreased (ANOVA: $P = 0.003$), while GSK-3ß activity increased (ANOVA $P = 0.002$) in depressed subjects. The study of PI3K and PTEN proteins indicated decrease in activity of PI3K and increase in PTEN levels in the same samples. Taken together, our data indicated that the signaling pathway through PI3K/GSK-3ß is blunted in depression. In cell culture experiments, the study showed activation of protein kinases by mood stabilizers (Lithium, Valproate), with a potentiation of cell proliferation and survival. These drugs induced an increase in Akt mRNA and protein levels to support potentiation of cell proliferation. In conclusion, these studies suggest a role of protein kinases and phosphatases, mostly known as oncogens, acting as downstream effectors of neurotransmitters and growth factors, in psychopathology and are supportive of the neuroplasticity hypothesis in affective disorders.

A. Introduction

The two main affective disorders, Major depressive disorders (MDD) and Bipolar disorder (BD) are among the most common and life-threatening diseases and represent a significant public health problem (Kupfer 2005; WHO report, 2001). Hundreds of millions of people worldwide are affected by mental illness and WHO estimates that 154 million people globally suffered from depression. However, in spite of intensive research, the exact neurobiological process leading to affective disruption are still limited. Recent hypotheses, yet to be developed, suggest changes in neuroplasticity of limbic and non limbic circuits, supported by alteration in signal transduction and gene expression (Carlson et al. 2006; Pittenger and Duman, 2007). The neuroplasticity is understood as the ability of the brain to undergo functionally relevant adaptation following external and internal stimuli (Fuchs et al 2004). However not all neural plasticity is beneficial as demonstrated by altered structure and function in the brain of patients with mood disorders (Drevets, 2000; Rajkowska, 2000). This study is in keeping with this evidence of compromised neuroplasticity and altered signaling pathways, by undertaking the study of the PI3K-PKB-GSK3 biochemical pathway, a signaling axis whose elements are involved in apoptosis, cellular growth and proliferation (Franke et al. 1997; Yamada and Araki, 2001; Datta et al, 2006). While some of these proteins are known as oncogens, most of them are well documented as either facilitators or negative regulators of signals in cell proliferation, neuroprotection or cell migration. We have therefore undertaken this study by focusing on their possible alteration in major depression and suicide subjects. For that, postmortem brains tissues have been used in the first part of this study. Secondly, a key question arising from the forgoing is whether there is evidence that antidepressant and mood stabilizers drugs can have an effect on the signaling pathways. Thus, the second issue of this research has focused on the actions of current mood stabilizers on these kinases. Given the complexity of mood disorders, investigating the targets of current drug treatment could shed more light on the psychopathological mechanisms.

1. Major Depressive Disorder and Neurobiology

The four-decade old theory of depression stipulated that depression can result from the dysfunction of several neurotransmitters or metabolites (Owens and Nemeroff, 1994; Heninger et al. 1996). Studies have mainly focused on alteration of serotonin and monoamines and their receptors (Pandey et al. 1995) or on the dysregulation of the hypothalamo-pituitary-axis (HPA) (Charlton, 1989). Further, it was suggested that longterm therapeutic antidepressant treatments are mediated by postsynaptic intracellular signaling pathways (Bezchlibnyk and Young, 2002). Afterwards, molecular and cellular studies have been undertaken to elucidate the intracellular machinery through which the monoamines ultimately control the function of the neurons, and new approaches to understanding major depression have focused on the regulation of the key pathways involved in cellular survival and plasticity (Duamn et al. 1997; Fuchs et al. 2006; Pettenger and Duman, 2007). These signaling pathways are of several categories and include for instance, signaling networks controlled by receptor-coupled second messengers (e.g. cAMP or Inositol triphosphate), and usually activated by classic neurotransmitters such as monoamines or amino acids (Duman et al. 1997; Pandey et al. 2003; Berridge et al. 1995). Other categories which have recently drawn attention, are biochemical pathways controlled by membrane receptors which contain tyrosine kinases and usually regulated by neurotrophic factors and cytokines; their activation leads to the intracellular cascades encompassing protein or lipid kinases, phosphatases, transcription factors etc (Dwivedi et al. 2003). These signaling networks which influence apoptosis or cell atrophy, cell proliferation, neuron growth, could support the morphological changes observed in the brains of depressed subjects (Drevets 2000; Rajkowska 2000; Manji et al. 2001) and in animal experiments (Magarios et al. 1996; McEwen et al. 2000).

In adults rodents, it was observed that chronic stress or glucocorticoids administration results in remodeling of the apical dentrites of the CA3 pyramidal neurons, which is manifested in a decrease of number and length of dendrites and in the ongoing rate of cell birth and cell proliferation in the dentate gyrus (McEwen et al. 2000, Gould et al. 1997). Moreover, it was shown that this stress or hormone effect was reversed by chronic, but not acute, treatment with the atypical antidepressant tianepine (Czeh et al 2001). In human studies, changes were observed in limbic and cortical regions implicated in the affective and cognitive impairment (Manji et al. 2001; Sheline et al. 1996). For instance, *postmortem* studies have observed a reduction in neuronal size, a reduction in glyal number and size in prefontal cortex, a decrease in cortical thickness and basal ganglia volume in patients with depression (Manji et al 2001a; Rajkowska et al 1999; Rajkowska 2000; Cotter et al. 2001). Functiuonal imaging studies also revealed that prefrontal cortical, ventral striatal and hippocampus volume is decreased (Bremer et al 2000). Thus, the emerging picture is that major depression is associated with cellular loss and volumes decrease. Sheline and collegues observed that the decrease of the hippocampus voulme was directely correlated to the disease duration (Sheline et al. 2003). Therefore, signaling pathways involved either in cell atrophy, cell proliferation or cell death are first-class candidates for further major depression studies.

1.1. PI3K, PTEN, PKB and GSK3 Proteins: Potential Targets

The phosphatidyl inositol 3-kinase (*PI3K*) - Protein kinase B (*PKB* also called *Akt*) is a signaling cascade originally described in insulin signaling, and then in multiple roles of cell physiology like cycle progression where it could underlie tumor-promoting effects through inactivation of GSK-3β (van der Heide et al. 2003; Linang and Sligerland, 2003). Later on, clinical and basic studies have shown that members of this signaling axis are altered in affective disorders (Hsiung et al 2003; Dwivedi et al 2003), and are modified by mood stabilizers drug treatment (Manji et al. 2001b); Carlson et al 2006; Mora et al 2002). The PI3K and Akt kinase activities are altered in suicide depression (Hsiung et al 2003; Karege et al 2007). Moreover, mood stabilizers like, Lithium, Olanzapine, Clozapine can stimulate the phosphorylation of Akt and PI3K. Moreover GSK3 activity is inhibited by Lithium and Valproate and can be regulated by serotonin (5HT) signaling (Li et al. 2006). Actually, these proteins are part of an important axis of cell signaling network regulated by growth factors, neurotrophins or insulin, and involved in cell proliferation, apoptosis or tumorigenesis (Dudek et al 1997; Ma et al 2007; Cantrel 2001).

In this study we have undertaken the assay of three protein and lipid kinases, namely the phosphatidyl inositol 3-kinase (*PI3K*), the protein kinase B (*PKB also called Akt*) and the glycogen synthase kinase-3 (*GSK3*) as well as a phosphatase, *PTEN*, in *postmortem* brain of depressed subjects. In the first part of the study, we focused to the *PKB* and *GSK*-3ß proteins, by measuring protein levels and enzymatic activities in tissue lysates from ventral prefrontal cortex.

1.2. PKB/Akt and GSK-3αβ

The *PKB/Akt*, is a serine/threonine kinase, which has recently emerged as a critical enzyme, the major role of which is to facilitate growth factor-mediated cell survival and to block apoptotic cell death (Lawlor and Alessi 2001; Cheng et al 2003). Previously, the PKB/Akt activity was shown to be blunted in the occipital cortex with a significant decrease in Akt activity and its upstream effectors (Hsiung et al 2003). We have undertaken this study to further explore this signaling pathway and characterize Akt regulation and its downstream effector, glycogen synthase kinase-3ß in depressed suicide subjects (Pap and Cooper, 1998). Glycogen synthase kinase-3ß (GSK-3ß) is a serine/threonine protein kinase, originally identified as a negative regulator of glycogen synthesis (Grimes and Jope 2001). It plays a key role in phosphorylation and regulation of metabolic enzymes and many transcription factors (Grimes and Jope, 2001). This diverse array of substrates gives to the enzyme the capacity to modulate diverse cellular processes such as embryogenesis, apoptosis and cell survival (Hetman et al. 2000; Jope and Bijur 2002; Grimes and Jope 2001). It was recently shown that GSK-3 plays a role in the transformation of epidermal cell (Ma et al 2007). GSK-3 exists in two highly homologous forms, GSK-3α and GSK-3ß, which are negatively regulated by phosphorylation on serine-21 and serine-9, respectively, a modification catalysed by several kinases, among them PKB (Cross et al 1995; Fang et al 2000; Goode et al 1992). Futhermore, it was observed that *5HT* signaling regulates the phosphorylation of *GSK*-3ß in mammalian brains *in vivo,* a mechanism thought to be achieved by a balance between opposing actions of 5HT receptors subtypes (Li et al. 2004). Based on the animal model of depression, it has been suggested that inhibition of *GSK*-3ß might contribute to antidepressant activity, thus increasing our interest in the role of *GSK*-3ß in major depression disorder (*MDD*) and suicidality (Kaidanovich-Beilin 2004; O'Brien et al 2004).

The interest in *GSK*-3ß in psychiatry has been highlighted by the discovery of its inhibition by lithium, which has further been found to inhibit both forms of *GSK-3* (Klein and Melton 1996; Stambolic et al 1996). Indeed, some studes have indicated that prophylactic administration of lithium to psychiatric patients could reduce the rates of suicide (Müller-Oerlinghausen et al 2003; Dunner et al. 2004). The aim of this preliminary study, therefore, was to examine the regulation of Akt/GSK-3ß signaling in completed suicide with documented major depression disorder (*MDD*). Results indicated a depression-associated alteration of the *PKB/GSK-3*, with a decrease in *PKB* and an increase in *GSK*-3ß activitiy in depressed subjects, suggesting an attenuation of this signaling pathway (Karege et al. 2007).

1.3. PI3K and PTEN

On the basis of these first results, we decided to go forward and explore the upstream elements which might be involved in this blunting. We have hypothesized that changes in Akt signaling are probably due to alterations in upstream lipid kinase and phosphatase which control the levels of the second messengers PtdIns (3,4,5)-P3. We therefore measured *PI3K* activity and protein levels and *PTEN* protein levels, as well. The PI3K is a heterodimer lipid and protein serine kinase comprising a 85-kDa regulatory subunit and a 110-kDa catalytic subunit (Cantley 2002). It has been demonstrated that this enzyme is required for the growth factor-dependent survival of a wide variety of cultured cells, including neurons (Dudek et al 1997). Biochemically, The PI3K enzyme phosphorylates at the D-3 position of the inositol ring of phosphoinosites to generate the lipid second messengers: PI-3-phosphate, PI-3,4-biphosphate, and PI-3,4,5 triphosphate (Whitman et al 1988). The lipid kinase is involved in regulation of multiple physiological functions (cell survival and proliferation, growth, cell motility, apoptosis, etc.) under activation by a variety of stimuli (quoted above). Activation of PI3K and the subsequent phosphorylation of the phosphoinosites, leading to the phospholipid products, are currently thought to stimulate the serine-threonine protein kinase Akt/PKB (Lawlor and Alessi 2001; Brazil et al 2004). Akt is activated by directly binding the PI3K-generated phosphoinositols (3,4,5-P_3) to their PH domain (James 1996; Franke et al 1997), and as mentioned above, the main downstream substrate of Akt is GSK-3ß .

Also upstream of this signaling pathway lies PTEN, another enzyme which has recently drawn attention, because of its crucial role in cell proliferation, in neuronal survival and in mediating Akt activation in a number of cell functions (Maehama and Dixon 1998; Yamada and Araki 2001). PTEN, an acronym of phosphatase and tensin homolog deleted on chromosome ten, is one of the frequently mutated tumor suppressors in human cancer, which functions primarily as a lipid phosphatase to regulate crucial signal transduction pathways and whose the main target is the phosphatidylinositols (3,4,5)-P_3. (Maehama and Dixon 1998; Yamada and Araki 2001). PTEN has also been reported as a modulator of cell signaling, growth and apoptosis and, by reducing PIP_3 levels, PTEN opposes the actions of PI3K-dependent pathways and behaves as a negative regulator of Akt. Although substantial progress has been made in understanding its role in suppressing malignacy, much less is known about its involvement in neurobiological processes. However, recent studies have reported critical alteration in Alzheimer's diseased patients and in depressed suicide victims (Griffin et al 2005; Hsiung et al 2003). This first part of our study aims to explore the regulation of this PI3K-PKB-GSK-3β in suicide major depressed subjects. The second part relies to mood satbilizers currently used to treat bipolar disorder and their molecular targets.

2. Bipolar Disorders and Mood Stabilizers

Bipolar disorder (BD) is a severe, chronic and debilitating disease characterized by cyclic periods of mania and depression which affect the lives and functioning of millions worldwide (Kupfer 2005). The cost associated with disability and premature death (suicide is estimated at 15% among BD) represents, as for the MDD, an economical burdern of several billions in the world and a number of health consequences co-occur with the disorder (Ciechanowski et al 2000; Musselman et al 1998). Unfortunately, the pathophysiologic mechanisms underlying this disorder are not completely understood, in spite of recent progress in medication. Growing evidence, however, suggest that mechanisms of neural plasticity and cellular resilience, including impairment of neurotrophic/neuroprotective signaling, appear to underline disturbances in the key affective circuits and significant progerss indevelopment of mood stabilizers have been performed (Manji and Duman, 2001; Carlson et al. 2006; Zarate et al., 2006)

Mood stabilizers are medications used to treat bipolar disorder, and most of current drugs have emerged as a neuroprotective agents against insult and apoptotic stimuli. This has reinforced the idea of a neurogenerative component in BD (Chen et al. 1999a; Manji et al. 1999). In this study, we have undertaken to investigate three types of mood stabilizers due to their common cellular effects (i.e. cell proliferation and cell survival) and molecular actions (PKB/Akt gene expression and protein phosphorylation). These three drugs are Lithium, the mainstay and oldest mood stabilizer, the Valproate (VA) and an atypical psychotropic drug, Clozapine, have been recently subject of intensive research, for both their clinical and cellular effects (Lai et al 2006; Mora et al. 2002; Li and El-Mallahk, 2000). All three, have been reported to affect the PKB and GSK-3ß proteins and a number of reports and reviews have been dedicated on the issue (Mora et al 2002; Shin et al 2006). This research has tried to bring together the three druds, for a comparison aim, in the same assay to find out possible differences and shared effects in their biochemical mechanisms on the above proteins.

2.1. Mood Stabilizers Used in this Study

-Lithium is a small cation, used in many aspects of Medecine for 150 years (Johnson and Amdisen, 1983) and later introduced in Psychiatry by John Cade, following its therapeutic actions in then called maniaco-depression disease (Cade et al., 1949). For its biochemical mechanisms, it was firstly shown to interfere with sodium channels, but modern research has focused on its effects on intracellular signaling (Berridge et al. 1982; Jope and Williams 1994). More recently, Lithium has been shown to be direct inhibitor of GSK-3ß (Stambolic et al. 1996; Klein and Melton, 1996). On neuronal aspect, lithium has emerged as a neuroprotective agent against insult and apoptic stimuli, thus contributing to support the hypothesis of a neurogenerative component in BD (Manji et al. 1999).

-Valproate (VA) most known for its anti-epileptic effects, was proposed as a mood stabilizer nearly 40 years ago and has been validated in many randomized trials in BD (Bowden et al., 1994). VA is also effective in accute mania, and as a prophylaxis for reccurrent mania and depression (Johannessen, 2000). On cellular level, numerous studies have reported neuroprotective effects against excitotoxic insults, low [K+]-induced cell death, or glutamate-induced oxidative stress (Mora et al. 1999; Shao et al. 2005).

- Clozapine (CZ), is a tricyclic dibenzodiazepine derivative, previously reported to inhibit GSK-3ß by phosphorylation at serine-9 and to exert neurotrophic actions in cell culture (Kang et al. 2004; Kozlovsky et al. 2006). On clinical level, Clozapine is the representative of atypical antipsychotic agents, whose major indication is the treatment of resistant schizophrenia, but also with mood-stabilizing effects (Ciapperelli et al. 2003; Frye et al. 1998).

Therefore, these three drugs may exert similar cellular actions, as neuroprotectors and may activate the same signaling axis, namey PKB-GSK-3ß, which could be the convergent point of their biochemical actions leading to subsequent neuronal changes and therapeutic effects. We have therefore undertaken this study to examine their effects on the expression and regulation of PKB/GSK-3ß signaling pathway. This intracellular pathway, as shown above, is thought to mediate many of downstream events controlled by phosphatidylinositol-3-kinase (PI3K), such as cell proliferation and survival (Lwalor and Alessi, 2001). The study of protein and lipid kinases and protein phosphatases involved in cell proliferation and anti-apoptotic processes are particularly relevant because of the above mentioned studies which demonstrated abnormal structural changes, including neuronal atrophy and changes in cell number in the brains of affective disorder victims (Knable et al 2004; Rajkowska 2000).

B. Methods and Materials

1. Collection and Dissection of *postmortem* Brains

Human brains were collected at autopsy from the Institute of Forensic Medicine, Geneva, Switzerland. This study was approved by the Research and Ethics Review Board of the department of Psychiatry, Faculty of Medicine in Geneva. This brain collection was obtained following legal procedures by the Republic and Canton of Geneva, which, in some cases requires family consent.

After the brain was removed from the cranium, it was examined for neuropathologic abnormalities and then blocks of tissue from the ventral prefrontal cortex (*VPFC*) (Brodmann's area 11) were dissected from the right hemisphere by forensic physicians and skilled technicians. The right hemisphere was used because of the reported association with suicidal tendencies and because of its functional insufficiency (Weinberg 2000). The samples were immediately stored at –80°C. During the fine dissection, blocks of tissue were thawed from -80°C to -20°C (*approx.* 30 mins) to make them easier to dissect, and then specimens were carefully dissected on a glass surface with dry ice, in accordance with the Brodmann' Atlas. Only grey matter samples (0.5-1 cm coronal slices) were carefully isolated and no white matter was included. Then samples were either stored at -80°C until used or immediately homogenized in appropriate buffers using a tissue grinder.

2. Subjects

Table 1 shows the characteristics of the subjects placed in four different groups, depending on the diagnosis. The suicide victims consisted of a well-defined population of twenty subjects (10 M/10 F) carefully determined by forensic physicians. Retrospective searching for *antemortem* clinical diagnosis and drug treatment was performed through reviews of medical histories of not only psychiatric units, but also internal and surgical units. Where possible, psychological autopsy was performed. Only subjects with or without the clinical diagnosis of DSM-IV criteria were included, and therefore the suicide group consisted of two subgroups: 10 suicide victims with documented *MDD* and a subgroup of 10 without documented *MDD*. The mean age was 45±15 and 48±14 *years* and the *postmortem* interval (PMI) from death to storage was 26±19 and 28±23 *hours*, for suicide with *MDD* and suicide without *MDD*, respectively. The causes of death were hanging (n = 5); drowning (n = 2); self-inflected gunshot wound (n = 3); jump (n = 3), carbon monoxide poisoning (n = 2), asphixia (n = 2) and multiple trauma (n = 3). Data were collected from the first hospitalization or ambulatory follow-up till the death of the subject. Subjects were considered as drug free if they did not take any antidepressants, antipsychotics or anticonvulsants during at least a year before death. Toxicological data were obtained from analysis of urine and blood samples. Moreover subjects with comorbidity of other psychiatric diseases or with positive toxicology (antidepressants, antipsychotics or anticonvuslsants) were excluded and only a few subjects with traces of alcohol (n = 2) or benzodiazepines (n = 2) were included in the study.

Table 1. Population

Diagnosis groups and demographic features of suicide victims and non-suicide subjects.

	Suicide victims		**Non-suicide subjects**	
	with MDD	**without MDD**	**without MDD**	**with MDD**
Subjects (M/F)	10 (6/4)	10 (5/5)	10 (5/5)	10 (5/5)
Age (yrs±sd)	45±15	48±14	42±15	45±16
PMI (hrs±sd)	26±19	28±23	27±16	26±15

MDD : Major Depression Disorder ; PMI : postmortem interval. M, male, F, female. Data are expressed as mean±sd. The group of non-suicide subjects with no MDD were used as controls.

The non-suicide subjects consisted of 20 subjects (10 M/10 F), who died from a determined cause, and only subjects whose *antemortem* medical records were available were included. According to psychological autopsy and retrospective searching for *antemortem* clinical diagnosis (*DSM-IV*), the non-suicide group consisted of two categories: a subgroup of 10 non-suicide subjects with documented *MDD* and a subgroup of 10 subjects without any antecedents of *MDD* or any other psychiatric disease. The latter group was used as a control group. The mean age was 42±15 and 45±16 years and the *postmortem* interval (PMI) was 27±16 and 26±15 hours for depressed and non-depressed subjects, respectively. Subjects with other psychiatric or neurological disorders or positive toxicology were excluded from this control group. With respect to the suicide group, the ages and genders of non-suicide subjects were balanced, and no statistically significant difference was observed in *postmortem* intervals (*PMI*) between the two groups. The causes of death varied and included myocardial infarction (n = 4); bronchopneumonia (n = 1), homicide, gunshot wound (n = 2); road traffic accident (n = 6); hypothermia (n = 1), various somatic diseases (n = 3), accidental drowning (n = 1) and CO poisoning (n = 2).

3. Cell Culture, Drug Treatment and Cell Proliferation

The human-derived neuroblastoma cells, the SH-SY5Y, purchased from ECACC (Salsibury, UK), were cultured in a humidified atmosphere at 95% air/5% CO2 at 37ºC in RPMI-1640 medium supplemented with 10% (v/v) fetal calf serum (FCS) with 100 U/ml penicillin and 100 μg/ml streptomycin (Invitrogen, Basel, CH). Depending on the experiment, cells were plated at various densities on 100 mm-dishes or six-well plates and where indicated, serum concentration could be decreased or withdrawn, for few hours before adding drugs and 10% FCS-supplemeted medium. This allowed to make cells quiescent and synchronize their growth during subsequent activation. In preliminary tests, different drugs levels were used in dose-response assays, but eventually only optimal drug doses (1mM lithium, 0.6 mM VA, 10 μM clozapine) were used in different experiments. Cells were counted in each 24-period to determine their proliferation, or survival level. For the survival study, cells were seeded at 10^5 cells/well in a six-well plate in complete medium. Four hours after seeding, the medium was changed to serum-free RPMI-1640 and the drugs were added. Two controls without drugs (serm-free and serum-feed medium) were performed. The medium and drugs were changed every two-three days and fresh drugs were added. Phase contrast images were acquired from three random fields from the coverslips at different times up to ten days, and cells viability was assessed by the trypan blue staining method.

4. Protein Quantification with the Immunoblot Method

Protein levels, enzyme activities, mRNA levels measured in this stuty, and the source of antibodies used as well, are summarized on Table 2. Cells or coronal brain slices (*approx.* 50 mg fresh weight) were homogenized in an extraction buffer of 50 mM Tris-HCl (pH 7.4), containing 150 mM NaCl, 1% Triton X-100, 1% sodium decolate, 0.01% SDS, 5mM EDTA, 1 mM phenylmethylsulfonyl fluoride (*PMSF*), aprotinin, leupeptin and pepstatin, 10μg/ml each. Homogenates were incubated for 20 mins at 4°C with gentle rocking, centrifuged (10'000 g x 10 mins, 4°C) and the supernatant was used for the assay samples. The loading buffer contained 0.125 mM Tris (pH 6.8), 20% glycerol, 10% mercaptoethanol, 4% SDS, and 0.02% bromophenol blue. Samples were heated at 95°C for 10 mins before gel loading. Protein concentrations were previously determined by the *BCA* kit (Pearse Chemical, Rockford, USA). Equal amounts of a soluble fraction of sample proteins (50 μg for GSK-3ß, β-actin and PI3K, 25μg for Akt-1 and PTEN) were electrophoresed on 10% (w/v ratio) SDS-polyacrylamide gel with a mini protean system (Bio-Rad Laboratories AG, Reinach, Switzerland) on pre-cast ready gels (Criterion, Bio-Rad). The samples were electrotransferred overnight onto PVDF membranes (GE Healthcare, Otelfingen, Switzerland), then blocked for 1h at room temperature in TBST solution with 5% (w/v) non-fat milk powder and bovine serum albumin, 2%. The blots were separately incubated overnight at 4°C with each of the following primary antibodies: anti-GSK-3αβ which recognonizes both alpha and beta isoforms (dilution 1:2000), (Santa-Cruz Biotechnology, Santa Cruz, CA) or anti-PTEN (1:1000), anti-PI-K (p85 regulatory subunit: *1:1000*) or anti-Akt-1 (*1:1000*) all from Cell Signaling Technology (Beverly, MA). After washing, membranes were incubated with an anti-rabbit IgG labelled with HRP membranes (GE Healthcare, Otelfingen, Switzerland), for 1h at room temperature. The membranes were washed and developed with the

Chemiluminescence ECL Western blotting membranes (GE Healthcare, Otelfingen, Switzerland), followed by exposition of the membranes with radiographic films (Hyperfilm ECL; GE Healthcare, Otelfingen, Switzerland). The autoradiograph exposure times were adjusted to the protein levels. In most cases, the PVDF membranes were stripped with Reprob Plus (Chemicon, Temecula, CA) and reincubated with ß-actin antibody (*1:1000 dilution*; Abcam Ltd. Cambridge, UK) as reference protein. The different proteins detected by their specific antibodies were identified by their molecular sizes (GSK-3α and ß bands were located at 52 and 47 kDa, respectively; the PTEN at 47kDa, Akt-1 migrated at 60 kDa, PI3K at 85 kDa and ß-actin was found at 46 kDa). Protein electrophoresis was run at different times, or on different gels, due to the very close molecular weights.

Table 2. Summary of proteins levels, enzymatic activity and mRNA levels measured, both in brain tisuue and cell culture, and the methods used as well.

Molecules	Brain tissue	Cell Culture	Methods	Source Antibodies
PI3-K protein levels	v	v	WB	CST
PI3-K enzy activity	v	-	^{33}P-transfer	-
Akt-1 levels	v	v	WB	CST
Akt-1 activity	v	-	^{33}P-transfer	-
Akt-1 mRNA	-	v	qRT-PCR	-
pAkt levels	-	v	WB	CST
PTEN levels	v	-	WB	CST
GSK3αß levels	v	v	WB	Santa Cruz
GSK3ß levels	v	v	WB	Santa Cruz
p-GSK3β	-	v	WB	Santa Cruz
GSK3ß activity	v	-	^{33}P-transfer	-
GSKαβ activity	v	-	^{33}P-transfer	-
ß-actin protein*	v	v	WB	Abcam
GAPDH mRNA*	-	v	qRT-PCR	-

* : used as controls ; WB : Western immunoblot ; CST : Cell Signaling Technology. See text for the detailed methods either for cell culture, brain dissection and laboratory methods.

Quantification of the immunoreactivities was performed by densitometric scanning using an image analysis system (Molecular Analyst, BioRad). The optical density (*OD*) obtained from each band was normalized against the corresponding ß-actin band. In pre-test experiments, the procedure was standardized with increasing levels of protein (15-100 µg/well) to determine the linear range, and standard curves were repeatedly performed. In some cases, recombinant proteins (Sigma Chemicals, Buchs, CH) were used in preliminary tests and the zone of linear range was determined. Test-retest stability in different samples assayed twice with different antibodies showed a significant correlation between tests.

5. Preparation of Brain Lysate, and Immunoprecipitation for Kinase Assays

Enzyme activities of selected proteins (Table 2) were measured in the brain lysates prepared as follows: tissue sections (*approx.* 50 mg: 1-2 mm depth) were homogenized in Phosphosafe ™ Extraction buffer (Novagen, Merck BioSciences) in 1/5 ratio (*W/V*). This buffer offers significant advantages over a classical lysis buffer by preserving the phosphorylation state of proteins, and maintaining kinase activity (Hayes et al 2004). The lysates were sonicated in Eppendorf tubes for 10 secs on ice and centrifuged at 14'000 g for 15 mins. The supernatant was saved for further use, and an aliquot of 50 μl was taken out for protein determination, with the Micro BCA method (Pierce Chemical Co., Rockford, IL).

6. Total GSK3 and GSK-3β Activity

The lithium-inhibitable activities of GSK-3α/ß or GSK-3ß alone were determined. To measure the GSK-3ß activity, kinase assay was performed on the above lysate after immunoprecipitation: 15 μl of primary GSK-3ß antibody (dilution:1/500) (Sigma Chemical,, St-Louis, USA) were added to 200 μl of brain lysates (50 μg prot.) and incubated with gentle rocking at 4°C, overnight. Then, Protein A agarose beads (20 μl of 50% bead slurry) and samples were mixed and incubated at 4°C with gentle rocking for 3h. The samples were microcentrifuged for 30 secs at 4°C and then washed twice with 500 μl of lysis buffer and twice with 500 μl of kinase buffer (25 mM *MOPS*, pH 7.2 containing 25 mM ß-glycerophosphate, 5 mM EGTA, 1 mM Na-Orthovanadate, 1 mM DTT). For both lithium-inhibitable activity of GSK-3ß and total GSK-3α/ß, the procedures were the same: assay was performed at 30°C for 30 mins, in the presence or absence of 30 mM lithium, according to the Upstate protocol (Upstate Biotechnology, Lake Placid, NY). In brief, 25 μl of the following mixture (5 μl per component) were prepared: GSK-3 kinase buffer; GSK3ß substrate peptide (40 μM) derived from glycogen synthase (Upstate Biotechnology); 30 mM LiCl (in inhibited set samples) or H_2O (in non-inhibited set samples); brain lyasate (10 μg prot.) or IP-GSK-3ß and a mix of 100 μM ATP and 15 mM $MgSO_4$ (to which 1000-3000 cpm/μl of ^{33}P-ATP were added). In control assays, recombinant GSK-3ß replaced lysates. After 30 mins at 30°C, the reaction was stopped and 20 μl were spotted onto prelabelled P81 phosphocellulose paper. The paper was immediately washed three times in 50 ml of 75 mM phosphoric acid and once in 50 ml acetone. ^{33}P incorporation into GSK-3ß substrate was quantitated by liquid scintillation spectrometry (Beckman-Coulter, Nyon, CH). The lithium-inhibitable activity was estimated as the difference between the two set sample values (with or without lithium) and ranged from 40% to 70% of total phosphorylation (non-lithium inhibited). By using the specific activity of the ^{33}P-ATP, the enzymatic activity was expressed in the amount of pmol of ^{33}P/min/mg prot.

7. PKB/Akt Phosphorylation Activity

Akt and GSK-3αβ kinase activity were measured in the brain lysates prepared as above. The lysate was immunoprecipitated with primary specific Akt antibody (Cell Signaling Technology). The immunoprecipitated Akt-1 activity (IP-Akt) was determined by phosphorylating a specific and optimal Akt substrate (AKtide-2T from Calbiochem) in the above assay buffer. For the procedure of immunoprecipitation, 15 μl of primary Akt-1 antibody (dilution:1/1000) (Sigma Chemical,, St-Louis, USA) were added to 200 μl of brain lysates (50 μg prot.) and incubated with gentle rocking at 4°C, overnight. Then, Protein A agarose beads (20 μl of 50% bead slurry) and samples were mixed and incubated at 4°C with gentle rocking for 3h, as for GSK-3ß. The Akt kinase activity was assayed in the presence of 20 μM AKtide-2T. In brief, 25 μl of the following mixture (5 μl per component) were prepared: MOPS buffer; IP-Akt (10 μg prot.); Akt substrate (20 μM AKtide); a mix of 100 μM ATP and 15 mM $MgSO_4$ (to which 1000-3000 cpm/μl of ^{33}P-ATP were added) and H_2O or 5μM staurosporine (negative controls). Positive control with recombinant Akt-1 (10-50 ng) (Calbiochem, Darmstadt, Germany) was performed. After 30 mins at 30°C, the reaction was stopped and 20 μl were spotted onto prelabelled P81 phosphocellulose paper. The paper was immediately washed three times in 50 ml of 75 mM phosphoric acid and once in 50 ml acetone. ^{33}P incorporation into Akt substrate was quantitated by liquid scintillation spectrometry (Beckman-Coulter, Nyon, CH). The enzymatic activity was expressed in the amount of pmol of ^{33}P/min/mg prot.

PI3K is a lipid kinase and preparation was slightly different: 500 μg of brain tissue were prepared in tha same buffer as for GSK-3ß (see above) then immunopreprécipatated by using a specific polyclonal antibody (5μg/sample: Cell Sigaling Technology) and binding on ProteinA/G-agarose beads (30μl) for 2h at 4oC. The complex was isolated by centrifugation, washed with a HEPES buffer (25 mM HEPES-150 mM NaCl 1mM EDTA, 1mM EGTA, 1% Triton X-100, 1mM sodium vanadate and a cocktail of protease inhibitors, final pH 7.5) and then suspended in 35 μl of kinase buffer (30 mM HEPES and 30 mM MgCl2, pH 7.5) containing 20μg of phosphatidylinositol. The kinase reaction was initiated by the addition of 5μl of 400 μM ATP (to which 5μCi of ^{33}P-ATP were added), incubated for 15 min at 30ºC, and stopped with 100 μl 1N HCl. Lipids were extracted according to the method of Whitman and collegues, were analysed by radioactivity and converted in *pmol,* using the specific activity of ^{33}P-ATP (Whitman et al (1998). The enzymatic activity was expressed in *pmol/min/mg prot.*

8. Gene Expression: Cell RNA Extraction and Quantitative RT-PCR

Approximately $5x10^6$ cells were collected after different periods of drug treatment, washed in PBS and then submitted to lysis with the RNAqueous kit from Ambion (Austin, Texas). Total RNA concentration was determined from a spectrophotometer optical density measurement (260 and 280 nm) and the purity of RNA was tested on a Bioanalyzer Agilent Technologies instrument (Basel, CH) which yield two sharp bands of 18S and 28S RNA. For all samples tested, the ratio between the spectrophotometer readings at 260 and 280 nm (OD 260/OD280) was between 1.8 and 2.0. Reverse transcription reactions were carried out on 500 ng of total RNA using the High Capacity cDNA Archive enzyme (Applied Biosystems) at 37ºC for 2 hrs.

For real time PCR, the ABI Prism 7900 HT Sequence Detection System instrument and the Assay-on-demands Gene Expression Products (TaqMan MGB probes, FAM dye-labeled) from Applied Biosystems (Applera Europe, Rotkreuz) were used to quantify mRNA for target gene and endogenous controls (ß-actin or GAPDH). The thermal conditions were given by the supplier : an initial step (2 min at 50 °C and 10 mins at 95 °C) was followed by 40 cycles of amplification (15 sec at 95 °C and 1 min at 60oC). All samples were run in triplicate. Data were analysed by the SDS 2.2 software with the comparative Ct method, where mRNA levels of target genes were normalized to the corresponding ß-actin or GAPDH genes expression

(Livak and Schmittgen, 2001). The drug-naive treated samples at time 0h were used as calibrator samples and results were expressed in percetange of change with respect to control.

9. Statistical Analyses

Data analyses were performed with a statistical software (StatView V, Brain Power, Calabas, CA). The results were reported as mean±SD. For Western blots assays, the values of optical density were expressed as the ratio to ß-actin protein. The statistics between groups were performed with parametric *ANOVA* or non-parametric Friedman test with significant difference set at $p<0.05$. *Ad hoc* Fisher or Mann-Whitney tests were used for multiple comparisons and significance was set at $p<0.01$. To test the potential effect of confounding variable of gender on various measures, a two-way *ANOVA* (gender and diagnosis) was used. In depressed suicide subjects, although postmortem interval (PMI) and age were matched as carefully as possible, and no difference between control and suicide victims was found (Table 1), the influence of the two parameters on protein levels and kinase activity were examined. For that, ANCOVA was used with PMI and age as covariates. The effect of the method of suicide on protein variations was tested by comparing violent and non-violent methods of death and an unpaired t test was used with significance set at $p<0.05$.

For the cell culture tests, data from immunoblotting and from qRT-PCR were expressed in relative values (percentage or ratios ±sd) with respect to housekeeping genes (GAPDH or ß-actin). Due to transformed values (expressed as ratio or percentage), non parametric Friedman tests for ANOVA was used for variance between groups, and multiple comparisons between controls (drug-untreated) and drug-treated cells was performed with *post hoc* Mann-Whitney U-test. The significance was set at $p<0.05$.

C. Results

1. PKB/Akt and GSK3αβ Protein Levels

For all immunoblot assays, preliminary tests were performed to establish the linear range and accurate amount of proteins for loading in routine assays. The linear range was obtained between 5 and 50 μg, and 25 or 50 μg per sample were loaded in duplicate in routine experiments.

Figure 1a shows representative Western blots of immunolabelling of the total levels of GSK-3α/ß, Akt-1 and ß-actin proteins from suicide victims (with and without *MDD*) and non-suicide subjects (with or without *MDD*), all in duplicate. The two forms of GSK-3 (GSK-3α and GSK-3ß) were co-revealed and migrated at 52 and 47 kDa, respectively and Akt-1 migrated at 60 kDa. The identity of proteins was ascertained by molecular size markers and the use of respective anti-peptides.

The relative optical density (ROD) of these protein bands, expressed as a ratio to ß-actin are displayed in Figs 2b-c. For the three proteins, there was no change in different diagnostic groups either for GSK-3α (*ANOVA*: $F_{3,36} = 0.583; p = 0.630$), and for GSK-3ß (*ANOVA was* $F_{3,36} = 0.971; p = 0.410$) (Fig 2) or for Akt-1 (*ANOVA*: $F_{3,36} = 0.618; p = 0.608$).

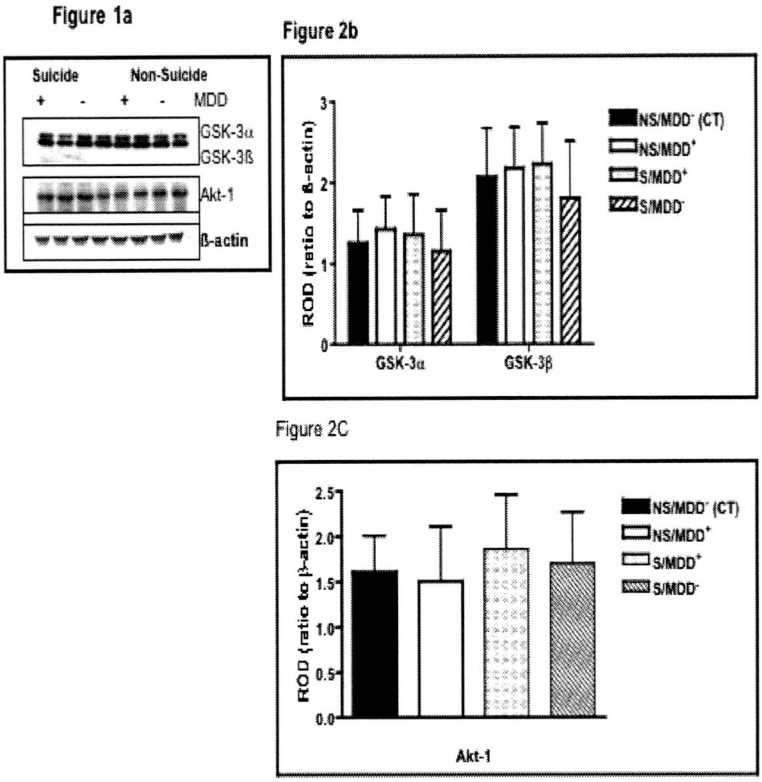

a): Representative immunoblots of the total levels of GSK-3α/ß, Akt-1 and ß-actin proteins extracted from the ventral prefrontal cortex of suicide victims with (+) and without (-) an *antemortem* diagnosis of major depression disorder (*MDD*) and non-suicide subjects with (+) or without (-) an *antemortem MDD*. Samples, loaded in duplicate, were separated by electrophoresis on acrylamide gel, probed with respective antibodies and detected by the Chemiluminescence ECL method.

b): Relative protein levels of total GSK-3α and GSK-3ß expressed in ratio to ß-actin, in suicide victims with (*S/MDD*$^{+}$) and without (*S/MDD*$^{-}$) major depression disorder, and in non-suicide subjects with (*NS/MDD*$^{+}$) and without (*NS/MDD*$^{-}$) (*CT*: controls) major depression disorder. One-way ANOVA yielded no statistical significant difference between groups. Statistical analysis of measures for GSK-3α (*ANOVA*: $F_{3,36} = 0.583$; $p = 0.630$), and for GSK-3ß (*ANOVA:* $F_{3,36} = 0.971$; $p = 0.410$).

c): Relative protein levels of Akt, expressed in ratio to ß-actin, in suicide victims with (*S/MDD*$^{+}$) and without (*S/MDD*$^{-}$) major depression disorder, and in non-suicide subjects with (*NS/MDD*$^{+}$) and without (*NS/MDD*$^{-}$) (*CT*: controls) major depression disorder. One-way ANOVA yielded no statistical significant difference between groups. Statistical analysis of measures for Akt-1 (*ANOVA:* $F_{3,36} = 0.618$; $p = 0.608$

Figure 1.[+]

[+] Licence: Figure 1a-c and Tables 1 and 3 are reproduced from Karege et al 2007 with the courtesy of Elsevier Limited licence: Number 1747061100696)

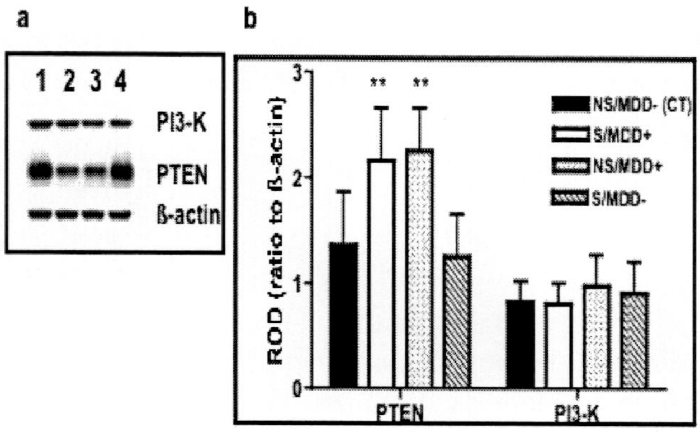

a) Representative immunoblots of the total levels of phosphatidylinositol 3-kinase (PI3-K), phosphatase and tensin homolog deleted on chromosome ten (PTEN) and ß-actin proteins extracted from the ventral prefrontal cortex of suicide victims with (line 1) and without (line 2) an *antemortem* diagnosis of major depression disorder (MDD) and non-suicide subjects with (line 4) or without (line 3: Controls) an *antemortem* MDD. Samples were separated by electrophoresis on polyacrylamide gel (PAGE), probed with respective antibodies and detected using the Chemiluminescence ECL method. See text for details.
b) Relative levels of total PTEN and PI3K in suicide victims with (S/D+) and without (S/D-) major depression disorder, and in non-suicide subjects with (C/D+) and without (controls) major depression disorder. Data are expressed as a ratio to β-actin OD. Each subgroup included 10 subjects. For PTEN measures, *ANOVA* yielded: $F(3,36)=10.59$, $p<0.001$ and *Post hoc* Fisher tests with respect to controls: ** $p < 0.002$, for both groups. No statistically significant difference was obtained for PI3K protein levels: *ANOVA yilded* $F(3,36)=0.604;p<0.798$).

Figure 2.

Table 3. Kinase activity of brain GSK-3, GSK-3ß and Akt-1 in diagnostic groups of suicide and non-suicide subjects.

	Non-suicide subjects		Suicide victims	
	No *MDD*	with *MDD*	No *MDD*	with *MDD*
Kinase activity measures				
Tot GSK-3α/ß	14.3±5.1	16.9±5.9	14.9±4.7	17.5±5.5
GSK-3ß	7.6±2.3	11.6±2.5**	8.1±2.4	12.2±2.8**
PKB/Akt	27.3±9.0	18.8±5.1**	25.8±6.3	17.8±4.9**

Mean vales (±SD) of lithium inhibitable total GSK-3α/ß and GSK-3ß alone, and Akt activity in the four subgroups: suicide victims with (MDD) and without MDD (no MDD) and non-suicide subjects with (MDD) and without (no MDD) major depression disorder. Values are expressed in *pmol/min/mg* protein of brain tissue. Statistical analysis: no significant differences in tot GSK-3 measures between different groups. For Akt measures, one-way ANOVA: $F_{3,36} = 5.372$; $p = 0.003$ and *post hoc Fisher test:* ** $p = 0.002$. For GSK-3ß measures, one-way ANOVA: $F_{3,36} = 8.567$; $p = 0.002$: and *post hoc Fisher* test: ** $p = 0.002$

Table 3 displays the mean values of the kinase activity (expressed in *pmol/min/mg prot.*) of total GSK-3α/ß, IP-GSK-3ß alone and IP-Akt in different diagnostic groups. With respect to total activity of GSK-3α/ß, no significant change was observed between diagnostic groups (*ANOVA:* $F_{3,36} = 0.847$; $p = 0.477$), while when GSK-3ß was assessed alone after immunoprecipitation, significant changes in different diagnostic groups were observed (ANOVA: $F_{3,36} = 8.567$; $p = 0.002$). *Post hoc* Fisher test for multiple comparisons showed significant increases in depressed suicide (*p<002*) and depressed non-suicide ($p<0.002$) subjects with respect to the control group (non-suicide non-depressed subjets). The group of suicide without documented major depression was not statistically different from the controls ($p = 0.697$).

With respect to Akt-1 kinase activity, significant changes between diagnostic groups were observed (*ANOVA:* $F_{3,36} = 5.372$; $p<0.003$). *Post hoc* Fisher test for multiple comparisons showed a significant decrease in Akt-1 activity in suicide depressed group ($p<0.002$) and in non-suicide depressed subjects ($p = 0.006$). The suicide group with non-documented history of depression was not different from the control group ($p = 0.613$).

Tests on the influence of potential confounding factors of human PMI and age yielded a non-significant effect on all proteins, whatever the subgroup. The effect of gender, as analysed by a two-way ANOVA (gender and diagnosis), was not significant. Likewise, there was no difference between violent and non-violent methods of suicide.

2. PI3K and PTEN

Figure 2a shows representative Western blots of immunolabeling of PI3K and PTEN and ß-actin proteins from suicide victims with (line 1: S/MDD+) and without (line 2: S/MDD-) major depression and non-suicide subjects without (line 3: NS/MDD-) or with (line 4: NS/MDD+) major depression. The identity of proteins was ascertained by molecular size markers and, in case of doubt, by the use of respective anti-peptides (not shown).

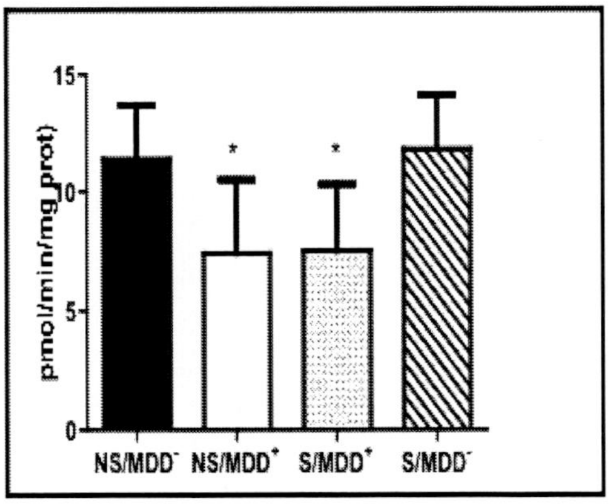

Figure 3. PI3K activitiy was measured in tissue lysates from VPFC of *postmortem* depressed suicide and non suicide subjects. Lysates were immunoprecipitated by specific antibodies and the complexes were separated on Protein G/A Agarose beads, then isolated, washed, and finally incubated for 15 min at 30ºC in presence of their substrates (10μM PI as a substrate of PI3K) in kinase buffer containing 5μCi (^{33}P-ATP). ANOVA yilded F3,36= 9.20, $p<0.001$. Post hoc Fisher test indicated decreased activity in both shown subgroups. *<0.05

The relative optical density (ROD) of these protein bands, expressed in mean values (±sd) of the ratio to ß-actin OD, are displayed in Figs 2b. Regarding PI3K protein levels, the mean values of ROD were not statistically different in the four diagnostic groups, as can be observed in Fig 2b (ANOVA: NS). For PTEN measures, ANOVA yielded statistically significant differences among diagnostic groups [$F(3,36) = 10.59$, $p < 0.001$)]. *Post hoc* Fisher tests for multiple comparisons indicated that PTEN measures were significantly higher in two diagnostic groups: the suicide victims with depression (S/MDD+) ($p < 0.002$) and the non-suicide subjects with a well-diagnosed history of depression [(NS/MDD+) ($p < 0.002$)] in comparison with the control group.

Figure 3 displays values of brain PI3K kinase activity (expressed in pmol/min/mg prot.) in different diagnostic groups. With respect to this kinase activity, significant variations between diagnostic groups were [(*ANOVA:* $F(3,36)=9.20$, $p<0.001$)]. *Post hoc* Fisher'tests indicated decreased activity in both depressed suicide (S/MDD+) and depressed non suicide group (NS/MDD+) ($p<0.01$), comnpared with the group of controls (NS/MDD-) ($p<0.01$).

Tests of the influence of potential confounding factors of human PMI and age yielded non-significant effect on all proteins, whatever the subgroup. The effect of gender, analysed by a two-way ANOVA (gender and diagnosis), was not significant. Likewise, there was no difference between violent and non-violent methods of suicide.

3. Effects of Drugs on Cell Proliferation

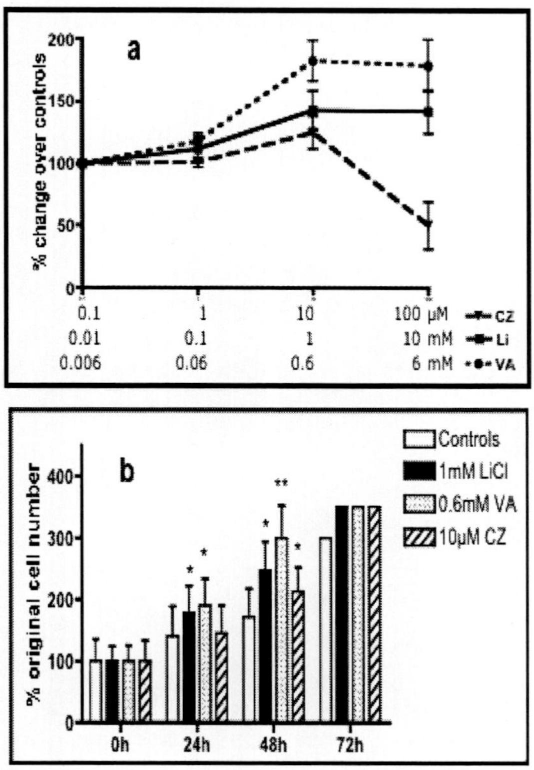

Figure 4. a) : Dose-response study of the the SH-SY5Y cells proliferation in RPMI-1640 with 10% FCS and complemented with mood stabilizers for a period of 48h, before cells confluence. Results, expressed in percentage change of controls (no-added drug), are means of 4-6 independent assays. Note CZ at 100 µM was toxic. 4b: Relative drug-induced cells growth expressed as percent (±sd) over basal (control cells) of 1mM LiCl, 0.6mM VA or 10µM clozapine-treated cells. Results are means±sd of 4-6 independent assays and non-parametric ANOVA was : Chi-2=23.6; p<0.001 and *post hoc* Mann-Whitney tests:. *p<0.05.; ** p<0.01.

To assess the effects of mood stabilzers, cells were cultured in RPMI-1640 with FCS (10%) and supplemented with drugs at increasing concentrations and cells were counted at different time points. Figure 4a displays dose-response assays of the three mood stabilizers on cell proliferation, after 48 h of drug treatment, before the cell confluence and data were expressed in percent change of non-drug supplemented controls. The three drugs significantly increased the FCS-induced cell growth and the optimal levels were found at 10µM, 1mM and 0.6mM for CZ, Lithium and VA, respectively. Higher concentrations were either no more efficient (Li and VA) or were toxic (CZ). To determine the time course effect of drugs, in comparison with non-drugs supplemented cells, cells were treated with drugs at their opimal cocentrations, and results were expressed with respect to the original cell number (time 0h). To show how vary the control, the refernce used here, is the original number of cells for each group. Figure 4b shows the time-course of cell proliferation in RPMI-1640 with 10% FCS and complemented with drugs, for up 72h (cell confluence). LiCl (1mM), VA (0.6mM) and CZ (10µM) addition resulted in a time-dependent significant potentiation of cell proliferation, compared with non-treated cells (non-parametric Friedman test : *chi-square*= *23.6*; $p<0.001$). While the action of lithium and VA has became significant after 24h, the effect of CZ was belated and significant after 48h (Mann-Whitney test for Li + and VA vs control; $p<0.01$ and $p<0.01$, respectively) It was established that the increase in cell proiliferation results from an increase in cell division confirmed by significant DNA neosythesis after 24h (result not shown).

4. Cytoprotective Actions in Serum-Starved Cells

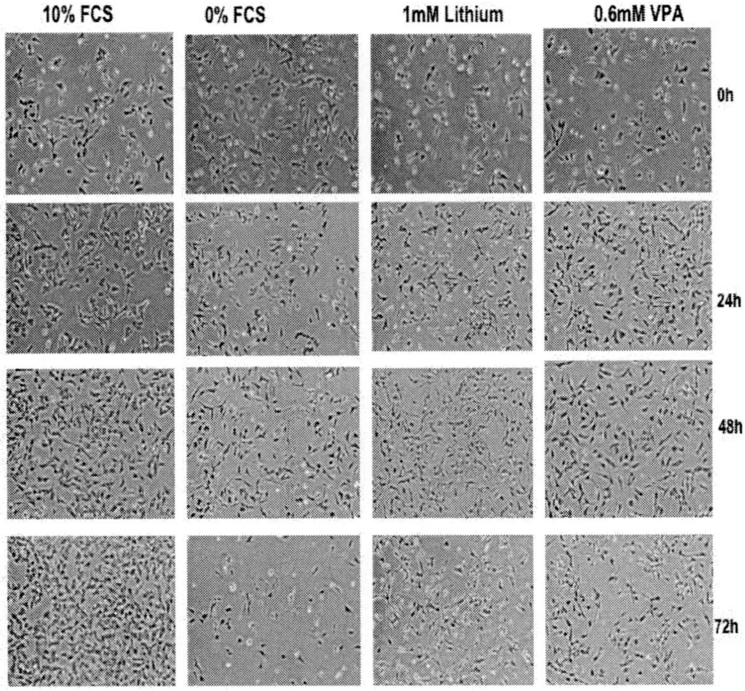

(a)

Figure 5. Continued on next page.

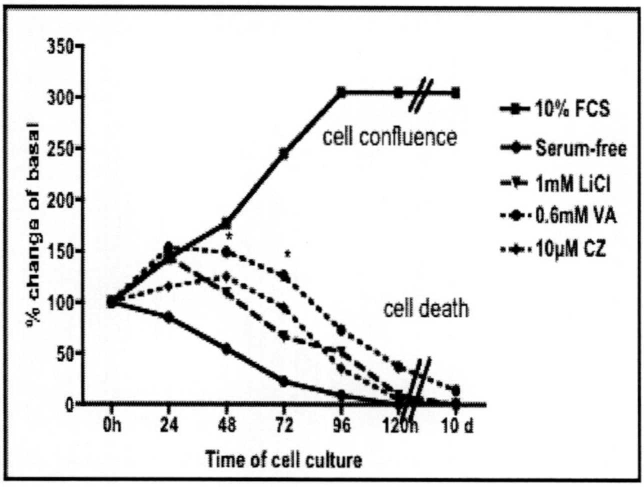

(b)

Figure 5. The protective effect of mood stabilizers against serum-starving induced cell death. See Fig 1 for concentrations. 5a: Phase contrast images of valproate or lithium-treated and control cells, for a period up to 72 h in serum-deprivated medium. 5b: Relative cell death or cell proliferation (in serum-fed cells) with respect to the starting cell number. Results are means (±(sd) of 4-6 independent assays. Friedman test for ANOVA was significant at periods between 24h and 72h (Chi-2 = 27.6; $p<0.01$). Mann-Whiney U-test tests performed with respect to serum-free cells were significant for VA and Lithium, at 48h and 72h. * $p<0.05$.

Figures 5 (Fig 5a for contrast images and Fig 5b for cell counting) show the drugs's effect on SH-SY5Y cells cultured in serum-free RPMI-1640 solution. As controls, serum-fed cells and serum-free without drug additon were also included in the assay. Serum deprivated cells underwent rapid and progressive death, which can be observed after few hours of serum withdrawal and significant after 24h, compared with serum-fed cells. The addition of drugs at their optimal concentrations : lithium (1mM), VA (0.6 mM) and CZ (10µM) on serum-free medium delayed cell death for a period of 48 to 72h, depending on the drug used (Friedman test of ANOVA chi square = 27.6; $p<0.01$) (Fig 5b). The effect was significant after 48h for VA and lithium (Mann-Whitney U-test; $p<0.05$ and $p<0.01$, respectively). At 120 h, there was still 50% of VA-treated cells, while in serum-free samples almost all cells have died and 20-30% remainned alive in lithium and CZ-treated samples. After 10 days, 5-10% of VA-treated cells were still alive (Fig 5b).

5. Immediate and Late Biochemical Effects on Akt/PKB and GSK-3ß

For the molecular assays, two approaches were used: the study of immediate effects on the phosphorylation of Akt at serine-473 and and GSK-3ß (Figure 6) the late effects on Akt protein and mRNA expression (Figure 7). Figure 6a displays representative immunoblots of ^{437}Ser-pAkt and ^{9}Ser-pGSK-3ß and their native forms after 30 mins of drug stimulation. The effect could last up to 6 h, or more but gradually declined (not shown). As indicated in relative values of optical densities (ROD), lithium and Valproate, but not clozapine, mediate

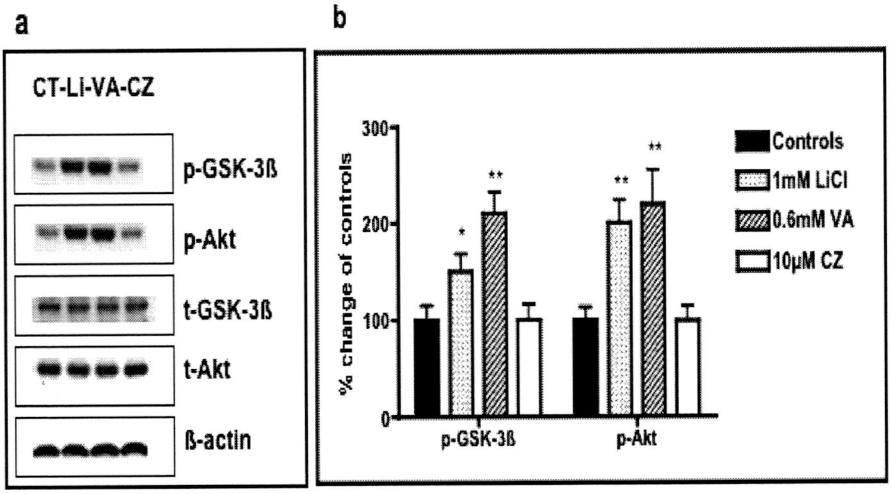

Figure 6. Rapid effect of mood stabilizers on phosphorylation of Akt-1 and GSK-3ß. 6a: Representative immunoblots of ^{437}ser-pAkt and ^{9}ser-pGSK-3ß and their native forms after 30 mins of drug stimulation. ß-actin was assessed as reference protein. 6b: Relative values of optical densities (ROD) of p-GSK-3ß and p-Akt immunoblots expressed as percent of non treated samples. Data expressed in percent of controls are means (±sd) of 4-6 independent assays and statistical analysis with Mann-Whitney U-test yielded significant change in lithium (1mM) and VA (0.6mM), but not clozapine (10µM) -treated cells. Respective U-tests were: * $p<0.05$ and ** $p<0.01$.

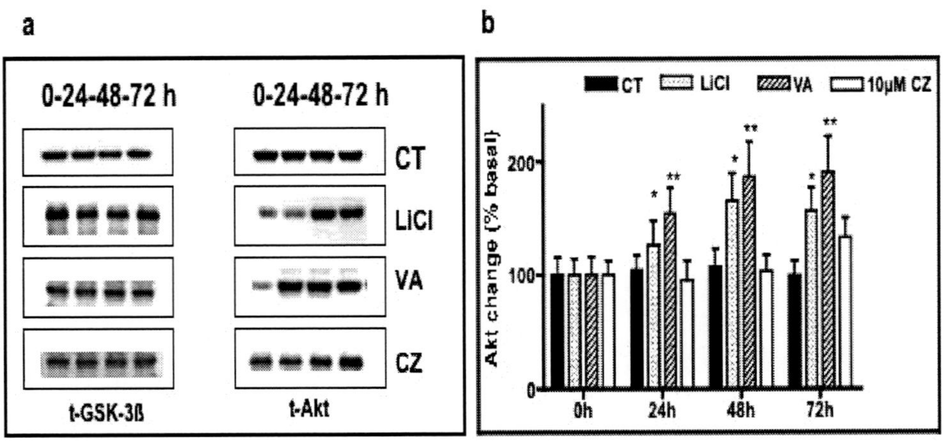

Figure 7. Late effect of mood stabilizers on total Akt and total GSK-3ß protein expression. a: representative immunoblots of t-GSK-3ß and t-Akt proteins during mood stabilizers treatment up to 72h. b: Relative optical densities (ROD) t-Akt proteins during 72h. Values are expressed as percent of increase over the time 0h. Results are means (±sd) of 4-6 independent assays and non parametric Friedman test for ANOVA was performed (chi-2 = 13.5; $p<0.05$) and *post hoc* Mann-Whitney tests were significant: *$p<0.05$; **, $p<0.01$.

an rapid phosphorylation of Akt and GSK-3ß (Fig 6b). Respective U-tests were significant for lithium effect on both GSK-3ß and Akt (*p<0.05* and *0.01*, respectively) and VA on GSK-3ß and Akt (*p<0.01* and *0.01*), respectively. There was no immediate phosphorylation with CZ, but a delayed effect was observed after 6h (data not shown. Analysis of total Akt and total GSK-3ß over a period of 72h of drug treatment revealed that Akt protein levels, but not GSK-3ß protein, significantly increased in the 72h-period (Figures 7a-b). Friedman test of ANOVA yielded for Akt changes: Chi-2=13,5; $p<0.05$. Multiple comparison analysis indicated that changes were significant after 24h for VA-treatment ($p<0.01$), 48h for lithium ($p<0.05$), and subsequent days. Only a trend to significant change was observed for clozapine at 72h ($p<0.08$) (Fig 7b). For GSK-3ß protein, no significant change, whatever the drug, was observed during this period (only immunoblots are shown).

6. Effects on Akt-1 Gene Expression

A time-course of Akt gene expression under drug treatment was undertaken. Figure 8 presents results from a real time qRT-PCR study for Akt-1 mRNA expression. Lithium and VA-treatment induced a time-dependent increase in Akt mRNA levels (Friedman test of ANOVA: *Chi-2=19.8* $p<0.001$ and *chi-2=20.2;* $p<0.001$, respectively). The change in Li- and VA-treated cells were significant after 12h, both for lithium (Mann-Whitney test: $p<0.05$) and VA (Mann-Whitney test: $p<0.01$). For the clozapine-treated cells, the global changes were not significant, but at 24h Akt mRNA was moderately increased compared with the original amount (*t-test:* $p<0.05$). For all drugs, no effect was observed on GSK-3ß mRNA during the same period.

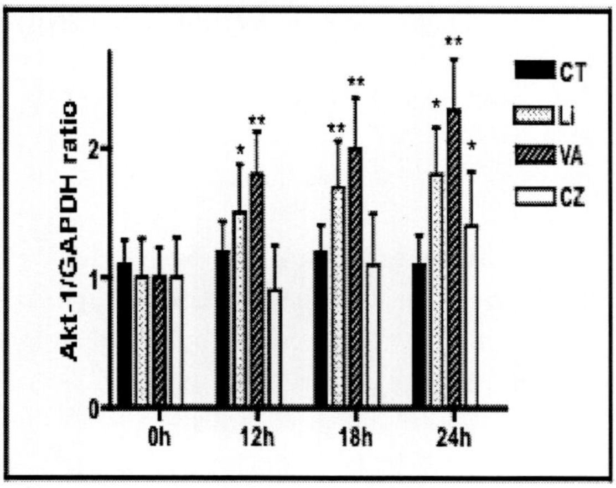

Figure 8. Time course of Akt-1 mRNA expression in neuroblastoma SH-SY5Y cells grown in a culture medium supplemented with mood stabilizers at their respective therapeutic doses. Akt-1 mRNA was assayed by the real time quantitative RT-PCR method and results were expressed in ratio values with respect to a house-keeping gene (i.e. GAPDH), as an endogenous gene and were calibrated to time 0h samples. Values are means (±sd) of 4-6 independent assays and non parametric Friedman test of ANOVA yieled significant changes for lithium (chi-2 =19.8 $p<0.001$) and valproate (chi-2= 20.2 ; $p<0.001$) but not clozapine (chi-2= 3.1 ; NS). *Post hoc* Mann-Whitney tests *vs* calibrator samples were performed at different times : *$p<0.05$; ** $p<0.01$.

D. Discussion

The first part of this study was aimed to investigate the alteration of the PI3K-Akt-GSK3αβ signaling axis in postmortem brain of suicide depressed subjects and non depressed controls. Protein levels and kinase activities were extracted from and assessed in tissue lysates from ventral prefrontal cortex (*VPFC*). The second part was dedicated to the study of mood stabilizers effects on both activation and expression of Akt-1 and GSK-3ß proteins in neuroblastoma SH-SY5Y cells culture. Effect of these drugs on cell proliferation and cell survival were assessed, as well. The following discussion will bring closer both parts for an integration of the conclusions as neuroplasticity is concerned.

1. The Study of Brain Protein Kinases : PI3K, PTEN, Akt/PKB and GSK3

As regards to the *postmortem* study, only drug-free subjects with a well-established diagnosis of suicide depression were included. However, by using brain tissues, one first difficulty to overcome was the use of posmortem samples where both suicide depressed and suicide non depressed subjects were mixed. The latter subjects deceased by suicide but without history of MDD. It seems that these people commited suicide for some reason other than depression. There was also a group of subjects with a long history of MDD but deceased by mean other than suicide. To disentangle all that, we performed designed matrics of four diagnosed groups of suicide and non -suicide subjects with or without a history of major depression, to decipher changes in protein levels and/or kinase activity of PI3K, PTEN, Akt-1 and GSK-3 in postmortem *VPFC*'s. With respect to non depressed control subjects, the main changes observed in this study can be summarized as follows:

1. The PI3K enzymatic activity was decreased in MDD, without change in protein levels.
2. The GSK3ß activity was increased in MDD but the total enzymatic activity (α+β isoforms) was not changed, nor were the total GSK3αβ protein levels.
3. The Akt-1 enzymatic activity was decreased, in the two groups of depressed subjects, but not in protein levels.
4. The PTEN protein levels were increased in these groups of depressed subjects. We did not measured the PTEN activity.

A part PTEN, where the activity was not assessed, all the changes were observed only in the enzymatic activities, but not in protein levels. Moreover, the changes were observed in both suicide and non suicide depressed subjects. To summarize the whole results, this study indicates that the PI3K-Akt-GSK3ß pathway signaling activity is blunted in major depression. These changes were associated with a history of *MDD*, rather than with suicide *per se*. This is the first time that a study gathers the four proteins in the same protocol encompassing suicide and MDD, and shows such a specific dysregulation, by indicating a blunting of the signaling axis in MDD. Actually all these four factors PI3K, PTEN, Akt-1 or GSK-3β, constitute a series of sequentially activated protein or lipid kinases and phosphatases.

When considering these proteins individually, this study is consistent with previous studies on postmortem brains of mood disorder subjects, which showed no difference in

GSK-3 protein levels or overall activity in patients with BD or MDD comapred with controls (Beasley et al 2002 ; Kozlovsky et al 2001 ; Lessort et al 1999). In a postmortem study of GSK-3ß levels and phosphorylation of tau protein, Lesort and collegues observed a tendency to modification of tau phosphorylation states in bipolar disorder (Lesort et al 1999). Later on, Hsiung and collegues demonstrated a decrease in Akt and PI3K activity without any change in protein levels, in the occipital cortex of depressed suicide victims (Hsiung et al 2003). In a recent study on depressed teenage suicide, Pandey and collegues have reported a decrease in ser9-pGSK-3β and β-catenin protein levels in PFC and the hippocampus and therefore suggesting an increased GSK-3β function (Pandey et al. 2006). The present study is therefore consonnant with those observations, which reported reduced PI3K, Akt or GSK3ß kinase function in affective disorders (Lessort et al. 1999 ; Hsiung et al, 2003 ; Pandey et al. 2006). The increase in PTEN protein levels is also consistent with a previous study where a similar alteration was reported in suicide depressed subjects (Hsiung et al 2003), confriming its role in blunting of PI3K-Akt signaling.

Now to come back to the neuroplasticity hypothesis, these proteins have been formerly described as regulators for the survival of neurons and neuroplasticity in general (D'Mello et al, 1997; Dudek et al, 1997; Yamada and Araki, 2001; Jope and Bijur, 2002). Accordingly, these abnormalties in the expression and function of proteins may be relevant to neural plasticity in major depression. There are a number of reports on alterations of molecules involved in neuroplasticity and cell survival (Duman et al, 1997; Dwivedi et al, 2003; Karege et al, 2005; Zarate et al, 2006). It is now accepted that neurogenesis and synaptic remodeling are influenced by neurotrophic factors and various kinases participate in phosphorylation and functional characterisation of critical proteins (Dwivedi et al. 2003 ; Chen and Manji 2006). By using designed matrics of four diagnosed groups of suicide and non-suicide subjects with or without a history of major depression, the study also attempted to reveal changes specific to suicide and depression diagnostics. In the suicide studies, the relationship between MDD and suicide has always been problematic, and it is difficult to determine what effects are attribuable to the presence of a major depression as opposed to vulnerability to suicidal behavior. Many individuals in *postmortem* studies of one meet the criteria of the other. Suicide and suicide behavior are likely to involve specific neurochemical abnormalties that might only overlap with major depression (Mann et al. 2001). The data indicated that these proteins were not modified in the *VPFC* of suicide victims with no documented history of MDD. This suggests that rather than being an indicator of suicide, the compromised PI3K/Akt signaling pathway may be indicator of depression. In a recent study on gene ontology analysis, differentially expressed genes have revealed specific patterns in limbic system for suicide and depression (Sequiera et al. 2006). In this report, the authors suggested that suicide may be more associated with the regulation of transcription factors while depression is mainly related to second messenger systems like G-protein coupled receptors (Sequiera et al. 2006). Further evidence indicated that in the cAMP-PKA pathways, various catalytic and regulatory subunits were selectively decreased in *postmortem* suicide victims and changes were specific to depressed suicide, while in extra-cellular signal-regulated kinase (ERK1/2), alterations were observed in all suicide subjects irrespective of psychiatric diagnosis (Dwivedi et al, 2006). The difficulty lies in the fact that most studies on depressed suicide victims did not include depressed non-suicide subjects and/or non-depressed suicide victims, as this study did. Therefore, this report represents a significant advance by showing neurochemical signals specific to depression in depressed suicide subjects. Nevertheless, it is possible that some *antemortem* medical records might have ignored depressive episodes that could have occurred at the time of death. In such a case, the dysregulation could be probably associated with only chronic MDD, but not accute depression.

What could the remote cause of this PI3K-Akt dyregulation in the VPFC of depressed persons be? In other words, is there any regulation by cellular signals, such as monoamines or growth factors ? Firstly, a wealth of evidence from neuropsycho-neuroanatomy and functional imaging studies in animal and humans indicates that prefrontal cortex is a key component of cortico-striatal circuits that are thought to generate pathological emotional behavior and accompanying physiological disturbunce (Tork, 1999). The ventral prefrontal cortex is of particular interest because it is one of the brain regions whose activity positively correlates with the Hamilton Depression Rating Score, and a number of studies have indicate abornormalties in the structure and function of this region (Milak et al, 2005; Rajkowska, 2000). Interestingly, 5HT systems have been shown to have anatomical connections with the ventral PFC and to contain a high density of 5HT receptors sites including 5HT1a subtypes (Stockmeier et al, 1998). The role of serotonin receptors in Akt signaling has been highlighted in recent years (Cohen et al. 2005 ; Hsiung et al 2005 ; Johnson-Farley et al. 2005). Positron emission studies in major depression have shown reduced 5HT1a recptors in many brain regions (Drevets 1999 ; Drevets et al. 2007, for review of litterature). It has been shown that although the activation of 5HT1a receptors stimulates multiple signals, such as ERK1/2, PI3K/Akt and nuclear transcription factor-kappaB (NF-kappaB), only the PI3K/Akt activation is required for 5HT1a receptor-dependent cell survival (Hsiung et al. 2005). Further, evidence has recently indicated that 5HT1a receptors couple to activation of Akt, but not ERK, in cultured hippocampus neurons (Cohen et al, 2005). Alteration in 5HT1a-coupled G protein function has been demonstrated in depressed suicide and downstream cascade including PI3K and Akt (Hsiung et al 2003). Therefore, the blunting of the PI3K/Akt signaling pathway may be relevant to 5HT1a receptors mediating cell survival-associated responses.

Alternatively, the PI3K/Akt dysregulation could stem from some impaired neurotrophic factor systems, such as brain-derived neurotrophic factor (BDNF). Our group and others have indeed reported decreases in BDNF levels in hippocampus and prefrontal cortex of suicide depressed subjects (Karege et al 2005 ; Dwivedi et al 2005). In many systems, BDNF, like many other growth factors, was shown to activate the PI3K-Akt signaling axis (Mizuno et al. 2003 ; Schratt et al 2004). In cultured hippocampal neurons, along with insulin-like growth fctor (IGF), BDNF was found to activate the neuroprotective effect of Akt pathway through Akt-dependent inhibitory phosphorylation of GSK-3ß (Johnson-Farley et al 2006). Consisitently, it can be conceived that blunting of the brain PI3K-Akt-GSK3ß signaling axis could alternatively stem from decreased BDNF levels in major depression. By gathering these information, a simplified but not exclusive model of PI3K-Akt-GSK3 signaling was proposed, where our data and archive data on 5HT1a and BDNF signaling systems were integrated in a hypothetic model of signaling (Figure 9). Of course, many other signaling routes conducting to the neuroplasticity disrupture have been proposed so far, but there are not antagonists to this one (Duman et al. al 1997; Bachmann et al 2005, Hsiung et al 2003).

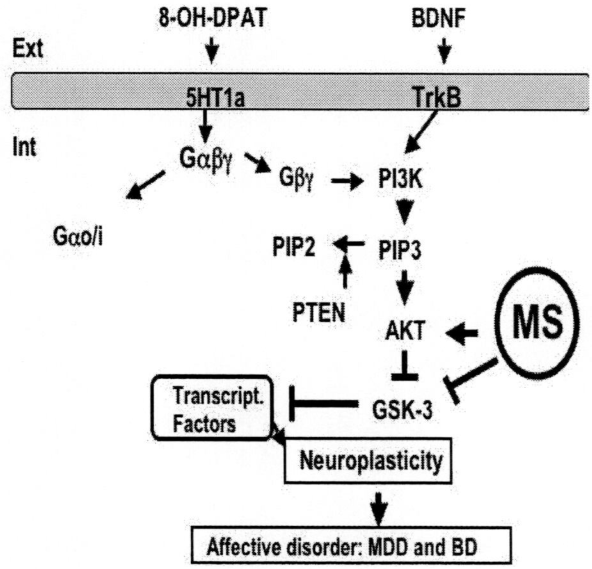

Figure 9. Simplified overview of PI3K-Akt-GSK3ß signaling axis that may be activated by 5HT1a receptrors and/or BDNF through its cognate receptor, TrkB, and rugulates the neuroplasticity in affective disorders. The model which is not exclusive (a number of other signalings axis, i.e. cAMP, were previously shown) was issued from the present study and from archive data.

Acronyms for Fig 9 :
MS : mood stabilizers ; 5HT1a : serotonin receptor, 1a-subtype ; PI3K : phosphatidyl-inositol 3-kinase, Akt, for Protein kinase B ; PIP3 : phosphaatidyl inositol triphosphate ; PIP2 : phosphatidyl inositol biphosphate ; GSK3 : glycogene synthse kinase 3 ; BDNF : brain-derived neurotrophic factor ; 8-OH-DPAT : 8-hydroxy-N,N-dipropyl-aminotetralin (8-OH-DPAT), is a 5HT1a receptor agonist ; TrkB : tyrosine kinase B, the BBNF receptor ; Gαβγ□: the three G protein-binding subunits (alpba, beta and gamma);

2. Effects of Mood Stabilizers on Akt and GSK3 Activation and Expression

The major goal of this second part was to determine whether drugs such as Valproate or Clozapine, could induce similar effects and activate the same molecular pathways, namely the Akt/GSK-3ß axis, with comparable efficacy and timing, as does Lithium, the mainstay and the oldest one. Their effects on cell proliferation and cell protection were also assessed. Different doses were tested and time-course experiments were performed. To compare with clinical situations, routine assays were performed at their respective optimal doses which correspond nearly to their therapeutically relevant concentrations, as reported by different authors (Bowden et al. 1994; Bowden et al. 1996; Perry et al. 1998; Sachs et al. 2004). The study yielded the following results:

a) On the cellular level:

1. The serum (FCS) stimulates the cell division and proliferation at a steady pace, in culture medium. Addition of the mood stabilizers to the culture medium results in speeding up the dose-dependent of the serum-sustained cell proliferation, by a

significant potentiation,. The effect was statistically significant after 24h (for Lithium and Valproate) and 48h (Clozapine).

2. The withdrawal of serum initiates the apoptotic cell death. Addition of mood stabilizers in the serum-free cell culture slowed down the cell death, delayed the apoptosis, and allowed the cells to survive few days more. This neuroprotective effect is more significant with VA and lithium than in CZ, and lasts several days.

b) On the molecular level, the following events were sequentially observed:

3. After 30 min, Lithium and Valproate, but not Clozapine, induce a significant phosphorylation of ^{9}Ser-GSK-3ß and ^{473}Ser-Akt, compared with drug-free serum feed controls.
4. At 12 h, Lithium and Valproate, not Clozapine, induce a significant increase in Akt-1 mRNA levels; no such a change was observed for GSK-3ß. Clozapine effect on Akt mRNA expression was significant after 24h.
5. After 24h, increase in Akt-1 protein levels, but not in GSK-3ß levels, was observed in Lithium and Valproate-treated cells.

These findings are globally consistent with previous works conducted on animal, and neuronal and non-neuronal cells which reported that psychotropic drugs, including Lithium, Valproate or other antipsychotics, can enhance cell viability and activate the Akt/GSK-3ß signaling pathway (Di Daniel et al. 2005; Kang et al. 2004; Chen and Manji, 1999; Manji and Duman 2001; Kozlovsky et al. 2006).

Interestingly, for the first time, it is here reported a drug-induced change in Akt-1 protein and mRNA levels during the proliferation period. In proliferating cells, increases in Akt mRNA and protein levels, but not in GSK-3ß, starts before the potentiation effect on the proliferation, and continues during the proliferation. It seems that the two molecular and cellular processes have a causal relationship. The molecular machinery of Akt synthesis (mRNA levels) precedes the drug-induced potentiation of cell proliferation. Akt was shown to be a key mediator of signal transduction process and mediates many of the survival signals (Lawlor and Alessi, 2001; Brunet et al. 2001). It was also previously demonstrated that Akt is not only involved in cell growth and survival but is also involved in glucose metabolism/uptake (Hajduch et al. 2001). Therefore, it becomes clear that addition of the mood-stabilizing drug has resulted in inducing more Akt mRNA and then protein levels to potentiate the ongoing cell proliferation. The increase in Akt, by preceding the increase in proliferation is probably the casue of this increase potentiation.

Moreover, the study observed differeneces among drugs, either on immediate effects or on gene expression. VA and Lithium were the most effective drug, while Clozapine was the less effective both for cellular and molecular activation. The drugs induced a rapid phosphorylation of GSK3ß and Akt-1. This cellular effect of GSK3ß was highlighted by others. Hetman et al have identified GSK3 as a critical downstream effector for the PI3K/Akt survival cascade in primary neurtons, bulding on previous work by Pap and Cooper (1998) in PC 12 cells (Pap and Cooper 1998). Both papers showed that GSK3ß is dephosphorylated and activated by apoptotic stimuli, and phosphorylated and inactivated in response to survival factors, in a PI3K-dependent manner. Actually, it has been shown that GSK-3ß is activated in two ways (Cohen and Frame, 2001). The first pathway is mediated by the PI3K-Akt signaling, following a tyrosine-kinase receptor stimulation (i.e. by insulin or growth factors) and resulting in ^{9}Ser-p-GSK3ß (Cross et al. 1995). The other major pathway is mediated by the Frizzled receptor and the extracellular signal Wnt (Zeng et al 1997). In this system GSK-3 works within a multimolecular complex composed of several molecules including Axin, APC. The de-activation of GSK-3ß by its phopshorylation results in the release of ß-catenin and its export into the nucleus to act as transcription factor (Grimes and Jope 2001). However in presence of upstream signals, like growth factors which activate the tyrosine kinase membrane receptors, the adopted pathway would be the PI3K-Akt signaling axis. Crowder and Friedman demonstrated that the GSK3 inhibitors may block apoptosis due solely to the inhibition of PI3K and Akt (Crowder and Friedman 2000). Thus, the effct of Lithium and Valproate acts through inhibiting dephosphorylation of Akt (Mora et al. 2002). The clinical relevance of such rapid cellular and molecular effects remain to be well understood. While in clinical context, these drugs are effective in chronic application, after several days (Manji and Duman, 2001), this study and others show that the action of drugs on phosphorylation of the proteins, is initiated more early. In fact, there are two phases, a rapid one and an late phase which requires synthesis of mRNA and traduction into proteins. This may suggest that longterm effects must probably be sensitized by accute effects. The study also showed that at therapeutic concentration, Clozapine was the tardiest drug, requiring many hours to significantly phosphorylate Akt protein and subsequently GSK-3ß protein (it took more than 6h: data not shown). Moreover, the increase in mRNA expression under Clozapine treatment, was moderate and late compared with Lithium and Valproate. Therefore, this result may suggest that the Clozapine effect on neuroblastoma SH-SY5Y cells is probably indirect and could result from complex changes occurring in the cell.

In contrast to phosphorylation of GSK-3ß which underwent changes, there was no drug-induced alteration neither in the mRNA levels, nor in the protein levels in cells culture, during the period of proliferation. Previous reports from different laboratories have indeed observed change either in enzymatic activity, or in the levels of the phosphorylated forms, but no report has observed change in total protein levels, per se. In brain tissues, the activity of GSK-3ß was also altered with respect to diagnosis, but not the protein levels. This lack of change in GSK3 protein levels, in contrast to the Akt levels suggest that probably the levels of GSK-3ß *per se*, are less important tha the capacity of the kinase to respond to drugs, its degree of activation is more determinant. But this is a speculation, more studies are warranted to determine the exact role of GSK-3ß in drug-inuced potentiation of the cell proliferation.

E. Conclusion

1. Signaling Axis and Mood Disorders

A growing body of evidence suggest that animal and human brains are provided of the ability of neuroplasticity or remodelling, and that dysfunction of these processes could contribute to the pathophysiology of mood disorders. Moreover, drug treatment or other efficient therapy could operate by induction of appropriate remodelling. Dysregulation in signaling networks responsable of cell survival or cell proliferation has been hypothesized and studies have shown this dysfunction in support of compromised neuroplasticity. This PI3K-PKB-GSK3ß pathway axis or its members taken individually, have been previously shown to regulate cellular processes of proliferation, neurprotection, cell mobility etc. In the present study we have observed an alteration of PI3K/PKB/GSK3ß signaling axis in brain of MDD. The increase of PTEN protein levels and the decrease of PI3K activity contribute to the blunting of the pathway axis. Our results are therefore consonant with this formerly described role of these protein and lipid kinases and phosphatases. This study was completed by *in vitro* study on neuronal-derived SH-SY5Y cell culture. The use of mood stablizers at clinical relevant levels, has allowed to delay serum-deprivated cell death and therefore to function as cytoprotectors, in one hand. In other hand the drugs have resulted in potentiation of serum-stimulated cell proliferation. On molecular level, these drugs activate Akt-1 protein and GSK3ß by a rapid phosphorylation, resulting in desactivating phosphorylation of GSK-3ß. This rapid action probably was aimed to sensitize the longterm time-dependent effect of drugs on increase in Akt gene and protein expression. The enhanced expression of Akt-1 mRNA precedes and accompanies the enhanced cell proliferation, thus confirming its role in anti-apoptotic and proliferation processes. It is reasonable to assume that the molecular and cellular dysfunctions which result in destabilization of mood play a role in the morphologic abnormalities observed in affective disorders. The relevant downstream effects of mood stabilizers, and antdepressant studies reported by others, on intracellular signaling and gene expression, undoubtedly will result in maintenance of this pathways.

To conclude, these two studies (postmortem and cell culture) present some novel studies. While some results should be considered preliminary given the small number of subjects, they provide platform for further investigation of the PI3K-Akt-GSK3ß signaling axis or its members, in affective disorders. Moreover, given the data on cell culture, the study might have relevance in the further development of new drugs in targetting the PI3K/Akt/GSK3 signal transduction.

2. Mood Disorder and Cancer

As previously mentioned, most of these proteins are known for their role in different types of cancers, including breask cancer, prostate or other sporadic cancers. Their implication in on mood disorders, raises the question of the relationship between cancer and depression (MDD or BD). Altough this topic is beyond of the aim of the study, it is nonetheless worthwhile to mention some of the outstanding studies that have investigated not only the high prevalance of depression in patients with cancer but, more importantly, the issue of depression as a risk factor in initiation and progerssion of cancer. There is an abundant litterature on the association and mutual effects between these pathologies (Persky et al, 1987 ; McDaniel et al. 1995 ; Penninx et al. 1998 ; McDaniel et al. 1995 ; Linkins et al 1990 ; Friedman 1994). The psychological distress and depression that are prevalent in some patients with cancer represent complex psychosocial and medical factors. And studies that examined the severity of cancer, as risk of depression demonstrate to be substantially higher as the severity of some cancer types increases (McDaniel et al 1995). In other hand, the idea that depression is an etiologic factor in the development of cancer is widespread in the gneral population and among healthy professionals (Croyle RT, 1998). The prevailing hypothesis for such an association is that depresion impairs immune function, which in turn predisposes a person to the initiation and progession of neoplastic disease (Ader et al. 1995). Moreover, the observed alteration in immunity in depressed persons suggests the exsistence of a link bettween depression and cancer (Herber and Cohen, 1992). A number of longitudinal epidemiological studies have been conducted to investigate their relationship. Although there is no agreement on this subject, numerous studies have reported evidence of an increased incidence of cancer in depressed persons (Linkins et al 1990 ; Friedman 1994), but others failed to observe such an incidence (Dalton et al. 2002). The pionner study of Persky and collegues suggested that psychologic depression might promote the development of cancer and spread of malignant neoplasms, and later on an epidemiological study of the National Institute of Aging in Betesda (MD, USA) concluded that when present at least 6 yeras, depression is associated with increased risk of cancer (Persky et al. 1987 ; Penninx et al. 1998). As regards to therapy, psychological intervention have resulted, for instance, in increased survival rates in patients with cancer (McDaniel et al. 1995). It was demonstrated that women with metastatic breast cancer who received weekly group pychotherapy survived an average of 18 months longer than did control patients who were randomly assigned to routine care (Mc Daniel et al 1995). There is also improvement in immune function (increase in granular lymphocytes and NK cells). In bipolar disorder, Tamoxifen©, a well-known anti-estrogen drug commonly used to treat women breast cancer, has shown surprising success in treating bipolar disorder (Zarate et al. 2007). Interestingly, this drug was further shown to target a protein kinase involved in cell siganling, the protein kinase C (PKC), awhich is also dysregulated in affective disorders (Nishizuka 1988 ; Pandey et al 2005). Could some phosphorylation processes be the crossroads between cancer and mood disorders ? This question deserves to be raised. Depression and cancer account for among the most severe human diseases. A comprehensive study of mechanisms shared by affective disorder and cancer in a population of subjects affected by these two pathogies is therefore urgently needed.

References

Ader R, Cohen N, Felten D (1995). Psychoneuroimmunology: interaction between the nervous system and the immune system. *Lancet* **345**:99–103

Bachmann RF, Schloesser RJ, Gould TD, Manji HK (2005). Mood stabilizers target cellular plasticity and resilience cascades: implications for the development of novel therapeutics. *Mol Neurobiol* **32** :173-202.

Beasley C, Cotter D, Everall I (2002). An investigation of the Wnt-signalling pathway in the prefrontal cortex in schizophrenia, bipolar disorder and major depressive disorder. *Schizophr Res* **58**:63-7.

Berridge MJ, Downes CP, Hanley MR (1982). Lithium amplifies agonist-dependent phosphatidylinositol responses in brain and salivary glands. *Biochem J* **206**: 587-95.

Belmaker RH, Bersudsky Y, Benjamin J, Agam G, et al 1995. Manipulation of inositol-linked second messenger systems as a therapeutic strategy in psychiatry. *Adv Biochem Psychopharmacol.* **49**:67-84.

Bezchlibnyk Y and Young T, 2002. The neurobiology ob bipolar disorder : focus on signal transduction pathways and the regulation o gene expression. *Can J Psychiatry* **47**: 135-148.

Bowden CL, Brugger AM, Swann AC, Calabresse J R, Janicak PG (1994). Efficacy of divalproex vs lithium and placebo in the treatment of mania. The Depakote Mania Study group. *J Am Med Association* **271** : 918-924.

Bowden CL (1996). Dosing strategies and time course of response to antimanic drugs. *J Clin Psychiatry* **57** (suppl 13): 4-9.

Brazil DP, Yang ZZ, Hemmings BA (2004): Advances in protein kinase B: AKTion on multiple fronts. *Trends Biochem Sci* **29**: 233-42.

Bremmer J, Narayan M, Anderson ER, Staib LH, Miller H, Charney DS 2000. Smaller hippocampal volume in major depression. *Am J Psychiatry* **157**: 115-117.

Brunet A, Datta SR, Greenberg M (2001). Transcription-dependent and –independent control of neuronal survival by the PI3K-Akt signaling pathway. *Current Opinion in Neurobiology* 11: 297-305.

Cade JF 1949 : .Lithium salts in the treatment of psychotic excitement. 1949. in *Bull World Health Organ*. 2000;78(4):518-20.

Cantley LC (2002): The phosphoinositide 3-kinase pathway. *Science* **296**: 1655-57.

Cantrell DA (2001): Phosphoinositide 3-kinase signaling pathways. *J Cell Science* **114**: 1439-45.

Carlson PJ, Singh JB, Zarate CA, Drevits WC and Manji HK 2006. Neural circuit and neuroplasticity in mood disorders : Insights for novel therapeutic targets. *NeuroRx* **3**. 22-41.

Charlton BG, Ferrier IN (1989). Hypothalamo-pituitary-adrenal axis abnormalities in depression: a review and a model. *Psychol Med.* **19**:331-6.

Chelecka-Franaszek E, Chuang D-M (1999). Lithium activates serine/threonine kinase Akt-1 and suppresses glutamine-induced inhibition of Akt-1 activity in neurons. *Proceedings of National Academy of Science USA* **9** : 8745- 8750.

Chen G, Hasanat KA, Bebchuk JM, Moore GJ, Glitz D, Manji HK (1999a). Regulation of signal transduction pathways and gene expression by mood stabilizers and antidepressants. *Psychosomatic Medecine* **61** : 599-617.

Chen G, Huang LD, Jiang YM, Manji HK (1999b): The mood stabilizing agent valproate inhibits the activity of glycogen synthase kinase-3. *J Neurochem.* **72**: 1327-1330.

Cheng AT (1995): Mental illness and suicide. A case-control study in east Taiwan. *Arch. Gen. Psychiatry* **52**: 594-603.

Cheng A, Wang S, Yang D, Xsao R, Mattson MP (2003): Calmodulin mediates brain-derived neurotrophic factor cell survival signaling upstream of Akt kinase in embryonic neocortical neurons. *J. Biol. Chem.* **278**: 7591-7599.

Chuang DM (2005). The Antiapoptotic Actions of Mood Stabilizers: Molecular Mechanisms and Therapeutic Potentials. *Ann New York Acad Sci* **1053** :195-204.

Ciapperelli A, Dell'Osso L, Bandettini di Poggio A, Carmassi C, Cecconi D, Cassano GB (2003). Clozapine in treatment-resistant patients with schizophrenia, schizoaffective disorder or psychotic bipolar disorder: A naturalistic 48-month follow-up study. *J Clin Psychiatry* **64** : 451-458.

Ciechanoviski PS, Katon WJ, Russo JE (2000). Depression and diabetes: impact of depressive syntoms on addherance function. And costs. *Arch Inter Med* 160: 3278-3285.

Cohen P, Frame S (2001). The renaissance of GSK3. *Nat Rev Mol Cell Biol.* **2(10):** 769-76.

Cohen DS, Johnson-Faley NN, Travkina T (2005) : 5HT receptors couple to activation of Akt, but not extracellular-regulated kinase (ERK), in cultured hippocampal neurons. *J Neurochem* **93** : 910-17.

Cotter D, Mackay D, Landau S, Kerwin R, Everall I 2001. Reduced glial cell density and neuronal size in the anterior cingulate cortex in major depressive disorder. *Arch Gen Psychiatry* 2001; **58**: 545–553.

Cross DA, Alessi DR, Cohen P, Andjelkovich M, and Hemmings BA (1995): Inhibition of glycogen synthase kinase-3 by insulin-mediated Protein kinase B. *Nature* **95**; 378: 785-789.

Croyle RT (1998). Depression as a risk factor for cancer: renewing a debate on the psychobiology of disease. *J Natl Cancer Inst* **90**:1856–7

Czeh B, Michaelis T, Watanabe T et al. (200)1. Stress-induced changes in cerebral metabolites, hippocampal volume, and cell proliferation are prevented by antidepressant treatment with tianeptine. *Proc Natl Acad Sci USA* **98**: 12796–12801.

Dalton SO, Mellemkjær L, Olsen JH et al (2002). Depression and Cancer Risk: A Register-based Study of Patients Hospitalized with Affective Disorders, Denmark,1969–1993. *Am J Epidemiol* **155** : 1088-95.

Datta SR, Brunet A, Greenberg M (2006). Cell survival: a play in three Akts. *Genes and Development* **13**: 2905-2927.

De Sarno P, Li X, Jope RS (2002). Regulation of Akt and glycogen synthase kinase-3ß phosphorylation by sodium valproate and lithium. *Neurpharmacology* **43**:1158-1164.

Di Daniel E, Mudge AW, Maycox PR (2005). Comparative analysis of the effects of four mood stabilizers in SH-SY5Y cells and in primary neurons. *Bipolar Disorder* **7**: 33-41.

D'Mello SR, Borodezt K, Stoltoff SP (1997). Insulin-like growth factor and potassium deplolarization maintain neuronal survival by distinct pathway : possible involvement of PI3-Kinase in IGF-1 signaling. *J Neurosci* **20** :326-331.

Drevets WC, Frank E, Price JC, Kupfer DJ, Holt D et al (1999). PET imaging of serotonin 1A receptor binding in depression. *Biol Psychiatry* **46** : 1375-87.

Drevets WC (2000). Neuroimaging studies of mood disorders. *Biol Psychiatry* 48 : 813-829.

Drevets WC, Thase ME, Moses-Kolko ELet al (2007) Serotonin-1A receptor imaging in recurrent depression: replication and literature review. *Nucl Med Biol.* **34**(7):865-77

Dudek H, Datta SR, Franke TF, Birnbaum MJ, yao R, Cooper GM, Segal RA, Kaplan DR, Greenberg ME (1997) : Regulation of neuronal survival by serine-threonine protein kinase Akt. *Science* **275** : 661-64.

Dunner DL (2004): Cortrelates of suicidal behavior and lithium in bipolar disorder. *J. Clin. Psychiatry* 2004; 65(suppl. 10): 5-10.

Duman RS, Heninger GR, Nestler EJ (1997). A molecular and cellular theory of depression. *Arch Gen Psychiatry* 54 : 597-605.

Dwivedi Y, Rizavi HS, Conley RR, Tamminga CA, Pandey GN (2003) : Altered gene expression of brain-derived neurotrophic factor and receptor tyrosine kinase B in postmortem brain of suicide subjects. *Arch Gen Psychiatry* **60** : 804-815.

Dwivedi Y (2006): Gene expression analysis of cell signaling molecules in postmortem brain of suicide victims. *Int J Neuropsychopaharmacol* **9** (suppl 1). SP47.05

Dwivedi Y, Mondal AC, Rizavi HS, Conley RR (2005): Suicide brain is associated with decreased expression of neurotrophins. *Biol. Psychiatry* **58** (4): 315-24.

Franke TF, Kaplan DR, Cantley LC, Toker A (1997). Direct regulation of the Akt protooncogene product by phosphoinisitol 3,4-biphosphate. *Science* **275**: 665-67.

Fang X, Yu SX, Lu Y, Bast RC, Woodgett JR, Mills GB (2000). Phosphorylation and inactivation of glycogen synthase kinase-3 by protein kinase A. *PNAS* **97**: 11960-11965.

Friedman GD (1994). Psychiatrically-diagnosed depression and subsequent cancer. *Cancer Epidemiol Biomarkers Prev* **3**:11–13

Frye MA, Ketter TA, Altshuller LL et al (1998). Clozapine in bipolar disorder: treatment implications for other atypical antipsychotics. *J of Affect Disorder* **48** : 91-104.

Fuchs B, Czéh B, Kole MHP, Michaelis T and Lucassen PJ. (2004). Alterations of neuroplasticity in depression : the hippocampus and beyond. *Eur Neuropsychopharmacol* **14** : S481-S490

Goode N, Hughes K, Woodget JR, Parker PJ (1992). Differential regulation of glycogen synthase kinase-3ß by Protein kinse C isotypes. *J. Biol. Chem.* **267**: 16878-16882.

Gould E, McEwen BS, Tanapat P, Galea LAM, Fuchs E. (1997). Neurogenesis in the dentate gyrus of the adult tree shrew is regulated by psychosocial stress and NMDA receptor activation. *J Neurosci* **17**: 2492–2498.

Grimes CA, Jope RS (2001). The multifaceted roles of glycogen synthase kinase 3ß in cellular signaling. *Prog. Neurobiol.* **65**: 391-426.

Griffin RJ, Moloney A, Kelliher M et al (2005). Activation of Akt/PKB, increased phosphorylation of Akt substrates and loss and altered distribution of Akt and PTEN are features of Alzheimer's disease pathology. *J Neurochem* **93**:105-117.

Gurvich N, Klein PS (2002). Lithium and Valproic acid: parallels and contrasts in diverse signaling contexts. *Pharmacology and Therapeutics* **96**: 45-66.

Hajduch E, Litherland GJ, Hundal HS (2001). Protein kinase B (PKB/Akt) – a key regulator of glucose transport? *FEBS Letters* **492** : 199-203.

Harwood AJ, Agam G (2003). Search for common mechanism for mood stabilizers as plasticity enhancers in the treatment of neuropsychiatric disorders. *J Clin Psychiatry* **64** (suppl 5) : 179-189.

Harwood AJ (2005). Lithium and bipolar mood disorder: the inositol-depletion hypothesis revisted. *Mol Psychiatry* **10** : 117-126.

Herbert TB, Cohen S (1993). Stress and immunity in humans: a meta-analytic review. *Psychosom Med* **55**:364–79

Hayes DL, Sawyer D (2004). Phosphosafe Extraction Reagent. *Innovations* **15**, 5.

Heninger GR, Delgado PL, Charney DS (1996). The revised monoamine theory of depression: a modulatory role for monoamines, based on new findings from monoamine depletion experiments in humans. *Pharmacopsychiatry.* **29**(1):2-11.

Hetman M, Cavanaugh JE, Kimelman D, Xia Z (2000). Role of glycogen synthase kinase-3beta in neuronal apoptosis induced by trophic withdrawal. *J Neurosci* **20** :2567-74.

Hirschfeld RM, Calabresse JR, Weissman MM (2003). Screening for bipolar disorder in the community. *J Clin Psychiatry* **64** : 53-59.

Hsiung SC, Adlersberg M, Arango V, Mann JJ, Tamir H, Liu KP (2003). Attenuated 5HT1a receptor signaling in brains of suicide victims: involvement of adenylyl cyclase, phosphatidylinositol 3-kinase, Akt and mitogen-activated protein kinase. *J Neurochem.* **87**: 182-194.

Hsiung SC, Tamir H, Franke TF, Liu KP (2005). Roles of extracellular signal-regulated kinase and Akt signaling in coordinating nuclear transcription factor-kappaB-dependent cell survival serotonin 1A receptor activation. *J Neurochem* **95**: 1653-66.

James SR, Downess CP, Gigg R, Grove SJ, Holmes AB, Alessi DR (1996). Specific binding of the Akt-1 protein kinase to phosphatidylinositol 3,4,5-triphosphate without subsequent activation. *Biochem J* **315**:709-13.

Johnson FN, Amdisen A (1983). The first era of lithium in medecine. A historical note. *Pharmacopsychiatria* **16** : 61-63.

Johnson-Farley NN, Tavrikina T, Cowen DS (2005). Cumulative activation of Akt and consequent inhibition of glycogen kinase-3 by brain-derived neurotrophic factor and insulin-like growth factor-1 in cultured hippocampal neurons. *J Pharmacol Exp Therapeutics* **316**: 1062-1069.

Johannessen CU (2000). Mechanisms of action of valproate: a commentatory. *Neurochem International* **37**: 103-110.

Jope RS, and Williams MB (1994). Lithium and brain signal transduction systems. *Biochem Pharmacol* **47**: 429-441.

Jope RS and Bijur GN (2002). Mood stabilizers, glycogen kinase-3ß and cell survival. *Mol. Psychiatry* **7:** S35-S45.

Kaidanovich-Beilin O, Milman A, Weizman A, Pick CG, Eldar-Finkelman H (2004). Rapid antidepressant-like activity of specific glycogen synthase kinase-3 inhibitor and its effect on ß-catenin in mouse hippocampus. *Biol. Psychiatry* **55**: 7881-4.

Kang UG, Seo MS, Roh MS, Kim Y, Yoon SC, Kim YS (2004). The effects of clozapine on the GSK-3-mediated signaling pathway. *FEBS Letters* **560** : 11-119.

Klein PS, Melton DA (1996). A molecular mechanism for the effect of lithium on development. *PNAS USA* **93** : 8455-8459.

Karege F, Vaudan G, Schwald M, Perroud N, La Harpe R (2005): Neurotrophin levels in postmortem brains of suicide of victims and the effects of *antemortem* diagnosis and psychotropic drugs. *Mol. Brain Research* **136**: 29-37.

Karege F, Perroud N, Burkhardt S, Schwald M et al. (2007). Alteration in Kinase activity but not in Protein Levels of Kinase B and Glycogen Synthase Kinase-3ß in ventral prefrontal cortex of depressed suicide victims. *Biol Psychiatry* **61**: 240-245.

Kelly TM, Mann JJ (1996). Validity of DSM-III-R diagnosis by psychological autopsy: a comparison with clinician antemortem diagnosis. *Acta. Psychiatr. Scand.* **94**: 337-43.

Klein PS, Melton DA (1996). A molecular mechanism for the effect of lithium on development. *PNAS USA* **93**: 8455-8459.

Knable MB, Barci MB, Webster MJ, Meador-Woodruff J, Torrey EF (2004). Molecular abnormalities of hippocampus in severe psychiatric illness: post-mortem findings from the Stanley Neuropathology Consortium. *Mol. Psychiatry* **9**: 609-20.

Kozlovsky N, Belmaker RH, Agam G (2001). Low GSK-3 activity in frontal cortex of schizophrenic patients. *Schizophr Res* **52**(:101-5.

Kozlovsky N, Amar S, Belmaker TH, Agam G (2006). Psychotropic drugs affect ser9-phosphorylated GSK-3ß protein levels in rident frontal cortex. *Int J Neuropsychopharmacol* **9**:337-342.

Kupfer D (2005). The increasing medical burden in bipolar diosrder. *JAMA* 283 : 2528-2530.

Lai JS, Zhao C, Warsh JJ, Li PP (2006). Cytoprotection by lithium and valproate varies between cell types and cellular stresses. *Eur J Pharmacol* **539** : 18-26.

Lawlor M, Alessi D (2001): PKB/Akt: a key mediator of cell proliferation, survival and insulin responses? *J. Cell Science* **114**: 2903-2910.

Lesort M, Greendorf A, Stockmeier C, Johnson GVW, Jope L (1999). Glycogen synthase kinase-3ß, ß-catenin, and tau in postmortem bipolar brain. *J. Neural. Transm.* **106**: 1217-1222.

Li R and El-Mallahk RS (2000). A novel evidence of different mechanisms of lithium and valproate neuroprotective action on human SY5Y neuroblatoma cells: caspase-3 dependency. *Neurosci Lett* **294**: 147-150.

Li X, Bijur GN, Jope RS (2002). Glycogen synthase kinase-3beta, mood stabilizers, and neuroprotection. *Bipolar Disorder* **4** : 137-44.

Li X, Zhu W, Roh MS, Friedman AB, Rosborough K, Jope RS (2004). *In vivo* regulation of glycogen synthase kinase-3ß (GSK3ß) by serotoninergic activity in mouse brain. *Neuropsychopharmacol.* **29**: 1426-1431.

Liang J and Slingerland M (2003). Multiple role of the PI3/PKB (Akt) Pathway in Cell Cycle Progression. *Cell Cycle* **2**: 339-345.

Linkins RW, Comstock GW (1990). Depressed mood and development of cancer. *Am J Epidemiol* **132**:962–72.

Livak KJ, Schmittgen TD (2001). Analysis of relative gene expression data using real-time quantitative PCR and the 2(delta-delta) Ct method. *Methods* **25**: 412-408.

Ma C, Wang J, Gao Y, Gao TW et al (2007). Role of glycogen synthase kinase 3-beta in the transformation of epidermal cells. *Cancer Res* **67** :7756-64.

Maehama T, Dixon JE (1998). The tumor suppressor, PTEN/MMCA1, dephosphorylates the lipid second messenger, phosphatidylinositol 3,4,5-triphosphates. *J Biol Chem* **273**: 13375-78

Magarinos AM, McEwen BS, Flugge G, Fuchs E. (1996.) Chronic psychosocial stress causes apical dendritic atrophy of hippocampal CA3 pyramidal neurons in subordinate tree shrews. *J Neurosci* **16**: 3534–3540

Manji HK, Moora GJ, Chen G (1999). Lithium at 50: have the neuroprotective effects of this unique cation been overlooked? *Biol Psychiatry* **46** : 949-940.

Manji HK, Drevets WC, Charney DS (2001a). The cellular neurobiology of depression. *Nat Med* **7**: 541–547.

Manji HK, Duman RS (2001b). Impairments of neuroplasticity and cellular resilience in severe mood disorders: implications for the development of novel therapeutics. *Psychopharmacol Bull* **35** : 5-49.

Mann JJ, Brent, DA Arango V (2001): The neurobiology and genetics of suicide and attempted suicide: a focus on the serotonergic system. *Neuropsychopharmacol* **24**: 467-77.

McEwen BS (2000). Effects of adverse experiences for brain structure and function. *Biol Psychiatry* **48**: 721–731.

Milak MS, Parsey RV, Keilp J, Oquendo MA (2005). Neuroanatomic correlates of psychopathologic components of major depressive disorder. *Arch Gen Psychiatry* **62**:397-408.

Mizuno M, Yamada K, Takei N et al. (2003). Phosphatidylinositol 3-kinase : a molcule mediating BDNF-dependent spatial memory formation. *Mol Psychiatry* **8** : 217-224

Mora A, Gonzalez-Polo RA, Fuentes JM, Soler G, Centeno F (1999). Different mechanisms of protection against apoptosis by valproate and lithium. *Eur J Biochemistry* **266** : 886-91.

Müller-Oerlinghausen B, Berghofer A, Ahrens B (2003). The antisuicidal and mortality-reducing effect of lithium prophylaxis: consequences for guidelines in clinical psychiatry. *Can. J. Psychiatry* **48**: 433-439.

Musselman DL, Evans DL, Nemeroff CB (1998). The relationship of depression to cardiovascular disease: epidemiology, biology, and treatment. *Arch Gen Psychiatry* **55**:580-92.

Nishizuka Y (1988). The molecular heterogeneity of protein kinase C and its implications for cellular regulation. *Nature*. **334**:661-665.

O'Brien WT, Harper AD, Jove F, Woodgett JR et al (2004). Glycogen synthase kinase-3ß haploinsufficiency mimics the behavioral and molecular effects of lithium. *J. Neurosci.* **24**: 6791-8.

Owens MJ and Numeroff CB (1994). Role of serotonin in the pathophysiology of depression: focus on the serotonin transporter, *Clin Chem.* **40**. 288-295

Pandey GN, Pandey SC, Dwivedi Y, et al (1995). Platelet serotonin-2A receptors: a potential biological marker for suicidal behavior. *Am J Psychiatry.* **152**:850-5.

Pandey GN, Dwivedi Y, Ren X, Rizavi HS et al C (2003). Altered expression and phosphorylation of myristoylated alanine-rich C kinase substrate (MARCKS) in postmortem brain of suicide victims with or without depression. *J Psychiatr Res.* **37** : 421-32.

Pandey GN, Dwivedi Y (2005). Focus on protein kinase A and protein kinase C, critical components of signal transduction system, in mood disorders and suicide. *Int J Neuropsychopharmacol* **8**:1-4.

Pandey GN, Ren X, Rizavi H et al (2006): GSK-3β and β-catenin in the *postmortem* brain of teenage suicide victims. The 11th European symposium on Suicide and Suicidal behavior: "Suicide: Interplay of genes and Environment". *Psychiatria Danubina* **18** (*suppl 1*).

Pap M and Cooper GM 1998). Role of glycogen synthase kinase-3 in the phosphatidylinositol 3-Kinase/Akt cell survival pathway. *J Biol Chem* **273**):19929-32.

Penninx BW, Guralnik JM, Pahor M et al (1998). Chronically depressed mood and cancer risk in older persons. *J Natl Cancer Inst* **90**:1888–93.

Perry PJ, Bever KA, Arndt S, Combs MD (1998). Relationship between patient variables and plasma clozapine concentrations: a dosing nomogram. *Biol Psychiatry* 44: 733-738.

Persky VW, Kempthorne-Rawson J, Shekelle RB (1987). Personality and risk of cancer: 20-year follow-up of the Western Electric Study. *Psychosom Med* **49**:435–49.

Pittenger C, Duman RS (2007). Stress, Depression and Neuroplasticity : A convergence of Mechanisms. *Neuropsychopharmaco Reviews* **1-22**

Rajkowska G, Miguel-Hidalgo JJ, Wei J et al (1999). Morphological evidence for neuronal and glial prefrontal cell pathology in major depression. *Biol Psychiatry* **45**: 1085-1098

Rajkowska G (2000): Postmortem studies in mood disorders indicate altered number of neurons and glyal cells. *Biol Psychiatry* **48**: 766-777.

Rich CL, Young D, Fowler R (1986): San Diego suicide study I. Young vs old subjects. *Arch Gen Psychiatry* **43**:577-82.

Sachs G, Chengappa KN, Suppes T et al (2004). Quetiapine with lithium or divalproex for the treatment of bipolar mania: a randomized, double-blind, placebo-controlled study. *Bipolar Disorder* **6** : 213-223.

Sassi RB, Soares JC (2001). Emerging therapeutics in bipolar mood disorder. *Expert Opinion on Therapeutic Targets* **5** : 587-599.

Sapolsky RM. (2000). Glucocorticoids and hippocampal atrophy in neuropsychiatric disorders. *Arch Gen Psychiatry* **57**: 925–935.

Schratt GM, Nigh, Chen WG et al (2004). BDNF reguates the translation of a select group of mRNA by mamalian target of rapamycin-phosphatidylinositol 3-kinase dependent patway during neuronal development. *J Neurosci* **24**, 7366-7377.

Sequeiera A, Kemplan T, Canetti L, Benkelfat C et al. (2006). Patterns of expression in the limbic system of suicides with and without depression. *Int J Neuropsychopaharmacol* **9** (*suppl 1*). SP47.04.

Shao L, Young T, Wang JF (2005). Chronic treatment with mood stabilizers lithium and valproate prevents excitotoxicity by inhibiting oxidative stress in rat cerebral cortical cells. *Biol Psychiatry* **58**:879-884.

Sheline Y, Wany P, Gado MH, Csernansky JG, Vannier MW (1996). Hippocampal atrophy in recurrent major depression. *PNAS USA* **93**: 3908–3913.

Sheline YI (2003). Neuroimaging studies of mood disorder effects on the brain. *Biol Psychiatry* **54** : 338-52.

Stambolic V, Ruel L, Woodgett JR (1996). Lithium inhibits glycogen synthase kinase-3 activity and mimics wingless signaling in intact cells. *Current Biol.* **6**: 1664-68.

Stockmeier C, Shapiro IA, Dilley GF, Kolli TN, Friedman I, Rajkowska G (1998). Increase in serotonin 1a autoreceptors in the midbrain of suicide victims with major depression-postmortem evidence for decreased serotonon activity. *J Neurosci* **18**:7394-401.

Stockmeier CA Stockmeier C, Mahajan, GR et al (2004): Cellular changes in the postmortem hippocampus in major depression. *Biol Psychiatry* **56**:640-50.

Tanriover G, Demis N, Pestereli E, Demir R, Kayisli UA (2005). Pten-mediated Akt activation in human neocortex during prenatal development. *Histochem Cell Biol* **123**: 393-406.

Touhara K, Hawes BE, Van Biesen T, Lefkowitz RJ (1995). G protein βγ subunits stimulate phosphorylation of Sch adapter protein. *Proc Natl Acad Sci USA* **92**: 9284-87.

Törk I (1990). Anatomy of the serotonin system. *Ann NY Acad Sci* **29**:165-81.

van der Heide LP, Ramakers GMJ, Smidt MP (2006). Insulin signaling in the central nervous system: Learning to survive. *Prog Neurobiol* **79**: 205-221.

Weinberg I (2000). The prisoners of despair: right hemisphere deficiency and suicide. *Neurosci. Biobehav. Rev.* **24**: 799-815.

Whitman M, Downes CP, Keeler M, Keeler T, Cantley L (1998). Type I phosphatidylinositol kinase makes a novel inositolphospholipid, phosphatidylinositol 3-phosphate. *Nature* **232**: 644-46.

World Health Organisation (2001). Chapiter 2. Burden of Mental and Behavior disorder. In the WHO report 2001 : Mental Health, New Understandings. New Hope. http://www.who.int/whr/ pdf 19-34.

Yamada KM, Araki M (2001). Tumor suppressor PTEN: modualtor of cell signaling, growth, migration and apoptosis. *J Cell Science* **114**: 2375-82.

Yatham LN, Goldstein JM, Vieta E et al (2005). Atypical antipsychotics in Bipolar depression: potential mechanisms of action. *J Clin Psychiatry* **66**: 40-48.

Zarate CA Jr, Singh J, Manji HK (2006). Cellular plasticity cascades: targets for the development of novel therapeutics for bipolar disorder. *Biol Psychiatry* **59**:1006-20.

Zarate CA Jr, Singh JB, Carlson PJ et al 2007. Efficacy of a protein kinase C inhibitor (tamoxifen) in the treatment of acute mania: a pilot study. *Bipolar Disorder*. **9**:561-70.

Zeng L, Fagotto F, Zhang Tet al 1997. The mouse Fused locus encodes Axin, an inhibitor of the Wnt signaling pathway that regulates embryonic axis formation. *Cell* **90** : 181-92.

In: Oncogene Proteins: Structure, Functions and Analyses ISBN: 978-1-60456-259-0
Editors: P.B. Murphy and J.R. Clarke, pp. 55-85

Chapter II

Oncogene Proteins: Targets for Cancer Immuno-cell Therapy

Hiroshi Fujiwara and Masaki Yasukawa
Department of Bioregulatory Medicine, Ehime University Graduate School of Medicine, Ehime, Japan

Abstract

Immunotherapy against cancers is a strategy to eradicate residual tumor cells by means of the tumor-specific cytotoxic T lymphocytes (CTLs) or tumor-related antigens specific monoclonal antibodies (moAb), in tumor-bearing individuals. This new proceeding may be a promising alternative to chemo-radiotherapy which intrinsically embraces organ toxicities. An ideal target of cancer immunotherapy is a protein which is exclusively and highly produced by tumor cells, but not by normal cells. Furthermore, the protein, which is closely relevant to tumorigenesis or biological property of tumor itself, is more desirable. From these points of view, oncogene proteins potentially can be ideal candidates for cancer immunotherapy. For example, immunogenicities of WT1, BCR-ABL, those are well-known oncogene proteins of leukemias, and several phase I/II clinical trials of vaccination by using peptides derived from those proteins as anticancer vaccines have already been conducted. Although excellent clinical trials and elaborative basic researches have been accumulating the lines of evidence to show substantial therapeutic benefits, several limitations of cancer vaccine strategy itself have been simultaneously uncovered. Those drawbacks of cancer immunotherapy comprise the invalidation of tumor-specific CTLs by regulatory T-lymphocytes and myeloid derived suppressor cells, and the immune evasion of tumor cells by down-modulation of surface HLA class I molecules. In addition, deterioration of intracellular antigen-processing machinery, and epigenetic alteration of target antigen as immunoediting have been also demonstated. In this chapter, focusing on three representative oncogene proteins ,WT1, BCR-ABL, HER2/neu, authors are reviewing the present state of cancer immunotherapy, especially the vaccine strategy to generate tumor-specific CTLs in tumor-bearing individuals, with describing some of our recent works.
(259 words)

1. Introduction

Oncogene protein is defined as the causative factor of cancer and is generally a chimeric protein which is exclusively produced by tumor cells, or is an overexpressed endogenous protein in tumor cells compared to normal cells. In general, both properties are closely correlated to growth activity and apoptosis-resistance property of tumor cells (1). Recent progress in molecular biology and immunology has provided the more profound comprehension of tumor immunology (2-5). Accumulation of lines of evidence in allogeneic hematopoietic stem cell transplantation (allo-HSCT) to successfully control hematological malignancies via allogeneic T lymphocytes from engrafted donor immune cells has rendered us encouraged to treat cancers by means of immune reaction, especially by induction and potentiation of tumor-specific cytotoxic T lymphocytes (CTLs) (6,7). Hitherto, many tumor antigens and tumor-associated antigens have been explored as candidates of target for cancer immunotherapy. Some of those antigens have already been conducted to clinical trials as tumor vaccines for the treatment of hematological malignancies and solid tumors. Despite the successful generation of target antigen-specific CTLs in vivo, the clinical efficacy has proven to be marginal, especially for solid tumors (8,9). The ideal target antigen for cancer immunotherapy is the protein which is exclusively or highly produced, processed and sufficiently presented by tumor cells, but not by normal cells. Simultaneously the target molecule should be the factor which is crucial for carcinogenesis or biological properties of tumor cells, and those properties are the very ones of oncogene protein themselves. Thus, cancer immunotherapy targeting oncogene proteins as tumor antigens seems to be effective. Herein, with some of our own experiences, authors are reviewing the present status of cancer immunotherapy, especially focusing on vaccine therapy with representative oncogene products, which have already been conducted to clinical trials. Those are including Wilms' tumor gene 1 (WT1), Breakpoint cluster region-Ableson gene product (BCR-ABL), and human epidermal growth factor receptor-2 (Her2/neu).

2. How do CTL Exert Cytocidal Effect to Cancer Cells?

Tumor immunology is consisted of both humoral and cellular immunology. Therapeutically, the anti-cancer humoral immunology is integrated into the anti-cancer monoclonal antibody. Herein the frameworks of cellular immunology against cancer cells is briefly reviewed, on which tumor vaccine strategy rationally relies. (Figure 1)

a) *Expression of target antigenic epitopes on tumor cells*

For the recognition and destruction of tumor cells by CD8+ CTL, target protein in the cytoplasm must be processed and the immunogenic epitope derived from the protein is presented on tumor cell surface through HLA class I molecule (10). Antigen processing machinery of tumor cell comprises proteosomes in the cytosol, endoplasmic reticulum (ER), transporters associated with antigen processing (TAP) which translocate generated peptides to bind newly synthesized HLA-class I heavy chain and β_2-microglobulin heterodimer, thereby optimized epitope will be presented in HLA-class I molecule groove (11,12). It is known that quite a few peptides are optimal to bind to HLA-class I molecule (13,14), and the efficacy of APM varies to each tumor cell, and that will be affected by various kinds of immune stress to tumor cells (15,16,17). These lines of evidence indicate that expression of antigenic epitope derived from oncogene protein by tumor cells is basically under hard constrains.

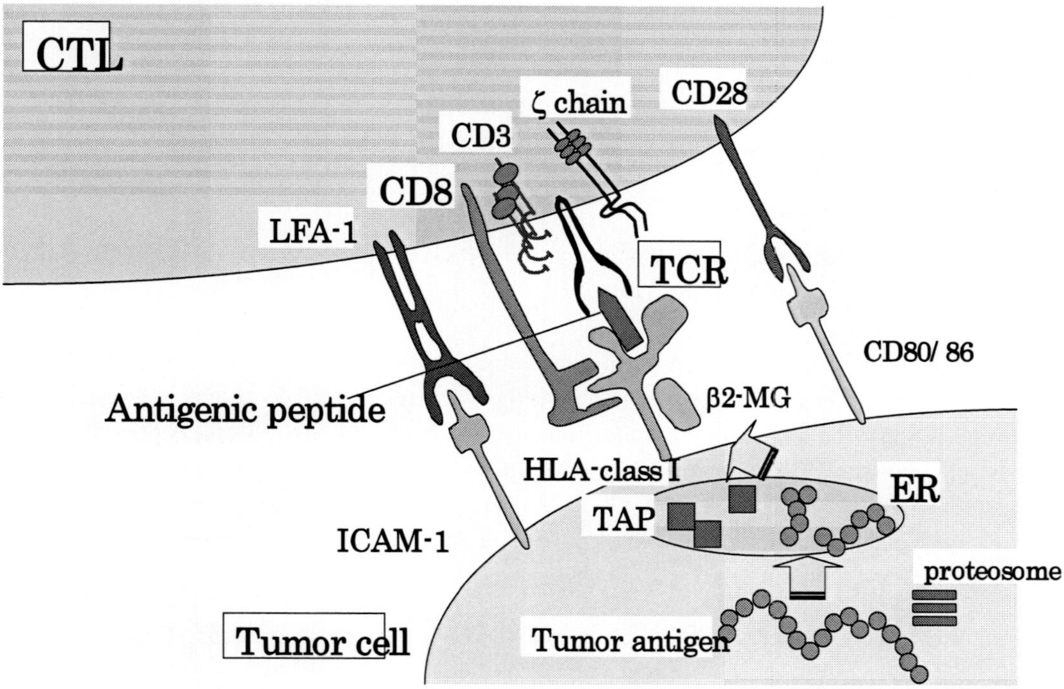

Figure 1. Basic concept of Tumor target recognition by CTL. Indeed CD4+ CTL actually exist, here the target tumor recognition by CD8+CTL is focused. Antigens which selectively expressed bytumor cells or overexpressed by tumor cells are considered as tumor antigens, which include oncogene proteins. Oncogene proteins are processed and presented by HLA-class I molecule on the tumor cell surface. That antigenic epitope is recognized by TCR on CTL assisted by coreceptors which compose "immunological synapse". After the recognition by TCR, CTL is activated to prolifeate and release IFN-γ and TNF-α, degranulate cytotoxic granules (perforin and granzyme.). Details are described in the text. ER: endoplasmic reticulum, TAP: Transportes associated with antigen processing. β2 MG: beta2 microglobulin, LFA-1: Leukocyte-function associated antigen-1, ICAM-1; Intercelluar adhesion molecule-1.

b) *Recognition by epitope-specific CTL/ Immunological synapse*

The target epitope recognition by epitope-specific CTL has recently been well characterized (18). Epitope-specific T-cell receptor (TCR) on CTL plays the central role in recognition of target epitope. The initial recognition is physically based on the structural interaction between the epitope presented in combination with HLA class I α-chain and TCR, which is termed "affinity" (19- 22). To potentiate that recognition by CTL, that initial epitope-TCR complex will be reinforced by assembled molecules (CD8, adhesion molecules, costimulatory molecules) to form "immunological synapse", which is termed "avidity" (23-28). This will provide the enough time for each CTL to efficiently exert their proliferation and cytotoxicity in response to the target (29,30).

c) *Cytocidal degranulation by CTL*

Finally, accepting the signal from the immunological synapse, CTL exert their target-specific cytocidal effect to induce apoptosis via degranulation of specific molecules, mainly perforin and granzyme B, and distinctively via FasL in the context of Fas-expression on target cells (31-34). Recently the resistant mechanism against this cytotoxic degranulation has been uncovered, Cathepsin B for self-protection of CTL, proteinase inhibitor 9 for immune evasion by tumor cells are some of those (35,36).

3. Oncogene proteins ; Targets of cancer immunotherapy
: WT1, BCR-ABL, and HER2/neu, the Representatives for Cancer Immunotherapy

Candidate antigens for cancer immunotherapy are validated for their applicability as vaccines, when those candidates are restricted to cancer cells, are immunogenic to elicit CTL with high avidity for target epitope, are functionally important for biological phenotypes of cancer cells, desirably are expressed on cancer stem cells, finally are applicable to wide range of cancers or cancer-bearing individuals. Based on those properties, clinical trials as cancer vaccine have been conducted in several oncogene proteins. Here, as representatives among them ; authors are reviewing WT1, BCR-ABL, and HER2/neu (Table 1).

Table 1. Summarized characteristics of WT1, BCR-ABL and HER2/neu.

Characteristics	WT1	BCR-ABL	HER2/neu
< General >			
Genotype	Wild type	Fusion chimera	Wild type
Expression pattern	Overexpression	Exclusive expression	Overexpression
Distribution	Various kinds of solid tumors, Leukemias and myeloma	Only leukemia	Breast, ovarian and several other solid tumors
< Immunogenicity >			
Autoantibody	yes	Not reported	yes
CTL in patient	yes	yes	yes
< Clinical trial >			
Vaccination	yes	yes	yes
Monoclonal antibody	no	no	yes

a) Wilms Tumor Gene 1 (WT1)

<Introduction>

WT1, one of zinc finger transcription factor which is closely relevant to primary development of urogenital system in embryo, was first identified as tumor suppressor (37, 38). Although it still remains to be determined that WT1 is more closely implicated in initiation or progression of cancers (39, 40), lines of evidence that wild-type WT1 is overexpressed in various kinds of cancers have been accumulated (41-50), and the overexpression of WT1 by cancer cells has the clinical prognostic value (51,52,53) have been accumulated. Thus the oncogenic aspect of WT1 has become emphasized, especially in leukemias(54). For example, the silencing of wt1 gene by siRNA effectively produced apoptosis of leukemia cell lines (55). Besides having uncovered its immunogenic epitopes to elicit CTL with high avidity, the observation that WT1 is expressed by leukemic stem cells (56, 57) has accelerated efforts to explore its clinical applicability for immunotherapy against cancers. Thus far, there have already been published several reports of phase I/II clinical study by means of WT1 vaccination.

<Laboratory And clinical Observations to Support the Applicability of WT1 for Cancer Immunotherapy>

First of all, to explore the property of WT1 as cancer vaccine, our laboratory and other some laboratories in the world have identified and characterized several antigenic epitopes of WT1 to elicit CD8+CTL in the context of HLA-A2, A24 and other HLA molecules, those generated CTL successfully lysed various kinds of cancer cell lines and primary cells *in vitro*, not only leukemias, but also various solid tumors (58-64). Furthermore, as a therapeutic setting for established caner, we successfully showed that the WT1-specific CTL could reject the implanted and established lung cancer in immunodeficient mice (65) (Figure 2). Simultaneously to aim to potentiate the WT1-specific CD8+ CTL efficacy, both epitopes for cytotoxic and helper CD4+ T-cells have been indentified (66,67). On the other hand, in addition to *in vitro* findings of the cellular immunogenicity of WT1, its humoral immunogenicity *in vivo*, in other words, the WT1-specific antibody production has been shown in patients with hematological malignancies, which has confirmed that WT1 actually exerts immunogenicity in patients with cancers (68,69).

Parallel to accumulation of laboratory findings, supportive data that WT-specific CTL is naturally induced in leukemia and cancer bearing individuals have also been accumulated (70-73). Furthermore, clinical observations in detail have indicated the possibility that WT1-specific CTL detected in the periphery of patients with leukemias who received allo-HSCT is closely correlated with GvL to provide longer disease remission.(74). Based on those laboratory and clinical findings to support the clinical usefulness of WT1 for cancer immunotherapy, several clinical trials of vaccination targeting WT1 have been conducted.

Most representatively, Oka et al. have reported the successful results of 26 patients bearing breast cancer (n=2), lung cancer (n=10), and leukemia (n=14) who received WT1 peptide vaccine in phase I clinical trial (ref.75:PNAS, 2004;101:13885-13890). In this report, all enrolled patients safely received WT1 peptide vaccine with Montanide ISA51 adjuvant, except severe leukocytopeina in a patient with myelodysplastic syndrome (MDS). In results,

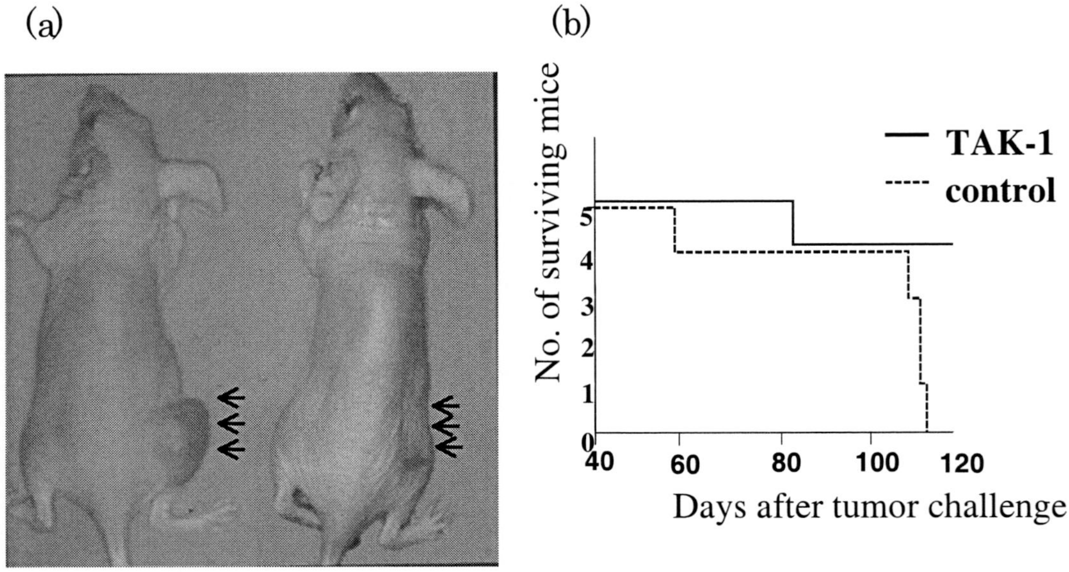

(from ref.65)

Figure 2. WT1-specific CD8+CTL clone therapeutically exert antitumor effect in vivo. HLA-A*2402-restricted CD8+T-cell clone (designated as TAK-1) which was established in our laboratory successfully demonstrated the antitumor effect in nude mice. (a) HLA-A*2402 positive human lung cancer cell line cells are inoculated in nude mice. After established the subcutaneous tumor, TAK-1 cells were adoptively transfused, and the tumor was successfully treated by those CTL. One of the representative mouse in CTL treated group is shown. (b) Survival benefit of the adoptive transfusion of WT-1specific CTL is successfully demonstrated. All mice in control group died within 120 days, while all but one in CTL-treated group were surviving during the observation periods. (Makita M *et al.* ref.65)

12 of 20 evaluated patients had clinical responses defined as the reduction of leukemic blast cell numbers, the reduction in size of tumor mass, and the reduction in serological tumor markers. Those clinical responses were positively correlated to the increased frequency of WT1-specific CTL in the periphery, which was evaluated by fluorescent peptide-MHC class I tetramer. Another German group have also conducted phase I/II trial and reported similar observation (76). The same Japanese group have also published the result of phase I/II trial of WT1 vaccine against solid tumors in 10 patients(77). In this report, the potential efficacy of WT1 vaccine against solid tumors has been validated, despite low number of patient with clinical effectiveness. Recently the same Osaka university group has proposed the very low-dose of WT1 peptide vaccination to avoid the severe suppression of normal hematopoiesis (78). We also conducted a phase I clinical study of WT1 vaccination with HLA-A*2402 restricted epitope for patients with leukemia and MDS, and experienced substantial clinical efficacy in accordance with evoked target-speicfic CTL response *in vivo* (Figure 3), which was confirmed by tetramer assay and by elispot assay to detect interferon-γ producing cells in response to target epitope (presented at 3rd International conference of WT1,in Berlin, 2007). (Figure#) As described above, WT1 is abundantly expressed in leukemia cells and leukemia stem cells, but not normal tissue, which provides the basis of safety and effectiveness of WT1 vaccine therapy for leukemias. Thus far, WT1 vaccination against solid tumors seems less

effective ,compared with leukemias. This less efficacy for solid tumors may be explained by the different expression level of antigenic epitope by each kind of cancer cells via several mechanisms which described in previous chapter.

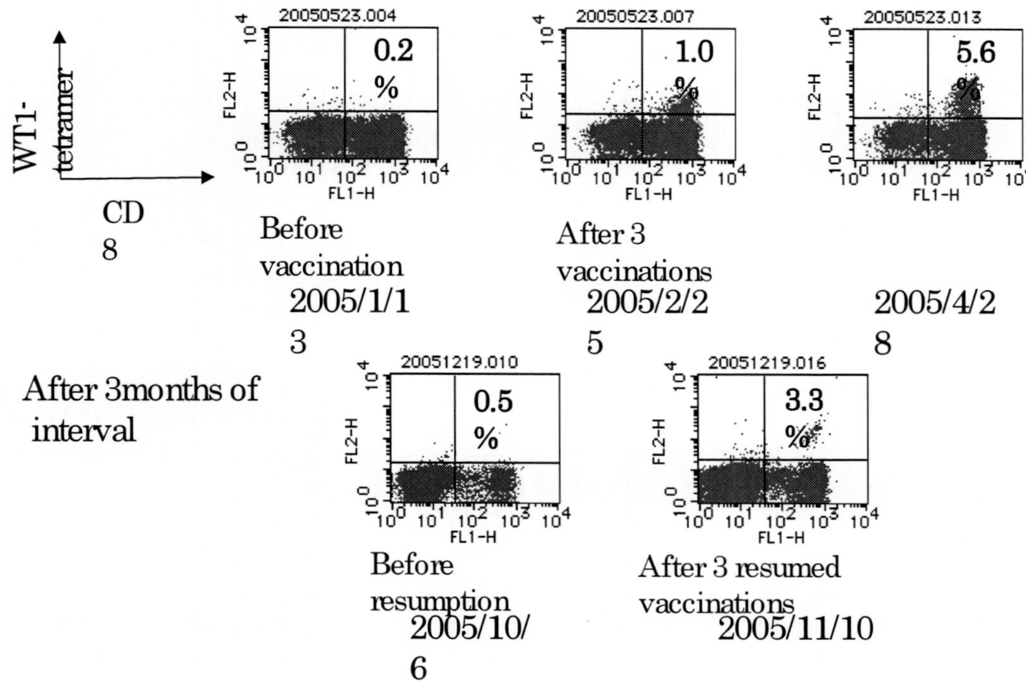

Figure 3. WT1-specific CD8+CTL was induced in the periphery of the patient with refractory acute myeloid leukemia who received WT1-peptide (WT1 $_{pp235\text{-}243}$:vaccination). Enrolled patient No.8 in phase I study of WT1 peptide vaccination with chemotherapy resistant acute myeloid leukemia was successfully controlled by WT1 peptide vaccination for longer than 2 years. In the periphery of this patient, WT1-specific CTL was detected by WT1-tetramer according to the number of vaccination (upper). Resumption after 3 months interval, WT1-peptide vaccination evoked the expansion of WT1-specific CTL again (Lower). It was shown that the induction of WT1-specific CTL was closely associated with the clinical efficacy of vaccination. (our unpublished data. Presented at 3rd International conference of WT1 in Berlin, 2007). The number listed in the upper right quadrant is the number of Tetramer positive cells (WT1-specific CTL).

To improve the clinical efficacy of WT1 vaccination therapy for cancers, developmental strategies have eagerly been under investigation. To use BCG-CWS as an adjuvant is one of them (79). Otherwise, to use the helper CD4+ T-cell directing epitope derived from WT1 in combination with the epitope for CTL induction (80,81), to innovate the hetroclitic epitope which is engineered for the higher binding affinity to target TCR, instead of the natural epitope (82), to use WT1 peptide-loaded DC(83), as a cellular vaccine, have been reported. Alternative to vaccination with WT1 peptide, the new strategy to use the target WT1 epitope-specific TCR gene-transfected artificial CTL for adoptive cell therapy has been innovated (84,85). This strategy will be able to provide abundant number of CTL against target-expressing cancer cells, which will be expected to exert the more stronger anticancer effect.

However this strategy requires pre-determined CTL clone or single TCR-gene, which presently centers on the main hindrance of this strategy for the wider application.

Taking all, although there still remains many obstacles to solve for the establishment of cancer immunotherapy targeting WT1 oncoprotein, accumulated lines of evidence strongly indicate that WT1 oncoprotein is a promising target of cancer immunotherapy.

b) Breakpoint Cluster Region-Abelson Gene Product (BCR-ABL)

<Introduction>

The reciprocal translocation between chromosome 9th and 22th: t(9;22)(q34;q11), known as Ph chromosome (86), which encodes Breakpoint Cluster Region (BCR) gene in 22th chromosome and Abelson (ABL) gene in 9th chromosome, leads the juxtapositional chimeric fusion gene *BCR-ABL*. The product protein of this fusion gene BCR-ABL has a constitutively elevated tyrosine kinase activity of ABL (87-91). This chimeric fusion gene produces various spliced isoforms, p190, p210 and p230, for example, then p190 is responsible for Ph^{+} acute lymphoblastic leukemia (ALL) and p210 is responsible for the leukemogenesis of chronic myelogenous leukemia (CML), respectively (92). In addition, clinically the chimeric gene transcript is regarded as a marker for minimal residual disease (MRD), and is also known to be an useful predictor for chemo-sensitivity and the consequent prognosis of patients with these leukemias who received chemotherapis and allo-HSCT (93-97). Based on those findings to describe that the tyrosine kinase activity of the chimeric gene is crucial for the pathogenesis of Ph-related leukemias, in recent years, the molecular agent targeting tyrosine kinase domain of bcr-abl gene, imatinib myslate has been successfully introduced into the treatment of CML and to some extent to Ph^{+}ALL (98, 99). On the other hand, the success of allogeneic stem cell transplantation for chronic phase of CML and donor lymphocyte infusion (DLI) for the treatment of relapse after transplant have produced lessons to indicate that CML cells are immunogenic and the engrafted donor immune cell derived CTL targeting CML cells are responsible for the durable disease remission (100-103). Furthermore, the BCR-ABL derived peptides were reported to be documented in elutriated peptides from CML cells (104). These accumulated lines of evidence have provided the concept that the chimeric oncoprotein BCR-ABL, especially its joining portion, is immunogenic and a potential candidate for the target of immunotherapy as well.

<Laboratory and Clinical Observations to Support the Applicability of BCR-ABL for Leukemia Immunotherapy>

Recently, the immunogenicity of BCR-ABL has been confirmed again by novel technique to use gene-modified DC to express the fusion protein by the transfection of target-gene encoding DNA, *in vitro* (105). Among 4 fusion variants of BCR-ABL, the p210 b3a2 variant seems the most promising source for providing the target epitopes which bind to HLA class I and class II molecules, to induce both of CD8+ and CD4+ CTLs (106-111). Despite the ability of b3a2 peptides to induce target-specific T cells *in vitro*, the cytocidal effect of b3a2 peptide-specific CTL widely varied, which may be due to the low affinity TCR of those b3a2 derived peptides-specific CTL. To circumvent this inconstancy of cytotoxicity of b3a2 targeting CTL, in these days, the exploration of target epitopes has been extended to other spliced variants, including epitopes of p190 for Ph1+ALL cells (112). On the other hand, clinical observations have also accumulated the lines of evidence to show BCR-ABL fusion protein derived epitopes are naturally processed and circulatory CTL are observed in the periphery of patients with CML who received allo-HSCT, imatinib myslate, or interferon-α(70,116,117).

Based on those laboratory and clinical observations, several clinical phase I/II vaccine trials have been conducted. Pinilla-Ibarz J and colleagues first reported the phase I clinical study of b3a2 peptide vaccination to 12 patients with chronic phase of CML (118). Administration of escalated doses of 5 peptides in combination vaccination with adjuvant QS-21 was safe and well-tolerated. In 3 of 6 patients who received 500μg of peptide vaccine, T cell response to the peptides was induced and delayed-type hypersensitivity (DTH) was also induced in 2 of those 3 patients. Those responses lasted up to 5 months after vaccination, while CTL were not generated even by 5 vaccinations. The same author group performed phase II study as well (119). In this phase II study, authors reported that target-responsive CD4+ and CD8+ T cells were generated in patients who received vaccination. Although the results appeared promising, the correlation between disease remission and vaccination failed to be determined, because enrolled patients were simultaneously treated by imatinib or IFN-α. As the more strongly supportive evidence, Bocchia M *et al.* well distinguished the clinical effect of vaccination by using p210-b3a2 peptides on CML from that of imatinib and/or IFN-α (120). In this paper, ten patients with CML treated by imatinib were comprised of 9 with stable cytogenetic disease (median 10% of Ph-chromosome in metaphases) and 1 with stable cytogenetic complete remission (CCR), and all of them received vaccinations. All patients showed improved cytogentic responses after six vaccinations, and 5 of 10 reached CCR. Three of these five patients had undetectable level of b3a2 gene transcript. IFN-α treated six CML patients with stable cytogenetic disease (median 10% of ph1-chromosome in metaphases) also received vaccination. Five of six patients showed improved cytogenetic responses, and 2 reached CCR. In total, peptide-specific delayed-type hypersensitivity was observed in 11 of 16 patients. Target-specific CD4+T cell proliferation was observed in 13 of 14 assessed cases, and target-epitope-induced IFN-γ production by CD4+T cell was observed in 5 of 5 assessed cases. The results suggested that vaccination with BCR-ABL b3a2 peptides could induce the clinical responses by generation of target-specific T cell responses in CML patients with low level of stable disease. On the other hand, Bornhauser M *et al.* reported that infusion of donor CD8+T cells stimulated with b3a2 peptide loaded donor DC, ex vivo, induced CCR in a patient with cytogenetic relapse after allo-HSCT with CD34+ cell selection. This is just a case report, but it suggested that targeting b3a2 peptides may be clinically acceptable and promissing (121). However, these results may simultaneously produce concerns that vaccination with BCR-ABL peptides themselves might be short for the induction of powerful CTL responses against CML, probably due to the immunological status of CML patients on vaccination. Alternative vaccinations with native peptide loaded DC, and heteroclitic peptides warrant further observations, to clear whether these strategies can overcome those problems or not (122-125).

At last, in the era of imatinib for the treatment of CML, the mutual correlation between imatinib and immunotherapy for the treatment of CML has become a new issue. As mentioned above, it is well demonstrated that imatinib is short to eradicate CML cells (ref.92. Lancet.2007;370:342-350), the combination with immunotherapy may be attractive, because of the combination of the two distinctive strategies which seem to have different mechanisms.

Regarding this issue, Brauer KM. *et al.* has reported that imatinib induced the down-regulation of immunogenic antigens on CML cells through the inhibition of BCR-ABL (126). The interference between immunotherapy, especially vaccination, and molecular target drugs including imatinib will come out to require further investigations to provide more profound comprehension.

c) Human Epidermal Growth Factor Receptor-2 (HER2) and neu (HER2/neu)

<Introduction>

Human epidermal growth factor receptor-2 and neu (HER2/neu) are the human and rodent homologues of an oncogenic growth factor receptor, which is located on the long arm of 17th chromosome, and is a member of *v-ervB* (avian erythroleukemia virus)-like oncogene family (127). The cytoplasmic domain of this HER2/neu protein is known to have tyrosine kinase activity like cellular epidermal growth factor receptor (EGFR) (128), the oncogenic signals via this cytoplasmic domain use mitogen-activated protein kinase (MAPK) and phosphatidylinositol-3-kinase (PI3K)/Akt pathway to modulate p27 and cyclin D1 (129). Contrary to constitutive expression in various normal epithelial cells (130), the overexpression of HER2 has been shown to be tumorigenic, in vitro (131, 132). In addition, the inhibition of HER2/neu in HER2/neu overexpressing tumors, by antisense (133), and by silencing with siRNA (134) induced tumor cell apoptosis. Hitherto, in human cancers, the overexpression of HER2, but not the mutation of tyrosine kinase activity containing cytoplasmic domain, is selectively documented in breast cancer and ovarian cancer (135). Overexpression of HER2/neu by either gene amplification or gene alteration other than amplification, has been observed in 25 to 30% of human primary breast and ovarian cancers, and that has simultaneously been pointed out to be closely correlated to their clinical behavior (136) Moreover, by mouse model, so-called, tumor stem cell of breast cancer has recently been shown among HER2/neu overexpressing cells (137). As recent evidence of its clinical sigenificance, the over expression of HER2/neu has been correlated with the efficacy of HER2-targeting monoclonal antibody "Trastuzumab (138, 139), via several underlying mechanisms, and HER2/neu overexprssing breast cancer has been reported to be sensitive to the HER2/neu tyrosine kinase inhibitor " Gefinitib". In addition, the loss of the efficacy of Gefinib has also been reported to be correlated the mutation of HER2/neu gene (140). As Trastuzumab, the selectively overexpressed oncogene protein HER2/neu by tumor cells can be one of ideal target for cancer immunotherapy. In the next paragraph, preclinical and clinical observations of immunotherapeutic approach to HER2/neu by using cellular or humoral immune responses are reviewed.

<Laboratory and Clinical Observations to Support the Applicability of HER2/nue for Cancer Immune-Cell Therapy>

Monoclonal antibody therapy is preceding in the clinical setting of immunotherapy targeting HER2/neu, Trastuzumab (Herceptin) is the representative. The discovery of overexpression of HER2/neu in human breast cancer and ovarian cancer, which is closely correlated to biological characteristics of cancer cells, rationally triggered the attempts to develop the monoclonal antibody (moAb) against the extracellular portion of HER2/neu for the abrogation of tumorigenic signaling. Among many developed moAbs, Trastuzumab is selected for the clinical setting. Trastuzumab (Herceptin) is a humanized mouse moAb against HER2/neu, which exerts anti-tumor effect by antibody-dependent cellular cytotoxicity (ADCC) or complement-dependent cytotoxicity (CDC) (141, 142, 143), in addition to pharmacological effects on receptor expression level, signaling pathway and so on, those are not much mentioned here. Clinical antitumor effect of Trastuzumab has been demonstrated by several clinical trials, the clinical efficacy of Trastuzumab is not curable and selective to HER2/neu overexpressing tumors approximately about 30% (ref.138, ref.139). Single agent treatment was selectively effective to HER2/neu overexpressing breast cancers, while less effective to HER2/neu overexpressing ovarian cancers (144) The most important impact in these clinical studies is that patients with early stage of HER2/neu overexpressing breast cancer can potentially be cured by Trastuzumab as an adjuvant therapy in combination with chemotherapy after surgical resection (145,146). Recently the second moAb, the inhibitor for HER dimerization, Pertuzumab is clinically under investigation (147). As one of other attempts to elaborate the antitumor efficacy of moAb to aim to improve ADCC activity mediated by NK cells, defucosylated Trastuzumab has been proposed for clinical setting (148).

Overexpression of tumor antigen HER2/neu can simultaneously provide another immunological strategy, the vaccination by means of the induction of target epitope-specific and tumor-destructing CTL. At present HER2/neu targeting vaccination is vigorously under development. In 1993 by Ioannides CG *et al.*, the 10 aminoacids (aa) peptide motif of HER2/neu $_{971\text{-}980\ \mathrm{aa}}$ which contains the anchoring motif to HLA-A2 moleclue (Rohtbard's epitope motif) was firstly shown to induce and expand target specific CTL from the malignant ascites of patient with ovarian cancer. This observation directly indicated that the peptide derived from HER2/neu is a target candidate for CTL which is naturally processed on HER2/neu overexpressing tumor cells (149). Since then, along with the excellent murine experiment that double transgenic mouse with human HLA-A2.1 and CD8 gene could show the HER2/neu peptide binding to HLA-A2 and the CTL induction ability, *in vivo* (150), the lines of evidence which indicate various HLA-A2 and A24 restricted peptides derived from HER2/neu can induce epitope-specific CTL from human PBMC, and can lyse the autologous ovarian cancer cells, metastatic breast cancer cells, pancreatic cancer cells, and renal cell cancer cells and colon cancer cells, *in vitro* (151-158). And peptides for helper CD4+ T cells which restricted by HLA-DR have also been identified (159). Finally Sotiropoulou PA *et al.* has directly shown the circulating HER2/nue derived peptides specific CTL precursors in the periphery of patients with HER2/nue overexpressing breast, ovarian, lung, colorectal, and prostate cancers. In this report, those CTL could recognized the target peptides and could lyse the autologuos tumor cells at the same time. This report has concluded the potential efficacy of HE2/nue targeting CTL in the clinical setting (160). Regarding to the HER2/neu derived peptides, recently Gritzapis AD *et al.* has shown that by using human HLA-A2.1-transgenic mouse, directly eluted peptides from primary human HER2/neu overexpressing and HLA-A2 positive cancer cells can induce target-specific and primary tumor destructive CTL in vivo, and the CTL activity induced by eluates of primary cancer cells were stronger than that induced by previously known single epitope. This finding has suggested that polyclonal CTL against multiple HER2/neu epitopes of primary tumors may be more potent than the single epitope vaccinaiton (161). As to clinical trials with HER2/neu peptide vaccination, a phase I trial with E75 peptide in combination with GM-CSF by Murry JL *et al.* showed the clinical safety and epitope-specific CTL induction ability (162). Recently Peoples GE and colleague have reported the clinical efficacy of vaccination with E75 peptide of HER2/neu against HLA-A2 positive patients with disease-free, node-positive breast cancer (high risk for recurrence) (163). In this report, authors mentioned that vaccination withE75 in combination with GM-CSF could induce peptide-specific T-cell reaction in patients, and with median follow-up period of 22 months, the disease-free survival in the vaccinated group was 85.7% compared to 59.8% in the unvaccinated group, with a recurrence rate of 8% compared to 21% of control group, respectively. These clinical observations indicated that HER2/neu peptide vaccination may be effective and prevent a recurrence in patients with so-called minimal residual disease state of breast cancer.

On the other hand, to overcome the limitations of single peptide vaccination, for example, to avoid the constraint of HLA restriction, DC has been introduced to circumvent those limitations and potentiate the vaccination efficacy against HER2/neu overexpressing cancers. Hitherto, target peptides loaded DC, DNA or RNA gene-modified DC to express HER2/nue have been proposed. Target peptides loaded DC in clinical grade has already been conducted in phase I/II study and could show to induce immunological reaction in patients with some clinical responses (164-167). Importantly, Czerniecki BJ *et al.* have reported that 14 patients with early staged breast cancer of ductal carcinoma insitu (DCIS)were successfully treated by DC with combined peptides directing both for HLA-class I and II, and surgical specimens of 7 out of 11 evaluable patients demonstrated the reduction of HER2/neu expression in residual DCIS, this finding suggested the vaccine efficacy for early stage of high risk cancer (168). Gene modified DC vaccine to express HER2/neu and/or accessory molecules is expected to potentiate DC vaccine efficacy by overcoming HLA-restriction, and concomitant provision of helper epitope, while so far, gene-modified DC vaccine strategy still remains experimental (169-173) At last, interestingly viewing from the point of HER2/neu targeting adoptive cell transfer, CTL induction from human umbilical cord blood with peptide loaded DC as naïve response induction, which may be expected to retain high proliferative activity, has been reported as a promising source of CTL (174).

4. Emerging Problems to Attenuate the Efficacy of Cancer Vaccine Therapy

Presently large number of clinical trials of cancer vaccine have been conducted, and the almost all results were of little clinical effect, or of marginal effect, at best (ref.9: Nat Med.2004;10:909-915.). Integration of laboratory and clinical lines of information about the "immune escape of tumor cells from host defense", various kinds of mechanisms for the immune evasion have been uncovered, and those still remain to be clarified (Figure 4).

First, as regarding the epitope expression by tumor cells, two distinct mechanisms of down-regulation of target-epitope expression have been reported. One is the down-regulation of antigen-presenting molecules inside of tumor cells which are supposed to present target-epitope for vaccine-induced CTL. These antigen-presenting molecules includes HLA class I molecule (175,176), and transporter for antigen presentation (TAP) (177). The other is "immunoediting", the epigenetic gene alteration to adversely affect the target protein expression of tumor cells, which occurs under the particular immune-stress triggered by the vaccination (178).

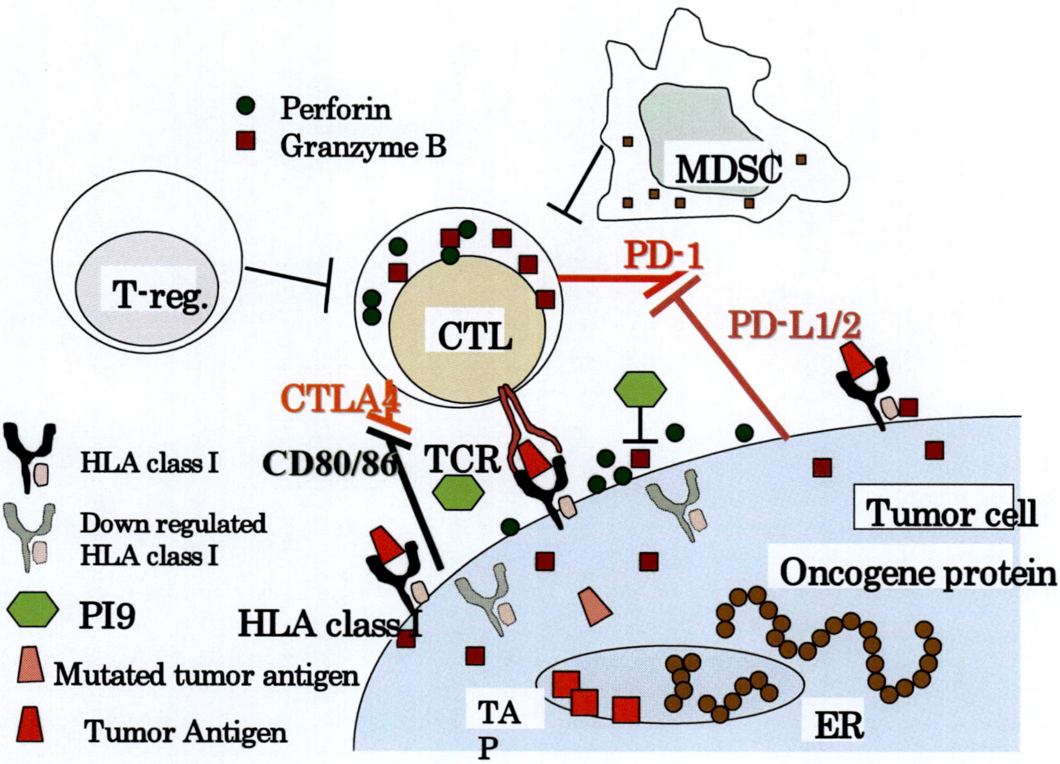

Figure 4. Factors adversely affect the CTL activity against cancer cells. Under immunological stress evoked by vaccination, multiple factors can adversely affect the CTL activity against cancer cells. As for cancer cells, the downregulation of HLA class I molecules, downregulation of tumor antigen processing machinery including TAP, epigenetic gene alteration of tumor antigen (as for immunoediting). As for CTL itself, inhibitory signals via CTLA-4, and PD-1 receptor. The ligands for PD-1 (PD-L1/2) can be provided by cancer cell itself. The inhibitory environmental factors are regulatory T-cell (T-reg.) and myeloid derived suppressor cells (MDSC). Finally, the anticytotoxic granule factor proteinase inhibitor 9(PI9) provided by cancer cell can degrade cytotoxic granules from CTL. Details are described in the text.

Second, as to changes inside of CTL, these factors are divided into two portions. One is the changes in CTL themselves, so-called, the co-inhibitory receptors come to appear on the surface of CTL. These molecules include CTL-associated Antigen 4 (CTLA-4)(179), and programmed death-1 (PD-1) (180,181). Ligands for CTLA-4 are CD80 and CD86 on activated antigen presenting cells (APC) and tumor cells, the ligation to CTLA-4 induces the negative regulation of T cell response, and tolerance. On the other hand, ligands for PD-1 are PD-L1 and PD-L2, those are also expressed on activated APC, and some tumor cells. PD-1 ligation also induces the negative regulation of CTL, which is so-called “exhaustion” of CTL, and it has been reported that the blockade of PD-1/PD-L1 enhanced the CTL activity.

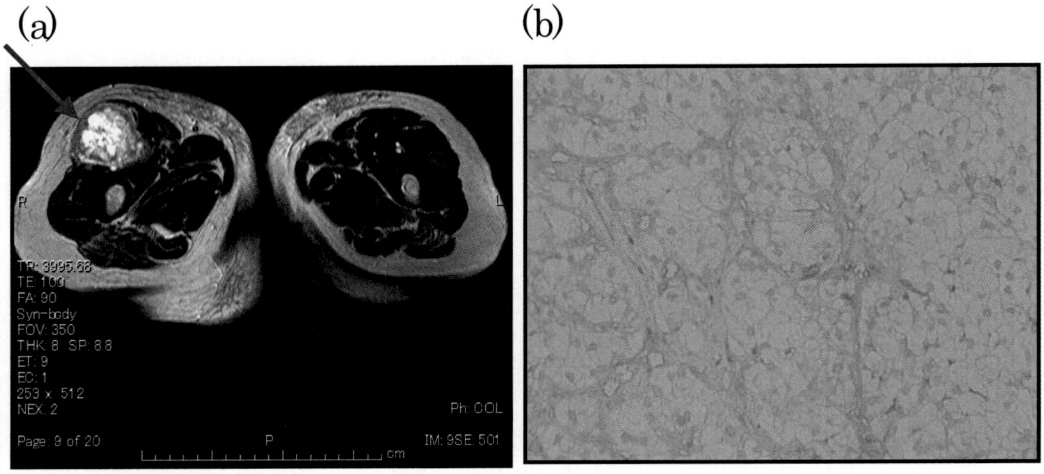

Figure 5. HLA-class I downregulation by progressed renal cell cancer cells. We experienced the disease progression case of renal cell cancer (RCC) who received vaccination by human telomerase reverse transcriptase peptide (hTERT). Under vaccination with hTERT, the patient had disease progression presenting intramuscular mass in right thigh, after durable stable disease of pulmonary lesions for 2 years. (a) Arrow indicates the recurrence of RCC in the right thigh. (b) Pathological examination revealed the downregulation of HLA class I molecules by immunohistochemistry. (unpublished data, presented at 3rd International conference of WT1 in Berlin, 2007).

Third, the regulatory environmental factors, irrelevant to the exhausted state of CTL, which inhibit CTL responses against tumor cells in the local area, are regulatory T lymphocytes (T-reg) and myeloid derived suppressor cells (MDSC)(182-186). Although functional mechanisms of T-reg. and MDSC are not fully elucidate, as to the clinical significance, in patients with hepatocellular carcinoma, the intratumoral balance of T-reg. and CTL has recently been demonstrated to be associated with the prognosis (187), and the phenotypic identification and functional evaluation of human MDSC in peripheral blood of melanoma patients with granulocyte-macrophage colony-stimulation factor-based antitumor vaccination have recently been described (188). Reasonably, to aim the enhancement of anti-tumor CTL activity, the selective depletion both of T-reg. and MDSC have already started to be explored in murine model (189, 190,191). On the other hand, the accumulating lines of information of innate immunity, especially toll-like receptors, has brought an another way to potentiate cancer immunity evoked by cancer vaccination via partially interfering those regulatory cells(192, 193).

5. Summary

Along with the progress in molecular biology and immunological techniques, cancer immunotherapy to aim the induction of cancer-reactive CTLs in cancer-bearing individuals has become real. Vaccine strategy to target tumor antigens is one of them. To elicit a long-lasting anticancer immunity in cancer-bearing individuals, vaccine strategy, instead of taking anticancer drugs for many years, looks more ideal. However, hitherto, the clinical efficacy of cancer vaccine trials have demonstrated to be rather disappointing.

One of explanations for this less clinical efficacy may be that large part of tumor-or tumor-associated antigens are not oncogene proteins, which are vital for the survival of cancer cells. From this point of view, oncogene protein is a both real and ideal target for cancer immunotherapy. As mentioned above, there still remains many obstacles to overcome in cancer immunotherapy, such as immune evasion of tumor cells by regulatory T cells, the immunoediting induced by vaccine-induced continuous immune stress, and so on (Table 2).

Table 2. Presently proposed strategies to improve immuno-cell therapy for cancer.

(1) To potentiate immunogenicity to induce CTL
- (a) Heteroclitic epitope to increase binding affinity to HLA molecule.
- (b) Peptide loaded or genetically modified DC as cellular vaccine.
- (c) Combination epitope vaccines directing both for CD4(helper) and CD8(cytotoxic).

(2) To make more numbers of CTL
- (a) Adoptively transfer of pre-characterized CTL in vitro.
    - (I) Precultured in ordinary methods, autologous or allogeneic CTL.
    - (II) Genetically characterized with target-specific TCR. Matured T lymphocyte, alternative source (cord blood.)

(3) To manipulate the immunosuppressive environment
- (a) To decrease the number or functionally suppress the T-reg.
- (b) To manipulate Toll-like receptors.

Nowadays the rapid progress in molecular biology and immunology continues to accelerate to develop innovative strategies. Thus, the future of cancer immunotherapy may be rather promising, even though of some marginal effect presently.

References

1. Introduction

[1] Tuttle TM. Oncogene products represent potential target of tumor vaccines. Cancer *Immunol Immunother*. 1996; 43:135-141.

[2] Kazansky DB. Intrathymic selection: new insight into tumor biology. *Adv Exp Med Biol.* 2007; 601: 133-134.

[3] Reiman JM, Kmieciak M, Manjili MH, Knutson KL. Tumor immunoediting and immunosculpting pathway to cancer progression. *Semin Cancer Biol.* 2007; 17: 275-287.

[4] van der Vliet HJ, Koon HB, Atkins MB, Balk SP, Exley MA. Exploiting regulatory T-cell populations for the immunotherapy of cancer. *J Immunother*. 2007; 30: 591-595.

[5] Martin-Orozco N, Dong C. Inhibitory costimulation and anti-tumor immunity. *Semin Cancer Biol.* 2007; 17: 288-298.

[6] Kolb HJ, Schmid C, Barrett AJ, Schendel DJ. Graft-versus-leukemia reaction in allogeneic chimeras. *Blood*. 2004; 103: 767-776.

[7] Sprangers B, Wijmeersch BV, Fevery S, Waer M, Billiau AD. Experimental and clinical approaches for optimization of the graft-versus-leukemia effect. *Nat Clin Pract Oncol.* 2007; 4: 404-414.
[8] Rosenberg SA. Progress in human tumor immunology and immunotherapy. *Nature.* 2001; 411: 380-384.
[9] Rosenberg SA, Yang JC, and Restifo N. Cancer immunotherapy: moving beyond current vaccines. *Nature Medicine.*2004;10:909-915.

2. How Do CTL Exert Cytocidal Effect to Cancer Cells?

Expression of Target Antigenic Epitope on Tumor Cells

[10] Pamer E, Cresswell P. Mechanisms of MHC class I-restricted antigen processing. *Annu Rev Immunol.* 1998; 16: 323-358.
[11] Momburg F, Roelse J, Howard JC, Butcher GW, Hammerling GJ, Neefjes JJ. Selectivity of MHC-encoded peptide transporters from human, mouse and rat. *Nature.* 1994; 367: 648-651.
[12] Rock KL, Gramm C, Rothstein L, Clark K, Stein R, Dick L, Hwang D, Goldberg AL. Inhibitors of the proteosome block the degradation of most cell proteins and the generation of peptides presented on MHC class I molecules. *Cell.* 1994; 78: 761-771.
[13] Lehner PJ. The calculus of immunity: quantitating antigen processing. *Immunity.* 2003; 18: 315-317.
[14] Park B, Lee S, Kim E, Cho K, Riddle SR, Cho S, and Ahn K. Redox regulation facilitates optimal peptide selection by MHC class I during antigen processing. *Cell.*2006;127:369-382.
[15] Eisenlohr LC, Huang L, Golovina TN. Rethinking peptide supply to MHC class-I molecules. *Nat Rev Immunol.* 2007; 7: 403-410.
[16] Chmbers B, Grufman P, Fredriksson V, Andersson K, Roseboom M, Laban S, Camps M, Wolpert EZ, Wiertz EJ, Offringa R, Ljunggren HG, van Hall T. Induction of protective CTL immunity against peptide transporter TAP-deficient tumors through dendritic cell vaccination. *Cancer Res.* 2007; 67: 8450-8455.
[17] Ferris RL, Whiteside TL, Ferrone S. Immune escape associated with functional defects in antigen-processing machinery in head and neck cancer. *Clin Cancer Res.* 2006; 12: 3890-3895.

Recognition by Epitope-Specific CTL / Immunological Synapse

[18] Davis MM, Bjorkman PJ. T-cell antigen receptor gene and T-cell recognition. *Nature.* 1988;334:395-402.
[19] Berg NN, Puente LG, Dawicki W, Ostergaard HL. Sustained TCR signaling is required for mitogen-activated protein kinase activation and degranulation by cytotoxic T lymphocytes. *Journal of Immunology.* 1998;161:2919-2924.
[20] Kalergis AM, Boucheron N, Doucey MA, Palmieri E, Goyarts EC, Vegh Z, Luescher IF, Nathenson SG. Efficient T cell activation requires an optimal dwell-time of interaction between the TCR and the pMHC complex. *Nat Immunol.* 2001; 2:229-234.

[21] Rosette C, Werlen G, Daniels MA, Holman PO, Alam SM, Travers PJ, Gascoigne NR, Palmer E, Jameson SC. The impact of duration versus extent of TCR occupancy on T cell activation: a revision of the kinetic proofreading model. *Immunity*. 2001; 15: 59-70.

[22] Slansky JE, Rattis FM, Boyd LF, Fahmy T, Jaffee EM, Schneck JP, Margulies DH, Pardoll DM. Enhanced antigen-specific antitumor immunity with altered peptide ligands that stabilize the MHC-peptide-TCR complex. *Immunity*. 2000; 13: 529-538.

[23] Purbhoo MA, Boulter JM, Price DA, Vuidepot AL, Hourigan CS, Dunbar PR, Olson K, Dawson SJ, Phillips RE, Jakobsen BK, Bell JI, Sewell AK. The human CD8 coreceptor effects cytotoxic T cell activation and antigen sensitivity primarily by mediating complete phosphorylation of the T cell receptor zeta chain. *J Biol Chem*. 2001; 276: 32786-32792.

[24] Arroyo AG, Campanero MR, Sanchez-Mateos P, Zapata JM, Ursa MA, del Pozo MA, Sanchez-Madrid F. Induction of tyrosine phosphorylation during ICAM-3 and LFA-1-mediated intercellular adhesion, and its regulation by the CD45 tyrosine phosphatase. *J Cell Biol.*. 1994;126:1277-1286.

[25] Friedman RS, Jacobelli J, Krummel MF. Surface-bound chemokines capture and prime T cells for synapse formation. *Nature Immunology*. 2006;7:1101-1108.

[26] Lee KH, Holdorf AD, Dustin ML, Chan AC, Allen PM, Shaw AS. T cell receptor signaling precedes immunological synapse formation. *Science*. 2002;295:1539-1542.

[27] Stinchombe JC, Bossi G, Booth S, Griffiths GM. The immunological synapse of CTL contains a secretory domain and membrane bridges. *Immunity*. 2001;15:751-761.

[28] Montoya MC, Sancho D, Vicente-Manzanares M, Sanchez-Madrid F. Cell adhesion and polarity during immune interactions. *Immunol Rev*.2002;186:68-82.

[29] Bossi G, Trambas C, Booth S, Clark R, Stinchcombe J,Griffiths GM. The secretory synapse:the secrets of a serial killer. *Immunol Rev*. 2002;189:152-160.

[30] Gonzalez PA, Carreno LJ, Figueroa CA, Kalergis AM. Modulation of immunological synapse by membrane-bound and soluble ligands. *Cytokine & Growth factor Rev*.2007;18:19-31.

Cytocidal Degranulation by CTL

[31] Barry M, Bleackley RC. Cytotoxic T lymphocytes: All roads lead to death. *Nat Rev Immunol.* 2002;2:401-409.

[32] Jenne DE, Tschopp J. Granzymes, a family of serine proteases released from granules of cytotoxic T lymphocytes upon T cell receptor stimulation. *Immunol Rev*.1988;103:53-71.

[33] Catalfamo M, Henkart PA. Perforin and granule exocytosis cytotoxicity pathway. *Cur Opin Immunol*. 2003;15:522-527.

[34] He JS, Ostergaard. CTLs contain and intracellular stores of FasL distinct from cytolytic granules. *J Immunol*. 2007;179:2339-23348

[35] Balaji KN, Schaschke N, Machleidt W, Catalfamo M, Henkart PA. Surface cathepsin B protects cytotoxic lymphocytes from self-destruction after degranulation. *J Exp Med.* 2002; 196: 493-503.

[36] Medema JP, de Jong J, Peltenburg LT, Verdegaal EM, Gorter A, Bres SA, Franken KL, Hahne M, Albar JP. Melief CJ, Offringa R. Blockade of the granzyme B/perforin pathway through overexpression of the serine protease inhibitor PI-9/SPI-6 constitutes a

mechanism for immune escape by tumors. *Proc Natl Acad Sci USA*. 2001;98:11515-11520.

3. Oncogene Proteins ; Targets of Cancer Immunotherapy: WT1, BCR-ABL, and HER2/neu, the Representatives for Cancer Immunotherapy Wilms' Tumor Gene 1 (WT1)

Wilms Tumor Gene 1 (WT1)

[37] Rivera MN, Haber DA. Wilms' tumor: connecting tumorigenesis and organ development in the kidney. *Nat Rev Cancer*. 2005;5:699-712.
[38] Ellisen LW, Carlesso N, Cheng T, Scadden DT, Haber DA. The Wilms' tumor suppressor WT1 directs stage-specific quiescence and differentiation of human hematopoietic progenitor cells. *EMBO J*. 2001;20:1897-1909.
[39] Burwell EA, McCarty GP, Simpson LA, Thompson KA, Loeb DM. Isoforms of Wilms' tumor suppressor gene 1 (WT1) have distinct effect on mammary epithelial cells. *Oncogene*. 2007;26:3423-3430.
[40] Han Y, San-Marina S, Yang L, Khoury H, Minden MD. The zinc finger domain of Wilms' tumor 1 suppressor gene (WT1) behaves as a dominant negative, leading to abrogation of WT1 oncogenic potential in breast cancer cells. *Breast Cancer Res*. 2007;9:R43, doi:10,1186/bcr1743.
[41] Oji Y, Miyoshi S, Maeda H, Hayashi S, Tamaki H, Nakatsuka S, Yao M, Takahashi E, Nakano Y, Hirabayashi h, Shintani Y, Oka Y, Tsuboi A, Hosen N, Asada M, Fujioka T, Murakami M, Kanato K, Motomura M, Kim EH, Kawakami M, Ikegame K, Ogawa H, Aozasa K, Kawase I, Sugiyama H. Overexpression of the Wilms' tumor gene WT1 in de novo lung cancers. *Int J Cancer*. 2002;200:297-303.
[42] Oji Y, Inohara H, Nakazawa N, Nakano Y, Akahashi S, Nakatsuka S, Koga S, Ikeba A, Abeno S, Honjo Y, Yamamoto Y, Iwai S, Yoshida K, Oka Y, Ogawa H, Yoshida J, Aozasa K, Kubo T, Sugiyama H. Overexpression of the Wilms' tumor gene WT1 in head and neck squamous cell carcinoma. *Cancer Sci*. 2003;94:523-529.
[43] Ueda T, Oji Y, Naka N, Nakano Y, Takahashi E, Koga S, Asada M, Ikeba A, Nakatsuka S, Abeno S, Hosen N, Tomita Y, Aozasa K, Tamai N, Myoui A, Yoshikawa H, Sugiyama H. Overexpression of the Wilms' tumor gene WT1 in human bone and soft-tissue sarcomas. *Cancer Sci*.2003;94:271-276.
[44] Oji Y, Yamamoto H, Nomura M, Nakano Y, Ikeba A, Nakatsuka S, Abeno S, Kiyotoh E, Jomgeow T, Sekimoto M, Nezu R, Yoshikawa Y, Inoue Y, Hosen N, Kawakami M, Tsuboi A, Oka Y, Ogawa H, Souda S, Aozasa K, Monden M, Sugiyama H. Overexpression of the Wilms' tumor gene WT1 in colorectal adenocarcinoma. *Cancer Sci*. 2003;94:712-717.
[45] Oji Y, Miyoshi Y, Koga S, Nakano Y, Ando A, Nakatsuka S, Ikeba A, Takahashi E, Sakaguchi N, Yokota A, Hosen N, Ikegame K, Kawakami M, Tsuboi A, Oka Y, Ogawa H, Aozasa K, Noguchi S, Sugiyama H. Overexpression of the Wilms' tumor gene WT1 in primary thyroid cancer. *Cancer Sci*. 2003;94:606-611.
[46] Oji Y, Nakamori S, Fujikawa M, Nakatsuka S, Yokota A, Tatsumi N, Abeno S, Ikeba A, Takashima S, Tsujie M, Yamamoto H, Sakon M, Nezu R, Kawano K, Nishida S,

Ikegame K, Kawakami M, Tsuboi A, Oka Y, Yoshikawa K, Aozasa K, Monden M, Sugiyama H. Overexpression of the Wilms' tumor gene WT1 in pancreatic ductal adenocarcinoma. *Cancer Sci.* 2004;95:583-587.

[47] Oji Y, Yano M, Nakano Y, Abeno S, Nakatsuka S, Ikeba A, Yasuda T, Fujiwara Y, Takiguchi S, Yamamoto H, Fujita S, Kanato K, Ito K, Jomgeow T, Kawakami M, Tsuboi A, Shirakata T, Nishida S, Hosen N, Oka Y, Aozasa K, Monden M, Sugiyama H. Overexpression of the Wilms' tumor gene WT1 in esophageal cancer. *Anticancer Res.* 2004;24:3103-3108.

[48] Oji Y, Suzuki T, Nakano Y, Maruno M, Nakatsuka S, Jomgeow T, Abeno S, Tatsumi N, Yokota A, Aoyagi S, Nakazawa T, Ito K, Kanato K, Shirakata T, Nishida S, Hosen N, Kawakami M, Tsuboi A, Oka Y, Aozasa K, Yoshimine T, Sugiyama H. Overexpression of the Wilms' tumor gene WT1 in primary astrocytic tumors. *Cancer Sci.* 2004;95:822-827

[49] Hylander B, Repasky E, Shrikant P, Intengan M, Beck A, Driscoll D, Singhal P, Lele S, Odunsi K. Expression of Wilms tumor gene (WT1) in epithelial ovarian cancer. *Gynecol Oncol.* 2006;101:12-17.

[50] Coosemans A, Nik SA, Caluwaerts S, Lambin S, Verbist G, Van Bree R, Schelfhout V, de Jonge E, Dalle I, Jacomen G, Cassiman JJ, Moerman P, Vergote I, Amant F. Upregulation of Wilms' tumor gene 1 (WT1) in uterine sarcomas. *Eur J Cancer.* 2007;43:1630-1637.

[51] Lapillonne H, Renneville A, Auvrignon A, Flamant C, Blaise A, Perot C, Lai JL, Ballerini P, Mazingue F, Fasola S, Dehee A, Bellman F, Adam M, Labopin M, Douay L, Leverger G, Preudhomme C, Landman-Parker J. High WT1 expression after induction therapy predicts high risk of relapse and death in pediatric acute myeloid leukemia. *J Clin Oncol.* 2006; 24:1507-1515.

[52] Sotobori T, Ueda T, Oji Y, Naka N, Araki N, Myoui A, Sugiyama H, Yoshikawa H. Prognostic significance of Wilms tumor gene 1 (WT1) mRNA expression in soft tissue sarcoma. *Cancer.* 2006;106:2233-2240.

[53] Srivastava A, Fuchs B, Zhang K, Ruan M, Halder C, Mahlum E, Weber K, Bolander ME, Sarkar G. High WT1 expression is associated with very poor survival of patients with osteogenic sarcoma metastasis. *Clin Cancer Res.* 2006;12:4237-4243.

[54] Yang L, Han Y, Saurez Saiz F and Minden MD. A tumor suppressor and oncogene: the WT1 story. *Leukemia.*2007;21:868-876.

[55] Glienke W, Maute L, Koehl U, Esser R, Milz E, Bergmann L. Effective treatment of leukemic cell lines with wt1 siRNA. Leukemia, 2007; doi:10.1038/sj.leu.2404878.

[56] Gao L, Bellantuono I, Elsasser A, Marley SB, Gordon MY, Goldman JM, Stauss HJ. Selective elimination of leukemic CD34(+) progenitor cells by cytotoxic T lymphocytes specific for WT1. *Blood.* 2000; 95:2198-2203.

[57] Hosen N, Shirakata T, Nishida S, Yanagihara N, Tsuboi A, Kawakami M, Oji Y, Oka Y, Okabe M, Tan B, Sugiyama H and Weissman IL. The Wilms' tumor gene WT1-GFP knock-in mouse reveals the dynamic regulation of WT1 expression in normal and leukemic hematopoiesis. *Leukemia.* 2007;21:1783-1791.

[58] Oka Y, Udaka K, Tsuboi A, Elisseeva OA, Ogawa H, Aozasa K, Kishimoto T, Sugiyama H. Cancer immunotherapy targeting Wilms' tumor gene 1 WT1 product. *J Immunol.* 2000;164:1873-1880.

[59] Ohminami H, Yasukawa M, Fujita S. HLA class I-restricted lysis of leukemia cells by a CD8(+) cytotoxic T-lymphocyte clone specific for WT1 peptide. *Blood.* 2000;95: 286-293.
[60] Azuma T, Makita M, Ninomiya K, Fujita S, Harada M, Yasukawa M. Identification of a novel WT1-derived peptide which induces human leukocyte antigen-A-24-restricted anti-leukemia cytotoxic T lymphocytes. *Br J Haematol.* 2002;116:601-603.
[61] Bellantuono I, Gao L, Parry S, Marley S, Dazzi F, Apperley J, Goldman JM, Stauss HJ. Two distinct HLA-A0201-presented epitopes of the Wilms tumor antigen 1 can function as targets for leukemia-reactive CTL. *Blood.*2002; 100;3835-3837.
[62] Koesters R, Linnebacher M, Coy JF, Germann A, Schwitalle Y, Findeisen P, von Knebel Doeberitz M. WT1 is a tumor-associated antigen in colon cancer that can be recognized by in vitro stimulated cytotoxic T cells. *Int J Cancer.* 2004;109:385-392.
[63] Azuma T, Otsuki T, Kuzushima K, Froelich CJ, Fujita S, and Yasukawa M. Myeloma cells are highly sensitive to the granule exocytosis pathway mediated by WT1-specific cytotoxic T lymphocytes. *Clin Cancer Res.*2004;10:7402-7412.
[64] Asemissen AM, Keilholz U, Tenzer S, Muller M, Walter S, Stevanovic S, Letsch A, Thiel E, Rammensee HG, Scheibenbogen C. Identification of a highly immunogenic HLA-A*01-binding T cell epitope of WT1. *Clin Cancer Res.* 2006;12:7476-7482.
[65] Makita M, Hiraki A, Azuma T, Tsuboi A, Oka y, Sugiyama H, Fujita S, Tanimoto M, Harada M, Yasukawa M. Antilung cancer effect of WT1-specific cytotoxic T lymphocytes. *Cin Cancer Res.* 2002;8:2626-2631.
[66] Guo Y, Niiya H, Azuma T, Uchida N, Yakushijin Y, Sakai I, Hato T, Takahashi M, Senju S, Nishimura Y, Yasukawa M. Direct recognition and lysis of leukemia cells by WT1-specific CD4+ T lymphocytes in an HLA class II-restricted manner. *Blood.* 2005;106:1415-1418.
[67] Kobayashi H, Nagato T, Aoki N, Sato K, Kimura S, Tateno M, Celis E. Defining MHC class II T helper epitopes for WT1 tumor antigen. *Cancer Immunol Immunother.* 2006;55:850-860.
[68] Elisseeva OA, Oka Y, Tsuboi A, Ogata K, Wu F, Kim EH, Soma T, Tamaki H, Kawakami M, Oji Y, Hosen N, Kubota T, Nakagawa M, Yamagami T, Hiraoka A, Tsukaguchi T, Udaka K, Ogawa H, Kishimoto T, Momura T, Sugiyama H. Humoral immune responses against Wilms tumor gene WT1 product in patients with hematopoietic malignancies. *Blood.* 2002; 99:3272-3279.
[69] Wu F, Oka Y, Tsuboi A, Elisseeva OA, Ogata K, Nakajima H, Fujiki F, Masuda T, Murakami M, Yoshihara S, Ikegame K, Hosen N, Kawakami M, Nakagawa M, Kubota T, Soma T, Yamagami T, Tsukaguchi M, Ogawa H, Oji Y, Hamaoka T, Kawase I, Sugiyama H. Th1-biased humoral immune responses against Wilms tumor gene WT1 product in the patients with hematopoietic malignancies. *Leukemia.* 2005;19:268-274.
[70] Rezvani K, Grube M, Brenchley JM, Sconocchia G, Fujiwara H, Price DA, Gostick E, Yamada K, Melenhorst J, Childs R, Hensel N, Douek DC, Barrett AJ. Functional leukemia-associated antigen-specific memory CD8+ T cells exist in healthy individuals and in patients with chronic myelogenous leukemia before and after stem cell transplantation. *Blood.* 2003;102:2892-2900.
[71] Gannage M, Abel M, Michallet AS, Delluc S, Lambert M, Giraudier S, Kratzer R, Niedermann G, Saveanu L, Guilhot F, Camoin L, Varet B, Buzyn A, Caillat-Zucman S. Ex vivo characterization of multiepitopic tumor-specific CD8+ T cells in patients with

chronic myelogenous leukemia: implications for vaccine development and adoptive cellular immunotherapy. *J Immunol.* 2005;174:8210-8218.

[72] Rezvani K, Brenchley JM, Price DA, Kilical Y, Gostick E, Sewell AK, Li J, Mielke S, Douek DC, Barrett AJ. T-cell responses directed against multiple HLA-A*0201-restricted epitopes derived from Wilms' tumor 1 protein in patients with leukemia and healthy donors: identification, quantification, and characterization. *Clin Cancer Res.* 2005;11:8799-8807.

[73] Gillmore R, Xue SA, Holler A, Kaeda J, Hadjiminas D, Healy V, Dina R, Parry SC, Bellantuono I, Ghani Y, Coombes RC, Waxman J, Stauss HJ. Detection of Wilms' tumor antigen-specific CTL in tumor-draining lymph nodes of patients with early breast cancer. *Clin Cancer Res*. 2006;12:34-42.

[74] Rezvani K, Yong AS, Savani BN, Mielke S, Keyvanfar K, Gostick E, Price DA, Douek DC, Barrett AJ. Graft-versus-leukemia effects associated with detectable Wilms tumor-1 specific T lymphocytes after allogeneic stem-cell transplantation for acute lymphoblastic leukemia. *Blood*. 2007;110:1924-1932.

[75] Oka Y, Tsuboi A, Taguchi T, Oaski T, Kyo T, Nakajima H, Elisseeva OA, Oji Y, Kawakami M, Ikegame K, Hosen N, Yoshihara S, Wu F, Fujiki F, Murakami M, Masuda T, Nishida S, Shirakata T, Nakatsuka S, Sasaki A, Udaka K, Dohy H, Aozasa K, Noguchi S, Kawase I, and Sugiyama H. Induction of WT1 (Wilms' tumor gene)-specific cytotoxic T lymphocytes by WT1 peptide vaccine and the resultant cancer regression. *Proc Natl Acad Sci USA*, 2004;101:13885-13890.

[76] Mailander V, Scheibenbogen C, Thiel E, Letsch A, Blau IW, Keilholz U. Complete remission in a patient with recurrent acute myeloid leukemia induced by vaccination with WT1 peptide in the absence of hematological or renal toxicity. *Leukemia*. 2004; 18:165-166.

[77] Morita S, Oka Y, Tsuboi A, Kawakami M, Maruno M, Izumoto S, Osaki T, Taguchi T, Ueda T, Myoui A, Nishida S, Shirakata T, Ohno S, Oji Y, Aozasa K, Hatazawa J, Udaka K, Yoshikawa H, Yoshimine T, Noguchi S, Kawase I, Nakatsuka S, Sugiyama H and Sakamoto J. A phase I/II trial of a WT1 (Wilms' Tumor gene) peptide vaccine in patients with patients with solid malignancies: safety assessment based on the phase I data. *Jpn J Clin Oncol*,2006;36:231-236.

[78] Kawakami M, Oka Y, Tsuboi A, Harada Y, Elisseeva OA, Furukawa Y, Tsukaguchi M, Shirakata T, Nishida S, Nakajima H, Morita S, Sakamoto J, Kawase I, Oji Y, Sugiyama H. Clinical and immunologic responses to very low-dose vaccination with WT1 peptide (5 microgram/body) in a patient with myelomonocytic leukemia. *Int J Hematol.* 2007;85:426-429.

[79] Nakajima H, Kawasaki K, Oka Y, Tsuboi A, Kawakami M, Ikegame K, Hoshida Y, Fujiki F, Nakano A, Masuda T, Wu F, Taniguchi Y, Yoshihara S, Elisseeva OA, Oji Y, Ogawa H, Azuma I, Kawase I, Aozasa K, Sugiyama H. WT1 peptide vaccination combined with BCG-CWS is more efficient for tumor eradication than WT1 peptide vaccination alone. *Cancer Immunol Immunother*. 2004;53:617-624.

[80] Fujiki F, Oka Y, Tsuboi A, Kawakami M, Kawakatsu M, Nakajima H, Elisseeva OA, Harada Y, Ito K, Li Z, Tatsumi N, Sakaguchi N, Fujioka T Masuda T, Yasukawa M, Udaka K, Kawase I, Oji Y, Sugiyama H. Identification and characterization of a WT1 (Wilms tumor gene) protein-derived HLA-DRB1*0405-restricted 16-mer helper epitope

that promotes the induction and activation of WT1-specific cytotoxic T lymphocytes. *J Immunother*. 2007;30:282-293.

[81] May RJ, Dao T, Pinilla-Ibarz J, Korontsvit T, Zakhaleva V, Zhang RH, Maslak P, Scheinberg DA. Peptide epitopes from Wilms' tumor 1 oncoprotein stimulate CD4+ and CD8+ T cells that recognize and kill human malignant mesothelioma tumor cells. *Clin Cancer Res*. 2007;13:4547-4555.

[82] Pinilla-Ibarz J, May RJ, Korontsvit T, Gomez M, Kappel B, Zakhaleva V, Zhang RH, Scheinberg DA. Improved human T-cell responses against synthetic HLA-0201 analog peptides derived from the WT1 oncoprotein. *Leukemia*. 2006;20:2025-2033.

[83] Ho WY, Nguyen HN, Wolfl M, Kuball J, Greenberg PD. In vitro methods for generating CD8+ T-cell clones for immunotherapy from the naïve repertoire. *J Immunol Methods*. 2006; 310: 40-52.

[84] Tsuji T, Yasukawa M, Matsuzaki J, Ohkuri T, Chamoto K, Wakita D, Azuma T, Niiya H, Miyoshi H, Kuzushima K, Oka Y, Sugiyama H, Ikeda H, Nishimura T. Generation of tumor-specific, HLA class I-restricted human Th1 and Tc1 cells by cell engineering with tumor peptide-specific T-cell receptor genes. *Blood*. 2005;106:470-476.

[85] Xue SA, Gao L, Hart D, Gillmore R, Qasim W, Thrasher A, Apperley J, Engels B, Uckert W, Morris E and Stauss H. Elimination of human leukemia cells in *NOD/SCID* mice by WT1-TCR gene-transduced human T cells. *Blood*. 2005;106:3062-3067.

Breakpoint Cluster Region-Abelson gene product (BCR-ABL fusion protein)

[86] Skorski T, Nieborowska-Skorska M, Nicolaides NC, Szczylik C, Iversen P, Iozzo RV, Zon G, Calabretta B. Suppression of Philadelphia1 leukemia cell growth in mice by BCR-ABL antisense oligonucleotide. *Proc Natl Acad Sci USA*. 1994;91:4504-4508.

[87] Ren R. Mechanisms of BCR-ABL in the pathogenesis of chronic myelogenous leukemia. *Nat Rev Cancer*. 2005;5:172-183.

[88] Koschmieder S, Gottgens B, Zhang P, Iwasaki-Arai J, Akashi K, Kutok JL, Dayaram T, Geary K, Green AR, Tenen DG , Huettner CS. Inducible chronic phase of myeloid leukemia with expansion of hematopoietic stem cells in a transgenic model of BCR-ABL leukemogenesis. *Blood*. 2005;105:324-334.

[89] Wendel HG, de Stanchina E, Cepero E, Ray S, Emig M, Fridman JS, Veach DR, Bornmann WG, Clarkson B, McCombie WR, Kogan SC, Hochhaus A, Lowe SW. Loss of p53 impedes the antileukemic response to BCR-ABL inhibition. *Proc Natl Acad Sci USA*. 2006;103:7444-7449.

[90] Van Etten RA. Oncogenic signaling: new insights and controversies from chronic myeloid leukemia. *J Exp Med*. 2007;204:461-465.

[91] Koptyra M, Falinski R, Nowicki MO, Stoklosa T, Majsterek I, Nieborowska-Skorska M, Blasiak J, Skorski T. BCR-ABL kinase induces self-mutagenesis via reactive oxygen species to encode imatinib resistance. *Blood*. 2006;108:319-327.

[92] Hehlmann R, Hochhaus A, Baccarani M on behalf of the European LeukemiaNet. Chronic myeloid leukemia. *Lancet*.2007;370:342-350.

[93] Langabeer SE, Gale RE, Harvey RC, Cook RW, Mackinnon S, Linch DS. Transcription-mediated amplification and hybridization protection assay to determine

BCR-ABL transcript level in patients with chronic myeloid leukemia. *Leukemia*. 2002; 16:393-399.

[94] Lange T, Deininger M, Brand R, Hegenbart U, Al-Ali H, Krahl R, Poenisch W, Uharek L, Leiblein S, Gentilini C, Petersdorf E, Storb RF, Niederwieser D. BCR-ABL transcripts are early predictors for hematological relapse in chronic myeloid leukemia after hematopoietic cell transplantation with reduced intensity conditioning. *Leukemia*. 2004, 18:1468-1475.

[95] Rapoport AP, Levine BL, Badros A, Meisenberg B, Ruehle K, Nandi A, Rollins S, Natt S, Ratterree B, Westphal S, Mann D, June CH. Molecular remission of CML after autotransplantation followed by adoptive transfer of costimulated autologus T cells. *Bone Marrow Transplant*. 2004;33: 53-60.

[96] Kaeda J, O'Shea D, Szydlo RM, Olavarria E, Dazzi F, Marin D, Saunders S, Khorashad JS, Cross NC, Goldman JM, Apperley JF. Serial measurement of BCR-ABL transcripts in the peripheral blood after allogeneic stem cell transplantation for chronic myeloid leukemia: an attempt to define patients who may not require further therapy. *Blood*. 2006;107:4171-4176.

[97] Spinelli O, Peruta B, Tosi M, Guerini V, Salvi A, Zanotti MC, Oldani E, Grassi S, Intermesoli T, Mico C, Rossi G, Fabris P, Lambertenghi-Deliliers G, Angelucci E, Barbui T, Bassan R, Rambaldi A. Clearance of minimal residual disease after allogeneic stem cell transplantation and the prediction of the clinical outcome of adult patients with high-risk acute lymphoblastic leukemia. *Haematologica*. 2007; 92: 612-618.

[98] Kantarjian HM, Giles F, Quintas-Cardama A, and Cortes J. Important therapeutic targets in chronic myelogenous leukemia. *Clin Cancer Res*. 2007; 13:1089-1097.

[99] Schiffer CA. BCR-ABL tyrosine kinase inhibitors for chronic myelogenous leukemia. *N Engl J Med*. 2007;357:258-265.

[100] Barrett A, Jiang YZ. Immune responses to chronic myeloid leukemia. *Bone Marrow Transplant*. 1992; 9:305-311.

[101] Kolb HJ, Schattenberg A, Goldman JM, Hertenstein B, Jacobsen N, Arcese W, Ljungman P, Ferrant A, Verdonck L, Niederwieser D, van Rhee F, Mittermueller J, de Witte T, Holler E, Ansari H; European Group for Blood and Marrow Transplantation Working Party Chronic Leukemia. Graft-versus-leukemia effect of donor lymphocyte transfusions in marrow grafted patients. *Blood*.1995;86:2041-2050.

[102] Falkenburg JH, Wafelman AR, Joosten P, Smit WM, van Bergen CA, BOngaerts JE, Kluin-Nelemans HC, Fibbe WE, Willemze R. Complete remission of accelerated phase chronic myeloid leukemia by treatment with leukemia-reactive cytotoxic T lymphocytes. *Blood*. 1999;94:1201-1208.

[103] Krause DS, Van Etten RA. Adoptive immunotherapy of BCR-ABL-induced chronic myeloid leukemia-like myeloproliferative disease in a murine model. *Blood*. 2004;104:4236-4244.

[104] Papadopoulos KP, Suciu-Foca N, Hesdorffer CS, Tugulea S, Maffei A, Harris PE. Naturally processed tissue- and differentiation stage-specific autologous peptides bound by HLA class I and II molecules of chronic myeloid leukemia blasts. *Blood*. 1997; 90:4938-4946.

[105] Sun JY, Krouse RS, Forman SJ, Senitzer D, Sniecinski I, Chatterjee S, Wong KK Jr. Immunogenicity of a p210 (BCR-ABL) fusion domain candidate DNA vaccine targeted

to dendritic cells by a recombinant adeno-associated virus vector in vitro. *Cancer Res.* 2002; 62: 3175-3183.

[106] Bocchia M, Wentworth PA, Southwood S, Sidney J, McGraw K, Scheinberg DA, Sette A. Specific binding of leukemia oncogene fusion protein peptides to HLA class I molecules. *Blood.* 1995; 85: 2680-2684.

[107] Bosch GJ, Joosten AM, Kessler JH, Melief CJ, Leeksma OC. Recognition of BCR-ABL positive leukemic blasts by human CD4+ T cells elicited by primary in vitro immunization with a BCR-ABL breakpoint peptide. *Blood.* 1996; 88: 3522-3527.

[108] Yotonda P, Firat H, Garcia-Pons F, Garcia Z, Gourru G, Vernant JP, Lemonnier FA, Leblond V, Langlade-Demoyen P. Cytotoxic T cell response against the chimeric p210 BCR-ABL protein in patients with chronic myelogenous leukemia. *J Clin Invest.* 1998; 101: 2290-2296.

[109] Osman Y, Takahashi M, Zheng Z, Koike T, Toba K, Liu A, Furukawa T, Aoki S, Aizawa Y. Generation of bcr-abl specific cytotoxic T-lymphocytes by using dendritic cells pulsed with bcr-abl (b3a2) peptide: its applicability for donor leukocyte transfusion in marrow grafted CML patients. *Leukemia.* 1999; 13: 166-174.

[110] Yasukawa M, Ohminani H, Kaneko S, Yakushijin Y, Nishimura Y, Inokuchi K, Miyakuni T, Nakao S, Kishi K, Kubonishi I, Dan K, Fujita S. CD4(+) cytotoxic T-cell clones specific for bcr-able b3a2 fusion peptide augment colony formation by chronic myelogenous leukemia cells in a b3a2-specific and HLA-DR-restricted manner. *Blood.* 1998; 92:3355-3361.

[111] Norbury LC, Clark RE, Christmas SE. b3a2 BCR-ABL fusion peptides as targets for cytotoxic T cells in chronic myeloid leukemia. *Br J Haematol.* 2000; 109: 616-621.

[112] Tanaka Y, Takahashi T, Nieda M, Masuda S, Kashiwase K, Ogawa S, Chiba S, Juji T, Hirai H. Generation of HLA-DRB1*1501-restricted p190 minor bcr-abl (e1a2)-specific CD4+ T lymphocytes. *Br J Haematol.* 2000; 109: 435-437.

[113] Kessler JH, Bres-Vloemans SA, van Veelen PA, de Ru A, Huijbers IJG, Camps M, Mulder A, Offringa R, Drijfhout JW, Leeksma OC, Ossendrop F, Melief CJM. BCR-ABL fusion regions as a source of multiple leukemia-specific $CD8^+$ T-cell epitopes. *Leukemia.* 2006; 20: 1738-1750.

[114] Crough T, Nieda M, Morton J, Bashford J, Durrant S, Nicol AJ. Donor-derived b2a2-specific T cells for immunotherapy of patients with chronic myeloid leukemia. *J Immunother.* 2002; 25:469-475.

[115] Volpe G, Cignetti A, Panuzzo C, Kuka M, Vitaggio K, Brancacio M, Perrone G, Rinaldi M, Prato G, Fava M, Geuna M, Pautasso M, Casnici C, Signori E, Tonon G, Tarone G, Marelli O, Fazio VM, Saglio G. Alternative BCR/ABL splice variants in Philadelphia chromosome-positive leukemias result in novel tumor-specific fusion proteins that may represent potential targets for immunotherapy approach. *Cancer Res.* 2007; 67:5300-5307.

[116] Butt NM, Rojas JM, Wang I,Christmas SE, Abu-Eisha HM, Clark RE. Circulating bcr-abl-specific CD8+ T cells in chronic myeloid leukemia patients and healthy subjects. *Haematologica.* 2005; 90:1315-1323.

[117] Gannage M, Abel M, Michallet AS, Delluc S, Lambert M, Giraudier S, Kratzer R, Niedermann G, Saveanu L, Guilhot F, Camoin L, Varet B, Buzyn A, Caillat-Zucman S. Ex vivo characterization of multiepitopic tumor- specific CD8 T cells in patients with

chronic myeloid leukemia: implications for vaccine development and adoptive cellular immunotherapy. *J Immunol.* 2005; 174: 8210-8218.

[118] Pinilla-Ibarz J, Cathcart K, Korontsvit T, Soignet S, Bocchia M, Caggiano J, Lai L, Jimenez J, Kolitz J, Scheinberg DA. Vaccination of patients with chronic myelogenous leukemia with bcr-abl oncogene breakpoint fusion peptides generates specific immune responses. *Blood.* 2000; 95:1781-1787.

[119] Cathcart K, Pinilla-Ibarz J, Korontsvit T, Schwartz J, Zakhaleva V, Papadopoulos EB, Scheinberg DA. A multivalent bcr-abl fusion peptide vaccination trial in patients with chronic myeloid leukemia. *Blood.* 2004; 103: 1037-1042.

[120] Bocchia M, Gentili S, Abruzzese E, Fanelli A, Tabilio A, Amabile M, Forconi F, Gozzetti A, Raspadori D, Amadori S, Lauria F. Effect of a p210 multipeptide vaccination associated with imatinib or interferon in patients with chronic myeloid leukemia and persistent residual disease: a multicentre observational trial. *Lancet.* 2005; 365: 657-662.

[121] Bornhauser M, Thiede C, Babatz J, Schetelig J, Illmer T, Kian A, Platzbecker U, Herr W, Rieber EP, Ehninger G, Schmitz M. Infusion of bcr/abl peptide-reactive donor T cells to achieve molecular remission of chronic myeloid leukemia after CD34+ selected allogeneic hematopoietic cell transplantation. *Leukemia.* 2006; 20: 2055-2057.

[122] He L, Feng H, Raymond A, Kreeger M, Zeng Y, Graner M, Whitesell L, Katsanis E. Dendritic-cell-peptide immunization provides immunoprotection against bcr-abl-positive leukemia in mice. *Cancer Immunol Immunother.* 2001; 50:31-40.

[123] Takahashi T, Tanaka Y, Nieda M, Azuma T, Chiba S, Juji T, Shibata Y, Hirai H. Dendritic cell vaccination for patients with chronic myelogenous leukemia. *Leuk Res.* 2003; 27: 795-802.

[124] Litzow MR, Dietz AB, Bulur PA, Bulter GW, Gastineau DA, Hoering A, Fink SR, Letndre L, Padley DJ, Paternoster SF, Tefferi A, Vuk-Pavlovic S. Testing the safety of clinical-grade mature autologous myeloid DC in a phase I clinical immunotherapy trial of CML. *Cytotherapy.* 2006;8:290-298.

[125] Pinilla-Ibarz J, Korontsvit T, Zakhaleva V, Roberts W, Scheinberg DA. Synthetic peptide analogs derived from bcr/abl fusion proteins and the induction of heteroclitic human T-cell responses. *Haematologica.* 2005; 90: 1324-1332.

[126] Brauer KM, Werth D, von Schwarzenberg, Bringmann A, Kanz L, Grunebach F, Brossart P. BCR-ABL activity is critical for the immunogenicity of chronic myelogenous leukemia cells. *Cancer Res.* 2007; 67: 5489-5497.

Human Epidermal Growth Factor Receptor-2 (HER2) and neu (HER2/neu)

[127] Schechter AL, Stern DF, Vaidyanathan L, Decker SJ, Drebin JA, Greene MI, Weinberg RA. The neu oncogene: an erb-B-related gene encoding a 185,000-Mr tumor antigen. *Nature.* 1984; 312:513-516.

[128] Akiyama T, Sudo C, Ogawara H, Toyoshima K, Yamamoto T. The product of the human c-erbB-2 gene: a 185-kilodalton glycoprotein with tyrosine kinase activity. *Science.* 1986; 232: 1644-1646.

[129] Lenferink AE, Busse D, Flanagan WM, Yakes FM, Arteaga CL. ErbB2/neu kinase modulates cellular p27(Kip1) and cyclin D1 through multiple signaling pathways. *Cancer Res.*2001; 61: 6583-6591.

[130] Press MF, Cordon-Cardo C, Slamon DJ. Expression of the HER-2/neu proto-oncogene in normal human adult and fetal tissues. *Oncogene.* 1990; 5: 953-962.

[131] Di Fiore PP, Pierce JH, Kraus MH, Segatto O, King CR, Aaronson SA. erbB-2 is a potent oncogene when overexpressed in NIH/3T3 cells. *Science.* 1987; 237:178-182.

[132] Hudziak RM, Schlessinger J, Ullrich A. Increased expression of the putative growth factor p185HER2 causes transformation and tumorigenesis of NIH/3T3 cells. *Proc Natl Acad Sci USA*, 1987; 84; 7159-7163.

[133] Colomer R, Lupu R, Bacus SS, Gelmann EP. arbB-2 antisense oligonucleotides inhibit the proliferation of breast carcinoma cells with erbB-2 oncogene amplification. *Br J Cancer.* 1994; 70: 819-825.

[134] Faltus T, Yuan J, Zimmer B, Kramer A, Loibl S, Kaufmann M, Strebhardt K. Silencing of the HER2/neu gene by siRNA inhibits proliferation and induces apoptosis in HER2/neu-overexpressing breast cancer cells. *Neoplasia.* 2004; 6: 786-795.

[135] Slamon DJ, Clark GM, Wong SG, Levin WJ, Ullrich A, McGuire WL. Human breast cancer: correlation of relapse and survival with amplification of the HER-2/neu onocgene. *Scinece.* 1987; 235: 177-182.

[136] Slamon DJ, Godolphin W, Jones LA, Holt JA, Wong SG, Keith DE, Levin WJ, Stuart SG, Udove J, Ullrich A , et al. Studies of the HER-2/neu proto-oncogene in human breast and ovarian cancer. *Science.* 1989; 244: 707-712.

[137] Liu JC, Deng T, Lehal RS, Kim J, Zacksenhaus E. Identification of tumorsphere- and tumor-initiating cells in HER2/Neu-induced mammary tumors. *Cancer Res.* 2007; 67: 8671-8681.

[138] Vogel CL, Cobleigh MA, Tripathy D, Gutheil JC, Harris LN, Fehrenbacher L, Slamon DJ, Murphy M, Novotny WF, Burchmore M, Shak S, Stewart SJ, Press M. Efficacy and safety of trastuzumab as a single agent I first-line treatment of HER2-overexpressing metastatic breast cancer. *J Clin Oncol.* 2002; 20: 719-726.

[139] Mass RD, Press MF, Anderson S, Cobleigh MA, Vogel CL, Dybdal N, Leiberman G, Slammon DJ. Evaluation of clinical outcomes according to HER2 detection by fluorescence in situ hybridization in women with metastatic breast cancer treated with trastuzumab. *Clin Breast Cancer.* 2005;6:240-246

[140] Piechocki MP, Yoo GH, Dibbley SK, Lonardo F. Breast cancer expressing the activated HER2/neu is sensitive to gefinitib in vitro and in vivo and acquires resistance through a novel point mutation in the HER2/neu. *Cancer Res.* 2007; 67: 6825-6843.

[141] Carter P, Presta L, Gorman CM, Ridgway JB, Henner D, Wong WL, Rowland AM, Kotts C, Carver ME, Shepard HM. Humanization of an anti-p185 HER2 antibody for human cancer therapy. *Proc Natl Acad Sci USA.* 1992; 89: 4285-4289.

[142] Gennari R, Menard S, Fagnoni F, Ponchio L, Scelsi M, Tagliabue E, Castiglioni F, Villani L, Magalotti C, Gibelli N, Oliviero B, Ballardini B, Da Prada G, Zambelli A, Costa A. Pilot study of the mechanism of action preoperative trastuzumab in patients with primary operable breast tumors overexpressing HER2. *Clin Cancer Res.* 2004; 10: 5650-5655.

[143] Arnold L, Gelly M, Penault-Llorca F, Benoit L, Bonnetain F, Migeon C, Cabaret V, Fermeaux V, Bertheau P, Garnier J, Jeannin JF, Coudert B. Trastuzumab-based

treatment of HER2-positive breast cancer: an antibody-dependent cellular cytotoxicity mechanism? *Br J Cancer*. 2006; 94: 259-267.

[144] Bookman MA, Darcy KM, Clarke-Pearson D, Boothby RA, Horowitz IR. Evaluation of monoclonal humanized anti-HER2 antibody, trastuzumab, in patients with recurrent or refractory ovarian or primary peritoneal carcinoma with overexpression of HER2: a phase II trial of the Gynecologic Oncology Group. *J Clin Oncol.* 2003; 21: 283-290.

[145] Piccart-Gebhart MJ, Procter M, Leyland-Jones B, Goldhirsch A, Untch M, Smith I, Gianni L, Baselga J, Bell R, Jackisch C, Cameron D, Dowsett M, Barrios CH, Steger G, Huang CS, Andersson M, Inbar M, Lichinitser M, Lang I, Nitz U, Iwata H, Thomssen C, Lohrisch C, Suter TM, Ruschoff J, Suto T, Greatorex V, Ward C, Straehle C, McFadden E, Dolci MS, Gelber RD, Herceptin Adjuvant (HERA) Trial Study Team. Trastuzumab after adjuvant chemotherapy in HER2-positive breast cancer. *N Engl J Med.* 2005;353:1659-1672.

[146] Romond EH, Perez EA, Bryant J, Suman VJ, Geyer CE Jr., Davidson NE, Tan-Chiu E, Martino S, Paik S, Kaufman PA, Swain SM, Pisansky TM, Fehrenbacher L, Kutteh LA, Vogel VG, Visscher DW, Yothers G, Jenkins RB, Brown AM, Dakhil SR, Mamounas EP, Lingle WL, Klein PM, Ingle JN, Wolmark N. Trastuzumab plus adjuvant chemotherapy for operable HER2-positive breast cancer. *N Engl J Med.* 2005; 353:1673-1684.

[147] Adams CW, Allison DE, Flagella K, Presta L, Clarke J, Dybdal N, McKeever K, Sliwkowski MX. Humanization of a recombinant monoclonal antibody to produce a therapeutic HER dimerization inhibitor, pertuzumab. *Cancer Immunol Immunother.* 2006; 55: 717-727.

[148] Suzuki E, Niwa R, Saji S, Muta M, Hirose M, Iida S, Shiotsu Y, Satoh M, Shitara K, Kondo M, Toi M. A nonfucosylated anti-HER2 antibody augments antibody-dependent cellular cytotoxicity in breast cancer patients. Clin Cancer Res. 2007; 13: 1875-1882.

[149] Ioannides CG, Fisk B, Fan D, Biddison WE, Wharton JT, O'brian CA. Cytotoxic T cells isolated from ovarian malignant ascites recognized a peptide derived from the HER-2/neu proto-oncogene. *Cell Immunol.* 1993;151: 225-234.

[150] Lustgarten J, Theobald M, Labadie C, LaFace D, Peterson P, Disis ML, Cheever MA, Sherman LA. Identification of Her-2/Neu CTL epitopes using double transgenic mice expressing HLA-A2.1 and human CD.8. *Human Immunol.* 1997; 52: 109-118.

[151] Fisk B, Chesak B, Pollack MS, Wharton JT, Ioannides CG. Oligopeptide induction of a cytotoxic T lymphocyte responses to HER-2/neu proto-oncogene in vitro. *Cell Immunol.* 1994; 157: 415-427.

[152] Disis ML, Smith JW, Murphy AE, Chen W, Cheever MA. In vitro generation of human cytolytic T-cells specific for peptides derived from the HER-2/neu protooncogene protein. *Cancer Res.* 1994; 54: 1071-1076.

[153] Linehan DC, Peoples GE, Hess DT, Summerhayes IC, Parikh AS, Goedegebuure PS, Eberlein TJ. In vitro stimulation of ovarian tumor-associated lymphocytes with a peptide derived from HER2/neu induces cytotoxicity against autologous tumor. *Surg Oncol.* 1995 ; 4: 41-49.

[154] Fisk B, Blevins TL, Wharton JT, Ioannides CG. Identification of an immunodominant peptide of HER-2/neu protooncogene recognized by ovarian tumor-specific cytotoxic T lymphocyte lines. *J Exp Med.* 1995 ; 181: 2109-2117.

[155] Linehan DC, Goedegebuure PS, Peoples GE, Rogers SO, Eberlein TJ. Tumor-specific and HLA-A2-restricted cytolysis by tumor-associated lymphocytes in human metastatic breast cancer. *J Immunol.* 1995; 155: 4486-4491.

[156] Piper M, Goedegebuure PS, Linehan DC, Ganguly E, Douville CC, Eberlein TJ. The HER2/neu-derived peptide p654-662 is a tumor-associated antigen in human pancreatic cancer recognized by cytotoxic T lymphocytes. *Eur J Immunol.* 1997 ; 27: 1115-1123.

[157] Brossart P, Stuhler G, Flad T, Stevanovic S, Rammensee HG, Kanz L, Brugger W. Her-2/neu-derived peptides are tumor-associated antigens expressed by human renal cell and colon carcinoma lines and are recognized by in vitro induced specific cytotoxic T lymphocytes. *Cancer Res.* 1998 ; 58: 732-736.

[158] Azuma K, Shichijo S, Shomura H, Matsueda S, Fujii T, Itoh K. Identification of HER2/neu-derived peptides capable of inducing both cellular and humoral immune responses in HLA-A24 positive breast cancer patients. *Breast Cancer Res Treat.* 2004; 86:19-29.

[159] Perez SA, Sotiropoulou PA, Sotiriadou NN, Mamalaki A, Gritzapis AD, Echner H, Voelter W, Pawelec G, Papamichail M, Baxevanis CN. HER-2/neu-derived peptide 884-899 is expressed by human breast, colorectal and pancreatic adenocaricinmas and is recognized by in-vitro-induced specific CD4(+) T cell clones. *Cancer Immunol Immunother.* 2002 ; 50: 615-624.

[160] Sotiropoulou PA, Perez SA, Voelter V, Echner H, Missitzis I, Tsavaris NB, Papamichail M, Baxevanis CN. Natural CD8+ T-cell responses against MHC class I epitope of the HER-2/neu oncoprotein in patients with epithelial tumors. *Cancer Immunol Immunother.* 2003 ; 52: 771-779.

[161] Gritzapis AD, Sotiriadou NN, Papamichail M, Baxevanis CN. Generation of human tumor-specific CTLs in HLA-A2.1-transgenic mice using unfractionated peptides from eluates of human primary breast and ovarian tumors. *Cancer Immunol Immunother.* 2004; 53: 1027-1040.

[162] Murray JL, Gillogly ME, Przepiorka D, Brewer H, Ibrahim NK, BooserDJ, Hortobagyi GN, Kudelka AP, Grabstein KH, Cheever MA, Ioannides CG. Toxicity, immunogenicity, and induction of E75-specific tumor-lytic CTLs by HER-2 peptide E75 (369-377) combined with granulocyte macrophage colony-stimulating factor in HLA-A2+ patients with metastatic breast and ovarian cancer. *Clin Cancer Res.* 2002 ; 8: 3407-3418.

[163] People GE, Gurney JM, Hucman MT, Woll MM, Ryan GB, Storrer CE, Fisher C, Shriver CD, Ioannides CG, Ponniah S. Clinical trial results of a HER2/neu (E75) vaccine to prevent recurrence in high-risk breast cancer patients. *J Clin Oncol.* 2005; 23: 7536-7545.

[164] Brossart P, Wirths S, Stuhler G, Reichardt VL, Kanz L, Brugger W. Induction of cytotoxic T-lymphocyte responses in vivo after vaccinations with peptide-pulsed dendritic cells. *Blood.* 2000; 96: 3102-3108.

[165] Disis ML, Rinn K, Knutson KL, Davis D, Caron D, dela Rosa C, Schiffman K. Flt3 ligand as a vaccine adjuvant I association with HER-2/neu peptide-based vaccines in patients with HER-2/neu-overexpressing cancers. *Blood* 2000; 99: 2845-2850.

[166] Kono K, Takahashi A, Sugai H, Fujii H, Choudhury AR, Kiessling R, Matsumoto Y. Dendritic cells pulsed with HER-2/neu-derived peptides can induce specific T-cell responses in patients with gastric cancer. *Clin Cancer Res.* 2002; 8: 3394-3340.

[167] Dees EC, Mckinnon KP, Kuhns JJ, Chwastiak KA, Spark S, Myers M, Collins EJ, Frelinger JA, Van Deventer H, Collichio F, Carey LA, Brecher ME, Graham M, Earp HS, Serody JS. Dendritic cells can be rapidly expanded ex vivo and safely administered in patients with metastatic breast cancer. *Cancer Immunol Immunother*. 2004; 53: 777-785.

[168] Czerniecki BJ, Koski GK, Koldovsky U, Xu S, Cohen PA, Mick R, Nisenbaum H, Pasha T, Xu M, Fox KR, Weinstein S, Orel SG, Vonderheide R, Coukos G, DeMichele A, Araujo L, Spitz FR, Rosen M, Levine BL, June C, Zhang PJ. Targeting HER-2/neu in early breast cancer development using dendritic cells with staged interleukin-12 burst secretion. *Clin Cancer Res*. 2007; 67: 1842-1852.

[169] Chen Y, Emtage P, Zhu Q, Foley R, Muller W, Hitt M, Gauldie J, Wan Y. Induction of ErbB-2/neu-specific protective and therapeutic antitumor immunity using genetically modified dendritic cells: enhanced efficacy by cotransduction of gene encoding IL-12. *Gene Ther*. 2001; 8:316-323.

[170] zum Buschenfelde CM, Metzger J, Hermann C, Nicklisch N, Peschel C, Bernhard H. The generation of both T killer and Th cell clones specific for the tumor-associated antigen HER2 using retrovirally transduced dendritic cells. *J Immunol*. 2001; 167: 1712-1719.

[171] Muller MR, Grunebach F, Nencioni A, Brossart P. Transfection of dendritic cells with RNA induces CD4- and CD8-mediated T cell immunity against breast carcinomas and reveals the immunodominance of presented T cell epitopes. *J Immunol*. 2003; 170: 5892-5896.

[172] Ffferson CL, Schickli J, Ko BK, Kawano K, Mouzi S, Palese P, Garcia-Sastre A, Ioannides CG. Activation of tumor antigen-specific cytotoxic T lymphocytes (CTL) by human dendritic cells infected with an attenuated influenza A virus expressing a CTL epitope derived from the HER-2/neu proto-oncogene. *J Virol*, 2003; 77: 7411-7424.

[173] Sakai Y, Morrison BJ, Burke JD, Park JM, Terabe M, Janik JE, Forni G, Berzofsky JA, Morris JC. Vaccination by genetically modified dendritic cells expressing a truncated neu oncogen prevents development of breast cancer in transgenic mice. *Cancer Res,* 2004; 64: 8022-8028.

[174] Wang P, Munger CM, Joshi AD, Pirruccello SJ, Joshi SS. Cytotoxicity of cord blood derived Her2/neu-specific cytotoxic T lymphocytes against human breast cancer in vitro and in vivo. *Breast Cancer Res Treat*, 2004; 83:15-23.

4. Emerging Problems to Attenuate the Efficacy of Cancer Vaccine Therapy

[175] Romero JM, Aptsiauri N, Vazquez F, Cozar JM, Canton J, Cabrera T, Tallada M, Garrido F, Ruiz-Cabello F. Analysis of the expression of HLA class I, proinflammatory cytokines and chemokines in primary tumors from patients with localized and metastatic renal cell carcinoma. *Tissue Antigen*. 2006; 68: 303-310.

[176] Cabrera T, Lara E, Romero JM, Maleno I, Real LM, Ruiz-Cabello F, Valero P, Camacho FM, Garrido F. HLA class I expression in metastatic melanoma correlates with tumor development during autologous vaccination. *Cancer Immunol Immunother*. 2007; 56: 709-717.

[177] Pandha H, Rigg A, John J, Lemoine N. Loss of expression of antigen-presenting molecules in human pancreatic cancer and pancreatic cancer cell lines. *Clin Experimental Immunol.* 2007; 148: 127-135.
[178] Singh R, Paterson Y. Immunoediting sculpts tumor epitopes during immunotherapy. *Cancer Res.* 2007; 67: 1887-1892.
[179] Tuve S, Chen B-M, Liu Y, Cheng T-L, Toure P, Sow PS, Feng Q, Kiviat N, Strauss R, Ni S, Li Z-Y, Roffler SR, Lieber A. Combination of tumor site-located CTL-Associated Antigen-4 blockade and systemic regulatory T-cell depletion induces tumor-destructive immune response. *Cancer Res.* 2007; 67:5929-5939.
[180] He Y-F, Zhang G-M, Wang X-H, Zhang H, Yuan Y, Li D, Feng Z-H. Blocking programmed death-1 ligand-PD-1 interactions by local gene therapy results in enhancement of antitumor effect of secondary lymphoid tissue chemokine. *J Immunol.* 2004; 173: 4919-4928.
[181] Hirano F, Kaneko K, Tamura H, Dong H, Wang S, Ichikawa M, Rietz C, Flies DB, Lau JS, Zhu G, Tamada K, Chen L. Blockade of B7-H1 and PD-1 by monoclonal antibodies potentiates cancer therapeutic immunity. Cancer Res. 2005; 65: 1089-1096.
[182] Curiel TJ. T regs and rethinking cancer immunotherapy. *J Clin Invest.* 2007; 117: 1167-1174.
[183] Kapp JA, Honjo K, Kapp LM, Goldsmith K, Bucy RP. Antigen, in the presence of TGF-β, induces up-regulation of Fox$P3^{gfp+}$ in $CD4^+$ TCR transgenic T cells that mediate linked suppression of $CD8^+$ T cell responses. *J Immunol.* 2007; 179: 2105-2114.
[184] Wei S, Kryczek I, Edwards RP, Zou L, Szeliga W, Banerjee M, Cost M, Cheng P, Chang A, Redman B, Herberman RB, Zou W. Interleukin-2 administration alters the CD4+FOXP3+ T-cell pool and tumor trafficking in patients with ovarian carcinoma. *Cancer Res.* 2007; 67: 7487-7494.
[185] Nagaraj S, Gupta K, Pisarev V, Kinarsky L, Sherman S, Kang L, Herber D, Schneck J, Gabrilovich DI. Altered recognition of antigen is a mechanism of CD8+ T cell tolerance in cancer. *Nat Med.* 2007; 13: 828-835.
[186] Talmadge J. Pathways mediating the expansion and immunosuppressive activity of myeloid-derived suppressor cells and their relevance to cancer therapy. *Clin Cancer Res.* 2007; 13: 5243-5248.
[187] Gao Q, Qiu SJ, Fan J, Zhou J, Wang XY, Xiao YS, Xu Y, Li YW, Tang ZY. Intratumoral balance of regulatory and cytotoxic T cells is associated with prognosis of hepatocellular carcinoma after resection. *J Clin Oncol.* 2007; 25: 2586-2593.
[188] Filipazzi P, Valenti R, Huber V, Pilla L, Canese P, Iero M, Castelli C, Mariani L, Parmiani G, Rivoltini L. Identification of a new subset of myeloid suppressor cells in peripheral blood of melanoma patients with modulation by a granulocyte-macrophage colony-stimulation factor-based antitumor vaccine. *J Clin Oncol.* 2007; 25:2546-2553.
[189] Tuve S, Chen B-M, Liu Y, Cheng T-L, Toure P, Sow PS, Feng Q, Kiviat N, Strauss R, Ni S, Li Z-Y, Roffler SR, Lieber A. Combination of tumor site-located CTL-Associated Antigen-4 blockade and systemic regulatory T-cell depletion induces tumor-destructive immune response. *Cancer Res.* 2007; 67:5929-5939. (Again)
[190] Ko H-J, Kim Y-J, Kim Y-S, Chang W-S, Ko S-Y, Chang S-Y, Sakaguchi S, Kang C-Y. A combination of chemoimmunotherapies can efficiently break self-tolerance and

induce antitumor immunity in a tolerogenic murine tumor model. *Cancer Res*. 2007; 67:7477-7486.

[191] Pan P-Y, Wang GX, Yin B, Ozao J, Ku T, Divino CM, Chen S-H. Reversion of immune tolerance in advanced malignancy: modulation of myeloid derived suppressor cell development by blockade of SCF function. *Blood*. 2007; DOI 10.1182/Blood-2007-04-086835.

5. Summary

[192] Rong-Fu Wang. Regulatory T cells and Toll-like receptors in Cancer Therapy. *Cancer Res*.2006; 66:4978-4990.

[193] Paulos CM, Kaiser A, Wrzesinski C, Hinrichs CS, Cassard L, Boni A, Muranski P, Sanchez-Perez L, Palmer DC, Yu Z, Antony PA, Gattinoni L, Rosenberg SA, Restifo NP. Toll-like receptors in Tumor Immunotherapy. *Clin Cancer Res*.2007; 13:5280-5289.

In: Oncogene Proteins: Structure, Functions and Analyses ISBN: 978-1-60456-259-0
Editors: P.B. Murphy and J.R. Clarke, pp. 87-115

Chapter III

PI3K/Akt/mTOR Signaling Pathway: Implications for Human Cancer

Xiaolin Wan [*] ***and Lee J. Helman***

Molecular Oncology Section, Pediatric Oncology Branch, Center for Cancer Research, National Cancer Institute, National Institutes of Health, Bethesda, Maryland, USA

Abstract

Aberrant intracellular signaling resulting from activation of oncogenic proteins, inactivation of tumor suppressors, and gene amplification of critical proteins are a key feature of human cancer. Phosphoinositol-3-kinase (PI3K)/Akt/mammalian target of rapamycin (mTOR) pathway is frequently dysregulated in many human malignancies and impacts a numbers of cellular processes including cell growth, survival, angiogenesis, migration, and cellular metabolism by mediating extracellular (mitogenic growth factor and stress) and intracellular (altered tyrosine receptor kinases and ras) signaling. Great progress has been made in understanding the regulation and roles of PI3K/Akt/mTOR signaling over the past ten years, creating an opportunity to manipulate this pathway therapeutically for cancer treatment. In this chapter, we discuss PI3K/Akt/mTOR pathway components and functions in tumorigenesis and tumor progression, as well as its alteration in cancer and relationship with different therapeutic approaches. Moreover, we summarize the current status of a variety of pharmacological agents that target this pathway in preclinical and clinical studies.

Introduction

Cancer has been viewed as the outcome of a multi-stage evolving genetic disease that is driven by progressive genetic abnormalities including activation of oncogenic proteins, inactivation of tumor suppressor genes, chromosomal abnormalities, and gene amplification of critical targets (1, 2). A consequence of these genetic abnormalities is the dysregulation of

[*] E-mail address: xiaolinw@mail.nih.gov. Tel: 301-451-7013; Fax: 301-451-7010; Corresponding author: Xiaolin Wan, Molecular Oncology Section, Pediatric Oncology Branch, Bldg. 10, Rm. CRC-1W-3750, NCI, National Institutes of Health, Bethesda, MD 20892-1928

various cell-signaling pathways that control cell function. The PI3K/Akt pathway is targeted by these genetic abnormalities more frequently than any other pathway in human cancer (3). Furthermore, this pathway is stimulated by many growth factors, with resultant activation of the pathway. mTOR is a central integrator of various extracellular and intracellular signals including growth factors, hormones, energy, nutrients, and stress. Among the signal inputs, growth factor- and homone-induced mTOR is the best characterized and is mediated via the PI3K/Akt signaling pathway. The best understood functions of mTOR are related to the control of protein synthesis through the inactivation of the 4E-binding protein1 (4E-BP1) and activation of p70S6K1 (S6K1). The PI3K/Akt/mTOR signaling pathway is frequently disturbed in many human cancers and controls many cellular processes including survival, cell cycle progression, transcription, translation, and angiogenesis. This pathway plays a key role not only in tumor development but also in the tumor's potential response to cancer therapy. Many of the new-targeted agents have been specifically designed to target the PI3K/Akt/mTOR pathway. This chapter focuses on the PI3K/Akt/mTOR pathway components and functions there its alterations in tumorigenesis and tumor progression. Moreover, we review the progress in the development of agents designed to target this pathway.

Overview of PI3K/Akt/mTOR

PI3Ks

PI3Ks, a large family of lipid signaling kinases, have been classified into three groups (class I, II, and III), according to their sequence homology, substrate preference, molecular structure and model of regulation (4, 5) (Fig. 1). Class I PI3Ks are heterodimeric proteins, which include class IA and class IB. In regulating cell growth and in tumorigenesis, class IA is the most important PI3K protein class and the majority of research focuses on these isoforms. The class IA PI3K subgroup consists of three catalytic subunits, p110α, p110β and p110γ, that form heterodimers with one of five regulatory subunits, p85α, p85β, p85γ, and p55α and p50α, which both produced by alternate splicing of the p85α. Class IA PI3Ks are often activated by growth factor receptor through a tyrosine-kinase-dependent mechanism (6). Upon growth factor stimulation, The Src homology 2 (SH2) domains of p85 bind to phosphorylated tyrosine in YXXM (Y, tyrosine; X, any amino acid; M, methinonine) motifs in receptor tyrosine kinases (RTKs) or their substrates adaptor proteins such as IRS-1)(7). Phosphotyrosine binding relieves the inhibition of p110 by p85 and recruits the PI3Ks to its substrate phosphatidylinositol 4,5-bisphosphate (PIP_2) at the plasma membrane (8). In addition, Ras-GTP can directly bind to a Ras-binding domain of the N-terminal region in p110, leading to PI3K activation (9). Upon activation, PI3Ks phosphorylate PIP2 to generate phosphatidylinositol 3,4,5-trisphosphate (PIP_3), which then recruit pleckstrin homology (PH) domain-containing proteins, such as the serine-threonine kinase Akt and phosphoinositide-dependent kinase 1 (PDK1) to the plasma membrane (10). The class IB PI3K consists of only one membrane, a heterodimer of a p110γ catalytic subunit and a p110 or p84 regulatory subunits. Since Class IB PI3Ks do not have p85 family regulatory subunits, they are not regulated by RTKs. Classes IB PI3Ks are activated by G-protein-coupled receptors (GPCRs)(11). Class II PI3Ks consists of a p110-like catalytic subunit that lacks adaptor subunits and preferentially uses phosphorylate phosphatidylinositol (PI) and phosphatidylinositol-4-phosphate (PI-4-P)(6). Three mammalian class II PI3K isoforms, PI3KC2α, PI3KC2β, and PI3KC2γ, have been identified (Fig. 1). Class III PI3Ks are heterodimeric enzymes consisting of adaptor (p150) and catalytic (Vps34) subunits. Recently, Vps34 was found to regulate autophagy (12, 13) and mammalian target of rapamycin (mTOR) activity in response to amino-acid availability (14, 15).

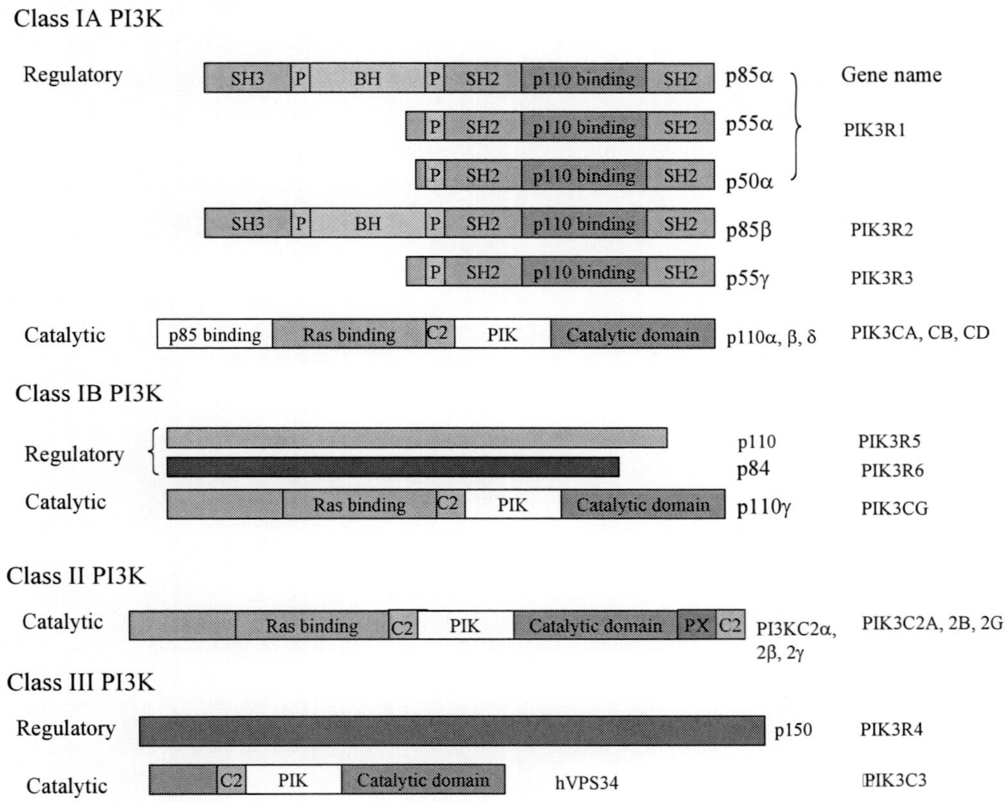

Figure 1. Schematic structure of the PI3K family members. Mammalian PI3Ks are composed of class IA, IB, class II and class III. The classes IA PI3Ks are heterodimers that consists of one p85 or p50-p55 regulatory subunit and one of three possible catalytic subunits (p110α, p110β or p110δ). The classes IB PI3Ks are heterodimers of either a p110 or p84 regulatory subunit and p110γ catalytic subunit. The classes II PI3Ks consist of only a p110-like catalytic subunit. The class III PI3K is thought to be a heterodimer of a p150 protein kinase and the mammalian homology of the yeast VPS34 (vacuolar protein-sorting defective 34). The regulatory subunits of class IA have a common core structure consisting of a p110 binding-domain flanked by two SH2 (Src-homology 2) domains, The two longer isoforms, p85α and p85β, have an extended N-terminal region containing SH3 domain and a Bcr/Rac GAP homology (BH) domain flanked by two proline-rich (P) regions. The catalytic subunits of class IA PI3K (p110α, β and γ, encoded by PIK3CA, CB, and CD) are composed of a p85-binding domain, a Ras-binding domain, a C2 domain (protein kinase C homology domain 2), a PIK (phosphatidylinositol kinase homology) domain and a C-terminal catalytic domain.

Akt

The serine/threonine kinase Akt, also known as protein kinase B (PKB), was originally identified as an oncogene transduced by Akt8 acute transforming retrovirus (16). In mammals, three isoforms of Akt encoded by three separate genes have been isolated and termed as Akt1/PKBα, Akt2/PKBβ, and Akt3/PKBγ, which are expressed in most tissues (17). Each member of the Akt family has a PH domain at the N-terminus, a kinase domain in the center sequence, and a serine/threonine-rich C-terminal region (regulatory domain). All Akt isoforms possess conserved threonine and serine residues (T308/S473 in Akt1, T309/S474 in Akt2 and T305/S472 inAkt3) that together with PH domain are critical for Akt activation (Fig. 2). Although these isoforms display high sequence homology, there seem to have differently biological and physiological functions (17). Up until now, it is not fully understood the role of individual PKB/Akt isoforms in various cellular processes because of the inability of the available reagents to differentiate between the three isoforms.

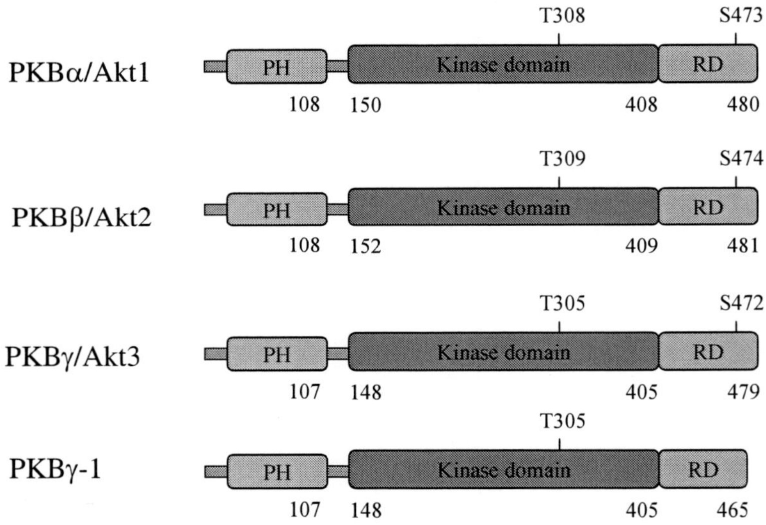

Figure 2. Domain structure of human Akt family membranes. Each Akt isoform consists of an N-terminal PH domain for binding PIP_3, a central kinase domain, and a C-terminal regulatory domain (RD). The two phosphorylation sites necessary for activation of Akt are shown.

mTOR

The mTOR, a serine/threonine protein kinase, contains 2549 amino acids and comprises several structural domains (Fig. 3A). The N terminus possesses ~20 tandem HEAT repeats of 37-43 amino acids (18), which are implicated in protein-protein interaction, such as Raptor (19). At the C terminus of the mTOR there are a FRAP, ataxia teleangiectasia mutated (ATM), transformation/transcription domain-associated protein (TRRAP)(FAT) domain followed by the FKBP-12, rapamycin binding domain (FRB), a catalytic kinase domain. The extreme C-terminal region of mTOR contains the FRAP, ATM, TRRAP, C-terminus (FATC) domain (20). FKBP with rapamycin can bind to the FRB domain and inhibit the activity of mTOR protein. The catalytic kinase domain in the C terminus is highly homologous to the lipid kinase domain of PI3K.

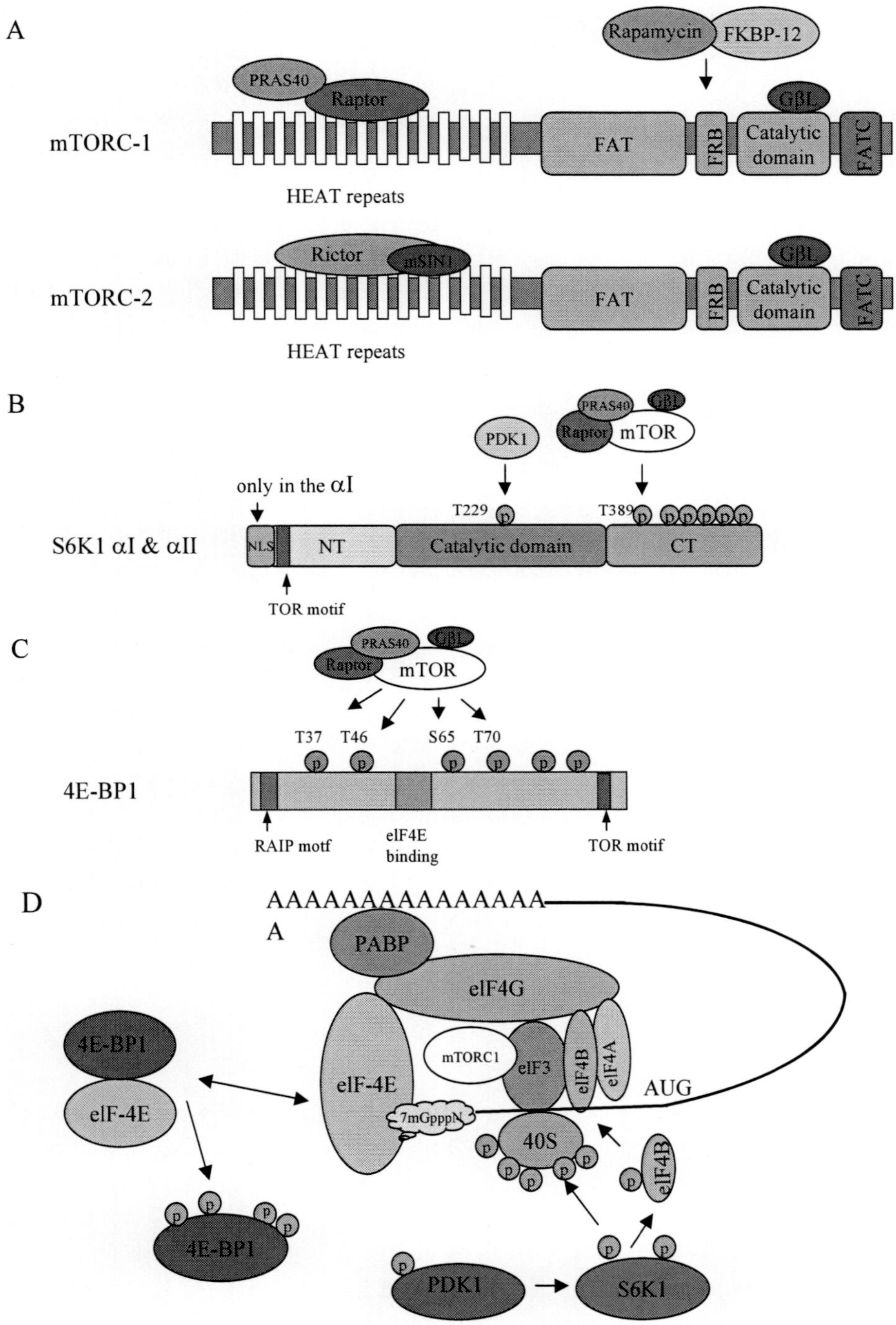

Figure 3. Summary of mTOR and its downstream targets S6K1 and 4E-BP1. A, schematic depiction of the basic features of mTOR structure and its complexes mTORC1 and mTORC2. B and C, the main features of S6K1 and 4E-BP1 structures. D, schematic illustration of preinitiation-complex assembly and eukaryotic translation initiation.

Recent studies have found that mTOR exists in two distinct complexes, mTOR complex 1 (mTORC1) and mTOR complex 2 (mTORC2)(Fig. 3A). mTORC1 is rapamycin sensitive and consists of mTOR, regulatory-associated protein of mTOR (raptor), G protein β subunit-like (GβL, also known as mLST8), and proline-rich Akt substrate 40 kDa (PRAS40)(21-23). mTORC2 is rapamycin insensitive and also contains mTOR and GβL, though it contains the adaptor protein rapamycin-insensitive companion of mTOR (rictor) and mammalian stress-activated protein kinase (SPARK)-interacting protein-1 (mSIN1) proteins instead of raptor and PRAS40 (24-26). The 159-kDa raptor containing three HEAT repeats in its amino-terminal domain is required for interaction with mTOR (21, 27). However, raptor also binds weakly to C-terminal of mTOR (21). Raptor may interact with mTOR at multiple contact sites, which is essential for positive regulation of mTOR activity and functions as scaffold for recruiting the mTOR substrates, eukaryote initiation factor 4E-binding protein 1 (4E-BP1) and ribosomal S6 kinase 1 (S6K1), mediated by a putative TOR signaling motif (27- 29). The TOS motif is a conserved five amino acid segment found in the N terminus of S6K1 (Phe-Asp-Ile-Asp-Leu) and in the C terminus of 4E-BP1 (Phe-Glu-Met-Asp-Ile), which are necessary for the phosphorylation of S6K1 and 4E-BP1 by mTORC1 *in vivo* (30)(Fig. 3B and C). PRAS40 is a 28-kDa proline-rich protein lacking named domains, but it is known to be phosphorylated near its C terminus by Akt and other kinases (31). Recent studies have found that PRAS40 is a raptor binding protein that interacts with mTORC1, but not mTORC2 and functions as a negative regulator of mTORC1 (23, 32). GβL, a 36-kDa protein with 7 WD40 repeats, binds to the kinase domain of mTOR and is present in both mTORC1 and mTORC2 (22, 28). However, deletion of GβL in mice seems not to affect mTORC1 activity or mTOR-raptor interaction, but it is required for the rictor-mTOR interaction (33). Rictor (also known as mAVO3) is a large protein with a molecular weight of 200 kDa) and shares conserved domains among eukaryotes, but its functions have so far not been elucidated (25). RNAi-mediated knockdown of rictor (but not raptor) results in loss of both actin polymerization and cell spreading (25, 34), and inhibition of Akt phosphorylation (24). mSIN1 was originally identified as a human protein that modulates yeast RAS function and has limited homology to AVO1, a component of TORC2 in yeast (35, 36). Recent studies have reported that mSIN1 is necessary for maintenance of rictor-mTOR complex integrity and regulation of Akt phosphorylation (37, 38).

S6K1 and 4E-BP1

Much of the current knowledge about mTOR in literature is associated with mTORC1, which regulates translation and cell growth by phosphorylating its two major downstream targets, S6K1 and 4E-BP1 in a rapamycin-sensitive manner. S6K1 is a member of the AGC family of serine/threonine protein kinase that phosphorylates the 40S ribosomal protein S6, thereby enhancing translation of mRNAs containing a 5' polypyrimidine tract (39, 40). In mammals, S6K1 has two isoforms, αI (85 kDa) and αII (70 kDa)(41, 42). αI contains an N-terminal 23 amino acid extension with a nuclear localization signal (NLS), which localizes it to nucleus (41, 42)(Fig. 3B). In contrast, αII is localized mostly in the cytoplasm (43). Current findings suggest that the two isoforms are controlled in a similar manner and are thus referred to here as S6K1 (44). The S6K1 are activated by phosphorylation at multiple residues (Fig. 3B). The important residues for S6K1 activation are Thr229 in the activation loop and

Thr389 in a conserved hydrophobic motif. S6K1 Thr389 was found to be the main rapamycin-sensitive residue and is directly phosphorylated by mTORC1 (45). Phosphorylation of Thr389 allows PDK1 to phosphorylate S6K1 at Thr229 and leads to full activation of S6K1 (45). Moreover, the N-terminal region of S6K1 contains a TOS motif that interacts with raptor to facilitate activation of S6K1 by mTORC1 (46).

Cap-dependent translation is mainly regulated by 4E-BPs, also known as PHAS (phosphorylated heat- and acid-stable), a family of three small acidic proteins (4E-BP1, 4E-BP2, and 4E-BP3)(29). Human 4E-BP1 (PHAS-1) contains 118 amino acids, with a RAIP motif (Arg-Ala-Ile-Pro) that lies within the first 24 residues of the N terminus (47), an elF4E-binding motif (YxxxxLϕ, where x is any amino acid and ϕ is a hydrophobic residue) located between residue 54 and 60 (48) and a TOR motif (Phe-Glu-Met-Asp-Leu) in the C terminus (30)(Fig. 3C). 4E-BP1 inhibits cap-dependent translation by binding to eIF4E and obstructing its interaction with eIF4G because both the 4E-BP1 and elF4E interact with elF4E through the elF4E-binding motif (49, 50). Thus, the 4E-BPs acts as molecular mimics of the elF4E-binding site in elF4G, and thus effectively compete with elF4G for elF4E binding. Binding of the 4E-BP1 to elF4E is reversible and is dependent on the phosphorylation status of 4E-BP1. Hypophosphorylated 4E-BPs binds avidly to eIF4E, whereas hyperphosphorylation of 4E-BP1 abrogates this interaction. Numerous types of extracellular stimuli, including serum, hormones, growth factors, mitogens, cytokines, and G-protein-coupled receptor agonists induces and increase 4E-BP1 phosphorylation, and a concomitant loss of eIF4E-binding activity. Conversely, nutrient or growth factor deprivation, or environmental stress induces 4E-BP1 dephosphorylation, accompanied by a dramatic increase in eIF4E-binding activity and an inhibition of translation (48, 51). Phosphorylation of 4E-BP1 is complex. 4E-BP1 contains at least six phosphorylation residues (Thr 37, Thr 46, Ser 65, Thr 70, Ser 83, and Ser 112)(52, 53), of these; four sites (Thr 37, Thr 46, Thr 65, and Thr 70) are known to be regulated by mTORC1 (53)(Fig. 3C). Phosphorylation of 4E-BP1 depends upon a TOR motif in the N terminus, which is required for 4E-BP1 to bind to raptor and recruits 4E-BP1 to mTOR kinase (53). Phosphorylation of Thr 37 and Thr 46 is thought to prime 4E-BP1 for sequential phosphorylation of Ser 65 and Thr 70, (54), which leads to 4E-BP1 dissociation from elF4E bound to the 7-methylguanosine cap structure (m7GpppN, where N is any nucleotide) at the 5'-end of nearly all transcribed mRNAs, allowing for recruitment of elF4G and elF4A, elF3, 40S ribosomal subunits to the cap, resulting in the assembly of the translation initiation complex (55). A recent study has proposed that elF3 acts as a platform to coordinate mTORC1 signaling with phosphorylation of S6K1 and 4EBP-1 and subsequent S6K1 phosphorylation of the 40 S ribosomal subunit S6 and eukaryotic initiation factor 4B (56)(Fig. 3D).

Activation and Regulation of PI3K/Akt/mTOR Signaling Pathway

Class I PI3Ks activation requires activated receptor tyrosine kinases (RTKs) or GTPase proteins (e.g. Ras). Binding of growth factors, cytokines and hormones to RTKs leads to dimerization of RTKs and autophosphorylation of the tyrosine residues of the YXXM consensus sequence in the kinase domain of the RTK, which results in binding of SH2 domain of p85 to phosphorylated tyrosine in YXXM motif, and recruits PI3K to membrane. The small GTPase Ras can also recruit and activate PI3K through direct binding to Ras-binding domain motif of the p110 catalytic subunit. At the membrane, PI3K phosphorylates PIP_2 at the 3' position on its inositol ring and converts PIP_2 to PIP_3. PIP_3 in turn recruits both PDK1 and PKB/Akt to the membrane via binding to their PH domains, where PDK1 phosphorylates and activates Akt at Thr308. Additional phosphorylation of Akt at Ser473 possibly by PDK2 results in full activation of Akt. The identity of PDK2 remains controversial. PDK1, integrin-linked kinase (ILK), and DNA-dependent protein kinase (DNA-PK) have been considered as potential candidates for PDK2 (57-59). Recent finding has reported that mTORC2 is required for phosphorylation of Akt at Ser 473 (24). The activity of PI3K is opposed by tensin homolog deleted on chromosome 10 (PTEN), which dephosphorylates PIP_3 to generate PIP_2, thus negative regulating PI3K signaling (Fig. 4).

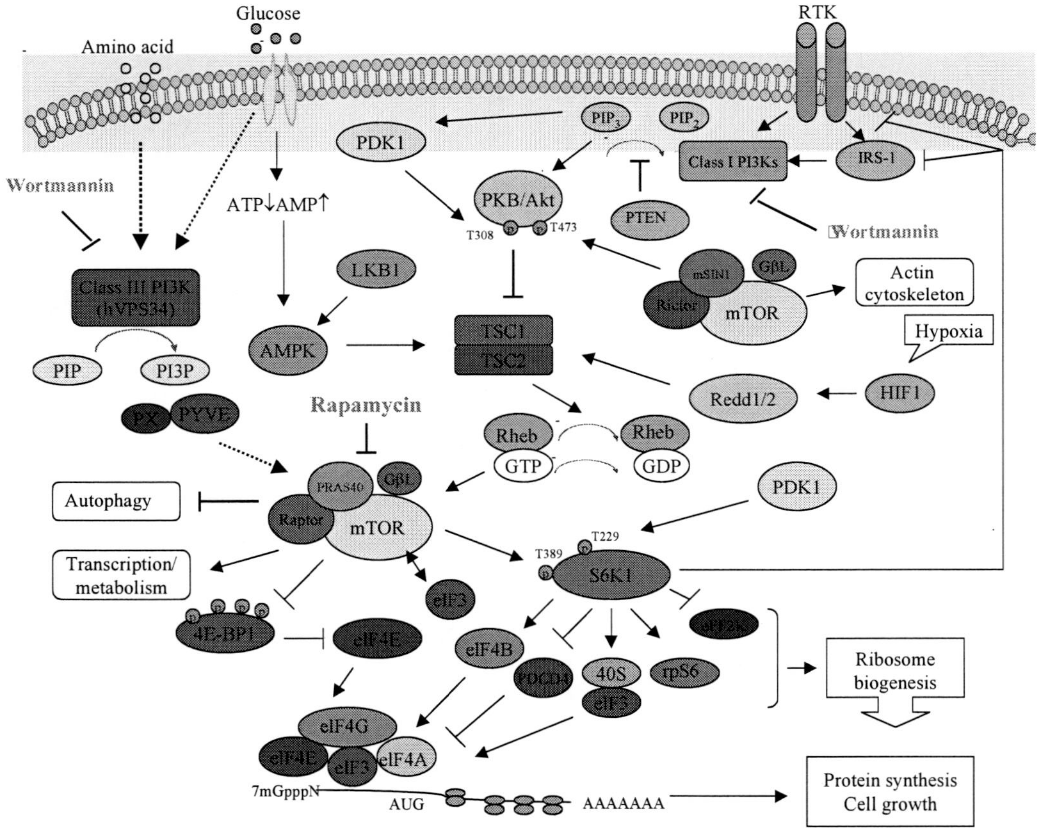

Figure 4. A model of PI3K/Akt/mTOR signaling pathway activation. This schematic outlines the possible mechanisms leading to activation of PI3K/Akt/mTOR pathway (see text).

Recent studies have placed the tumor suppressors tuberous sclerosis complex 1 (TSC1, also known as hamartin) and TSC2 (turberin) in the middle of an evolutionarily conserved signaling pathway that control mTOR, which provides a link between the growth factor-regulated PI3K-Akt and the nutrient-sensitive mTOR pathways (60). TSC-1 and TSC-2 form a heterodimer that inhibits the activity of Ras homolog enriched in brain (Rheb), a small GTPase required for mTOR activation (61). TSC2 is phosphorylated on multiple sites by various kinases including Akt, extracellular signal regulated kinase (ERK), AMP-activated protein kinase (AMPK), glycogen synthase kinase 3 (GSK3), and S6K1 (62). The mechanism by which Akt activates mTORC1 appears to be through direct phosphorylation of TSC2 at Ser939, 981, and Thr1462 resulting in dissociation and degradation of TSC1/2, thereby releasing the small GTPase Rheb and positively regulates mTOR signaling (63, 64). In addition, it has been reported that overexpression of Rheb induces phosphorylation of S6 and 4E-BP1, which both are downstream targets of mTOR (65, 66). Moreover, Rheb has been recently found to directly bind mTOR in a GTP-dependent manner to stimulate mTOR activity (67, 68). However, it remains unclear how GTP binding to Rheb activates mTORC1. Conversely, under energetic stress, or increasing AMP levels, the LKB1 phosphorylates AMPK at Thr 172 on its activation loop, which in turn phosphorylates TSC2 at Ser1345 and activates ability of TSC-2 to inhibit mTORC1 (69), but the underlying mechanism is not fully understood. LKB1is a tumor suppressor inactivated in Peutz-Jeghers syndrome, (70), which encodes a threonine kinase that has been shown genetically and biochemocally to be the direct activating kinase of the AMPK (71). When intracellular ATP levels drop and AMP levels rise, AMP directly binds the regulatory γ subunit of AMPK, resulting in a conformational shift that exposes the activation-loop threonine, which is then phosphorylated by LKB1 (72). In addition, Redd1 has been demonstrated to inhibit mTORC1 activity through the TSC1/2 in response to hypoxia, which is independent the LKB1/AMPK pathway (73)(Fig. 4).

Nutrients such as amino acids appear to regulate signaling through the mTOR kinase, mTORC1. However, there has been little understanding that how mTORC1 can sense amino acids. Recent studies have found that human vacuolar protein sorting 34 (hVPS34), a Class III PI3K producing PI3P, acts as a signaling intermediate between amino acids and mTORC1independently of TSC1/2 or Rheb (74, 75). Early studies had shown that treatment with the Class I PI3K inhibitor wortmannin resulted in blocking amino acids-mediated S6K1 activation (76, 77). Further study found that amino acids induce S6K1 activation even with knockdown of Class I PI3K by siRNA, whereas knockdown of hVPS34 blocked amino acids-induced S6K1 activation (78). Unlike Class I PI3K that produces PIP_3, the Class III PI3K uses unphosphorylated phosphatidylinositol (PIP) as a substrate to produce the PI3P, which mediates the recruitment of proteins containing FYVE (Fab1p, YOTB, Vac1p and EEA1) or PX (Phox homology) domains to endosomal membranes (79). The mechanism by which hVPS34 signals to mTORC1/S6K1 via the PI3P is not known. Recent evidence has shown that glucose deletion leads to inhibition of hVPS34 activity (75), suggesting a possible role of hVPS34 as a common nutrient sensor for activation of mTORC1/S6k1.

As indicated above, mTORC1 and mTORC2 are two structurally and functionally distinct protein complexes (Fig. 3A & 4). mTORC1 is activated by growth factors and nutrients via a Class I PI3K, involving Akt-mediated phosphorylation of TSC2 and activation of Rheb GTPase and through a poorly characterized Class III PI3K signaling network. Activation of mTORC1 results in phosphorylation of translational regulators S6K1 and 4E-BP1, which augments protein synthesis. The anti-fungal macrolide rapamycin does not directly inhibit mTOR, but rather binds to the intracellular immunophilin FK506-binding protein 12 (FKBP12) and the resultant complex binds to the FK-rapamycin domain of mTOR, in turn leading to inhibition of mTOR function in mTORC1, thereby preventing phosphorylation of S6K1 and 4E-BP1. Study of mTORC2 is in its infancy. Little is known regarding the regulation of mTORC2. The mechanism by which growth factors activate mTORC2 through

Class I PI3K remains unclear. Unlike mTORC1, it is not activated by nutrients, and rapamycin does not bind to mTORC2, and consequently, it is often called rapamycin-insensitive complex, although prolonged treatment disrupts mTORC2 assembly in certain cell lines (80). mTORC2 function is less defined than that of mTORC1 but is believed to be involved in actin cytoskeleton. mTORC2 is able to phosphorylate Akt at its hydrophobic motif Ser 473.

Dysregulation of the PI3K/Akt/mTOR Signaling Pathway in Human Cancer

Constitutive activation of PI3K/Akt/mTOR signaling is a common feature in many human cancers that have multiple alterations including both mutation and amplification of oncogenes and loss of tumor suppressor genes, leading to dysregulation of this pathway as specified in Table 1.

Table 1. Abnormalities of the PI3K/Alt/mTOR pathway in human cancer

Molecule	Dysfunction	Type of cancer	References
RTK	Receptor activation	Many tumor types	81-83
PIK3CA (p110α)	Mutations	Breast, colon, glioma,hepatocellular, ovarian, lung, and gastric	84
	Amplification	Head and neck, gastric, breast, cervical,thyroid, and sequamous cell	
PIK3R1(p85α)	Mutations	Colon and ovarian	85
Akt1/PKBα	Amplification	Gastric and glioma	86, 87
Akt2/PKBβ	Amplification	Ovarian, breast and pancreatic	88, 89
	Mutation	Colorectal	91
Akt3/PKBγ	Amplification	Hormone-independent breast and prostate	90
PDK1	Mutation	Colorectal	91
Rheb	Amplification	Breast, prostate, colon, ovarian, melanoma, and pancreatic	108
elF-4E	Amplification	In many human tumors	110-112
S6K1	Amplification	Breast and ovarian	112
PTEN	Mutations Loss of heterozygosity Methylation	Glioma, melanoma, prostate, breast, gastric, colorectal, leukaemia, and endometrial	118-120
TSC1/2	Mutation	Tuberous sclerosis	118
LKB1	Mutation	Peutz-Jeghers syndrome and Lung	118, 124. 125
PDCD4	Loss of expression	Lung and breast	128, 129

As described above, PI3K/Akt signaling is activated as a result of ligand-dependent activation of RTKs and /or G-protein-coupled receptor. RTKs are central components of cell signaling networks that control fundamental cellular activities. Currently, more than 50 mammalian RTKs have been identified such as the well-known insulin receptor (IR), insulin-like growth factor 1 receptor (IGF-1R), epidermal growth factor receptor (EGFR), ErbB, fibroblast growth factor receptor (FGFR), c-Kit, and Met (HGF receptor), which are frequently activated by overexpression and activating mutation in many types of cancer (81-83), resulting in activation of downstream signaling pathways such as the PI3K/Akt/mTOR pathway. Mutation or amplification of the PIK3CA (p110α) also commonly occurs in variety of human cancers (84), although PIK3R1 mutation is rare (85). Activation of Akt isoforms have been implicated in specific functions in cancer. Akt1 has been reported to be amplified in a single gastric carcinoma and glioblastoma (86, 87). Akt2 is amplified in ovarian, breast and pancreatic tumors (88, 89), and Akt3 is overexpressed in hormone-independent breast and prostate cancers (90). Recently, mutations of Akt2 and PDK1 were observed in colorectal cancer (91). Furthermore, activated Akt has been detected by immunohistochemical analysis of Akt phosphorylation at Ser 473 in various tumor types including melanoma, bile duct, head and neck, endometrial, breast, lung, colon, ovarian, pancreatic and prostate cancers (92-101). Moreover, active Akt has been functionally associated with a poor prognosis (96, 100-106) and resistance to chemotherapy and radiotherapy in many human cancers (93, 94, 97, 107). With regarding to Rheb functions as shown in Fig. 4, activation of mTORC1 requires the small GTPase Rheb. Rheb is overexpressed in several types of cancer including breast, prostate, colon, ovarian, melanoma, and pancreatic cancers (108).

Numerous downstream targets of mTOR are known to be activated and overexpressed in many human cancers, which are associated with oncogenesis and drug resistance. One of the best studies examples is elF-4E, which is overexpressed as a result of gene amplification in many human cancers and its overexpression has routinely been linked to progression of these cancers, including tumor recurrence, tumor metastasis, and poor prognosis (109-112). In contrast to elF-4E, 4E-BP1 functions as a tumor suppressor, which the levels of 4E-BP1 expression correlate inversely with tumor progression, and ovexexpression of 4E-BP1 resulted in reduction of the elF-4E overexpression-mediated tumorigenic effect and decrease in tumor size (111, 113). Furthermore, overexpression of 4E-BP1 has been shown to display a pro-apoptotic effect (114, 115).

Recently, our proteomic network analysis of microdissected human RMS specimens, which were obtained from the Children's Oncology Group Intergroup Rhabdomyosarcoma Study (IRS) IV, D9502 and D9803, with 12 year follow-up, demonstrated that high levels of Akt Ser473, 4E-BP1 Thr37/46, elF4G Ser1108, and S6K1 Thr389 and low level of 4EBP-1 expression in Stage III RMS patients were associated poor overall (OAS) and poor disease-free (DFS (116, 117). There is also a statistically significant association between survival of patients and status of Akt/mTOR pathway activation.

The PI3K/Akt/mTOR pathway is regulated by many tumor suppressors, including PTEN, TSC1/2, LKB1 and Programmed cell death protein 4 (PDCD4). Aberrant activation of the TSC1/2-Rheb-mTORC1 axis via loss of tumor suppressor protein function has been connected to the pathobiology of some tumor predisposition syndromes, such as Cowden's syndrome and Bannayan-Riley-Ruvalcaba syndrome (PTEN), Peutz-Jeghers syndrome (LKB1), and tuberous sclerosis (TSC1/2)(118). In addition to genetic syndrome-associated mutation, loss of PTEN function including mutation, deletion and/or promoter methylation has been frequently found in a variety of cancers with concomitant constitutive activation of the PI3K/Akt/mTOR pathway (119, 120). Furthermore, our study found that the levels of PTEN expression inversely correlate with Akt phosphorylation on Ser 473 (121). Moreover, tumors associated with PTEN inactivation are particularly sensitive to mTOR inhibitors (122, 123). Recent studies have reported that inactivating somatic mutation of LKB1 is more common in lung adenocarcinomas (124, 125). These observations have increased the interest towards signaling pathway of this tumor suppressor kinase. PDCD4 was recently identified as a novel tumor suppressor gene, which is involved in the control of translation via inhibiting the translation initiation factor elF4A (126, 127). PDCD4 is a substrate of S6K1, which is rapidly degraded upon phosphorylation by S6K1 (127). Consequently, loss of PDCD4 function is expected to promote cell growth and ultimately lead to cancer. Indeed, loss of PDCD4 expression has been strongly implicated in the development and progression of lung cancer and aggressive breast cancer (128, 129).

Overall, the PI3K/Akt/mTOR signaling pathway is dysregulated by a variety of mechanisms and many of its components are mutated or overexpressed in a large fraction of human cancers, which is intimately implicated in cancer development and progression.

Current Development of the PI3K/Akt/mTOR Pathway Inhibitors for Cancer Therapy

Considering this association of PI3K/Akt/mTOR dysregulation with human cancer progression, the rational development of inhibitors to specifically target the PI3K/Akt/mTOR pathway may represent a good strategy for the prevention and treatment of cancer. As shown in Fig. 4, there are many potential targets for modulation of the PI3K/Akt/mTOR signaling axis. This section will focus on discussion of four representative targets, the upstream mediators of PI3K/Akt signaling such as growth factor receptors, PI3K, Akt, and mTOR.

Inhibitors of PI3K/Akt/mTOR Upstream Mediator Signaling

As discussed above, growth factor RTKs are the most common upstream target of the PI3K/Akt/mTOR cascade, which have been approved to be one of the most amenable to therapeutic intervention for cancer therapy. For the sake of brevity, only some RTKs inhibitors are discussed below to exemplify the emerging paradigm for targeting the upstream mediators of the PI3K/Akt/mTOR pathway (Table 2).

The choice of RTKs as a target for cancer therapy is driven by a multitude of preclinical data including *in vitro* and *in vivo*, and clinical data supporting a powerful pro-survival signal of the receptor in tumor growth and development. A number of approaches have been explored to therapeutically target RTKs, but by far most successful strategies use humanized antibodies or small molecule tyrosine kinase inhibitors. Antibodies usually bind to the extracellular domains of receptor, resulting in inhibition of their activation, and promotion of receptor internalization and downregulation, whereas small molecules inhibit ATP binding to the receptor, thereby hindering autophosphorylation and preventing kinase activation.

The first used anti-IGF-1R antibody was αIR3, which has been known for more than twenty years (130). Extensive preclinical studies have demonstrated that IGF-1R antibodies suppress in vitro proliferation and/or *in vivo* tumor growth in a wide range of tumor types,

Table 2. Currently developmental Inhibitors of targeting RTKs for cancer therapy

Target	Compound	Class	Company	Trial status	References
IGF-1R	h7C10	Antibody	Merck	Phase I	132
	CP-751, 871	Antibody	Pfizer	Phase II	133
	IMC-A12	Antibody	Imclone	Phase I	134
	EM164	Antibody	ImmunoGen	Preclinical	136
	19D12	Anitbody	Schering-Plough	Preclinical	137
	PPP	Kinase inhibitor	Karolinska/Biovitrum	Preclinical	138
	BMS-554417/-536924	Kinase inhibitor	Bristol-Myers Squibb	Preclinical	139
	NVP-AEW541/-ADW742	Kinase inhibitor	Novarits	Preclinical	131,140
EGFR/HER-1	IMC-C225 (cetuximab)	Antibody	Imclone	Phase II/III	143, 144
	ABX-EGF (Panitumumab)	Antibody	Abgenix	Phase II/III	145
	EMD 72000 (Matuzumab)	Antibody	Meck KgaA	Phase I/II	145
	MDX-447	Antibody	Meck KagA	Phase I/II	145
	ZD1839 (Gefitinib)	Kinase inhibitor	AstraZeneca	Phase II/III	146, 147
	OSI-774 (Erlotinib)	Kinase inhibitor	OSI-Genetech-Roche	Approved,	148
	CI-1033 (Canertinib)	Kinase inhibitor	Pfizer	Phase I/II	149
	GW572016 (Lapatinib)	Kinase inhibitor	GlaxoSmithkline	Phase II/III	150
HER-2/Neu	Trastuzumab (Herceptin)	Antibody	Genentech	Approved	151,152
BCR-ABL, PDGFR/c-Kit	Imatinib (STI-571, Gleevec)	Kinase inhibitor	Novartis	Approved	153, 154
IGF-1R/HER-2	Insm-18	Kinase inhibitor	Insmed	Phase I/II	155
IGF-1R/EGFR	Di-diabody	Antibody	Imclone	Preclinical	156
FLT3/c-Kit/VEGFR	SU5416 (Semaxanib))	Kinase inhibitor	Sugen	Phase II	157, 158
FLT3/c-Kit/VEGFR/ PDGFR	PKC412 (Midostaurin)	Kinase inhibitor	Novartis	Phase II	159

such as breast, lung, colon, cervical, ovarian, prostate, pancreatic, neuroblastoma, melanoma, Ewing's sarcoma rhabdomyosarcoma (RMS), and osteosarocma (131-134). Our recent studies have demonstrated that h7C10 inhibited RMS cell growth *in vitro* (135) and RMS xenograft growth *in vivo* (unpublished data). Moreover, combining treatment with h7C10 and rapamycin led to blocking mTOR inhibition induced feedback activation of Akt as well enhancing the effect of rapamycin on the suppression of RMS cell growth and tumor growth. Several IGF-IR tyrosine kinase inhibitors (TKIs) have been developed like PPP, BMS-554417, and NVP-AEW541, which have been shown to inhibit tumor cell growth in vitro and in vivo in a variety of human cancers (131,138-142).

Human epidermal growth factor receptor (HER) family of tyrosine kinases consists of EGFR (HER1, ErbB1), HER2 (HER2/neu, ErbB2), HER3 (Erb3) and HER4 (ErbB4). IMC-C225 (cetuximab) is the most widely studied anti-HER1/EGFR antibody, which has demonstrated antitumor activity against a variety of cancer cell lines in preclinical studies (143, 144). It is currently approved as monotherapy or in combination with chemotherapy or radiation for the treatment of patients with metastatic colorectal cancer (mCRC) and sequamous cell carcinoma of the head and neck (SCCHN)(145). Other antibodies, such as ABX-EGF (Panitumumab), EMD-72000 (Matuzumab), and MDX-447, are currently undergoing evaluation in preclinical and clinical trials (145). Extensive studies have evaluated the efficacy of EGFR tyrosine kinase inhibitors such as ZD1839 (Gefitinib), OSI-774 (Erlotinib), CI-1033 (Canertinib), and GW572016 (Lapatinib)(146-149). The most successful inhibitor of RTKs used as an anticancer agent is imatinib, also known as Gleevec (USA),

Glivec (Europe), and STI-571, that has been shown remarkable clinical activity in leukemia and gastrointestinal stromal tumors (GIST)(153,154).

PI3K Inhibitors

The best-characterized inhibitors of PI3K are wortmannin and LY294002 (Table 3). Wotmannin is a fungal metabolite whose first identification can be traced back to the mid-1970 as an anti-inflammatory agent isolated from Penicillium wortmanni and later found to inhibit class I PI3K activity by binding covalently to a conserved lysine residue (Lys-802 of p110a and Lys-883 of p110y) in the ATP-binding cleft of the kinase (160, 161). Wortmannin also inhibits class III PI3K to reduce autophagy that acts as a tumor suppressor involved in type II programmed cell death (162). The flavonoid derivative LY294002 is a reversible inhibitor of PI3K, which competes with ATP for the ATP binding site of the catalytic subunit of p110 (163). In addition, recent studies found that LY294002 also inhibited casein kinase 2 (CK2) and Pim-1 serine/threonine kinase activity (164, 165). Although wortmannin and LY294002 have shown antitumor activity in *vitro* and *in vivo*, which are associated with inhibition of PI3K and downstream component of this pathway, suggesting that they might offer a valuable approach to cancer therapy, the use of these compounds in clinical applications is limited due to poor pharmacokinetics, low selectivity for specific kinase

Table 3. Developmental Inhibitors of targeting the PI3K/Akt/mTOR pathway for cancer therapy

Compound	Target(s)	Mechanism of action	Trial status	References
Wortmannin	Class I/III PI3K	Irreversible; binds p110α/γ	Preclinical	160-162
LY294002	Class I/II PI3K/CK2/Pim-1	Reversible; inhibits p110 ATP binding site	Preclinical	163-165
PI-103	Class I PI3K	Inhibits p110α/γ	Preclinical	166, 167
PIK-75	Class I PI3K	Inhibits p110α/γ	Preclinical	166, 167
IC87114	Class I PI3K	Inhibits p110 δ	Preclinical	166, 167
TGX115	Class I PI3K	Inhibits p110 β/δ	Preclinical	166, 168
UCN-01	PDK1	Inhibits ATP binding domain	Phase I/II	169, 170
BX-320/-795/-912	PDK1	Binds to PDK1 catalytic domain	Preclinical	171
OSU-03012	PDK1	Unknown	Preclinical	172, 173
Perifosine	Akt	Binds PH domain; Inhibits translocation	Phase I/II	174
AKTide-2T	Akt1	Binds to the substrat-binding site of kinase domain	Preclinical	175
API-2	Akt	Unknown	Phase I/II	176, 177
Akt-I-1	Akt1	Binds Akt at site other than PH damain	Preclinical	178, 179
Akt-I-2	Akt2	Binds Akt at site other than PH damain	Preclinical	178, 179
Akt-I-1/2	Akt1/Akt2	Binds Akt at site other than PH damain	Preclinical	178, 179
Rapamycin	mTOR	Inhibits mTOR by binding FKBP-12	Phase I/II	180, 181
CCI-779	mTOR	Inhibits mTOR by binding FKBP-12	Approved	117, 182
RAD001	mTOR	Inhibits mTOR by binding FKBP-12	Phase I/II	117, 182
AP23573	mTOR	Inhibits mTOR by binding FKBP-12	Phase I/II	117, 182

inhibition and unacceptable toxicity. Currently a number of initiatives are under evaluation that has identified isoform-selective ATP-competitive inhibitors of the PI3K with better pharmacodynamic. Some of these new PI3K inhibitors are shown in Table 3 (166-168).

PDK1 Inhibitors

The attractiveness of PDK1 as a target for cancer therapy is linked to its ability to phosphorylate and activate Akt. In addition to Akt, PDK1 has been shown to be a critical activator of other kinases in the AGC kinase superfamily that are important promoters of cancer progression, including protein kinase C (PKC), serum and glucocorticoid-regulated kinase (SGK), and S6K1. These kinases have a homologous region in their activation loops containing a consensus substrate recognition site for PDK1. Hence, compounds that target PDK1 may indirectly inhibit these other kinases. This off-target activity may be advantageous from a therapeutic standpoint but obviously complicates their development as compounds that selectively inhibit the PI3K/Akt/mTOR network. UCN-01, a staturosporine analogue originally isolated as a selective PKC inhibitor from the culture broth of Streptomyces sp, may represent the prototypic PDK1 inhibitor, which potently inhibits PDK1 *in vitro* and *in vivo* by forming a complex with the kinase domain of PDK1 (169, 170). UCN-01 is currently being explored in cancer patients in phase I/II clinical trials, both as single agent and in combination with chemotherapy. In addition to UCN-01, other PDK1 inhibitors such as BX-320 and celecoxib derivative OSU-03012, a selective cyclooxygenase-2 (COX-2) inhibitor, have been found to inhibit cell growth through targeting PDK1 (171, 172)(Table 3).

Akt Inhibitors

Rational design of Akt inhibitors is based on directly targeting three regions of this protein, PH domain, the ATP binding pocket in the kinase domain, and regulatory domain using competitive, pseudosubstrate, allosteric or other mechanisms. So far, no ATP-competitive inhibitors specific against Akt has been identified due to the fact that there is a high degree of homology in the ATP-binding pocket between Akt, PKA and PKC (173). An alternative approach to ATP-competitive inhibitors that might result in improved specificity would be to identify compounds that blocked PIP_3 binding and prevented Akt translocation to the plasma membrane. An example is the orally bioavailable alkylphospholipid perifosine that has now entered phase II clinical trials for various types of cancer (Table 3). As a phospholipid analogue, perifosine likely interferes with lipid signaling and thus impedes Akt translocation to the cell membrane preventing its activation (174). In contrast to ATP-competitive inhibitors, pseudosubstrate peptide inhibitors bind to the peptide/protein substrate sites of the catalytic domain and inhibit the kinase activity. A 14-mer peptide, ARKRERTYSFGHHA (AKTide-2T), has been identified that binds to the substrate-binding site of Akt1 and inhibits Akt1 activity (175). Akt/PKB signaling inhibitor-2 (API-2), also known as triciribine, was identified by screening a chemical library of 1920-compounds from the National Cancer Institute diversity set (176). API-2 is highly selective for Akt and does not inhibit PI3K and PDK1. As yet the mechanism for API-2 inhibition of Akt activation is unknown. API-2 has undergone limited phase I/II clinical trials in cancer patients with advanced solid tumors many years ago and it exhibited significant side effects at high dose including hyperglycemia, hypocalcemia, hepatotoxicity, thrombocytopenia and hyperglycemia, which hamper its further development in the clinic (177). However, this compound has been shown to be a non-isoforn-specific Akt inhibitor at much lower doses than previously tested (176), which provides possibility that lower doses may inhibit Akt activation and tumor growth without previous high-dose-associated side effects. The allosteric Akt inhibitors are particularly interesting, because they can target specifically Akt isoforms and might result in mitigating pronounced side effects. Recent studies have identified isoform-specific Akt inhibitors, named Akt-I-1, Akt-I-2, and Akt-I-1, 2, by a high-throughput kinase activity screen (178, 179). These inhibitors have been shown to have a great potential for pharmacologically targeting Akt1, Akt2 or both Akt1 and Akt2 for anticancer drug discovery, but these compounds have poor solubility and pharmacokinetics properties that precluded their evaluation in animal models (179).

mTOR Inhibitors

mTOR inhibitors are the most developed class of compounds for targeting the PI3K/Akt/mTOR signaling pathway. Rapamycin (sirolimus) is a natural product isolated from a strain of *Streptomyces hygroscopicus,* found in the soil of the Vai Atore region of Easter Island (Rapa Nui), for which rapamycin was named. Rapamycin is a macrolide antibiotic initially discovered as an anti-fungal agent in mid-1970s (180). Rapamycin was approved by the Food and Drug Administration (FDA) as an immunosuppressant for use after renal transplantation in 1990s. The potential for rapamycin to act as an antitumor agent was evaluated by the NCI screening program, which demonstrated potent inhibitory activity against several human cancers in vitro and in vivo. Rapamycin has also shown activity against many types of cancer in phase I/II clinical trials (181). However, rapamycin's poor aqueous solubility and chemical stability limited its clinical development as an anticancer agent, which led to efforts directed at synthesized analogs with more favorable pharmaceutical characteristics. So far, three analogues of rapamycin have been synthesized. These include the soluble ester analogue of rapamycin, CCI-779 (temsirolimus), an orally bioavailable derivative of rapamycin, RAD001 (everolimus) and a non-prodrug rapamycin analogue, the phosphorous-containing derivative AP23573 (Table. 3). Currently, these derivatives are being evaluated in cancer clinical trials. To date, clinical results have shown good tolerability of treatment with these derivatives in most reports and varying effectiveness of these mTOR inhibitors in a subset of patients (117). Moreover, CCI-779 (temsirolimus) has been approved by the FDA for the treatment of advanced renal cell carcinoma. Further studies to define the optimal dose/schedule of these inhibitors, patient selection, and combination strategies with other biological agents are clearly required to achieve maximal efficiency and minimal side effects.

Conclusion

Much progress had been made in our understanding of the diverse roles of the PI3K/Akt/mTOR pathway and its role in human cancer. This work has strongly implicated this pathway as a therapeutic target for cancer treatment. These research efforts have been facilitated greatly by the development of the specific inhibitors to target this pathway, such as rapamycin and its analogues that have shown promising clinical efficacy and low toxicity profiles in variety tumors in a subset of patients. However, the challenges associated with the transition of these inhibitors from the laboratory bench to the oncology clinics have underscored the fact that we still have much to learn about the intricacies of the PI3K/Akt/mTOR pathway as well as the integration of this pathway into the network of signaling cascades.

References

[1] Hanahan, D. & Weinberg, RA. (2000). The hallmarks of cancer. *Cell.***100**:57-70.

[2] Vogelstein, B. & Kinzler, KW. (2004). Cancer genes and the pathways they control. *Nat Med.*10:789-799.

[3] Hennessy, BT. Smith, DL. Ram, PT. Lu, Y. & Mills, GB. (2005). *Exploiting the PI3K/AKT pathway for cancer drug discovery.* **4**:988-1004.

[4] Cantley, LC. (2002). The phosphoinositide 3-kinase pathway. *Science.* **296**:1655-1657.

[5] Vanhaesebroeck, B. Leevers, SJ. Ahmadi, K. Timms, J. Katso, R. Driscoll, PC. Woscholski, R. Parker, PJ. & Waterfield MD. (2001). Synthesis and function of 3-phosphorylated inositol lipids. *Annu Rev Biochem.* **70**:535-602.

[6] Katso, R. Okkenhaug, K. Ahmadi, K. White, S. Timms, J. Waterfield, MD. (2001). Cellular function of phosphoinositide 3-kinases: implications for development, homeostasis, and cancer. *Annu Rev Cell Dev Biol.***17**:615-675.

[7] Songyang Z, Shoelson SE, Chaudhuri M, Gish G, Pawson T, Haser WG, King F, Roberts T, Ratnofsky S, Lechleider RJ, et al. (1993). SH2 domains recognize specific phosphopeptide sequences. *Cell.* **72**:767-78.

[8] Yu, J. Wjasow, C. & Blacker, JM. (1998). Regulation of the p85/p110alpha phosphatidylinositol 3'-kinase. Distinct roles for the n-terminal and c-terminal SH2 domains. *J Biol Chem.* **273**:30199-30203.

[9] Rodriguez-Viciana, P. Warne, PH. Dhand, R. Vanhaesebroeck, B. Gout, I. Fry, MJ. Waterfield, MD. & Downward, J. (1994). Phosphatidylinositol-3-OH kinase as a direct target of Ras. *Nature.* **370**:527-532.

[10] Corvera, S. & Czech, MP. (1998). Direct targets of phosphoinositide 3-kinase products in membrane traffic and signal transduction. *Trends Cell Biol***.** **8**:442-446.

[11] Wymann, MP. Björklöf, K. Calvez, R. Finan, P. Thomast, M. Trifilieff, A. Barbier, M. Altruda, F. Hirsch, E. & Laffargue, M. (2003). Phosphoinositide 3-kinase gamma: a key modulator in inflammation and allergy. *Biochem Soc Trans.* **31**:275-280.

[12] Wurmser, AE. & Emr, SD. (2002). Novel PtdIns(3)P-binding protein Etf1 functions as an effector of the Vps34 PtdIns 3-kinase in autophagy. *J Cell Biol.* **158**:761-772.

[13] Kihara, A. Noda, T. Ishihara, N. & Ohsumi, Y. (2001). Two distinct Vps34 phosphatidylinositol 3-kinase complexes function in autophagy and carboxypeptidase Y sorting in Saccharomyces cerevisiae. *J Cell Biol.***152**:519-530.

[14] Byfield, MP. Murray, JT. & Backer, JM. (2005). hVps34 is a nutrient-regulated lipid kinase required for activation of p70 S6 kinase. *J Biol Chem.* **280**:33076-33082.

[15] Nobukuni, T. Joaquin, M. Roccio, M. Dann, SG. Kim, SY. Gulati, P. Byfield, MP. Backer, JM. Natt, F. Bos, JL. Zwartkruis, FJ. & Thomas, G. (2005). Amino acids mediate mTOR/raptor signaling through activation of class 3 phosphatidylinositol 3OH-kinase. *Proc Natl Acad Sci U S A.* **102**:14238-14243.
[16] Staal, SP. (1987). Molecular cloning of the akt oncogene and its human homologues AKT1 and AKT2: amplification of AKT1 in a primary human gastric adenocarcinoma. *Proc Natl Acad Sci U S A.* **84**:5034-5037.
[17] Brazil, DP. Yang, ZZ. & Hemmings, BA. (2004). Advances in protein kinase B signalling: AKTion on multiple fronts. *Trends Biochem Sci.* **29**:233-242.
[18] Andrade, MA. & Bork, P. (1995). HEAT repeats in Huntington's disease protein. *Nat Genet.* **11**:115-116.
[19] Kin, D. Sarbassov, DD. Ali, SM. King, JE. Latek, RR, Erdjument-Bromage, H. Tempst, P. & Sabatini, DM. (2002). mTOR interacts with raptor to form a nutrient-sensitive complex that signals to the cell growth machinery. *Cell.* **110**:163-175.
[20] Soliman, GA. (2005). The mammalian target of rapamycin signaling network and gene regulation. *Curr Opin Lipidol.* **16**:317-323.
[21] Kim, DH. Sarbassov, DD. Ali, SM. King, JE. Latek, RR. Erdjument-Bromage, H. Tempost, P. & Sabatini, DM. (2002). mTOR interacts with raptor to form a nutrient-sensitive complex that signals to the cell growth machinery. *Cell.* **110**:163-175.
[22] Kim, DH. Sarbassov, DD. Ali, SM. King, JE. Latek, RR. Guntur, KV. Erdjument-Bromage, H. Tempost, P. & Sabatini, DM. (2003). GbetaL, a positive regulator of the rapamycin-sensitive pathway required for the nutrient-sensitive interaction between raptor and mTOR. *Mol Cell.* **11**:895-904.
[23] Sancak, Y. Thoreen, CC. Peterson, TR, Lindquist, RA. Kang, SA. Spooner E. Carr, SA. & Sabatini, DM. (2007). PRAS40 is an insulin-regulated inhibitor of the mTORC1 protein kinase. *Mol Cell.* **25**:903-915.
[24] Sarbassov, DD. Guertin, DA. Ali, SM. & Sabatini, DM. (2005). Phosphorylation and regulation of Akt/PKB by the rictor-mTOR complex. *Science.* **307**:1098-1101.
[25] Sarbassov, DD. Ali, SM. Kim, DH. Guertin, DA. Latek, RR. Erdjument-Bromage, H. Tempst, P. & Sabatini. DM. (2004). Rictor, a novel binding partner of mTOR, defines a rapamycin-insensitive and raptor-independent pathway that regulates the cytoskeleton. *Curr Biol.* **14**:1296-1302.
[26] Frias, MA. Thoreen, CC. Jaffe, JD. Schroder, W. Sculley, T. Carr, SA. & Sabatini, DM. (2006). mSin1 is necessary for Akt/PKB phosphorylation, and its isoforms define three distinct mTORC2s. *Curr Biol.***16**:1865-1870.
[27] Hara, K. Maruki, Y. Long, X. Yoshino, K. Oshiro, N. Hidayat, S. Tokunaga, C. Avruch, J. & Yonezawa, K. (2002). Raptor, a binding partner of target of rapamycin (TOR), mediates TOR action. *Cell.***110**:177-189.
[28] Bhaskar, PT. & Hay, N. (2007). The two TORCs and Akt. *Dev Cell.* **12**:487-502.
[29] Hay, N. & Souenberg. (2003). Upstream and downstream of mTOR. *Genes Dev.* **18**:1926-1945.
[30] Schalm, SS. & Blenis, J. (2002). Identification of a conserved motif required for mTOR signaling. *Curr Biol.* **16**;12:632-639.
[31] Kovacina, KS. Park, GY. Bae, SS. Guzzetta, AW. Schaefer, E. Birnbaum, MJ, & Roth, RA. (2003). Identification of a proline-rich Akt substrate as a 14-3-3 binding partner. *J Biol Chem.* **278**:10189-10194.

[32] Vander Haar, E. Lee, SI. Bandhakavi, S. Griffin, TJ. & Kim, DH. (2007). Insulin signalling to mTOR mediated by the Akt/PKB substrate PRAS40. *Nat Cell Biol.* **9**:316-323.

[33] Guertin, DA. Stevens, DM. Thoreen, CC. Burds, AA. Kalaany NY, Moffat, J. Brown, M. Fitzgerald, KJ. & Sabatini, DM. (2006). Ablation in mice of the mTORC components raptor, rictor, or mLST8 reveals that mTORC2 is required for signaling to Akt-FOXO and PKCalpha, but not S6K1. *Dev Cell.* **11**:859-871.

[34] Jacinto E, Loewith R, Schmidt A, Lin S, Ruegg MA, Hall A, & Hall MN. (2004). Mammalian TOR complex 2 controls the actin cytoskeleton and is rapamycin insensitive. *Nat Cell Biol.* **6:**1122-1128.

[35] Colicelli, J. Nicolette, C. Birchmeier, C. Rodgers, L. Riggs, M. & Wigler, M. (1991). Expression of three mammalian cDNAs that interfere with RAS function in Saccharomyces cerevisiae. *Proc Natl Acad Sci USA* **88**:2913–2917.

[36] Loewith, R. Jacinto, E. Wullschleger, S. Lorberg, A. Crespo, JL. Bonenfant, D. Oppliger, W. Jenoe, P. & Hall, MN. (2002). Two TOR complexes, only one of which is rapamycin sensitive, have distinct roles in cell growth control. *Mol Cell.* **10**:457–468.

[37] Frias, MA. Thoreen, CC. Jaffe, JD. Schroder, W. Sculley, T. Carr, SA. & Sabatini, DM. (2006). mSin1 is necessary for Akt/PKB phosphorylation, and its isoforms define three distinct mTORC2s. *Curr Biol.* **16**:1865-1870.

[38] Jacinto, E. Facchinetti, V. Liu, D. Soto, N. Wei, S. Jung, SY. Huang, Q. Qin, J. & Su, B. (2006). SIN1/MIP1 maintains rictor-mTOR complex integrity and regulates Akt phosphorylation and substrate specificity. *Cell.***127**:125-137.

[39] Jefferies, HB. Reinhard, C. Kozma, SC. & Thomas, G. (1994). Rapamycin selectively represses translation of the "polypyrimidine tract" mRNA family. *Proc Natl Acad Sci U S A.* **91**:4441-4445.

[40] Jefferies, HB. Fumagalli, S. Dennis, PB. Reinhard, C. Pearson, RB. & Thomas, G. (1997). Rapamycin suppresses 5'TOP mRNA translation through inhibition of p70s6k. *EMBO J.* **16**:3693-3704.

[41] Grove, JR. Banerjee, P. Balasubramanyam, A. Coffer, PJ. Price, DJ. Avruch, J. & Woodgett, JR. (1991). Cloning and expression of two human p70 S6 kinase polypeptides differing only at their amino termini. *Mol Cell Biol.* **11**:5541-5550.

[42] Reinhard, C. Thomas, G. & Kozma, SC. (1992). A single gene encodes two isoforms of the p70 S6 kinase: activation upon mitogenic stimulation. *Proc Natl Acad Sci U S A.* **89**:4052-4056.

[43] Reinhard, C. Fernandez, A. Lamb, NJ. & Thomas, G. (1994). Nuclear localization of p85s6k: functional requirement for entry into S phase. *EMBO J.* **13**:1557-1565.

[44] Um, SH. D'Alessio, D. & Thomas, G. (2006). Nutrient overload, insulin resistance, and ribosomal protein S6 kinase 1, S6K1. *Cell Metab.* **3**:393-402.

[45] Averous, J. & Proud, CG. (2006). When translation meets transformation: the mTOR story. *Oncogene.* **25**:6423-6435.

[46] Gingras, AC. Raught, B. & Sonenberg, N. (1999). eIF4 initiation factors: effectors of mRNA recruitment to ribosomes and regulators of translation. *Annu Rev Biochem.* **68**:913-963.

[47] Tee, AR. & Proud, CG. (2002). Caspase cleavage of initiation factor 4E-binding protein 1 yields a dominant inhibitor of cap-dependent translation and reveals a novel regulatory motif. *Mol Cell Biol.* **22**:1674-1683.

[48] Pause, A. Belsham, GJ. Gingras, AC. Donzé, O. Lin, TA. Lawrence, JC. Jr. & Sonenberg, N. (1994). Insulin-dependent stimulation of protein synthesis by phosphorylation of a regulator of 5'-cap function. *Nature.* **371**:762-767.

[49] Gingras, AC. Kennedy, SG. O'Leary, MA. Sonenberg, N. & Hay, N. (1998). 4E-BP1, a repressor of mRNA translation, is phosphorylated and inactivated by the Akt (PKB) signaling pathway. *Genes & Dev.* **12**:502-513.

[50] Sonenberg, N. & Gingras, AC. (1998). The mRNA 5' cap-binding protein eIF4E and control of cell growth. *Curr Opin Cell Biol.* **10**:268-275.

[51] Brunn, GJ. Fadden, P. Haystead, TA. & Lawrence, JC Jr. (1997). The mammalian target of rapamycin phosphorylates sites having a (Ser/Thr)-Pro motif and is activated by antibodies to a region near its COOH terminus. *J Biol Chem.* **272**:32547-32550.

[52] Heesom, KJ. Avison, MB. Diggle, TA. & Denton, RM. (1998). Insulin-stimulated kinase from rat fat cells that phosphorylates initiation factor 4E-binding protein 1 on the rapamycin-insensitive site (serine-111). *Biochem J.* **336**:39-48.

[53] Schalm, SS. Fingar, DC. Sabatini, DM. & Blenis, J. (2003). TOS motif-mediated raptor binding regulates 4E-BP1 multisite phosphorylation and function. *Curr Biol.* **13**:797-806.

[54] Gingras, AC. Raught, B. & Sonenberg, N. (2001). Regulation of translation initiation by FRAP/mTOR. *Genes Dev.***15**:807-826.

[55] Hay, N. & Sonenberg, N. (2004). Upstream and downstream of mTOR. *Genes Dev.* **18**:1926-1945.

[56] Holz, MK. Ballif, BA. Gygi, SP. & Blenis, J. (2005). mTOR and S6K1 mediate assembly of the translation preinitiation complex through dynamic protein interchange and ordered phosphorylation events. *Cell.* **123**:569-580.

[57] Balendran, A. Casamayor, A. Deak, M. Paterson, A. Gaffney, P. Currie, R. Downes, CP. & Alessi, DR. (1999). PDK1 acquires PDK2 activity in the presence of a synthetic peptide derived from the carboxyl terminus of PRK2. *Curr Biol.* **9**:393-404.

[58] Persad, S. Attwell, S. Gray, V. Mawji, N. Deng, JT. Leung, D. Yan, J. Sanghera, J. Walsh, MP. & Dedhar, S. (2001). Regulation of protein kinase B/Akt-serine 473 phosphorylation by integrin-linked kinase: critical roles for kinase activity and amino acids arginine 211 and serine 343. *J Biol Chem.* **276**:27462-27469.

[59] Feng, J. Park, J. Cron, P. Hess, D. & Hemmings, BA. (2004). Identification of a PKB/Akt hydrophobic motif Ser-473 kinase as DNA-dependent protein kinase. *J Biol Chem.* **279**:41189-41196.

[60] Fingar, DC. & Blenis, J. (2004). Target of rapamycin (TOR): an integrator of nutrient and growth factor signals and coordinator of cell growth and cell cycle progression. Oncogene. **23**:3151-3171.

[61] Choo, AY. Roux, PP. & Blenis, J. (2006). Mind the GAP: Wnt steps onto the mTORC1 train. *Cell.***126**:834-836.

[62] Li, Y. Corradetti, MN. Inoki, K. & Guan, KL. (2004). TSC2: filling the GAP in the mTOR signaling pathway. *Trends Biochem Sci.* **29**:32-38.

[63] Inoki, K. Li, Y. Zhu, T. Wu, J. & Guan, KL. (2002). TSC2 is phosphorylated and inhibited by Akt and suppresses mTOR signalling. *Nat Cell Biol.* **4**:648-657.

[64] Potter, CJ. Pedraza, LG. & Xu, T. (2002). Akt regulates growth by directly phosphorylating Tsc2. *Nat Cell Biol.* **4**:658-665.

[65] Castro, AF. Rebhun, JF. Clark, GJ. & Quilliam, LA. (2003). Rheb binds tuberous sclerosis complex 2 (TSC2) and promotes S6 kinase ctivation in a rapamycin- and farnesylation-dependent manner. *J Biol Chem.* **278**:32493-32496.

[66] Tabancay, AP Jr. Gau, CL. Machado, IM. Uhlmann, EJ. Gutmann, DH. Guo, L. & Tamanoi, F. (2003). Identification of dominant negative mutants of Rheb GTPase and their use to implicate the involvement of human Rheb in the activation of p70S6K. *J Biol Chem.* **278**:39921-39930.

[67] Long, X. Lin, Y. Ortiz-Vega, S. Yonezawa, K. & Avruch, J. (2005). Rheb binds and regulates the mTOR kinase. *Curr Biol.* **15**:702-713.

[68] Long, X. Ortiz-Vega, S. Lin, Y. & Avruch, J. (2005). Rheb binding to mammalian target of rapamycin (mTOR) is regulated by amino acid sufficiency. *J Biol Chem.* **280**:23433-23436.

[69] Inoki, K. Zhu, T. & Guan, KL. (2003). TSC2 mediates cellular energy response to control cell growth and survival. *Cell.***115**:577-590.

[70] Hemminki, A. Markie, D. Tomlinson, I. Avizienyte, E. Roth, S. Loukola, A. Bignell, G. Warren, W. Aminoff, M. Hoglund, P. Jarvinen, H. Kristo, P. Pelin, K. Ridanpaa, M. Salovaara, R. Toro, T. Bodmer, W. Olschwang, S. Olsen, AS. Stratton, MR. de la Chapelle, A. & Aaltonen, LA. (1998). A serine/threonine kinase gene defective in Peutz-Jeghers syndrome. *Nature.* **391**:184-187.

[71] Woods, A. Johnstone, SR. Dickerson, K. Leiper, FC. Fryer, LG. Neumann, D. Schlattner, U. Wallimann, T. Carlson, M. & Carling, D. (2003). LKB1 is the upstream kinase in the AMP-activated protein kinase cascade. *Curr Biol.***13**:2004-2008.

[72] Hardie, DG. (2005). New roles for the LKB1-->AMPK pathway. *Curr Opin Cell Biol.***17**:167-173.

[73] Brugarolas, J. Lei, K. Hurley, RL. Manning, BD. Reiling, JH. Hafen, E. Witters, LA. Ellisen, LW. & Kaelin, WG Jr. (2004). Regulation of mTOR function in response to hypoxia by REDD1 and the TSC1/TSC2 tumor suppressor complex. *Genes Dev.* **18**:2893-2904.

[74] Nobukuni, T. Joaquin, M. Roccio, M. Dann, SG. Kim, SY. Gulati, P. Byfield, MP. Backer, JM. Natt, F. Bos, JL. Zwartkruis, FJ. & Thomas, G. (2005). Amino acids mediate mTOR/raptor signaling through activation of class 3 phosphatidylinositol 3OH-kinase. *Proc Natl Acad Sci U S A.* **102**:14238-14243.

[75] Byfield, MP. Murray, JT. & Backer, JM. (2005). hVps34 is a nutrient-regulated lipid kinase required for activation of p70 S6 kinase. *J Biol Chem.* **280**:33076-33082.

[76] Hara, K. Yonezawa, K. Weng, QP. Kozlowski, MT. Belham, C. & Avruch, J. (1998). Amino acid sufficiency and mTOR regulate p70 S6 kinase and eIF-4E BP1 through a common effector mechanism. *J Biol Chem.* **273**:14484-14494.

[77] Iiboshi, Y. Papst, PJ. Kawasome, H. Hosoi, H,. Abraham, RT. Houghton, PJ. & Terada, N. (1999). Amino acid-dependent control of p70(s6k). Involvement of tRNA aminoacylation in the regulation. *J Biol Chem.* **274**:1092-1099.

[78] Nobukuni, T. Joaquin, M. Roccio, M. Dann, SG. Kim, SY. Gulati, P. Byfield, MP. Backer, JM. Natt, F. Bos, JL. Zwartkruis, FJ. & Thomas, G. (2005). Amino acids mediate mTOR/raptor signaling through activation of class 3 phosphatidylinositol 3OH-kinase. *Proc Natl Acad Sci U S A.* **102**:14238-14243.

[79] Lemmon, MA. (2003). Phosphoinositide recognition domains. *Traffic.* **4**:201-213.

[80] Sarbassov, DD. Ali, SM. Sengupta, S. Sheen, JH. Hsu, PP. Bagley, AF. Markhard, AL. & Sabatini, DM. (2006). Prolonged rapamycin treatment inhibits mTORC2 assembly and Akt/PKB. *Mol Cell.* **22**:159-168.

[81] Hunter, T. (1998). The Croonian Lecture 1997. The phosphorylation of proteins on tyrosine: its role in cell growth and disease. *Philos Trans R Soc Lond B Biol Sci.* **353**:583-605.

[82] Blume-Jensen, P. & Hunter, T. (2001). Oncogenic kinase signalling. *Nature.* **411**: 355-365.

[83] Baserga, R. (2005). The insulin-like growth factor-I receptor as a target for cancer therapy. *Expert Opin Ther Targets.* **9**:753-768.

[84] Engelman, JA. Luo, J. & Cantley, LC. (2006). The evolution of phosphatidylinositol 3-kinases as regulators of growth and metabolism. *Nat Rev Genet.***7**:606-619.

[85] Philp, AJ. Campbell, IG. Leet, C. Vincan, E. Rockman, SP. Whitehead, RH. Thomas, RJ. & Phillips, WA. (2001). The phosphatidylinositol 3'-kinase p85alpha gene is an oncogene in human ovarian and colon tumors. *Cancer Res.* **61**:7426-7429.

[86] Staal, SP. (1987). Molecular cloning of the akt oncogene and its human homologues AKT1 and AKT2: amplification of AKT1 in a primary human gastric adenocarcinoma. *Proc Natl Acad Sci U S A.* **84**:5034-5037.

[87] Knobbe, CB. & Reifenberger, G. (2003). Genetic alterations and aberrant expression of genes related to the phosphatidyl-inositol-3'-kinase/protein kinase B (Akt) signal transduction pathway in glioblastomas. *Brain Pathol.* **13**:507-518.

[88] Bellacosa, A. de Feo, D. Godwin, AK. Bell, DW. Cheng, JQ. Altomare, DA. Wan, M. Dubeau, L. Scambia, G. Masciullo, V. Ferrandina, G. Benedetti Panici, P. Mancuso, S. Neri, G. & Testa, JR. (1995). Molecular alterations of the AKT2 oncogene in ovarian and breast carcinomas. *Int J Cancer.* **64**:280-285.

[89] Cheng, JQ. Ruggeri, B. Klein, WM. Sonoda, G. Altomare, DA. Watson, DK. & Testa, JR. (1996). Amplification of AKT2 in human pancreatic cells and inhibition of AKT2 expression and tumorigenicity by antisense RNA. *Proc Natl Acad Sci U S A.* **93**:3636-3641.

[90] Nakatani, K. Thompson, DA. Barthel, A. Sakaue, H. Liu, W. Weigel, RJ. & Roth, RA. (1999). Up-regulation of Akt3 in estrogen receptor-deficient breast cancers and androgen-independent prostate cancer lines. *J Biol Chem.* **274**:21528-21532.

[91] Parsons, DW. Wang, TL. Samuels, Y. Bardelli, A. Cummins, JM. DeLong, L. Silliman, N. Ptak, J. Szabo, S. Willson, JK. Markowitz, S. Kinzler, KW. Vogelstein, B. Lengauer, C. & Velculescu, VE. (2005). Colorectal cancer: mutations in a signalling pathway. *Nature.* **436**:792.

[92] Alkan, S. & Izban, KF. (2002). Immunohistochemical localization of phosphorylated AKT in multiple myeloma. *Blood.* **99**:2278-2279.

[93] Tanno, S. Yanagawa, N. Habiro, A. Koizumi, K. Nakano, Y. Osanai, M. Mizukami, Y. Okumura, T. Testa, JR. & Kohgo, Y. (2004). Serine/threonine kinase AKT is frequently activated in human bile duct cancer and is associated with increased radioresistance. *Cancer Res.* **64**:3486-3490.

[94] Gupta, AK. McKenna, WG. Weber, CN. Feldman, MD. Goldsmith, JD. Mick, R. Machtay, M. Rosenthal, DI. Bakanauskas, VJ. Cerniglia, GJ. Bernhard, EJ. Weber, RS. & Muschel, RJ. (2002). Local recurrence in head and neck cancer: relationship to radiation resistance and signal transduction. *Clin Cancer Res.* 885-892.

[95] Kanamori, Y. Kigawa, J. Itamochi, H. Shimada, M. Takahashi, M. Kamazawa, S. Sato, S. Akeshima, R. & Terakawa, N. (2001). Correlation between loss of PTEN expression and Akt phosphorylation in endometrial carcinoma. *Clin Cancer Res.* **7**:892-895.

[96] Perez-Tenorio, & G. Stal, O; Southeast Sweden Breast Cancer Group. (2002). Activation of AKT/PKB in breast cancer predicts a worse outcome among endocrine treated patients. *Br J Cancer.* **86**:540-545.

[97] Brognard, J. Clark,AS. Ni, Y. & Dennis, PA. (2001). Akt/protein kinase B is constitutively active in non-small cell lung cancer cells and promotes cellular survival and resistance to chemotherapy and radiation. *Cancer Res.* **61**:3986-3997.

[98] Roy, HK. Olusola, BF. Clemens, DL. Karolski, WJ. Ratashak, A. Lynch, HT. & Smyrk, TC. (2002). AKT proto-oncogene overexpression is an early event during sporadic colon carcinogenesis. *Carcinogenesis.* **23**:201-205.

[99] Kurose, K. Zhou, XP. Araki, T. Cannistra, SA. Maher, ER. & Eng, C. (2001). Frequent loss of PTEN expression is linked to elevated phosphorylated Akt levels, but not associated with p27 and cyclin D1 expression, in primary epithelial ovarian carcinomas. *Am J Pathol.* **158**:2097-2106.

[100] Yamamoto, S. Tomita, Y. Hoshida, Y. Morooka, T. Nagano, H. Dono, K. Umeshita, K. Sakon, M. Ishikawa, O. Ohigashi, H. Nakamori, S. Monden, M. & Aozasa, K. (2004). Prognostic significance of activated Akt expression in pancreatic ductal adenocarcinoma. *Clin Cancer Res.* **10**:2846-2850.

[101] Kreisberg, JI. Malik, SN. Prihoda, TJ. Bedolla, RG. Troyer, DA. Kreisberg, S. & Ghosh, PM. (2004). Phosphorylation of Akt (Ser473) is an excellent predictor of poor clinical outcome in prostate cancer. *Cancer Res.* **64**:5232-5236.

[102] Ermoian, RP. Furniss, CS. Lamborn, KR. Basila, D. Berger, MS. Gottschalk, AR. Nicholas, MK. Stokoe, D. & Haas-Kogan, DA. (2002). Dysregulation of PTEN and protein kinase B is associated with glioma histology and patient survival. *Clin Cancer Res.* **8**:1100-1106.

[103] Dai, DL. Martinka, M. & Li, G. (2005). Prognostic significance of activated Akt expression in melanoma: a clinicopathologic study of 292 cases. *J Clin Oncol.* **23**:1473-1482.

[104] Min, YH. Cheong, JW. Kim, JY. Eom, JI. Lee, ST. Hahn, JS. Ko, YW. & Lee, MH. (2004). Cytoplasmic mislocalization of p27Kip1 protein is associated with constitutive phosphorylation of Akt or protein kinase B and poor prognosis in acute myelogenous leukemia. *Cancer Res.* **64**:5225-5231.

[105] Schlieman, MG. Fahy, BN. Ramsamooj, R. Beckett, L. & Bold, RJ. (2003). Incidence, mechanism and prognostic value of activated AKT in pancreas cancer. *Br J Cancer.* **89**:2110-2115.

[106] Nakanishi, K. Sakamoto, M. Yamasaki, S. Todo, S. & Hirohashi, S. (2005). Akt phosphorylation is a risk factor for early disease recurrence and poor prognosis in hepatocellular carcinoma. *Cancer.***103**:307-312.

[107] Clark, AS. West, K. Streicher, S. & Dennis, PA. (2002). Constitutive and inducible Akt activity promotes resistance to chemotherapy, trastuzumab, or tamoxifen in breast cancer cells. *Mol Cancer Ther.* **1**:707-717.

[108] Basso, AD. Mirza, A. Liu, G. Long, BJ. Bishop, WR. & Kirschmeier, P. (2005). The farnesyl transferase inhibitor (FTI) SCH66336 (lonafarnib) inhibits Rheb farnesylation

and mTOR signaling. Role in FTI enhancement of taxane and tamoxifen anti-tumor activity. *J Biol Chem.* **280**:31101-31108.

[109] Lazaris-Karatzas, A. Montine, KS. & Sonenberg, N. (1990). Malignant transformation by a eukaryotic initiation factor subunit that binds to mRNA 5' cap. *Nature.* **345**: 544-547.

[110] Mamane, Y. Petroulakis, E. Rong, L. Yoshida, K. Ler, LW. & Sonenberg, N. (2004). eIF4E--from translation to transformation. *Oncogene.* **23**:3172-3179.

[111] De Benedetti, A. & Graff, JR. (2004). eIF-4E expression and its role in malignancies and metastases. *Oncogene.* **23**:3189-3199.

[112] Mamane, Y. Petroulakis, E. LeBacquer, O. & Sonenberg, N. (2006). mTOR, translation initiation and cancer. *Oncogene.* **25**:6416-6422.

[113] Jiang, H. Coleman, J. Miskimins, R. & Miskimins,WK. (2003). Expression of constitutively active 4EBP-1 enhances p27Kip1 expression and inhibits proliferation of MCF7 breast cancer cells. *Cancer Cell Int .* **3**:2.

[114] Avdulov, S. Li, S. Michalek, V. Burrichter, D. Peterson, M. Perlman, DM. Manivel, JC. Sonenberg, N. Yee, D. Bitterman, PB. & Polunovsky, VA. (2004). Activation of translation complex eIF4F is essential for the genesis and maintenance of the malignant phenotype in human mammary epithelial cells. *Cancer Cell.* **5**:553-563.

[115] Polunovsky, VA. Gingras, AC. Sonenberg, N. Peterson, M. Tan, A. Rubins, JB. Manivel, JC. & Bitterman PB. (2000). Translational control of the antiapoptotic function of Ras. *J Biol Chem.* **275**:24776-24780.

[116] Petricoin, EF 3rd. Espina, V. Araujo, RP. Midura, B. Yeung, C. Wan, X. Eichler, GS. Johann, DJ Jr, Qualman, S. Tsokos, M. Krishnan, K. Helman, LJ. & Liotta, LA. (2007). Phosphoprotein pathway mapping: Akt/mammalian target of rapamycin activation is negatively associated with childhood rhabdomyosarcoma survival. *Cancer Res.* **67**:3431-3440.

[117] Wan, X. & Helman, LJ. (2007). The biology behind mTOR inhibition in sarcoma. *The Oncologist.* **12**:1007-1018

[118] Inoki, K. Corradetti, MN. & Guan, KL. (2005). Dysregulation of the TSC-mTOR pathway in human disease. *Nat Genet.* **37**:19-24.

[119] Cantley, LC. & N.G. Neel, NG. (1999). New insights into tumor suppression: PTEN suppresses tumor formation by restraining the phosphoinositide 3-kinase/AKT pathway. *Proc Natl Acad Sci U S A.* **96**:4240–4245.

[120] Steelman, LS. Bertrand, FE. & McCubrey, JA. (2004). The complexity of PTEN: mutation, marker and potential target for therapeutic intervention. *Expert Opin Ther Targets.* **8**:537-550.

[121] Wan, X. & Helman, LJ. (2002). Effect of insulin-like growth factor II on protecting myoblast cells against cisplatin-induced apoptosis through p70 S6 kinase pathway. *Neoplasia.* **4**:400-408.

[122] Neshat, MS. Mellinghoff, IK. Tran, C. Stiles, B. Thomas, G. Petersen, R. Frost, P. Gibbons, JJ. Wu, H. & Sawyers, CL. (2001). Enhanced sensitivity of PTEN-deficient tumors to inhibition of FRAP/mTOR. *Proc Natl Acad Sci U S A.* **98**:10314-10319.

[123] Podsypanina, K. Lee, RT. Politis, C. Hennessy, I. Crane, A. Puc, J. Neshat, M. Wang, H. Yang, L. Gibbons, J. Frost, P. Dreisbach, V. Blenis, J. Gaciong, Z. Fisher, P. Sawyers, C. Hedrick-Ellenson, L. & Parsons, R. (2001). An inhibitor of mTOR reduces

neoplasia and normalizes p70/S6 kinase activity in Pten+/- mice. *Proc Natl Acad Sci U S A.* **98**:10320-10325.

[124] Sanchez-Cespedes, M. Parrella, P. Esteller, M. Nomoto, S. Trink, B. Engles, JM. Westra, WH. Herman, JG. & Sidransky, D. (2002). Inactivation of LKB1/STK11 is a common event in adenocarcinomas of the lung. *Cancer Res.* **62**:3659–3662.

[125] Carretero, J. Medina, PP. Pio, R. Montuenga, LM. & Sanchez-Cespedes, M. (2004). Novel and natural knockout lung cancer cell lines for the LKB1/STK11 tumor suppressor gene. *Oncogene.* **23**:4037–4040.

[126] Yang, HS. Jansen, AP. Komar, AA. Zheng, X. Merrick, WC. Costes, S. Lockett, SJ. Sonenberg, N. & Colburn, NH. (2003). The transformation suppressor Pdcd4 is a novel eukaryotic translation initiation factor 4A binding protein that inhibits translation. *Mol Cell Biol.* **23**:26-37.

[127] Dorrello, NV. Peschiaroli, A. Guardavaccaro, D. Colburn, NH. Sherman, NE. & Pagano, M. (2006). S6K1- and betaTRCP-mediated degradation of PDCD4 promotes protein translation and cell growth. *Science.***314**:467-471.

[128] Chen, Y. Knosel, T. Kristiansen, G. Pietas, A. Garber, ME. Matsuhashi, S. Ozaki, I. & Petersen I. (2003). Loss of PDCD4 expression in human lung cancer correlates with tumour progression and prognosis. *J Pathol.* **200**:640-646.

[129] Afonja, O. Juste, D. Das, S. Matsuhashi, S. & Samuels, HH. (2004). Induction of PDCD4 tumor suppressor gene expression by RAR agonists, antiestrogen and HER-2/neu antagonist in breast cancer cells. Evidence for a role in apoptosis. *Oncogene.* **23**:8135-8145.

[130] Flier, JS. Usher, P, & Moses, AC. (133). Monoclonal antibody to the type I insulin-like growth factor (IGF-I) receptor blocks IGF-I receptor-mediated DNA synthesis: clarification of the mitogenic mechanisms of IGF-I and insulin in human skin fibroblasts. *Proc Natl Acad Sci U S A.* **83**:664-668.

[131] Mitsiades, CS. & Mitsiades, N. (2005). Treatment of hematologic malignancies and solid tumors by inhibiting IGF receptor signaling. *Expert Rev Anticancer Ther.* **5** 487-499.

[132] Goetsch, L. Gonzalez, A. Leger, O. Beck, A. Pauwels, PJ. Haeuw, JF. & Corvaia, N. (2005). A recombinant humanized anti-insulin-like growth factor receptor type I antibody (h7C10) enhances the antitumor activity of vinorelbine and anti-epidermal growth factor receptor therapy against human cancer xenografts. *Int J Cancer.***113**:316-328.

[133] Cohen, BD. Baker, DA. Soderstrom, C. Tkalcevic, G. Rossi, AM. Miller, PE. Tengowski, MW. Wang, F. Gualberto, A. Beebe, JS. & Moyer. JD. (2005). Combination therapy enhances the inhibition of tumor growth with the fully human anti-type 1 insulin-like growth factor receptor monoclonal antibody CP-751,871. *Clin Cancer Res.***11**:2063-2073.

[134] Burtrum, D. Zhu, Z. Lu, D. Anderson, DM. Prewett, M. Pereira, DS. Bassi, R. Abdullah, R. Hooper, AT. Koo, H. Jimenez, X. Johnson, D. Apblett, R. Kussie, P. Bohlen, P. Witte, L. Hicklin, DJ. & Ludwig, DL. (2003). A fully human monoclonal antibody to the insulin-like growth factor I receptor blocks ligand-dependent signaling and inhibits human tumor growth in vivo. *Cancer Res.* **63**:8912-8921.

[135] Wan, X. Harkavy, B. Shen, N. Grohar, P. & Helman, LJ. (2007). Rapamycin induces feedback activation of Akt signaling through an IGF-1R-dependent mechanism. *Oncogene.* **26**:1932-1940.

[136] Maloney, EK. McLaughlin, JL. Dagdigian, NE. Garrett, LM. Connors, KM. Zhou, XM. Blattler, WA. Chittenden, T. & Singh, R. (2003). An anti-insulin-like growth factor I receptor antibody that is a potent inhibitor of cancer cell proliferation. *Cancer Res.* **63**:5073-5083.

[137] Wang, Y. Hailey, J. Williams, D. Wang, Y. Lipari, P. Malkowski, M. Wang, X. Xie, L. Li, G. Saha, D. Ling, WL. Cannon-Carlson, S. Greenberg, R. Ramos, RA. Shields, R. Presta, L. Brams, P. Bishop, WR. & Pachter, JA. (2005). Inhibition of insulin-like growth factor-I receptor (IGF-IR) signaling and tumor cell growth by a fully human neutralizing anti-IGF-IR antibody. *Mol Cancer Ther.* **4**:1214-1221.

[138] Vasilcanu, D. Girnita, A. Girnita, L. Vasilcanu, R. Axelson, M. & Larsson, O. (2004). The cyclolignan PPP induces activation loop-specific inhibition of tyrosine phosphorylation of the insulin-like growth factor-1 receptor. Link to the phosphatidyl inositol-3 kinase/Akt apoptotic pathway. *Oncogene.* **23**:7854-7862.

[139] Haluska, P. Carboni, JM. Loegering, DA. Lee, FY. Wittman, M. Saulnier, MG. Frennesson, DB. Kalli, KR. Conover, CA. Attar, RM. Kaufmann, SH. Gottardis, M. & Erlichman, C. (2006). In vitro and in vivo antitumor effects of the dual insulin-like growth factor-I/insulin receptor inhibitor, BMS-554417. *Cancer Res.***66**:362-371.

[140] Garcia-Echeverria, C. Pearson, MA. Marti, A. Meyer, T. Mestan, J. Zimmermann, J. Gao, J. Brueggen, J. Capraro, HG. Cozens, R. Evans, DB. Fabbro, D. Furet, P. Porta, DG. Liebetanz, J. Martiny-Baron, G. Ruetz, S. & Hofmann, F. (2004). In vivo antitumor activity of NVP-AEW541-A novel, potent, and selective inhibitor of the IGF-IR kinase. *Cancer Cell.* **5**:231-239.

[141] Menu, E. Jernberg-Wiklund, H. Stromberg, T. De Raeve, H. Girnita, L. Larsson, O. Axelson, M. Asosingh, K. Nilsson, K. Van Camp, B. & Vanderkerken, K. (2006). Inhibiting the IGF-1 receptor tyrosine kinase with the cyclolignan PPP: an in vitro and in vivo study in the 5T33MM mouse model. *Blood.***107**:655-660.

[142] Manara, MC. Landuzzi, L. Nanni, P. Nicoletti, G. Zambelli, D. Lollini, PL. Nanni, C. Hofmann, F. Garcia-Echeverria, C. Picci, P. & Scotlandi, K. (2007). Preclinical in vivo study of new insulin-like growth factor-I receptor--specific inhibitor in Ewing's sarcoma. *Clin Cancer Res.* **13**:1322-1330.

[143] Kim, ES. Khuri, FR. & Herbst, RS. (2001). Epidermal growth factor receptor biology (IMC-C225). *Curr Opin Oncol.* **13**:506-513.

[144] Ciardiello, F. & Tortora, G. (2002). Anti-epidermal growth factor receptor drugs in cancer therapy. *Expert Opin Investig Drugs.***11**:755-768.

[145] Rocha-Lima, CM. Soares, HP. Raez, LE. & Singal, R. (2007). EGFR targeting of solid tumors. *Cancer Control.***14**:295-304.

[146] Polychronis, A. Sinnett, HD. Hadjiminas, D. Singhal, H. Mansi, JL. Shivapatham, D. Shousha, S. Jiang, J. Peston, D. Barrett, N. Vigushin, D. Morrison, K. Beresford, E.Ali, S. Slade, MJ. & Coombes, RC. (2005). Preoperative gefitinib versus gefitinib and anastrozole in postmenopausal patients with oestrogen-receptor positive and epidermal-growth-factor-receptor-positive primary breast cancer: a double-blind placebo-controlled phase II randomised trial. *Lancet Oncol.* **6:**383-391.

[147] Baselga, J. Albanell, J. Ruiz, A. Lluch, A. Gascon, P. Guillem,V. Gonzalez, S. Sauleda, S. Marimon, I. Tabernero, JM. Koehler, MT. & Rojo, F. (2005). Phase II and tumor pharmacodynamic study of gefitinib in patients with advanced breast cancer. *J Clin Oncol.* **23**:5323-5333.

[148] Johnston, JB. Navaratnam, S. Pitz, MW. Maniate, JM. Wiechec, E. Baust, H. Gingerich, J. Skliris, GP. Murphy, LC. Los, M. (2006). Targeting the EGFR pathway for cancer therapy. *Curr Med Chem.***13**:3483-3492.

[149] Campos, S. Hamid, O. Seiden, MV. Oza, A. Plante, M. Potkul, RK. Lenehan, PF. Kaldjian, EP. Varterasian, ML. Jordan, C. Charbonneau, C. & Hirte, H. (2005). Multicenter, randomized phase II trial of oral CI-1033 for previously treated advanced ovarian cancer. *J Clin Oncol.* **23**:5597-5604.

[150] Burris, HA 3rd. Hurwitz, HI. Dees, EC. Dowlati, A. Blackwell, KL. O'Neil, B. Marcom, PK. Ellis, MJ. Overmoyer, B. Jones, SF. Harris, JL. Smith, DA. Koch, KM. Stead, A. Mangum, S. & Spector, NL. (2005). Phase I safety, pharmacokinetics, and clinical activity study of lapatinib (GW572016), a reversible dual inhibitor of epidermal growth factor receptor tyrosine kinases, in heavily pretreated patients with metastatic carcinomas. *J Clin Oncol.* **23**:5305-5313.

[151] Burstein, HJ. Harris, LN. Marcom, PK. Lambert-Falls, R. Havlin, K. Overmoyer, B. Friedlander, RJ Jr. Gargiulo, J. Strenger, R. Vogel, CL. Ryan, PD. Ellis, MJ. Nunes, RA. Bunnell, CA. Campos, SM. Hallor, M. Gelman, R. & Winer, EP. (2003). Trastuzumab and vinorelbine as first-line therapy for HER2-overexpressing metastatic breast cancer: multicenter phase II trial with clinical outcomes, analysis of serum tumor markers as predictive factors, and cardiac surveillance algorithm. *J Clin Oncol.* **21**:2889-2895.

[152] Hudis, CA. (2007). Trastuzumab--mechanism of action and use in clinical practice. *N Engl J Med.* **357**:39-51.

[153] Deininger, M. Buchdunger, E. & Druker, BJ. (2005). The development of imatinib as a therapeutic agent for chronic myeloid leukemia. *Blood.* **105**:2640-2653.

[154] Tibes, R, Trent, J. & Kurzrock, R. (2005). Tyrosine kinase inhibitors and the dawn of molecular cancer therapeutics. *Annu Rev Pharmacol Toxicol.* **45**:357-384.

[155] Youngren, JF. Gable, K. Penaranda, C. Maddux, BA. Zavodovskaya, M. Lobo, M. Campbell, M. Kerner, J. & Goldfine, ID. (2005). Nordihydroguaiaretic acid (NDGA) inhibits the IGF-1 and c-erbB2/HER2/neu receptors and suppresses growth in breast cancer cells. *Breast Cancer Res Treat.* **94**:37-46.

[156] Lu, D. Zhang, H. Koo, H. Tonra, J. Balderes, P. Prewett, M. Corcoran, E. Mangalampalli, V. Bassi, R. Anselma, D. Patel, D. Kang, X. Ludwig, DL. Hicklin, DJ. Bohlen, P. Witte, L. & Zhu, Z. (2005). A fully human recombinant IgG-like bispecific antibody to both the epidermal growth factor receptor and the insulin-like growth factor receptor for enhanced antitumor activity. *J Biol Chem.* **280**:19665-19672.

[157] Yee, KW. O'Farrell, AM. Smolich, BD. Cherrington, JM. McMahon, G. Wait, CL. McGreevey, LS. Griffith, DJ. & Heinrich,MC. (2002). SU5416 and SU5614 inhibit kinase activity of wild-type and mutant FLT3 receptor tyrosine kinase. Blood.**100**:2941-2949.

[158] Giles. FJ. Stopeck, AT. Silverman, LR. Lancet, JE. Cooper, MA. Hannah, AL. Cherrington, JM. O'Farrell, AM. Yuen, HA. Louie, SG. Hong, W. Cortes, JE. Verstovsek, S. Albitar, M. O'Brien, SM. Kantarjian, HM. & Karp, JE. (2003). SU5416,

a small molecule tyrosine kinase receptor inhibitor, has biologic activity in patients with refractory acute myeloid leukemia or myelodysplastic syndromes. *Blood*. **102**:795-801.

[159] Stone, RM. DeAngelo, DJ. Klimek, V. Galinsky, I. Estey, E. Nimer, SD. Grandin, W. Lebwohl, D. Wang, Y. Cohen, P. Fox, EA. Neuberg, D. Clark, J. Gilliland, DG. & Griffin, JD. (2005). Patients with acute myeloid leukemia and an activating mutation in FLT3 respond to a small-molecule FLT3 tyrosine kinase inhibitor, PKC412. *Blood*.**105**:54-60.

[160] Wiesinger, D. Gubler, HU. Haefliger, W. & Hauser, D. (1974). Antiinflammatory activity of the new mould metabolite 11-desacetoxy-wortmannin and of some of its derivatives. *Experientia*. **30**:135-136.

[161] Wymann, MP. Bulgarelli-Leva, G. Zvelebil, MJ. Pirola, L. Vanhaesebroeck, B. Waterfield, MD. & Panayotou, G. (1996). Wortmannin inactivates phosphoinositide 3-kinase by covalent modification of Lys-802, a residue involved in the phosphate transfer reaction. *Mol Cell Biol*. **16**:1722-1733.

[162] Shintani, T. & Klionsky, DJ. (2004). Autophagy in health and disease: a double-edged sword. *Science*. **306**:990-995.

[163] Vlahos, CJ. Matter, WF. Hui, KY. & Brown, RF. (1994). A specific inhibitor of phosphatidylinositol 3-kinase, 2-(4-morpholinyl)-8-phenyl-4H-1-benzopyran-4-one (LY294002). *J Biol Chem*. **269**:5241-5248.

[164] Tolloczko, B. Turkewitsch, P. Al-Chalabi, M. & Martin, JG. (2004). LY-294002 [2-(4-morpholinyl)-8-phenyl-4H-1-benzopyran-4-one] affects calcium signaling in airway smooth muscle cells independently of phosphoinositide 3-kinase inhibition. *J Pharmacol Exp Ther*. **311**:787-793.

[165] Amaravadi, R. & Thompson, CB. (2005). The survival kinases Akt and Pim as potential pharmacological targets. *J Clin Invest*.**115**:2618-2624.

[166] Knight, ZA. Gonzalez, B. Feldman, ME. Zunder, ER. Goldenberg, DD. Williams, O. Loewith, R. Stokoe, D. Balla, A. Toth, B. Balla, T. Weiss, WA. Williams, RL. & Shokat, KM. (2006). A pharmacological map of the PI3-K family defines a role for p110alpha in insulin signaling. *Cell*.**125**:733-747.

[167] Chaussade, C. Rewcastle, GW. Kendall, JD. Denny, WA. Cho, K. Gronning, LM. Chong, ML. Anagnostou, SH. Jackson, SP. Daniele, N. & Shepherd, PR. (2007). Evidence for functional redundancy of class IA PI3K isoforms in insulin signalling. *Biochem J*.**404**:449-458.

[168] Knight, ZA. Chiang, GG. Alaimo, PJ. Kenski, DM. Ho, CB. Coan, K. Abraham, RT. & Shokat, KM. (2004). Isoform-specific phosphoinositide 3-kinase inhibitors from an arylmorpholine scaffold. *Bioorg Med Chem*. **12**:4749-4759.

[169] Sato, S. Fujita, N. & Tsuruo, T. (2002). Interference with PDK1-Akt survival signaling pathway by UCN-01 (7-hydroxystaurosporine). *Oncogene*. **21**:1727-1738.

[170] Komander, D. Kular, GS. Bain, J. Elliott, M. Alessi, DR. & Van Aalten, DM. (2003). Structural basis for UCN-01 (7-hydroxystaurosporine) specificity and PDK1 (3-phosphoinositide-dependent protein kinase-1) inhibition. *Biochem J*. **375**:255-262.

[171] Feldman, RI. Wu, JM. Polokoff, MA. Kochanny, MJ. Dinter, H. Zhu, D. Biroc, SL. Alicke, B. Bryant, J. Yuan, S. Buckman, BO. Lentz, D. Ferrer, M. Whitlow, M. Adler, M. Finster, S. Chang, Z. & Arnaiz, DO. (2005). Novel small molecule inhibitors of 3-phosphoinositide-dependent kinase-1. *J Biol Chem*. **280**:19867-19874.

[172] Zhu, J. Huang, JW. Tseng, PH. Yang, YT. Fowble, J. Shiau, CW. Shaw, YJ. Kulp, SK. & Chen, CS. (2004). From the cyclooxygenase-2 inhibitor celecoxib to a novel class of 3-phosphoinositide-dependent protein kinase-1 inhibitors. *Cancer Res.* **64**:4309-4318.

[173] Yang J, Cron P, Thompson V, Good VM, Hess D, Hemmings BA, Barford D. (2002). Molecular mechanism for the regulation of protein kinase B/Akt by hydrophobic motif phosphorylation. *Mol Cell.***9**:1227-1240.

[174] Kondapaka, SB. Singh, SS. Dasmahapatra, GP. Sausville, EA. & Roy, KK. (2003). Perifosine, a novel alkylphospholipid, inhibits protein kinase B activation. *Mol Cancer Ther.* **2**:1093–1103.

[175] Obata, T. Yaffe, MB. Leparc, GG. Piro, ET. Maegawa, H. Kashiwagi, A. Kikkawa, R. & Cantley, LC. (2000). Peptide and protein library screening defines optimal substrate motifs for AKT/PKB. *J Biol Chem.* **275**:36108-36115.

[176] Yang, L. Dan, HC. Sun, M. Liu, Q. Sun, XM. Feldman, RI. Hamilton, AD. Polokoff, M. Nicosia, SV. Herlyn, M. Sebti, SM. & Cheng, JQ. (2004). Akt/protein kinase B signaling inhibitor-2, a selective small molecule inhibitor of Akt signaling with antitumor activity in cancer cells overexpressing *Akt. Cancer Res.* **64**:4394-4399.

[177] Stauffer, F. Holzer, P. & Garcia-Echeverria, C. (2005). Blocking the PI3K/PKB pathway in tumor cells. *Curr Med Chem Anticancer Agents.* **5**:449-462.

[178] Barnett, SF. Defeo-Jones, D. Fu, S. Hancock, PJ. Haskell, KM. Jones, RE. Kahana, JA. Kral, AM. Leander, K. Lee, LL. Malinowski, J. McAvoy, EM. Nahas, DD. Robinson, RG. & Huber, HE. (2005). Identification and characterization of pleckstrin-homology-domain-dependent and isoenzyme-specific Akt inhibitors. *Biochem J.* **385**:399-408.

[179] DeFeo-Jones, D. Barnett, SF. Fu, S. Hancock, PJ. Haskell, KM. Leander, KR. McAvoy, E. Robinson, RG. Duggan, ME. Lindsley, CW. Zhao, Z. Huber, HE. & Jones, RE. (2005). Tumor cell sensitization to apoptotic stimuli by selective inhibition of specific Akt/PKB family members. *Mol Cancer Ther.* **4**:271-279.

[180] Vezina, C. Kudelski, A. & Sehgal, SN. (1975). Rapamycin (AY-22,989), a new antifungal antibiotic. I. Taxonomy of the producing streptomycete and isolation of the active principle. *J Antibiot (Tokyo).* **28**:721-726.

[181] Granville, CA. Memmott, RM. Gills, JJ. & Dennis, PA. (2006). Handicapping the race to develop inhibitors of the phosphoinositide 3-kinase/Akt/mammalian target of rapamycin pathway. *Clin Cancer Res.* **12**:679-689.

[182] Smolewski, P. (2006). Investigating mammalian target of rapamycin inhibitors for their anticancer properties. *Expert Opin Investig Drugs.* **15**:1201-1227.

In: Oncogene Proteins: Structure, Functions and Analyses ISBN: 978-1-60456-259-0
Editors: P.B. Murphy and J.R. Clarke, pp. 117-138

Chapter IV

Modulation of Tumor Angiogenesis by Polyunsaturated Fatty Acids: what is the Mechanism?

Nenad Petrovic and Michael Murray
Pharmacogenomics and Drug Development group
Faculty of Pharmacy, University of Sydney, Australia

Abstract

In the last decade the role of dietary fats in cancer-related diseases has attracted considerable interest within the community. First reports of the beneficial role of ω-3 long-chain polyunsaturated fatty acids (ω-3 PUFA) in human health occurred in the mid-1970s with observations that the decrease in prevalence of cardiovascular diseases and cancer correlated with dietary intake of ω-3 PUFA. Current research interest is focused on the protective effects of oily fish-derived ω-3 PUFA, mostly eicosapentaenoic (EPA) and docosahexaenoic acid (DHA) that counteract the pro-carcinogenic effects of animal meat-derived ω-6 PUFA, typified by arachidonic acid. Both epidemiologic and experimental findings indicate that ω-3 PUFA, almost absent from typical Western diets today, exert protective effects against cancer progression. Although beneficial effects of ω-3 PUFA have been widely documented, the current and future use of these PUFA, as preventive or complementary treatments in cancer and other diseases, is hampered by the lack of mechanistic information. The ω-3 PUFA have been shown to be less potent than ω-6 PUFA in activation of important oncogenes such as protein kinase C, transcription factors *ras* and AP1 and could differentially modulate apoptosis mediated by NFκB and Bcl-2. It is now believed that multiple mechanisms contribute to the observed biological effects of ω-3 PUFA. PUFA were able to modulate cancer regulation through their roles as structural phospholipids of the cell membrane, affecting membrane fluidity, cellular signaling and cellular interactions. PUFA also exhibit numerous functional properties, either directly or after conversion to their respective lipid mediators. Both ω-6 and ω-3 PUFA compete for the same bio-conversion enzymes: cyclooxygenase (COX), lipoxygenase (LOX) and epoxygenase (CYP) enzymes to form different classes of eicosanoids (prostaglandins, thromboxanes, leukotrienes and epoxides). Consequently, when the supply of ω-3 PUFA is increased relative to ω-6 PUFA, the synthesis of the corresponding eicosanoids is also increased. These ω-3 PUFA-derived eicosanoids have generally been shown to be less inflammatory and to be weaker cancer-promoters than their ω-6 counterparts. One of the mechanisms by which ω-3 PUFA could play a significant role in carcinogenesis is in the inhibition of new tumor blood vessel formation, or angiogenesis, which is an important process affecting cancer growth and dissemination. Angiogenesis involves complex interplay between tumour and endothelial cells that culminates in proliferation, migration, and differentiation of new endothelium. This review is focused on a comprehensive analysis of published research studies aimed at characterization of the potential mechanisms by which ω-3 PUFA influence angiogenesis.

Introduction

The potential role(s) of dietary fats in overall human health and cancer development and progression is attracting considerable interest within the community. Major constituents of body fats include the long-chain fatty acids, (18-22 carbons in length) that are usually esterified with glycerol (triglycerides) or phosphatidyls (phospholipids) within cell membranes. Based on their structure, fatty acids are divided into three major groups: unsaturated (containing no double bonds), mono-unsaturated (one double bond) and polyunsaturated (two or more double bonds). When required, these fatty acids are released from their respective esters by lipases and phospholipases to perform important cellular and tissue functions. Due to their metabolic energy content, fatty acids (mostly the unsaturated fatty acids) are a major source of energy production. Apart from energy metabolism, fatty acids perform numerous structural roles and are constituents of cellular membranes (influencing membrane fluidity and function, including ion transport, receptor binding and electrical potentials) and lipid-containing tissue structures (e.g, nerve insulation membranes). In addition, fatty acids serve as precursors for the synthesis of numerous intracellular and extracellular bioactive lipids that modulate a variety of pathways that regulate important cellular functions (e.g., prostaglandins, thromboxanes and leukotrienes which modulate vascular tone, platelet aggregation, and inflammatory responses).

This review will analyze information that is currently available on the role of polyunsaturated fatty acids (PUFA) in the regulation of cancer progression. Some PUFA are synthesized by human tissues, and these are known as “nonessential PUFA”. In contrast, some of the PUFA are necessary (or essential) since they cannot be synthesized by mammalian cells and must be obtained from the diet. There are two known families of essential PUFA: ω-3 (or n-3) and ω-6 (or n-6). These numbers refer to the location of the first double bond from the methyl (omega or n) end of the molecule (Figure 1). Since mammalian enzymes cannot form double bonds in fatty acids before the 9th carbon from the omega end, both ω-3 and ω-6 PUFA are essential (1). The ω-3 and ω-6 PUFA cannot be inter-converted, but both can be elongated and desaturated to form other PUFA of the same series.

There are three major ω-3 PUFA: alpha-linolenic acid (ALA, 18:3n-3) found in green leafy vegetables, walnuts, canola, and flaxseed oils, eicosapentaenoic acid (EPA, 20:5n-3), and docosahexaenoic acid (DHA, 22:6n-3). The numbers are derived from the structures of these acids: total number of carbons followed by total number of double bonds and position of the first double bond. Cells are able to convert ALA to EPA and DHA. Extensive research indicates that ω-3 PUFA reduce inflammation and help prevent certain chronic diseases, such as heart disease and arthritis (2-5). These PUFA are highly concentrated in the brain and appear to be particularly important for cognitive and behavioral function. In fact, infants who do not get enough ω-3 PUFA from their mothers during pregnancy are at risk of developing vision and neural problems (6-8). In the ω-6 PUFA family the most important member is linoleic acid (LA, 18:2n-6) which is found in vegetable oils, such as corn or sunflower oils. The ω-6 PUFA AA and ω-3 PUFA EPA and DHA are obtained directly from the diet, bypassing the need for metabolic conversion from LA or ALA. Thus, AA is readily available in meats, while EPA and DHA are found in fish and fish oils.

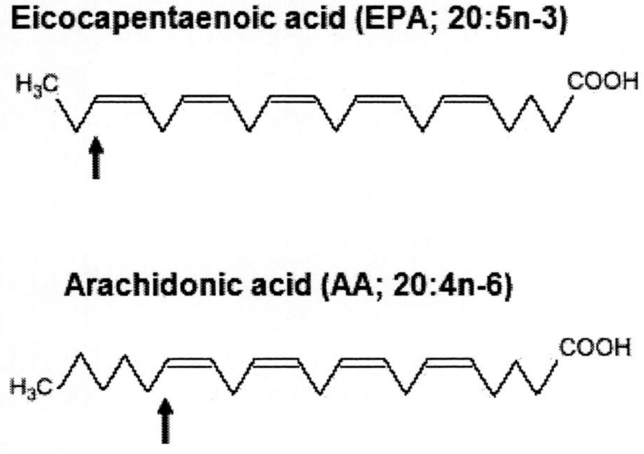

Figure 1. Chemical structures of eicosapentaenoic acid (EPA) and arachidonic acid (AA). The first number represents the total number of carbons followed by the total number of double bonds. The "n" refers to the position of the first double bond from the methyl end (indicated by arrows).

Epidemiological and Animal Studies

Interest in ω-3 PUFA began in the early 1970s with observations that the consumption of fish is beneficial to human health. Since then much has been learned about the role of PUFA in cardiovascular disease and cancer; the data suggest that ω-3 PUFA are protective, but that ω-6 PUFA are pro-inflammatory and cancer-promoting (9, 10). The dietary intake of ω-3 PUFA has generally declined in recent years while the dietary intake of ω-6 PUFA has substantially increased due to greater ingestion of vegetable oils and meat (11). Interest in both the scientific and general communities in the role of ω-3 PUFA in the prevention and treatment of cancer-related diseases has risen significantly (a recent search of the Internet using "omega-3" and "cancer" as the keywords resulted in 5.18 million hits!). It is now believed that as much as 30-40% of all kinds of cancer can be prevented, or at least suppressed, with a healthy lifestyle and dietary measures (12).

The ratio of dietary ω-6 to ω-3 was about 1.5:1 as recently as 200 years ago and it is now estimated at 16:1 in the average North American diet (13). It is now apparent that the current Western diet is deficient in ω-3 PUFA and contains an excess of ω-6 PUFA (11). Intake of ω-3 PUFA in United States (*per capita*) is approximately one fourth of that found in Norway and one fifteenth of that found in Japan (14).

There is ample epidemiological evidence that intake of ω-3 PUFA is protective in certain types of cancer, most notably of breast, prostate and colon. The incidence of these cancers is increasing in Japan (15) and in Alaskan natives as the cultures adopt Western diets, decrease fish intake and increase their dietary intake of ω-6 PUFA (16). Diets with ω-6/ω-3 PUFA ratios of ~1 are today believed to be optimal (17) and are similar to diets that are consumed by certain populations such as the Inuit (18), who have a lower incidence of cardiovascular disease. Current Western diets have ω-6/ω-3 ratios of approximately 30 (19), in some cases as high as 50 (20).

One of the cancers that have been most intensely investigated in regard to the effects of ω-3 PUFA is colorectal cancer, where significant dose-dependent reduction of risk has been found to be associated with increased consumption of ω-3 PUFA (21, 22). In a study by Hall *et al*, blood ω-3 PUFA levels were found to be inversely associated with colorectal cancer risk, while ω-6 PUFA did not significantly affect the disease (23). Indeed, the risk of colorectal cancer might also be estimated from erythrocyte ω-3 PUFA content (24). Interestingly, Japanese migrants to the USA who have adopted an American diet have an increased colon cancer incidence compared to their counterparts in Japan, where ω-3 PUFA consumption remains high (25). The same trend was observed from an analysis of dietary data from 24 European countries which suggested that a low ω-3/ω-6 PUFA ratio is a risk factor for colon cancer development (26).

Epidemiological studies also suggest that consumption of fish reduces the incidence of prostate cancer (27, 28). In one of the largest studies of its kind, involving 6,272 men with 30 years of follow-up, fatty fish consumption was found to be associated with a decreased risk of prostate cancer (28). Furthermore, serum levels of ω-3 PUFA are significantly lower in patients with benign prostatic hyperplasia and prostate cancer (1.5- to 3.3-fold), while ω-6 PUFA levels are higher (29, 30). In the "Health Professionals Follow-up Study" which followed 47,882 men for 12 years, higher consumption of fish was strongly associated with a reduced risk of metastatic prostate cancer (31). A positive correlation between the LA-to-DHA ratio and tumor volume was detected in prostate cancers whereas there was a negative correlation with the AA-to-DHA ratio (32).

ω-3 PUFA appear to influence the incidence and progression of a wide range of tumors. Decreased risk of non-Hodgkin's lymphoma and chronic lymphocytic lymphoma has been also associated with ω-3 PUFA intake (33). Effects of ω-3 PUFA on lymphocytic cell populations are also apparent in the intestinal immune system (34, 35) where these fatty acids alter the balance between CD4+ T-helper (Th1 and Th2) subsets by suppressing Th1 cell development which affects resistance to intracellular pathogens and allergic hypersensitivity.

In general, consumption of fish and fish oil appears to be associated with protection against breast cancer and female colorectal cancer (36). Epidemiological studies suggest that higher intake of fish correlates with lower breast cancer incidence and mortality (37). As in studies that have investigated colorectal cancer, erythrocyte PUFA content could be used as a marker of breast cancer risk (38).

Nevertheless, such findings have not yielded a consensus on the link between ω-3 PUFA intake and breast cancer protection (39-44). The potential explanation for these findings could be related to the difficulties in choosing an appropriate cohort and in designing a short-term treatment that would faithfully mimic long-term dietary intervention. For example, the study by Vatten *et al* (40) was performed in a cohort of Norwegian women who were already consuming much higher levels of ω-3 PUFA than their counterparts in other Western countries. On the other hand, dietary intake of fish or ω-3 PUFA has been clearly demonstrated to have an inverse association with breast cancer risk in Japanese subjects, in both large-scale case-control and prospective cohort studies (45-47) Supporting these findings was the observation in this population, whose dietary ω-3 PUFA content is the highest in the world, that the rate of mammary cancer has markedly increased in recent years. This could be related to the increased intake of Westernized diets, with high consumption of meat, animal fat and/or saturated fatty acids (48, 49).

Dietary increases in ω-3 PUFA have been investigated not only for cancer-protective and tumor growth-inhibition effects, but also as potential complementary treatments to enhance the response of tumors to traditional anti-cancer therapy (50-53). Thus, the level of DHA in breast adipose tissue of patients with complete or partial remission due to cytotoxic drugs was higher than in patients with no response (54). This supports the contention that nutritional strategies may be useful in conjunction with conventional chemotherapeutic approaches.

Epidemiological findings have been further substantiated by studies in experimental animals. Diets rich in ω-6 PUFA stimulated the growth and metastasis of human breast cancer cell lines that had been implanted in nude mice (55-56), whereas diets providing high levels of ω-3 PUFA exerted suppressive effects on tumor progression (57, 58). The elegant study of Xia *et al* (59) confirmed the anticancer benefits of increasing body ω-3 PUFA content. Expression of the *fat-1* transgene in mice, which encoded an ω-3 desaturase from *Caenorhabditis elegans* that converts ω-6 into ω-3 PUFA, reduced melanoma growth in mouse xenograft models (60).

As a consequence of epidemiological findings a number of *in vivo* studies in animals and *in vitro* studies have been performed in order to understand the molecular basis of PUFA effects on cancer development and progression (reviewed in 61 and 62).

Mechanisms of the Effects of ω-3 PUFA on Tumorigenesis

Formation of tumors in mammalian tissues is a multistep process that progresses through the acquisition of numerous genetic changes and leads to uncontrolled growth of transformed cells. The most critical of these genetic alterations are oncogenic activation, loss of telomerase activity, and induction of aneuploidy (63-65). Numerous regulatory pathways could therefore be potentially affected by ω-3 PUFA. What is surprising is that almost all major signaling pathways are reported to be targets for ω-3 PUFA treatment (some are listed in Table 1). It remains to be understood how relatively simple molecules such as PUFA, which are common dietary components, could modulate such a variety of unrelated cellular processes. Considered together, however, it now appears that multiple mechanisms likely contribute to the observed biological actions of ω-3 PUFA. The weight of evidence points to two major processes – the enzymatic formation of a range of eicosanoid mediators from PUFA substrates, and lipid peroxidation.

Table 1. Proposed mechanisms for suppression of cancer by ω-3 PUFA

Factors affected by ω-3 PUFA	References
Induction of PPAR-γ and p53 (increase in apoptosis)	66
Increase in lipid peroxidation and PPARs (increase in cell death)	67
Membrane lipid peroxidation (increase in apoptosis)	68, 69
Activation of sphingomyelinase (inhibition of cell growth)	70
Increase in BRCA1 and BRCA2 (increase in tumor suppression)	71, 72
Lipid raft function (modulation of signaling by lipid rafts)	73, 74
Phosphorylated Bad kinase (induction of apoptosis)	75
Decrease of protein kinase C expression and function (downregulation of TGFbetaRII receptor and cell growth)	76, 77
Activation of c-Jun N-terminal protein kinase (induction of mitochondrial damage)	78
Up-regulation of Bax/Bcl-2 kinase ratio (induction of apoptosis)	79
Suppression of p53-p21WAF1/CIP1, ERK1/2 MAPK, p27KIP1, BRCA1 and NF-κB kinases (inhibition of cell proliferation and survival)	80
Inhibition of AP-1 transcription factor activity (suppression of cell transformation)	81
Modulation of oncogenic Ras and Ras-p21 function (inhibition of progression of colon cancer)	82, 83
Inhibition of p27Kip1, CDK2, ERK1/ERK2, and retinoblastoma phosphorylation (inhibition of cell growth)	84
Activation of Stat5 (induction of cell differentiation)	85
Change in profile of synthesized eicosanoids (suppression of ω-6 eicosanoid synthesis)	86, 87, 88

Biotransformation of PUFA to Eicosanoid Mediators

Once released intracellularly ω-3 and ω-6 PUFA are rapidly converted into numerous bioactive lipid products termed eicosanoids (i.e., lipid mediators derived from 20-carbon fatty acids) (89). Figure 2 shows the important enzymatic pathways that mediate the formation of eicosanoid products and that exert a range of effects in tissues.

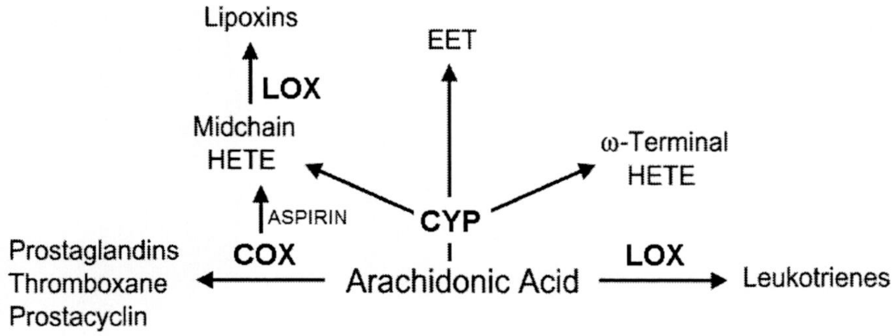

Figure 2. Major metabolic pathways for bioconversion of ω-6 PUFA arachidonic acid. Not all of the pathways operate in every cell type.

Three enzymatic pathways are important in eicosanoid formation: 1) the cyclooxygenase pathway (involving COX-1, a constitutive enzyme, and COX-2, an inducible enzyme) that leads to the synthesis of prostaglandins, prostacyclins, and thromboxanes (90); 2) the lipoxygenase (LOX) pathway that results in synthesis of hydroperoxides and leukotrienes (90); and, 3) the cytochrome P450 (CYP) monooxygenase pathway that generates epoxy- and hydroxy-fatty acids (91, 92). Eicosanoids have optimal biological activities at concentrations that are similar to or lower than the physiological levels of their PUFA precursors.

Further supporting evidence of an important role for eicosanoids in the regulation of cancer progression is provided by the beneficial effects of COX inhibitors in prevention of colon (93) and prostate (94) cancer. Thus, COX-derived prostanoids of ω-6 PUFA, such as AA, are strongly implicated in tumorigenesis. The mechanisms by which ω-3 PUFA may be beneficial in cancer have not been established but evidence is increasing that bioconversion to eicosanoids may be pertinent. ω-3 and ω-6 PUFA are alternative substrates for the same enzymes, and the nature of the substrate determines the specific eicosanoids produced in cells. ω-3 PUFA-derived eicosanoids possess generally weaker biological activity than their AA-derived counterparts. Whereas the ω-6 series prostaglandins produced from AA tend to be pro-inflammatory and to stimulate proliferation in most tissues, the ω-3-based prostaglandins are anti-inflammatory and decrease cancer growth (95). Thus, cellular increases in the ω-3 PUFA DHA and EPA have been shown to inhibit the synthesis of eicosanoids from AA by competition at the level of the COX enzyme (96). Utilization of alternate PUFA substrates to produce analogous series of eicosanoids is now well accepted. How these individual eicosanoids influence tumor progression is being evaluated increasingly in experimental studies.

The observed effects of ω-3 PUFA on tumorigenesis could also be partially explained in terms of the relative availability of ω-6 and ω-3 PUFA. Thus, animal and epidemiologic studies have suggested that the ratio of these PUFA types, rather than their absolute level, may be the most important factor that determines biological action (97). Prolonged intake of dietary ω-3 PUFA *in vivo* markedly enriches tissues with ω-3 PUFA and depletes ω-6 PUFA content. Thus, supplementation of patient diets with fish oil suppressed rectal epithelial cell proliferation and PGE_2 synthesis when the dietary ω-6:ω-3 ratio was 2:1 but not 4:1, even at the same absolute level of fish oil intake (98). In Japan and the Mediterranean, which are regions with reported low incidences of inflammatory conditions and the longest life expectancy, the dietary ratio of ω-6:ω-3 PUFA is about 3:1 (99). It has also been shown that Japanese that migrated to the US and acquired the local dietary habits, which resulted in an increased dietary ω-6:ω-3 PUFA ratio of 16:1, suffered from health problems that were similar to those in the local population (99). Thus, the nature of dietary PUFA intake has emerged as an environmental factor of critical importance in the progression of certain cancers. Enrichment of diets with ω-3 PUFA that enhance the production of beneficial ω-3-derived eicosanoids may prevent tumorigenesis.

Fatty Acid Binding Proteins and Peroxidation of PUFA

Phospholipids comprising esterified saturated fatty acids that regulate membrane fluidity are a significant component of cell membranes. Increasing the esterified PUFA content beyond physiological levels influences the optimal functioning of membrane components, including receptors, ion channels and lipid rafts. The transport of PUFA to cells is tightly controlled by specialized binding proteins, principally serum albumin and lipoproteins. Membrane incorporation of esterified PUFA is an enzymatic process that is also tightly regulated.

Lipid peroxidation is the degradation of cell membrane lipids mediated by reactive oxygen species (oxygen free radicals and inorganic and organic peroxides). This process proceeds by a free radical chain reaction mechanism and mostly affects PUFA, because they contain multiple double bonds. The consequence of uncontrolled lipid peroxidation at the cell membrane is usually rapid onset of apoptosis and cell death, which accounts for reports that ω-3 and ω-6 PUFA both induce apoptosis of cancer cells in culture. Since ω-6 PUFA are very similar to ω-3 PUFA in terms of their extent of unsaturation, their peroxidizability is also similar.

Most of the *in vitro* studies of PUFA effects in cell culture that have been undertaken to date have employed concentrations of PUFA that are much higher than could be achieved by dietary manipulations and that are likely to exceed the binding capacities of serum proteins. Thus, *in vitro* concentrations of PUFA (either ω-3 or ω-6) that exceed 50 μM have been found to induce cell toxicity and apoptosis (100-102). Such non-physiological increases in PUFA exert toxic effects through unregulated membrane lipid peroxidation (103). This nonspecific toxicity can obscure the more selective homeostatic effects of ω-3 and ω-6 PUFA seen at physiological PUFA concentrations. If the binding capacity of fatty acid-binding proteins is exceeded, non-specific toxicity predominates and gives rise to reports that ω-3 and ω-6 PUFA are similarly toxic to cells. Before general agreement about the role of the change in cell membrane fluidity and initiation of lipid peroxidation in the biological effects of ω-3 PUFA can be reached, *in vitro* cell culture systems should more faithfully represent *in vivo* conditions.

Angiogenesis: Role in Tumor Progression

A number of processes related to cancer progression are influenced by ω-3 PUFA and their eicosanoid derivatives. During embryonic development blood vessels are created *de novo* from endothelial-cell precursors (angioblasts) in a process called vasculogenesis. In adult organisms, the circulatory system may be differentiated for the formation of new blood vessels from pre-existing capillaries by the process of angiogenesis (104). Such growth of the vasculature is usually quite limited and apart from growing tumors, relatively few adult tissues require ongoing angiogenesis. These include female reproductive organs, organs that are actively growing, or regenerating tissue following injury (105). While vasculogenesis and angiogenesis have been considered to be distinct processes, they share some similarities. Recent evidence suggests that endothelial progenitors may contribute to vessel growth both in the embryo and in ischemic, malignant or inflamed tissues in the adult (106, 107). However, the percentage of endothelial-precursor-cell incorporation is generally low and depends on the nature of the tumor, supporting the concept that the majority of tumor neo-vascularization seems to occur via angiogenesis (106).

Following transformation, newly formed tumors require a constant supply of nutrients and oxygen for growth and maintaining an adequate supply of these factors rapidly becomes growth-limiting. In early disease (tumors that are smaller than 1-2 mm in diameter), tumor cells are oxygenated by simple diffusion in a phase defined as avascular. Tumors of this size can remain dormant for many months after reaching a steady state between proliferation and apoptosis (108).

The demand for oxygen and nutrients varies widely among tumors and is often dependent upon tumor type, their stage of growth, and their location. These requirements can change over the course of tumor progression (109). Tumor cells characteristically show a reduced oxygen dependency compared to normal cells, and are thought to sustain their metabolism under relatively anaerobic conditions by increasing glycolysis to maintain ATP production (110). However, despite their increased tolerance of hypoxia, tumors remain strictly dependent on an adequate oxygen supply for their exponential growth. During initial chronic oxygen deprivation, and as growth progresses and the lack of oxygen becomes acute, the induction of the expression of a battery of genes allows tumors to overcome hypoxia by increasing their blood supply through angiogenesis. This neovascularization signals the change from a relatively dormant state to a tumor capable of rapid growth and is characterized by a set of physiological events termed the "angiogenic switch" that is provoked by a change in the balance of pro- and anti-angiogenic molecules (105).

In addition to physiological stimuli, factors influencing the angiogenic switch include genetic mutation, metabolic and mechanical stress, inflammation, activation of oncogenes, and/or the inactivation of tumor-suppressor genes (111, 112). Factors affecting the onset of the angiogenic switch may be tumor and tissue-specific and may also change during tumor progression (113, 114). The switch can occur at different stages of tumor growth, again depending on the type of the tumor and the surrounding tissue. The process of angiogenesis begins with perivascular detachment and vessel dilation, followed by angiogenic sprouting, new vessel formation and maturation, and the recruitment of perivascular cells. New blood vessel formation will continue as long as the tumor grows in order to supply oxygen and nutrients to the hypoxic and necrotic areas of the tumor. The induction of angiogenesis is a crucial process that impacts not only malignant tumor growth, but also tumor invasiveness and metastasis (115).

ω-3 PUFA have been found to suppress both tumor and pathological retinal angiogenesis in laboratory animals (116-118) and some preliminary findings indicate that these PUFA could be used as supplementary anti-angiogenic agents in the treatment of benign vascular tumors (119). Suppression of pathological angiogenesis by ω-3 PUFA and/or their active metabolites could be an important adjunct to existing anti-cancer therapies. However, before the implementation of such ω-3 PUFA-based approaches, a better understanding of the mechanisms by which ω-3 PUFA affect tumor angiogenesis and metastasis is warranted.

Angiogenesis can be divided into three distinct phases (Figure 3): 1) Disassembly of existing blood vessel in the vicinity of the tissue affected by hypoxia into individual activated endothelial cells; 2) replication and migration of endothelial cells toward the hypoxic tumor and 3) Assembly and differentiation of endothelial cells into the new blood vessel and formation of connections between primary vessel and tumor. Each of these phases has been shown to be affected by ω-3 PUFA. Activation and proliferation of endothelial cells (mediated by vascular growth factor (VEGF)) was inhibited by EPA, which down-regulated the expression of the VEGF receptor Flk-1 on endothelial cells (120). In the second stage of angiogenesis, the migration of activated endothelial cells toward growing tumors was inhibited by ω-3 PUFA (121).

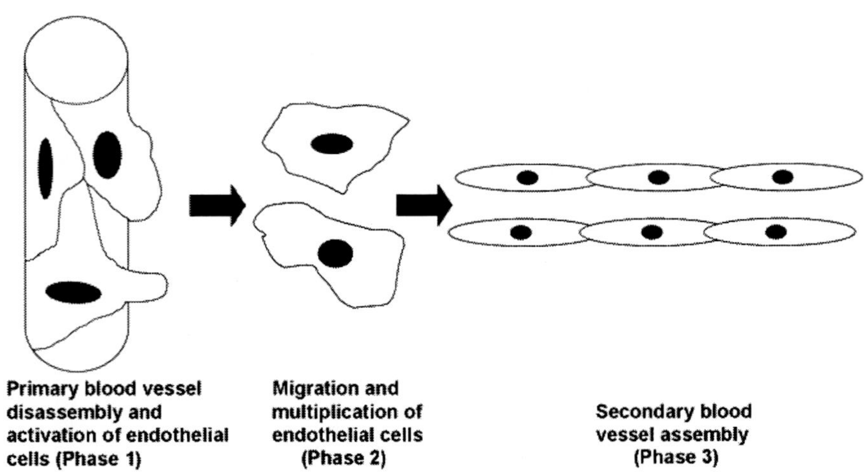

Figure 3. Major phases of angiogenesis.

The final stage of angiogenesis can be analyzed using the *in vitro* "tube formation assay." Here, activated endothelial cells are placed on the surface of an artificial basement membrane substitute where they differentiate and interact to form intricate networks of connected cells (or tubes). The ω-3 PUFA EPA and DHA are inhibitory in the tube formation assay using primary endothelial cells (122, 123). ω-3 PUFA offer additional anti-angiogenic effects, including down-regulation of a number of endothelium-specific pro-atherogenic and pro-inflammatory genes (124).

Regulation of Angiogenesis by ω-3 PUFA

Although numerous mechanisms have been proposed for the suppression of angiogenesis by ω-3 PUFA these usually involve direct effects on the physical properties of the endothelial cell membrane which increases apoptosis and/or modifies caveolae/lipid raft function (125-128). As already suggested, many of those studies have used unrealistically high concentrations of free PUFA. To the present the participation of ω-3 PUFA-derived eicosanoid metabolites has not been widely considered.

The absence of studies that have investigated biotransformation products of ω-3 PUFA as potential modulators of angiogenesis is surprising in view of the well-documented role of ω-6 eicosanoids in angiogenic regulation (129). Thus, in hypoxia angiogenesis is stimulated along with COX-2 expression and prostanoid synthesis (130, 131). The pro-angiogenic effects of COX appear to be mediated principally by prostaglandin E_2 (PGE_2), thromboxane A_2 (TXA_2), and prostaglandin I_2 (PGI_2) (139-142). Pro-angiogenic actions include induction of VEGF (132, 133), activation of endothelial migration and tube formation (134, 135), and increased expression of the anti-apoptotic protein Bcl-2 and the survival kinase Akt (136, 137).

Other eicosanoid classes also have the capacity to regulate angiogenesis. Thus, epoxides generated by the action of family 2 CYPs on PUFA have been implicated in hypoxia-induced endothelial cell migration and angiogenesis (138-142). Similarly, products of the action of LOX-5, LOX-12 or LOX-15, depending on cell type, are also potent activators of angiogenesis (143-145). There is evidence that at least some of the effects of ω-6 PUFA-derived eicosanoids are mediated via G-protein-activated receptors (146). However, signaling paradigms by which specific ω-3/ω-6 eicosanoid metabolites regulate angiogenesis and other cellular processes remain to be established. Interestingly, however, Wada *et al.* recently demonstrated that ω-3 EPA-derived eicosanoids were less potent than their ω-6 AA-derived analogues as ligands for prostanoid receptors (147).

Recently, new classes of potential angiogenic modulators have been identified, including the lipoxins and other mediators formed by the dual action of biotransformation enzymes, such as aspirin-treated COX (Figure 2) and LOX, or CYP and LOX enzymes (148-149). Although the capacity of COX to generate prostanoids is decreased when the enzyme has been acetylated by aspirin, 15-epi-lipoxins are still produced in endothelial cells (150). These epi-lipoxins may be converted to lipoxins by LOX enzymes.

There are a number of outstanding issues whose significance is presently unclear. In contrast to other eicosanoids, lipoxins derived from both ω-6 and ω-3 PUFA (termed resolvins) (116) have been reported to possess similar anti-angiogenic properties in *in vitro* systems (151). Accordingly, it is difficult to account for the well-recognized differential effects of ω-3/ω-6 PUFA and their eicosanoid products on angiogenesis in terms of resolvins. It is pertinent that PUFA precursors must undergo complex substrate transfer between endothelial cells (which induce COX-2 when subjected to hypoxia) and activated leukocytes (that express LOX enzymes); this is termed transcellular eicosanoid biosynthesis (150). At this point it is quite unclear how the interplay of these pathways (or those involving microsomal CYPs and leukocyte LOX) influences resolvin formation *in situ*.

The hypothesis that eicosanoids may be major mediators of the regulation of angiogenesis by PUFA relies on the different biological activities of eicosanoids derived from ω-6 and ω-3 PUFA substrates. As indicated previously, many reports have suggested that ω-6 and ω-3 PUFA compete for enzymes involved in PUFA biotransformation and that metabolism of ω-6 PUFA may be suppressed by ω-3 PUFA. Thus, increased intake of EPA by patients suppresses the synthesis of AA-derived eicosanoids and increases the formation of the analogous EPA-derived mediators (152). This is in accord with the observed importance of the ω-6/ω-3 PUFA dietary ratio that was outlined previously.

Variability in physiological effects between ω-3 and ω-6 derived lipid mediators could be potentially explained by findings reported by Malkowski *et al* (153) and Thuresson *et al* (154) where the low rate of oxygenation of EPA compared to AA by COX-1 was attributed to the "strained" conformation of EPA at the COX-1 active site (Figure 4). Such conformational differences between ω-6 and ω-3 PUFA (further influencing structure of all lipid mediators derived from those PUFA) might be the cause for their less efficient interaction with enzymes and receptors.

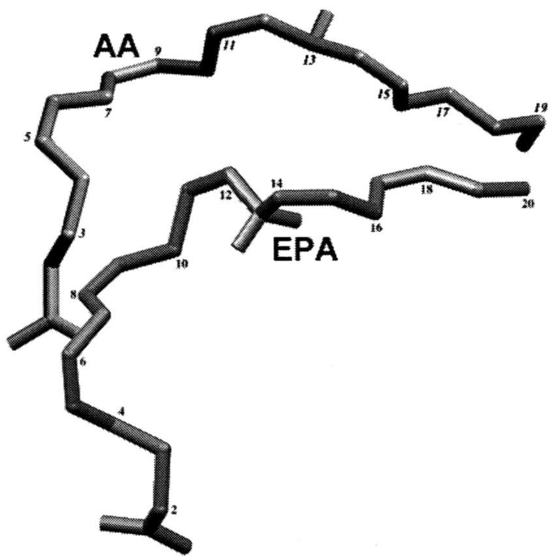

Figure 4. Conformational differences of AA and EPA when bound to the COX-1 active site (adapted from references 153 and 154).

Conclusion

Current dietary trends, particularly in the developed world, include increased consumption of ω-6 PUFA, and this has led to an imbalance of the ω-3/ω-6 PUFA ratio in favor of ω-6 PUFA. The consequences of this imbalance are now becoming clearer and include exaggerated inflammatory responses in tissues, as well as enhancement of tumor development and angiogenesis. There is increasing evidence that many of these effects are due to eicosanoid products rather than to the parent PUFA themselves.

There appear to be health benefits that could arise from readjustments to dietary ω-3 PUFA intake. These would include direct effects of the ω-3 PUFA themselves, indirect benefits from the formation of ω-3-derived eicosanoid mediators and to the development of new anti-angiogenic and anti-cancer therapies based on these lipid molecules.

References

[1] Lichtenstein AH, Kennedy E, Barrier P, Danford D, Ernst ND, Grundy SM, Leveille GA, Van Horn L, Williams CL, Booth SL. Dietary fat consumption and health. *Nutr Rev*. 1998; 56: S3-S19

[2] Colhoun HM. Eicosapentaenoic acid for prevention of major coronary events. *Lancet*. 2007; 370: 215

[3] Jacobson TA. Beyond lipids: the role of omega-3 fatty acids from fish oil in the prevention of coronary heart disease. *Curr Atheroscler Rep*. 2007; 9: 145-53

[4] Goldberg RJ, Katz J. A meta-analysis of the analgesic effects of omega-3 polyunsaturated fatty acid supplementation for inflammatory joint pain. *Pain*. 2007; 129: 210-223

[5] Lau CS. Collateral benefits of fish oil therapy for rheumatoid arthritis. *J Rheumatol*. 2006; 33: 1931-1933

[6] Coleman H, Chew E. Nutritional supplementation in age-related macular degeneration. *Curr Opin Ophthalmol*. 2007; 18: 220-223

[7] Moriguchi T, Greiner RS, Salem N Jr. Behavioral deficits associated with dietary induction of decreased brain docosahexaenoic acid concentration. *J Neurochem*. 2000; 75: 2563-2573

[8] Salem N Jr, Moriguchi T, Greiner RS, McBride K, Ahmad A, Catalan JN, Slotnick B. Alterations in brain function after loss of docosahexaenoate due to dietary restriction of n-3 fatty acids. *J Mol Neurosci*. 2001;16: 299-307

[9] Simopoulos AP. The importance of the ratio of omega-6/omega-3 essential fatty acids. Biomed Pharmacother. 2002; 56: 365-379

[10] Connor WE. Importance of n-3 fatty acids in health and disease. *Am J Clin Nutr*. 2000; 71: 171S-175S

[11] Simopoulos AP. Omega-3 fatty acids in health and disease and in growth and development. *Am J Clin Nutr*. 1991; 54: 438-463

[12] Divisi D, Di Tommaso S, Salvemini S, Garramone M, Crisci R. Diet and cancer. *Acta Biomed*. 2006; 77: 118-123

[13] Simopoulos AP. Evolutionary aspects of diet, the omega-6/omega-3 ratio and genetic variation: nutritional implications for chronic diseases. *Biomed Pharmacother*. 2006; 60: 502-507

[14] Sugano M, Hirahara F. Polyunsaturated fatty acids in the food chain in Japan. *Am J Clin Nutr*. 2000; 71: 189S-196S2

[15] Wynder EL, Fujita Y, Harris RE, Hirayama T, Hirayama T. Comparative epidemiology of cancer between the United States and Japan: a second look. *Cancer*. 1991; 67: 746-763

[16] Ziegler RG, Hoover RL, Pike MC, Hildesheim A, Nomura AMY, West DW, Wu-Williams AH, Kolonel LN, Horn-Ross PI, Rosenthal JF, Hyer MB. Migration patterns and breast cancer risk in Asian-American women. *J Natl Cancer Inst*. 1993; 85: 1819-1827

[17] Simopoulos AP, Leaf A, Salem N. Jr. Essentiality of and recommended dietary intakes for omega-6 and omega-3 fatty acids. *Ann. Nutr. Metab*. 1999; 43: 127–130

[18] Dewailly E, Blanchet C, Lemieux S, Sauve L, Gingras S, Ayotte P, Holub BJ. n-3 fatty acids and cardiovascular disease risk factors among the Inuit of Nunavik. *Am. J. Clin. Nutr*. 2001; 74: 464-473

[19] Simopoulos AP. Essential fatty acids in health and chronic disease. *Am. J. Clin. Nutr.* 1999; 70: 560S–569S

[20] Weber PC. 1989. Are we what we eat? Fatty acids in nutrition and in cell membranes: cell functions and disorders induced by dietary conditions. *Proceedings of the International Conference on Fish Lipids and Their Influence on Human Health*, Svanoy, Norway

[21] Theodoratou E, McNeill G, Cetnarskyj R, Farrington SM, Tenesa A, Barnetson R, Porteous M, Dunlop M, Campbell H. Dietary fatty acids and colorectal cancer: a case-control study. *Am J Epidemiol*. 2007; 166: 181-195

[22] Byers T. Nutrition and cancer among American Indians and Alaska Natives. *Cancer* 1996; 78: 1612–1616

[23] Hall MN, Campos H, Li H, Sesso HD, Stampfer MJ, Willett WC, Ma J. Blood levels of long-chain polyunsaturated fatty acids, aspirin, and the risk of colorectal cancer. *Cancer Epidemiol Biomarkers Prev*. 2007; 16: 314-321

[24] Kuriki K, Wakai K, Hirose K, Matsuo K, Ito H, Suzuki T, Saito T, Kanemitsu Y, Hirai T, Kato T, Tatematsu M, Tajima K. Risk of colorectal cancer is linked to erythrocyte compositions of fatty acids as biomarkers for dietary intakes of fish, fat, and fatty acids. *Cancer Epidemiol Biomarkers Prev*. 2006; 15: 1791-1798

[25] Bingham SA. Epidemiology of colorectal cancer. In: Strain J, Sadler M, Caballero B, editors. *Encyclopedia of human nutrition*, vol. 1. New York: Academic Press; 1998; pp 230-235

[26] Caygill C, Hill M. Fish n-3 fatty acids and human colorectal and breast cancer mortality. *Eur J Cancer Prev*. 1995; 4: 329-332

[27] Norrish AE, Skeaff CM, Arribas GL, Sharpe SJ, Jackson RT. Prostate cancer risk and consumption of fish oils: a dietary biomarker-based case-control study. *Br J Cancer*. 1999; 81: 1238-1242

[28] Terry P, Lichtenstein P, Feychting M, Ahlbom A, Wolk A. Fatty fish consumption and risk of prostate cancer. *Lancet*. 2001; 357: 1764-1766

[29] Yang YJ, Lee SH, Hong SJ, Chung BC. Comparison of fatty acid profiles in the serum of patients with prostate cancer and benign prostatic hyperplasia. *Clin Biochem*. 1999; 32:405-409

[30] Freeman VL, Meydani M, Yong S, Pyle J, Flanigan RC, Waters WB, Wojcik EM. Prostatic levels of fatty acids and the histopathology of localized prostate cancer. *J Urol*. 2000; 164: 2168-2172

[31] Augustsson K, Michaud DS, Rimm EB, Leitzmann MF, Stampfer MJ, Willett WC, Giovannucci E. A prospective study of intake of fish and marine fatty acids and prostate cancer. *Cancer Epidemiol Biomarkers Prev*. 2003; 12: 64-67

[32] Ritch CR, Wan RL, Stephens LB, Taxy JB, Huo D, Gong EM, Zagaja GP, Brendler CB. Dietary fatty acids correlate with prostate cancer biopsy grade and volume in Jamaican men. *J Urol*. 2007; 177: 97-101

[33] Chang ET, Balter KM, Torrang A, Smedby KE, Melbye M, Sundstrom C, Glimelius B, Adami HO. Nutrient intake and risk of non-Hodgkin's lymphoma. *Am J Epidemiol.* 2006; 164: 1222-1232

[34] Chapkin RS, McMurray DN, Lupton JR. Colon cancer, fatty acids and anti-inflammatory compounds. Curr Opin Gastroenterol. 2007; 23: 48-54
[35] Hosack Fowler K, Chapkin RS, McMurray DN. Effects of purified dietary n-3 ethyl esters on murine T lymphocyte function. *J Immunol.* 1993; 151: 5186-5197
[36] Caygill CP, Charlett A, Hill MJ. Fat, fish oil and cancer. *Br J Cancer.* 1996; 74: 159-164
[37] Sasaki S, Horacsek M, Kesteloot H. An ecological study of the relationship between dietary fat intake and breast cancer mortality. *Prev Med.* 1993; 22: 187-202
[38] Kuriki K, Hirose K, Wakai K, Matsuo K, Ito H, Suzuki T, Hiraki A, Saito T, Iwata H, Tatematsu M, Tajima K. Breast cancer risk and erythrocyte compositions of n-3 highly unsaturated fatty acids in Japanese. *Int J Cancer.* 2007; 121: 377-385
[39] Holmes MD, Hunter DJ, Colditz GA, Stampfer MJ, Hankinson SE, Speizer FE, Rosner B, Willett WC. Association of dietary intake of fat and fatty acids with risk of breast cancer. *JAMA.* 1999; 281: 914-920
[40] Vatten LJ, Solvoll K and Loken EB. Frequency of meat and fish intake and risk of breast cancer in a prospective study of 14,500 Norwegian women. *Int J Cancer.* 1990; 46: 12-15
[41] Prentice RL, Caan B, Chlebowski RT, Patterson R, Kuller LH, Ockene JK, Margolis KL, Limacher MC, Manson JE, Parker LM, Paskett E, Phillips L Robbins J, Rossouw JE, Sarto GE, Shikany JM, Stefanick ML, Thomson CA, Van Horn L, Vitolins MZ, Wactawski-Wende J, Wallace RB, Wassertheil-Smoller S, Whitlock E, Yano K, Adams-Campbell L, Anderson GL, Assaf AR, Beresford SA, Black HR, Brunner RL, Brzyski RG, Ford L, Gass M, Hays J, Heber D, Heiss G, Hendrix SL, Hsia J, Hubbell FA, Jackson RD, Johnson KC, Kotchen JM, LaCroix AZ, Lane DS, Langer RD, Lasser NL, Henderson MM. Low-fat dietary pattern and risk of invasive breast cancer: the Women's Health Initiative Randomized Controlled Dietary Modification Trial. *JAMA.* 2006; 295: 629-42
[42] MacLean CH, Newberry SJ, Mojica WA, Khanna P, Issa AM, Suttorp MJ, Lim YW, Traina SB, Hilton L, Garland R, Morton SC. Effects of omega-3 fatty acids on cancer risk: a systematic review. *JAMA.* 2006; 295: 403-415
[43] Hooper L, Thompson RL, Harrison RA, Summerbell CD, Ness AR, Moore HJ, Worthington HV, Durrington PN, Higgins JPT, Capps NE, Riemersma RA, Ebrahim SBJ, Davey-Smith G. Risks and benefits of omega 3 fats for mortality, cardiovascular disease, and cancer: systematic review. *BMJ.* 2006; 332: 752-60
[44] Engeset D, Alsaker E, Lund E, Welch A, Khaw KT, Clavel-Chapelon F, Thiebaut A, Chajes V, Key TJ, Allen NE Amiano P, Dorronsoro M, Tjønneland A, Stripp C, Peeters PH, van Gils CH, Chirlaque MD, Nagel G, Linseisen J, Ocké MC, Bueno-de-Mesquita HB, Sacerdote C, Tumino R, Ardanaz E, Sánchez MJ, Panico S, Palli D, Trichopoulou A, Kalapothaki V, Benetou V, Quirós JR, Agudo A, Overvad K, Bjerregaard L, Wirfält E, Schulz M, Boeing H, Slimani N, Riboli E. Fish consumption and breast cancer risk. The European Prospective Investigation into Cancer and Nutrition (EPIC). *Int J Cancer.* 2006; 119: 175-82
[45] Hirose K, Tajima K, Hamajima N, Inoue M, Takezaki T, Kuroishi T, Yoshida M, Tokudome S. A large-scale, hospital-based case-control study of risk factors of breast cancer according to menopausal status. *Jpn J Cancer Res.* 1995; 86: 146-154

[46] Hirose K, Takezaki T, Hamajima N, Miura S, Tajima K. Dietary factors protective against breast cancer in Japanese premenopausal and postmenopausal women. *Int J Cancer*. 2003; 107: 276-282

[47] Wakai K, Tamakoshi K, Date C, Fukui M, Suzuki S, Lin Y, Niwa Y, Nishio K, Yatsuya H, Kondo T, Tokudome S, Yamamoto A, Toyoshima H, Tamakoshi A; JACC Study Group. Dietary intakes of fat and fatty acids and risk of breast cancer: a prospective study in Japan. *Cancer Sci*. 2005; 96: 590-599

[48] Zhang J, Temme EH, Kesteloot H. Fish consumption is inversely associated with male lung cancer mortality in countries with high levels of cigarette smoking or animal fat consumption. *Int J Epidemiol*. 2000; 29: 615-621

[49] Kuriki K, Tokudome S, Tajima K. Association between type II diabetes and colon cancer among Japanese with reference to changes in food intake. *Asian Pac J Cancer Prev*. 2004; 5: 28-35

[50] Hardman WE. Omega-3 fatty acids to augment cancer therapy. *J Nutr*. 2002; 132 (11 Suppl): 3508S-3512S

[51] Pardini RS. Nutritional intervention with omega-3 fatty acids enhances tumor response to anti-neoplastic agents. *Chem Biol Interact*. 2006; 162: 89-105

[52] Ding WQ, Liu B, Vaught JL, Palmiter RD, Lind SE. Clioquinol and docosahexaenoic acid act synergistically to kill tumor cells. *Mol Cancer Ther*. 2006; 5: 1864-1872

[53] Colas S, Maheo K, Denis F, Goupille C, Hoinard C, Champeroux P, Tranquart F, Bougnoux P. Sensitization by dietary docosahexaenoic acid of rat mammary carcinoma to anthracycline: a role for tumor vascularization. *Clin Cancer Res*. 2006; 12: 5879-5886

[54] Bougnoux P, Germain E, Chajes V, Hubert B, Lhuillery C, Le Flock O, Body G and Calais G. Cytotoxic drugs efficacy correlates with adipose tissue docosahexaenoic acid level in locally advanced breast carcinoma. *Br J Cancer*. 1999; 79: 1765-1769

[55] Rose DP and Connolly J. Omega-3 fatty acids as cancer chemopreventive agents. *Pharmacol Ther*. 1999; 83: 217–244

[56] Rose DP, Rayburn J, Hatala MA and Connolly JM. Effects of dietary fish oil on fatty acids and eicosanoids in metastasizing human breast cancer cells. *Nutr Cancer*. 1994; 22: 131-141

[57] Senzaki H, Iwamoto S, Ogura E, Kiyozuka Y, Arita S, Kurebayashi J, Takada H, Hioki K and Tsubura A. Dietary effects of fatty acids on growth and metastasis of KPL-1 human breast cancer cells in vivo and in vitro. *Anticancer Res*. 1998; 18: 1621-1627

[58] Hubbard N, Lim D and Erickson K. Alteration of murine mammary tumorigenesis by dietary enrichment with n-3 fatty acids in fish oil. Cancer Lett. 1998; 124: 1-7

[59] Xia S, Lu Y, Wang J, He C, Hong S, Serhan CN, Kang JX. Melanoma growth is reduced in fat-1 transgenic mice: impact of omega-6/omega-3 essential fatty acids. *Proc Natl Acad Sci USA*. 2006; 103: 12499-12504

[60] Kang JX, Wang J, Wu L, Kang ZB. Transgenic mice: fat-1 mice convert n-6 to n-3 fatty acids. *Nature*. 2004; 427: 504

[61] Larsson SC, Kumlin M, Ingelman-Sundberg M., Wolk A. Dietary long-chain n-3 fatty acids for the prevention of cancer: a review of potential mechanisms. *Am J Clin Nutr*. 2004; 79: 935-945

[62] Hardman WE. (n-3) fatty acids and cancer therapy. J Nutr. 2004; 134: 3427S-3430S

[63] Duesberg P, Rasnick D, Li R, Winters L, Rausch C, Hehlmann R. How aneuploidy may cause cancer and genetic instability. *Anticancer Res*. 1999; 19; 4887-4906

[64] Hahn WC, Weinberg RA. Rules for making human tumor cells. *N Engl J Med*. 2002; 347: 1593-1603

[65] Pelengaris S, Khan M, Evan G. c-MYC: more than just a matter of life and death. *Nat Rev Cancer*. 2002; 2: 764-776

[66] Zand H, Rhimipour A, Bakhshayesh M, Shafiee M, Nour Mohammadi I, Salimi S. Involvement of PPAR-gamma and p53 in DHA-induced apoptosis in Reh cells. *Mol Cell Biochem*. 2007; 304: 71-77

[67] Trombetta A, Maggiora M, Martinasso G, Cotogni P, Canuto RA, Muzio G. Arachidonic and docosahexaenoic acids reduce the growth of A549 human lung-tumor cells increasing lipid peroxidation and PPARs. Chem Biol Interact. 2007; 165: 239-25

[68] Leonardi F, Attorri L, Benedetto RD, Biase AD, Sanchez M, Tregno FP, Nardini M, Salvati S. Docosahexaenoic acid supplementation induces dose and time dependent oxidative changes in C6 glioma cells. *Free Radic Res*. 2007; 41: 748-756

[69] Li M, Kong ZM, Liu ZL. Antioxidant enzyme activities and lipid peroxidation induced by eicosapentaenoic acid (EPA) in PC12 cells. *Cell Biol Toxicol*. 2006; 22: 331-337

[70] Wu M, Harvey KA, Ruzmetov N, Welch ZR, Sech L, Jackson K, Stillwell W, Zaloga GP, Siddiqui RA. Omega-3 polyunsaturated fatty acids attenuate breast cancer growth through activation of a neutral sphingomyelinase-mediated pathway. *Int J Cancer*. 2005; 117: 340-348

[71] Jourdan ML, Maheo K, Barascu A, Goupille C, De Latour MP, Bougnoux P, Rio PG. Increased BRCA1 protein in mammary tumours of rats fed marine omega-3 fatty acids. *Oncol Rep*. 2007; 17: 713-719

[72] Bernard-Gallon DJ, Vissac-Sabatier C, Antoine-Vincent D, Rio PG, Maurizis JC, Fustier P, Bignon YJ. Differential effects of n-3 and n-6 polyunsaturated fatty acids on BRCA1 and BRCA2 gene expression in breast cell lines. *Br J Nutr*. 2002; 87: 281-289

[73] Siddiqui RA, Harvey KA, Zaloga GP, Stillwell W. Modulation of lipid rafts by Omega-3 fatty acids in inflammation and cancer: implications for use of lipids during nutrition support. *Nutr Clin Pract*. 2007; 22: 74-88

[74] Schley PD, Brindley DN, Field CJ. (n-3) PUFA alter raft lipid composition and decrease epidermal growth factor receptor levels in lipid rafts of human breast cancer cells. *J Nutr*. 2007; 137: 548-553

[75] Berquin IM, Min Y, Wu R, Wu J, Perry D, Cline JM, Thomas MJ, Thornburg T, Kulik G, Smith A, Edwards IJ, D' Agostino R, Zhang H, Wu H, Kang JX, Chen YQ. Modulation of prostate cancer genetic risk by omega-3 and omega-6 fatty acids. *J Clin Invest*. 2007; 117: 1866-1875

[76] Murray NR, Weems C, Chen L, Leon J, Yu W, Davidson LA, Jamieson L, Chapkin RS, Thompson EA, Fields AP. Protein kinase C beta II and TGF beta RII in omega-3 fatty acid-mediated inhibition of colon carcinogenesis. *J Cell Biol*. 2002; 157: 915-920

[77] Moore NG, Wang-Johanning F, Chang PL, Johanning GL. Omega-3 fatty acids decrease protein kinase expression in human breast cancer cells. Breast *Cancer Res Treat*. 2001; 67:279-283

[78] Lee M, Bae MA. Docosahexaenoic acid induces apoptosis in CYP2E1-containing HepG2 cells by activating the c-Jun N-terminal protein kinase related mitochondrial damage. *J Nutr Biochem*. 2007; 18: 348-354

[79] Chiu LC, Wong EY, Ooi VE. Docosahexaenoic acid from a cultured microalga inhibits cell growth and induces apoptosis by upregulating Bax/Bcl-2 ratio in human breast carcinoma MCF-7 cells. *Ann N Y Acad Sci*. 2004; 1030: 361-368

[80] Menendez JA, Mehmi I, Atlas E, Colomer R, Lupu R. Novel signaling molecules implicated in tumor-associated fatty acid synthase-dependent breast cancer cell proliferation and survival: Role of exogenous dietary fatty acids, p53-p21WAF1/CIP1, ERK1/2 MAPK, p27KIP1, BRCA1, and NF-kappaB. *Int J Oncol*. 2004; 24: 591-608

[81] Liu G, Bibus DM, Bode AM, Ma W-Y, Holman RT and Dong Z. Omega 3 but not omega 6 fatty acids inhibit AP-1 activity and cell transformation in JB6 cells. *Proc Natl Acad Sci USA*. 2001; 98: 7510-7515

[82] Collett ED, Davidson LA, Fan YY, Lupton JR, Chapkin RS. n-6 and n-3 polyunsaturated fatty acids differentially modulate oncogenic Ras activation in colonocytes. *Am J Physiol Cell Physiol*. 2001; 280: C1066-C1075

[83] Singh J, Hamid R, Reddy BS. Dietary fat and colon cancer: modulating effect of types and amount of dietary fat on ras-p21 function during promotion and progression stages of colon cancer. *Cancer Res*. 1997; 57: 253-258

[84] Khan NA, Nishimura K, Aires V, Yamashita T, Oaxaca-Castillo D, Kashiwagi K, Igarashi K. Docosahexaenoic acid inhibits cancer cell growth via p27Kip1, CDK2, ERK1/ERK2, and retinoblastoma phosphorylation. *J Lipid Res*. 2006; 47: 2306-2313

[85] Liu YE, Pu W, Wang J, Kang JX, Shi YE. Activation of Stat5 and induction of a pregnancy-like mammary gland differentiation by eicosapentaenoic and docosapentaenoic omega-3 fatty acids. *FEBS J*. 2007; 274: 3351-62

[86] Horia E, Watkins BA. Complementary actions of docosahexaenoic acid and genistein on COX-2, PGE2 and invasiveness in MDA-MB-231 breast cancer cells. *Carcinogenesis*. 2007; 28: 809-815

[87] Brown MD, Hart CA, Gazi E, Bagley S, Clarke NW. Promotion of prostatic metastatic migration towards human bone marrow stoma by Omega 6 and its inhibition by Omega 3 PUFAs. *Br J Cancer*. 2006; 94: 842-853

[88] Kobayashi N, Barnard RJ, Henning SM, Elashoff D, Reddy ST, Cohen P, Leung P, Hong-Gonzalez J, Freedland SJ, Said J, Gui D, Seeram NP, Popoviciu LM, Bagga D, Heber D, Glaspy JA, Aronson WJ. Effect of altering dietary omega-6/omega-3 fatty acid ratios on prostate cancer membrane composition, cyclooxygenase-2, and prostaglandin E2. *Clin Cancer Res*. 2006; 12: 4662-4670

[89] Funk CD (2001) Prostaglandins and leukotrienes: advances in eicosanoid biology. Science 294: 1871-1875

[90] Ziboh VA, Cho Y, Mani I and Xi S. Biological significance of essential fatty acids/prostanoids/lipoxygenase-derived monohydroxy fatty acids in the skin. *Arch Pharm Res*. 2002; 25: 747-758

[91] Capdevila JH, Harris RC and Falck JR. Microsomal cytochrome P450 and eicosanoid metabolism. *Cell Mol Life Sci*. 2002; 59: 780-789

[92] Oliw EH Oxygenation of polyunsaturated fatty acids by cytochrome P450 monooxygenases. *Prog Lipid Res*. 1994; 33: 329-354

[93] Giardiello FM, Hamilton SR, Krush AJ, Piantadosi S, Hylind LM, Celano P, Booker SV, Robinson CR, Offerhaus GJ. Treatment of colonic and rectal adenomas with sulindac in familial adenomatous polyposis. *N Engl J Med*. 1993; 328: 1313-1316

[94] Jacobs EJ, Rodriguez C, Mondul AM, Connell CJ, Henley SJ, Calle EE, Thun MJ. A large cohort study of aspirin and other nonsteroidal anti-inflammatory drugs and prostate cancer incidence. *J. Natl. Cancer Inst.* 2005; 97: 975-980

[95] Rose DP and Connolly JM. Effects of fatty acids and inhibitors of eicosanoid synthesis on the growth of a human breast cancer cell line in culture. *Cancer Res.* 1990; 50: 7139-7144

[96] Rose DP and Connolly JM. Regulation of tumor angiogenesis by dietary fatty acids and eicosanoids. *Nutr Cancer.* 2000; 37: 119-127

[97] Maillard V, Bougnoux P, Ferrari P, Jourdan ML, Pinault M, Lavillonniere F, Body G, Le Floch O, Chajes V. N-3 and n-6 fatty acids in breast adipose tissue and relative risk of breast cancer in a case-control study in Tours, France. *Int J Cancer* 2002; 98: 78-83

[98] Bartram HP, Gostner A, Reddy BS, Rao CV, Scheppach W, Dusel G, Richter A, Richter F, Kasper H. Missing anti-proliferative effect of fish oil on rectal epithelium in healthy volunteers consuming a high-fat diet: potential role of the n-3:n-6 fatty acid ratio. *Eur J Cancer Prev.* 1995; 4: 231-237

[99] Sugano M, Hirahara F. Polyunsaturated fatty acids in the food chain in Japan. *Am J Clin Nutr.* 2000; 71: 189S-196S

[100] Chiu LC, Wong EY, Ooi VE. Docosahexaenoic acid modulates different genes in cell cycle and apoptosis to control growth of human leukemia HL-60 cells. *Int J Oncology.* 2004; 25:737-474

[101] Colquhoun A, Schumacher RI. Gamma-linolenic acid and eicosapentaenoic acid induce modifications in mitochondrial metabolism, reactive oxygen species generation, lipid peroxidation and apoptosis in Walker 256 rat carcinosarcoma cells. *Biochim Biophys Acta.* 2001; 1533: 207-219

[102] Rizzo MT, Regazzi E, Garau D, Akard L, Dugan M, Boswell HS, Rizzoli V, Carlo-Stella C. Induction of apoptosis by arachidonic acid in chronic myeloid leukemia cells. *Cancer Res.* 1999; 59: 5047-5053

[103] Guichardant M, Bacot S, Moliere P, Lagarde M. Hydroxy-alkenals from the peroxidation of n-3 and n-6 fatty acids and urinary metabolites. *Prostaglandins Leukot Essent Fatty Acids.* 2006; 75: 179-182

[104] Carmeliet P. Mechanisms of angiogenesis and arteriogenesis. *Nat Med.* 2000; 6: 389-395

[105] Hanahan D, Folkman J. Patterns and emerging mechanisms of the angiogenic switch during tumorigenesis. *Cell.* 1996; 86: 353-364

[106] Lyden D, Hattori K, Dias S, Costa C, Blaikie P, Butros L, Chadburn A, Heissig B, Marks W, Witte L, Wu Y, Hicklin D, Zhu Z, Hackett NR, Crystal RG, Moore MA, Hajjar KA, Manova K, Benezra R, Rafii S. Impaired recruitment of bone-marrow-derived endothelial and hematopoietic precursor cells blocks tumor angiogenesis and growth. *Nat Med.* 2001; 7: 1194-1201

[107] Rafii S, Heissig B, Hattori K. Efficient mobilization and recruitment of marrow-derived endothelial and hematopoietic stem cells by adenoviral vectors expressing angiogenic factors. *Gene Ther.* 2002; 9: 631-641

[108] Black WC, Welch HG. Advances in diagnostic imaging and overestimations of disease prevalence and the benefits of therapy. *N Engl J Med.* 1993; 328: 1237-1243

[109] Hlatky L, Hahnfeldt P, Folkman J. Clinical application of antiangiogenic therapy: microvessel density, what it does and doesn't tell us. *J Natl Cancer Inst.* 2002; 94: 883-893
[110] Papetti M, Herman IM. Mechanisms of normal and tumor-derived angiogenesis. *Am J Physiol Cell Physiol.* 2002; 282: C947-C970
[111] Kerbel R, Folkman J. Clinical translation of angiogenesis inhibitors. *Nat Rev Cancer.* 2002; 2: 727-739
[112] Carmeliet P, Jain RK. Angiogenesis in cancer and other diseases. *Nature.* 2000; 407: 249-257
[113] Folkman J. Looking for a good endothelial address. *Cancer Cell.* 2002; 1: 113-115
[114] Dor Y, Porat R, Keshet E. Vascular endothelial growth factor and vascular adjustments to perturbations in oxygen homeostasis. *Am J Physiol Cell Physiol.* 2001; 280: C1367-C1374
[115] Hanahan D, Weinberg RA. The hallmarks of cancer. *Cell,* 2000; 100: 57-70
[116] Connor KM, Sangiovanni JP, Lofqvist C, Aderman CM, Chen J, Higuchi A, Hong S, Pravda EA, Majchrzak S, Carper D, Hellstrom A, Kang JX, Chew EY, Salem N Jr, Serhan CN, Smith LE. Increased dietary intake of omega-3-polyunsaturated fatty acids reduces pathological retinal angiogenesis. *Nat Med.* 2007; 13: 868-873
[117] Hardman WE, Sun L, Short N, Cameron IL. Dietary omega-3 fatty acids and ionizing irradiation on human breast cancer xenograft growth and angiogenesis. *Cancer Cell Int.* 2005; 5: 12
[118] Wen B, Deutsch E, Opolon P, Auperin A, Frascogna V, Connault E, Bourhis J. n-3 polyunsaturated fatty acids decrease mucosal/epidermal reactions and enhance antitumour effect of ionising radiation with inhibition of tumour angiogenesis. *Br J Cancer.* 2003; 89: 1102-1107
[119] Sterescu AE, Rousseau-Harsany E, Farrell C, Powell J, David M, Dubois J. The potential efficacy of omega-3 fatty acids as anti-angiogenic agents in benign vascular tumors of infancy. *Med Hypotheses.* 2006; 66: 1121-1124
[120] Yang SP, Morita I, Murota SI. Eicosapentaenoic acid attenuates vascular endothelial growth factor-induced proliferation via inhibiting Flk-1 receptor expression in bovine carotid artery endothelial cells. *J Cell Physiol.* 1998; 176: 342-349
[121] Tsuzuki T, Shibata A, Kawakami Y, Nakagawa K, Miyazawa T. Conjugated eicosapentaenoic acid inhibits vascular endothelial growth factor-induced angiogenesis by suppressing the migration of human umbilical vein endothelial cells. *J Nutr.* 2007; 137: 641-646
[122] Kanayasu T, Morita I, Nakao-Hayashi J, Asuwa N, Fujisawa C, Ishii T, Ito H, Murota S. Eicosapentaenoic acid inhibits tube formation of vascular endothelial cells in vitro. *Lipids.* 1991; 26: 271-276
[123] Tsuji M, Murota SI, Morita I. Docosapentaenoic acid (22:5, n-3) suppressed tube-forming activity in endothelial cells induced by vascular endothelial growth factor. *Prostaglandins Leukot Essent Fatty Acids.* 2003; 68: 337-342
[124] De Caterina R, Massaro M. Omega-3 fatty acids and the regulation of expression of endothelial pro-atherogenic and pro-inflammatory genes. *J Membr Biol.* 2005; 206: 103-116

[125] Jensen LD, Hansen AJ, Lundbaek JA. Regulation of endothelial cell migration by amphiphiles - are changes in cell membrane physical properties involved? *Angiogenesis*. 2007; 10: 13-22

[126] Artwohl M, Roden M, Waldhausl W, Freudenthaler A, Baumgartner-Parzer SM. Free fatty acids trigger apoptosis and inhibit cell cycle progression in human vascular endothelial cells. *FASEB J.* 2004; 18: 146-148

[127] Kim HJ, Vosseler CA, Weber PC, Erl W. Docosahexaenoic acid induces apoptosis in proliferating human endothelial cells. *J Cell Physiol.* 2005; 204: 881-888

[128] Chen W, Jump DB, Esselman WJ, Busik JV. Inhibition of cytokine signaling in human retinal endothelial cells through modification of caveolae/lipid rafts by docosahexaenoic acid. *Invest Ophthalmol Vis Sci.* 2007; 48: 18-26

[129] Mezentsev A, Seta F, Dunn MW, Ono N, Falck JR, Laniado-Schwartzman M. Eicosanoid regulation of vascular endothelial growth factor expression and angiogenesis in microvessel endothelial cells. *J Biol Chem.* 2002; 21: 18670-18676

[130] Saed GM, Munkarah AR, Abu-Soud HM, Diamond MP. Hypoxia upregulates cyclooxygenase-2 and prostaglandin E(2) levels in human peritoneal fibroblasts. *Fertil Steril.* 2005; 83: Suppl 1:1216-9121

[131] Pichiule P, Chavez JC, LaManna JC. Hypoxic regulation of angiopoietin-2 expression in endothelial cells. *J Biol Chem.* 2004; 279: 12171-12180

[132] Höpper MM, Voelkel NF, Bates TO, Allard JD, Horan M, Shepherd D, Tuder RM. Prostaglandins induce vascular endothelial growth factor in a human monocytic cell line and rat lungs via cAMP. *Am J Respir Cell Mol Biol.* 1997; 17: 748-756

[133] Pai R, Szabo IL, Soreghan BA, Atay S, Kawanaka H, Tarnawski AS. PGE2 stimulates VEGF expression in endothelial cells via ERK2/JNK1 signaling pathways, *Biochem Biophys Res Commun.* 2001; 286: 923-928

[134] Daniel TO, Liu H, Morrow JD, Crews BC, Marnett LJ. Thromboxane A2 is a mediator of cyclooxygenase-2-dependent endothelial migration and angiogenesis, *Cancer Res.* 1999; 59: 4574-4577

[135] Nie D, Lamberti M, Zacharek A, Li L, Szekeres K, Tang K, Chen Y, Honn KV. Thromboxane A(2) regulation of endothelial cell migration, angiogenesis, and tumor metastasis. *Biochem Biophys Res Commun.* 2000; 267: 245-251

[136] Nor JE, Christensen J, Liu J, Peters M, Mooney DJ, Strieter RM, Polverini PJ. Up-Regulation of Bcl-2 in microvascular endothelial cells enhances intratumoral angiogenesis and accelerates tumor growth. *Cancer Res.* 2001; 61: 2183-2188

[137] Dimmeler S, Zeiher AM. Akt takes center stage in angiogenesis signaling, *Circ. Res.* 2000; 86: 4-5

[138] Michaelis UR, Fisslthaler B, Barbosa-Sicard E, Falck JR, Fleming I, Busse R. Cytochrome P450 epoxygenases 2C8 and 2C9 are implicated in hypoxia-induced endothelial cell migration and angiogenesis. *J Cell Sci.* 2005; 118: 5489-5498

[139] Michaelis UR, Fisslthaler B, Medhora M, Harder D, Fleming I, Busse R. Cytochrome P450 2C9-derived epoxyeicosatrienoic acids induce angiogenesis via cross-talk with the epidermal growth factor receptor (EGFR). *FASEB J.* 2003; 17: 770-772

[140] Pozzi A, Macias-Perez I, Abair T, Wei S, Su Y, Zent R, Falck JR, Capdevila JH. Characterization of 5,6- and 8,9-epoxyeicosatrienoic acids (5,6- and 8,9-EET) as potent in vivo angiogenic lipids. *J Biol Chem.* 2005; 280: 27138-27146

[141] Zhang B, Cao H, Rao GN. Fibroblast growth factor-2 is a downstream mediator of phosphatidylinositol 3-kinase-Akt signaling in 14,15-epoxyeicosatrienoic acid-induced angiogenesis. *J Biol Chem.* 2006; 281: 905-914

[142] Wang Y, Wei X, Xiao X, Hui R, Card JW, Carey MA, Wang DW, Zeldin DC. Arachidonic acid epoxygenase metabolites stimulate endothelial cell growth and angiogenesis via mitogen-activated protein kinase and phosphatidylinositol 3-kinase/Akt signaling pathways. *J Pharmacol Exp Ther.* 2005; 314: 522-532

[143] Romano M, Claria J. Cyclooxygenase-2 and 5-lipoxygenase converging functions on cell proliferation and tumor angiogenesis: implications for cancer therapy. *FASEB J.* 2003; 17: 1986-1995

[144] Nie D, Tang K, Diglio C, Honn KV. Eicosanoid regulation of angiogenesis: role of endothelial arachidonate 12-lipoxygenase. *Blood.* 2000; 95: 2304-2311

[145] Harats D, Ben-Shushan D, Cohen H, Gonen A, Barshack I, Goldberg I, Greenberger S, Hodish I, Harari A, Varda-Bloom N, Levanon K, Grossman E, Chaitidis P, Kuhn H, Shaish A. Inhibition of carcinogenesis in transgenic mouse models over-expressing 15-lipoxygenase in the vascular wall under the control of murine preproendothelin-1 promoter. *Cancer Lett.* 2005; 229: 127-134

[146] Liu X, Lazenby AJ, Siegal GP. Signal transduction cross-talk during colorectal tumorigenesis. *Adv Anat Pathol.* 2006; 13: 270-274

[147] Wada M, DeLong CJ, Hong YH, Rieke CJ, Song I, Sidhu RS, Yuan C, Warnock M, Schmaier AH, Yokoyama C, Smyth EM, Wilson SJ, FitzGerald GA, Garavito RM, Sui DX, Regan JW, Smith WL. Enzymes and receptors of prostaglandin pathways with arachidonic acid-derived versus eicosapentaenoic acid-derived substrates and products. *J Biol Chem.* 2007; 282: 22254-22266

[148] Fierro IM. Angiogenesis and lipoxins. *Prostaglandins Leukot Essent Fatty Acids.* 2005; 73: 271-275

[149] Fierro IM, Kutok JL, Serhan CN. Novel lipid mediator regulators of endothelial cell proliferation and migration: aspirin-triggered-15R-lipoxin A(4) and lipoxin A(4). *J Pharmacol Exp Ther.* 2002; 300: 385-392

[150] Clària J, Serhan CN. Aspirin triggers previously undescribed bioactive eicosanoids by human endothelial cell–leukocyte interactions, *Proc Natl Acad Sci USA* 1995; 92: 9475-9479

[151] Cezar-de-Mello PF, Nascimento-Silva V, Villela CG, Fierro IM. Aspirin-triggered Lipoxin A4 inhibition of VEGF-induced endothelial cell migration involves actin polymerization and focal adhesion assembly. *Oncogene.* 2006; 25: 122-129

[152] Fischer S, Weber PC. Prostaglandin I3 formed in vivo in man after dietary eicosapentaenoic acid. *Nature* 1984; 307: 165-168

[153] Malkowski MG, Thuresson ED, Lakkides KM, Rieke CJ, Micielli R, Smith WL, Garavito RM. Structure of eicosapentaenoic and linoleic acids in the cyclooxygenase site of prostaglandin endoperoxide H synthase-1. *J Biol Chem.* 2001; 276: 37547-37555

[154] Thuresson ED, Lakkides KM, Rieke CJ, Sun Y, Wingerd BA, Micielli R, Mulichak AM, Malkowski MG, Garavito RM, Smith WL. Prostaglandin endoperoxide H synthase-1: the functions of cyclooxygenase active site residues in the binding, positioning, and oxygenation of arachidonic acid. *J Biol Chem.* 2001; 276:10347-1057

In: Oncogene Proteins: Structure, Functions and Analyses ISBN: 978-1-60456-259-0
Editors: P.B. Murphy and J.R. Clarke, pp. 139-159

Chapter V

Survivin Splice Variants: New Players in Cancer*

Janardhan Sampath [1]**, *Seiji Fukuda*** [2] ***and Louis M. Pelus*** [1**]

[1] Department of Microbiology and Immunology, Walther Oncology Center, Indiana University School of Medicine, Indianapolis, IN, 46202.

[2] Department of Pediatrics, Shimane University School of Medicine, Shimane, Japan

Abstract

The IAP family member Survivin, is overexpressed in most cancers, with low or no expression in most differentiated adult tissues, making it an attractive target for cancer treatment. Survivin expression is a poor prognostic marker in a number of cancers. Clinical studies are progressing evaluating targeted Survivin knockdown using anti-sense oligonucleotides, immunotherapy using Survivin primed dentritic cells and disruption of Survivin stability by antagonism of Survivin interaction with Hsp90. Strategies based upon ribozymes directed against Survivin mRNA, or dominant-negative cDNA to block Survivin function are in pre-clinical stages. Survivin, like many genes, is alternately spliced and a number of new splice variants have recently been identified. In some cancers, expression of Survivin splice variants correlates with steroid receptor loss and p53, suggesting that like the wild-type form, some Survivin splice variants may have prognostic relevance. This chapter will review the current understanding of the function of Survivin splice variants and their expression and sub-cellular localization in neoplastic tissues, as well as critically evaluate the potential effect of current Survivin directed therapies on alternatively spliced Survivin isoforms.

Keywords: Survivin, splice variant, cancer expression, targeted therapy

---

* Supported by grants HL69669 and HL79654 to L.M.P from the National Institutes of Health.

** Correspondence should be addressed to: Louis M. Pelus, Ph.D., Department of Microbiology and Immunology, Indiana University School of Medicine, 950 West Walnut St, Indianapolis, IN 46202

Introduction

Survivin is a member of the highly conserved inhibitor of apoptosis (IAP) family of endogenous caspase inhibitors. It is highly over expressed in most transformed cells, solid tumors and leukemias and lymphomas, primarily during G_2/M phase of the cell cycle [1] and plays a vital role in cell survival and proliferation. Survivin was originally shown to be developmentally regulated; expressed in fetal tissues with no detectable expression in adult differentiated tissues except the thymus and placenta. Cancer specific expression with limited normal tissue expression and a role in loss of apoptosis and cell cycle regulation in cancer cells has made Survivin an attractive therapeutic target. However, more recently, it has become apparent that Survivin is expressed in some normal cells, particularly hematopoietic stem and progenitor cells (HSPC) [2-4], vascular endothelial cells [5,6], keratinocyte [7] and neural [8,9] stem cells and some mature blood cells [10-14] where it plays a developmental role [11,15]. Nevertheless, despite expression in normal stem cells, the levels of Survivin are much lower than in transformed cells justifying the rationale for development of anti-Survivin based therapies.

Survivin contains an amino terminal BIR domain, a ~75 amino acid domain characteristic of all IAP proteins, and a carboxy terminal coiled coil domain. Multiple alignments of Survivin protein from various species clearly show that both the BIR and the coiled coil domains are highly conserved with high amino acid identity [16]. Although the cross species function of Survivin has not been totally characterized, the presence of a homologous Survivin gene in multiple species suggests that it likely plays a vital role. Currently, there is considerable effort in developing targeted therapeutic drugs that effectively inhibit Survivin function or production. Small and large molecule drugs directed against Survivin and immunotherapy using Survivin primed dendritic cells are in clinical trials. However, in light of the role(s) that splice variants may play in Survivin function, it is currently unknown if any of the targeted therapies under development will affect function and/or expression of alternatively spliced Survivins. A clear understanding of the relative expression of Survivin isoforms and the interplay of the pro- and anti-apoptotic functions of Survivin splice variants is required before successful targeted anti-Survivin therapies can be fully developed. In this chapter, we will highlight this emerging aspect of Survivin biology.

Survivin and Splice Variants

Transcriptional Regulation of Survivin and Its Splice Variants

The human Survivin gene, localizes to chromosome 17q25 and its basal transcription is regulated by multiple Sp1 binding sites [17,18] and a TATA less promoter [17]. Consistent with cell cycle regulation of Survivin, one cell cycle homology domain (CHR) and 3 cell cycle dependent elements (CDEs) have been identified in the 5' flanking region [17]. A GC rich region was also found in the promoter region of Survivin, however this region was not found to be methylated in normal adult tissues [17]. E2F binding sites have been identified proximal to the CDE/CHR elements and the E2F family of transcriptional activators (E2F1 and 3) and repressors (E2F4 and 5) in association with retinoblastoma family members, pRB and p130, variably interact at this site in a cell cycle dependent fashion, activating or repressing Survivin transcription [19]. The tumor suppressor, wild-type p53, represses Survivin expression, although the precise mechanism by which this occurs is controversial [20,21]. Since p53 and pathways that regulate pRB function are frequently mutated in cancers, these studies highlight two ways by which Survivin could be aberrantly expressed in cancers. Besides the loss of repression by tumor suppressors, several transcription factors that are activated in many cancers have been shown to transcriptionally activate Survivin. The Bcr-abl fusion protein found primarily in chronic myeloid leukemia up regulates Survivin transcription, possibly through increased activity of AP1 and/or E2F transcriptional activators [22]. Ras, a transforming oncogene that is activated in leukemia and solid tumors, also activates Survivin transcription [23]. Recently, estrogen receptor alpha (ERα) has been shown to interact with p53 and block its repression of Survivin [24]. Insulin like growth factor-1 (IGF-1), a survival factor required for growth of multiple myelomas, can activate NFκB activity through the PI-3K/AKT pathway resulting in up regulation of Survivin [25]. Similarly, Survivin has been shown to be up-regulated in adult T-cell leukemia induced by the human T-cell leukemia virus-1 (HTLV1) through NFκB [26]. The oncogene c-myc activates Survivin transcription in breast cancer cells [27] and DEC1, a transcription factor over expressed in a number of different cancers, activates Survivin expression in cancer cells predominantly by interacting with Sp1 binding sites on the Survivin promoter [28]. Stat3 another transcription factor highly activated in breast, head and neck, endometrial and cervical cancers also up regulates Survivin expression [29-31].

Flt3, a type III tyrosine kinase receptor expressed mainly in the primitive hematopoietic compartment and internal tandem duplication (ITD) mutations of the Flt3 gene that constitutively activate the Flt3 tyrosine kinase are frequently found in patients with AML and associated with poor prognosis [32]. Incubation of primary human $CD34^+$ cells, the cell population that contains hematopoietic stem and progenitor cells (HSPC), with Flt3 ligand, induces Survivin expression, indicating that Survivin lies down stream of Flt3 signaling pathways in $CD34^+$ cells. Consistent with these findings, ITD-Flt3 increases expression of Survivin in the mouse hematopoietic cell line BaF3, resulting in reduced active caspase-3 levels and prolonged cell survival and growth [33]. Over expression of human wild-type Flt3 does not induce growth factor independent proliferation of primary mouse bone marrow granulocyte and macrophage progenitor cells (CFU-GM), however, ectopic expression of ITD-Flt3 mutants cloned from patients with AML significantly enhanced survival of CFU-GM and c-kit^+, Sca-1^+, lin^- cells, the population of mouse cells that contains HSPC, under conditions of growth factor deprivation and allowed for factor independent proliferation. Conditional deletion of Survivin significantly reduced growth factor independent survival of these cells induced by ITD-Flt3 [34]. These findings indicate that ITD-Flt3 up regulates Survivin expression independent of hematopoietic growth factors and that the ITD-Flt3/Survivin axis provides a survival and growth advantage in transformed hematopoietic cells.

All the Survivin splice variants that have been described thus far have the same invariant start codon as well as the subsequent 25 amino acids [16], therefore it is reasonable to assume that transcription factors that regulate Survivin WT would also regulate all the splice variants. Bcr-abl has been shown to increase the expression of Survivin 2B and ΔEx3, similar to Survivin WT [22]; however the effect of over expression of Bcr-abl on other variants is currently unknown. Similarly, the effect of increased activity of other transcription factors described above on Survivin splice variants is unknown.

Insulin like growth factor-1 (IGF-1) treatment of prostate cancer cells was found to increase the level of Survivin WT protein; however neither the mRNA level nor the stability of Survivin WT protein was determined to be altered in these cells indicating that the observed effect most likely is due to an alteration in translation. Treatment of cells with rapamycin, an mTOR (mammalian target of rapamycin) inhibitor or a siRNA against p70S6K1, a downstream target of mTOR, decreased the level of Survivin WT protein. Conversely, over expression of p70S6K1 resulted in increased Survivin WT protein demonstrating that the increased Survivin WT protein observed after IGF-1 treatment in these cells is due to alterations in translation through the AKT/mTOR/p70S6K1 pathway [35].

Structure of Human Survivin and Predictions of the Structure of Survivin Splice Variants

Wild-type (WT) human Survivin is encoded by 4 exons and is 142 amino acids long. Cryptic exons that are a part of the Survivin WT introns are retained in some of the splice variants, i.e., Survivin 2α, 2B and 3B, and in other variants, some of the known exons are skipped, e.g. Survivin ΔEx3. A number of new variants identified either through a search of the EST database [36] or by RT-PCR [37] have recently been described (Figure 1); however their functions have not been extensively characterized. Clustal W [38] analysis of all the human Survivin splice variants shows that the first N-terminal 73 amino acids are highly conserved in all isoforms except the new variants Survivin ΔptEx2, ΔptEx1/2, ΔptEx1/2G/T and Δpt Ex2/3 [16]. Interestingly, the predicted coiled coil domain (AA $_{97-141}$ of Survivin WT) is conserved only in Survivin WT and 2B whereas the BIR domain (AA $_{16-87}$ of Survivin WT) is disrupted in all the splice variants except Survivin WT and 3B.

The BIR domain (AA $_{1-88}$) of Survivin consists of three helices and three anti-parallel β sheets and residues C57, C60, H77 and C84 form a zinc finger that stabilizes the BIR domain. Crystal structure shows that human Survivin is a bow tie shaped homodimer [39-41] with the dimer interface resulting from hydrophobic interactions involving the N-terminus, specifically AA $_{6-10}$, the linker region involving AA $_{89-97}$ and the N-terminus of the coiled coil domain [39].

Survivin splice variant amino acid sequence alignment indicates that only the 3B variant retains the wild-type BIR domain. The C-terminal α-helix of the BIR domain is interrupted by new sequences from intron 2B in the 2B and 2B-3B variants and Jpred [42] predicts that amino acids derived from these new sequences might form a β-sheet and a α-helix. The 4 zinc binding amino acids (C57, C60, H77 and C84) are highly conserved in the Survivin 2B and 2B-3B variants except that the last two conserved amino acids (H77 and C84) are pushed further away from the other two zinc coordinating amino acids by the presence of the β-sheet and α-helix. The presence of these extra amino acids in 2B and 2B-3B interrupts not only the α3 helix of the BIR but also affects zinc coordination resulting in a non-functional BIR domain. The essential role of zinc coordination in the function of Survivin is highlighted by the C84A mutant [43,44] that loses anti-apoptotic function. As a result, the Survivin 2B and 2B-3B variants would not be predicted to retain anti-apoptotic function, which is consistent with data reported to date [45-47]. The BIR domain and the zinc coordinating amino acids are intact in the Survivin 3B variant, suggesting that this variant may be anti-apoptotic, which is also consistent with recent evidence [48]. The AA $_{64-87}$ region shown to interact with Smac/Diablo is conserved in Survivin 3B, 2B and 2B-3B variants, although interrupted by the 23 new amino acids in the 2B and 2B-3B splice variants, suggesting that the 3B variant should be able to interact with Smac/Diablo. In the case of the ΔEx3 variant, the C-terminus is different from wild-type Survivin and the BIR domain is interrupted by deletion of exon 3, resulting in a frame shift and the amino acids required for zinc coordination are not conserved. These observations suggest that the ΔEx3 variant would not be anti-apoptotic. However, numerous reports show that this variant retains anti-apoptotic function [45-47]. Since the BIR domain and zinc coordinating amino acids are not conserved in all the other splice variants they are predicted not to be anti-apoptotic.

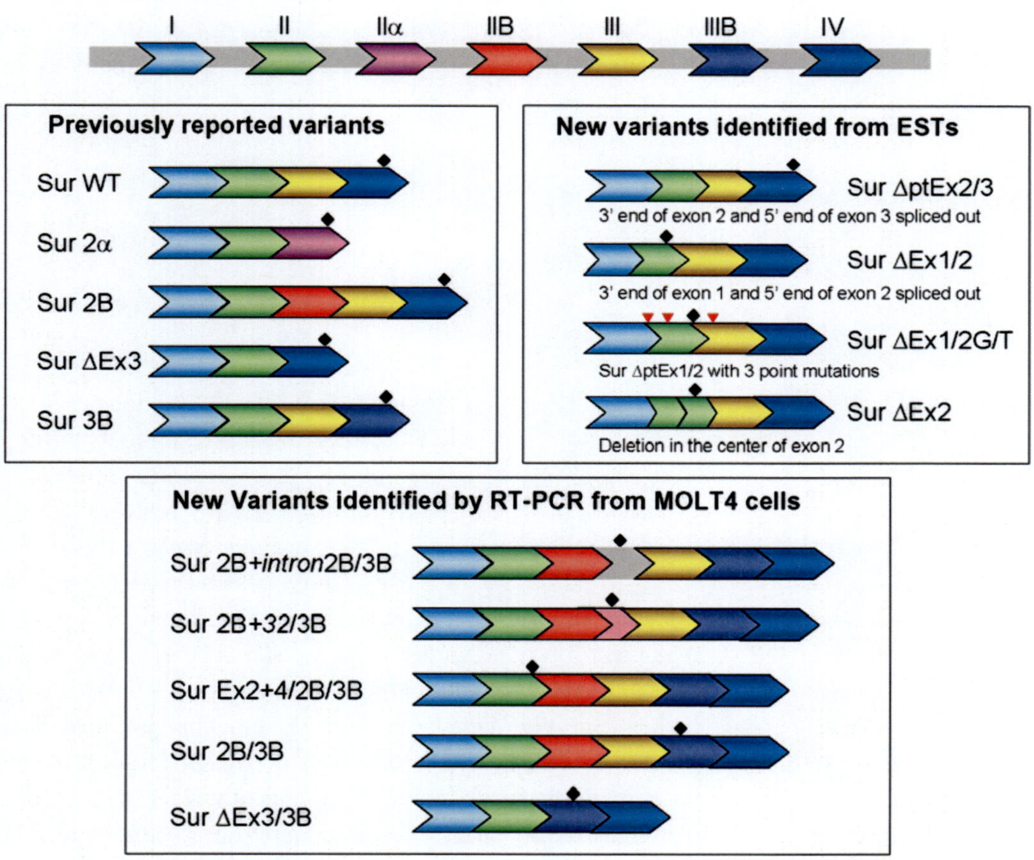

Figure 1. Gene structure of Survivin. Boxes represent exons and the gray line represents the introns and 3' and 5; UTR. The splice of variants that have been previously reported and studied in some detail are shown in the upper left box (Sur WT, Sur2α, Sur2B, Sur ΔEx3 and Sur3B). New variants recently identified in the EST database are shown in the upper right box. Variants identified by RT-PCR using Molt4 RNA are shown in the lower box. The black filled diamond signifies the presence of a "Stop" codon sequence. The red triangles indicate point mutations.

Interestingly, all the Survivin splice variants that have been identified except Survivin 2B and 3B lack all the regions that have been shown to be essential for dimerization, suggesting that only the 2B and 3B variants could potentially interact with wild-type Survivin. Recently, Survivin 2B [45,49] and 3B [48] have been shown to interact with Survivin WT, however, Survivin ΔEx3 and 2α, which lack two of the three regions required for dimerization have also been shown to interact with Survivin WT [45,49,50]. It is possible that the interaction is not direct but mediated by other adaptor protein(s). However, *in vitro* interaction studies using a GST-Survivin fusion protein show that it can interact with *in vitro* transcribed and translated Survivin splice variants 2B and ΔEx3 [49], suggesting that the presence of one of the three interaction domains required for dimerization is enough for this interaction. Recent *in vitro* and *in vivo* evidence show Survivin 3B to be the only variant capable of interacting with Survivin WT, although a weak interaction was observed with the 2B variant under less stringent conditions [48]. Survivin 2α is the smallest known variant of Survivin and its BIR domain is truncated, lacking the C-terminal helix. Considering the fact that the C-terminal truncated mutant of Survivin WT (containing AA_{1-99}) was found not to fold properly [39], it raises the possibility that Survivin 2α might be an unstable protein.

Inhibition of Apoptosis by Survivin and Its Splice Variants

Survivin is highly expressed in most cancer cell lines [1] and its expression correlates with resistance to apoptosis induced by chemotherapeutic agents [51-53] and death inducers [51,53-59]. Cytoplasmic Survivin interacts with caspases 3 [51] and 7 [60], however Survivin can also translocate to the nucleus, where it interacts with Cdk4 releasing $p21^{CIP1/WAF1}$, which phosphorylates Rb resulting in cell cycle entry [58] and subsequently translocates to the mitochondria and inhibits caspase 3. Survivin also localizes inside the mitochondria of transformed cells and primary cancers, but not in normal human cells [61]. Disruption of the binding of $^{79}K - K^{90}$ in the BIR domain with the chaperone Hsp90 results in degradation of Survivin with increased apoptosis and cell cycle arrest [62,63]. In transformed cells, highest expression of Survivin is seen in G_2/M phase, with very low/no expression in G_1. In general, higher cytosolic Survivin localization is observed compared to nuclear localization. Cytosolic Survivin localizes to the centrosomes, interphase microtubules, spindle poles and mitotic spindle microtubules at metaphase and anaphase and midbodies at telophase, consistent with a role in cytokinesis [64]. Nuclear Survivin localizes to the kinetochores of metaphase chromosomes and to the central spindle midzone during anaphase in cells undergoing mitosis [64]. Wild-type Survivin interacts with the aryl hydrocarbon receptor-interacting protein (AIP) through the D142 residue in the C-terminus coiled coil domain [65]. Mutation of D142 accelerates degradation of Survivin that is blocked by proteasome inhibition, suggesting that this interaction regulates Survivin stability. Down regulation of AIP by RNAi or antagonizing the interaction increases Survivin degradation resulting in enhanced sensitivity to UVB and etoposide [65]. Survivin 2B is the only variant that retains the coiled coil domain including D142; therefore, it might interact with AIP.

Interaction of Survivin and Splice Variants with Chromosomal Passenger Complex Proteins

Survivin localizes to centromeres at metaphase and spindle midzone in anaphase like other chromosomal passenger complex (CPC) proteins [66,67] and interacts with Aurora B kinase, Borealin, INCEP, and other CPC proteins [68-71]. Disruption of Survivin, Aurora-B or INCEP results in defects in mitosis and cytokinesis (reviewed in [72]), due to improper localization of Survivin and Aurora-B during metaphase and anaphase [68]. The C-terminal helical coiled coil domain of Survivin WT containing the T117 residue that is phosphorylated by Aurora-B kinase is involved in interacting with INCEP. Mutation of Survivin at T117 prevents proper localization during mitosis due to its inability to interact with INCEP [73]. Mutation of ^{70}DD^{71} to AA in the BIR domain of Survivin WT also results in improper localization with an increase in multinucleated cells due to loss of interaction with Aurora-B, indicating that the BIR domain of Survivin interacts with Aurora-B [74]. However, the precise region within the BIR domain responsible for the interaction remains unknown. This region of the BIR domain has also been shown to be involved in Smac/Diablo interaction [39]. Consistent with the requirement for an intact BIR domain and coiled coil domain for proper co-localization with other CPC proteins, stably expressed C-terminal GFP fusions of Survivin 2B that contain a disrupted BIR but an intact coiled coil domain and Survivin ΔEx3 that lacks both an intact BIR and coiled coil domain both failed to interact with endogenous Borealin and Aurora-B [49]. However, Survivin 2B might interact with INCEP, since it retains the C-terminal coiled coil domain.

Nuclear Export of Survivin and Splice Variants

Although a nuclear localization sequence (NLS) has not been identified in Survivin, a nuclear export sequence (NES) has been identified and Survivin WT interacts with Crm1, a protein involved in nuclear protein export [47,75]. While Survivin WT is present almost exclusively in the cytoplasm, a L98A mutation that retains Crm1 interaction results in predominantly nuclear localization [47] and a T97A/E mutation [47] or NES deletion, in which the Crm1 interaction is lost, results in equal localization in both the nuclear and cytoplasmic compartments [75]. Cytoplasmic localization of Survivin is essential for its anti-apoptotic function, with the L98A mutant showing increased sensitivity to γ-radiation as well as TRAIL [47]. Although the NES mutant can interact with other CPC proteins and can homodimerize, it fails to localize to the centromere during mitosis, suggesting that interaction of Survivin with Crm1 is essential for centromere localization [75]. Intriguingly, the T97A/E mutant that fails to interact with Crm1 still localizes to centromeres during mitosis [47]. Furthermore, since the NES deletion mutant does not rescue the increase in multinucleated cells or block inhibition of cell proliferation, Survivin-Crm1 interaction seems to be required for CPC function [75].

Survivin 2B, 3B and 2B-3B variants all have the NES sequence and Survivin 2B is present in the cytoplasm and nucleus, similar to Survivin WT; however the sub-cellular localization of Survivin 3B and 2B-3B are currently unknown. Closer analysis of Survivin 2B and ΔEx3 variants suggest that the anti-apoptotic function is not solely determined by cytoplasmic localization, since Survivin ΔEx3, which lacks an intact BIR domain as well as a

NES and localizes exclusively to the nucleus in actively proliferating cells, still retains its anti-apoptotic function [45,76]. However, Survivin 2B, with an NES and a disrupted BIR domain lacks anti-apoptotic function [45-47]. The presence of a NES alone does not appear to be sufficient for centromere localization, since Survivin 2B does not associate with the centromere in spite of having a NES. Moreover, the presence of a NLS alone is not sufficient for nuclear localization since deletion of the NLS in Survivin ΔEx3 results in cytoplasmic as well as nuclear accumulation [76]. A NLS has also been identified in Survivin Δpt2/3, which only differs from Survivin ΔEx3 by 4 amino acids [36]. Interestingly, a nucleolar localization sequence (NoLS) as well as a degradation signal has been identified in Survivin ΔEx3 [77], however the functional significance of this is unknown.

Expression of Survivin and Its Splice Variants in Cancer

Survivin is over-expressed in most cancers including solid tumors, leukemia and lymphoma. Expression of Survivin WT in cancers and normal tissues has recently been reviewed [78], however much less is known about the expression patterns of splice variants in cancer tissues and what functional role they may play.

Solid Tumors

In samples from soft tissue sarcoma patients, Survivin WT was the predominant transcript observed. Survivin ΔEx3 was detected in ~25% of samples, while Survivin 2B was not detected. Expression of Survivin WT inversely correlated with survival [79]. In a second cohort of 76 patients, Survivin WT was also the predominant form identified, although Survivin ΔEx3 and 2B levels were also detected [80]. Expression of Survivin WT, ΔEx3 and 2B correlated significantly with overall survival and a significant association between ΔEx3 expression and disease stage was observed. However, these conclusions may be somewhat misleading. Since each of the PCR products are of different size and the reactions were carried out in the same tubes, the amplification efficiency of the products will be based on their size, thus, the longest product would be amplified the least. In fact, the 2B variant, which is the longest product, was not amplified at all in the first study.

Using specific reverse primers but a common forward primer and a forward and a reverse probe for each variant, Survivin WT was detected in 24 brain cancer samples tested, except for one sample of pilocytic astrocytoma. Survivin 2B and ΔEx3 were not detectable in normal brain, but trace amounts of Survivin WT were detected. Malignant tumors expressed higher levels of Survivin WT and ΔEx3 than benign tumors and expression in gliomas was significantly higher than in non-gliomas. Survivin 2B expression was no different between malignant and benign tumors or between gliomas and non-gliomas [81]. In colorectal carcinomas, Survivin WT, ΔEx3 and 2B were higher in tumors than in normal tissues and the ratio of Survivin 2B to Survivin WT was significantly higher in tumors than in normal tissues. Overall, higher levels of the splice variant Survivin 2B correlated with favorable prognosis [82]. Analysis of the sequences of primers and probes used in the quantitative RT-

PCR reactions [81,82], indicates that the primer/probe combinations used for detecting Survivin WT would also detect the Survivin 3B variant (identical size product). Similarly, along with Survivin 2B, the recently identified 2B+intron2B/3B variant [37] would also be detected. Therefore, it is possible that the interpretation of higher expression of Survivin WT in malignant tumors and gliomas over benign tumors and non gliomas might not be reliable. The suggestion of increased expression of Survivin 2B in benign tumors and non-gliomas should be revisited with the use of more specific primers and probes.

In a study of 30 patients with gastric cancer, Survivin WT, 2B and ΔEx3 were all detected, with Survivin WT being the dominant form. A significant decrease in Survivin 2B expression in patients with stage III + IV disease was observed, while Survivin WT and ΔEx3 transcript levels remained constant [83]. In a second study of 77 patients, all the tumors showed Survivin WT expression with 79% of the tumors expressing 2B and 65% expressing ΔEx3. Based on semi-quantitative RT-PCR, it was concluded that expression of Survivin 2B inversely correlated with tumor grade, depth of tumor and invasion. Expression of Survivin WT correlated with poor prognosis and an inverse correlation was found between Survivin ΔEx3 expression and apoptosis. No correlation between Survivin 2B and ΔEx3 and prognosis was observed [84]. Since a single primer set that detects all three Survivin variants was used for PCR, conclusions regarding the role of Survivin 2B expression might not be accurate, based upon amplification efficiency issues.

Expression of Survivin WT was detected in 100%, 2B in 63% and ΔEx3 in 69% of hepatocellular carcinoma samples. Tumors with more malignant histology contained higher levels of Survivin WT and tumors with higher proliferative index contained elevated levels of Survivin WT and ΔEx3. No correlation between high Survivin 2B expression and proliferative index was found [85]. The primers and probes were only slightly different than those described earlier; therefore the conclusions concerning Survivin WT and 2B need further validation. Survivin WT is the dominant transcript detected in renal cell carcinomas, but a significant decrease in Survivin 2B at late stages was noted [86]. The aggressive clear cell RCC was found to have similar levels of Survivin as the chromophilic/papillary RCC. Again, as in the previous studies described, the primers and probes used to detect the wild-type and 2B isoforms can also detect other variants. Survivin WT, 2B and ΔEx3 variants were detected in 41 patients with cervical cancer; however no correlation was found between Survivin expression and stage of tumor, histology, age or menopausal state [87]. One set of primers was used to detect the expression of all three variants and closer analysis of the primers indicates that the 5'end of the primers can form hairpin loops and both primers can form dimers and cross dimers, with very low ΔG. Thus the conclusions drawn from this study may not be reliable.

In lung cancer patients, semi-quantitative RT-PCR analysis indicated that Survivin 2B expression correlated with better prognosis and ΔEx3 with poor prognosis. Although the primers used in this study were designed to specifically detect Survivin 2B, in light of recent findings [36,37], the primers would also detect Survivin2B/3B as well as Survivin Ex2+4/2B/3B. As a result, the conclusion of better prognosis associated with Survivin 2B expression might really be detecting these new variants. Survivin WT was the predominant form detected, followed by 2B then ΔEx3 in 19 medulloblastoma samples [88]. Although the ΔEx3 specific primers and probes did not cross react with any other variant, the 2B specific primer and probes could detect the newly identified Survivin Ex2+4/2B/3B and 2B/3B variants along with Survivin 2B. The primers/probes would also detect the Survivin 2B+32/3B variant. In a second study of 20 tumor samples and normal brain tissues, Survivin ΔEx3 was the only variant found to be present in normal adult cerebrum; however Survivin WT, 2B and ΔEx3 were all detected in fetal brain tissue [89]. No significant correlation was identified between the expression of the splice variants and clinical staging or histology, however a significant association was identified between expression of Survivin ΔEx3 and poor outcome.

Breast Cancer

The expression of Survivin splice variants has been most extensively evaluated in patients with breast cancer and warrants a detailed discussion. In two early studies using PCR primers that would detect all variants, Survivin WT was the most dominant transcript found in tumor biopsies, followed by ΔEx3 then 2B. Interestingly, Survivin ΔEx3 was also detected in one tumor in the absence of Survivin WT and 2B expression. No Survivin transcripts were detected in normal breast tissue. Expression of wild-type Survivin or any of the alternatively spliced variants did not correlate with disease-free or overall survival [90]. Survivin WT was higher in nodal metastases than in primary cancers and Survivin ΔEx3 was higher in ductal than lobular cancers. No correlation was found between estrogen or progesterone receptor status and the expression of variants [91]. In a study of 131 patients with invasive breast cancers [92] Survivin WT was detected in 94% of tumors, 2B in 70%, ΔEx3 in 80%, and Survivin 3B was detected in 33% of the samples. High grade carcinomas and tumors with mutant p53 had higher expression of Survivin 3B. Treatment reduced Survivin WT and 2B expression but did not affect ΔEx3 and 3B expression. Analysis of the specific primers used shows significant homology between the forward 3B primer and several newly described Survivin variants and the reverse primer matches these variants perfectly, therefore the PCR conditions might not be optimal and detection of Survivin 3B could be an underestimate.

Two recent studies using more specific primers and probes are particularly revealing. In 275 breast cancers [93], expression of all Survivin variants correlated well with the loss of estrogen and progesterone receptors. Survivin ΔEx3 and 2α associated with ductal carcinoma and young age. Total Survivin and ΔEx3 were highest in advanced stage tumors. Lower expression of Survivin 2B was associated with high grade tumors and interestingly, Survivin 2α and 3B predicted poor relapse-free survival. The primers and probes used in this study were specifically designed to pick only the intended variant, however the primer/probe pairs for Survivin 2B and 3B would likely detect several of the newly described variants. Therefore, the conclusions drawn in this study with respect to Survivin 2B and 3B may require reanalysis using more specific reagents. In a separate study of 162 breast carcinoma samples, Survivin WT and 2α were most highly expressed. Expression of tetramerization mutant p53 significantly correlated with increased expression of Survivin WT, ΔEx3 and 3B but not with Survivin 2α or 2B [94]. However, the primers and probes used to detect 2B would likely detect Survivin 2B+intron2B/3B in addition to 2B. Furthermore, the observed high Survivin 2α levels may be a consequence of small amplicon size resulting in high amplification efficiency compared to the other variants. The primers and probes for

Survivin ΔEx3 and 3B detect specifically only those variants but not 2B, which might explain why only their expression correlated with presence of mutant p53.

A significant correlation between wild-type Survivin and/or splice variant expression and the presence of mutant p53 and absence of estrogen and progesterone receptors was seen in the last two breast cancer studies. The tumor suppressor, p53, can suppress the expression of Survivin in vitro [20,21] and p53 is mutated in most cancers (reviewed in [95]), thus expression of Survivin WT, ΔEx3 and 3B fits well with the existing knowledge on p53 and breast cancer. The correlation between absence of estrogen receptor (ER) and progesterone receptor and Survivin expression [93] is consistent with the positive regulation of the progesterone receptor by ER [96]. In addition, patients with ER negative tumors have a better prognosis than ER positive tumors, making the correlation of Survivin expression with ER negative tumors an attractive hypothesis.

Leukemia and Lymphoma

Similar to solid tumors, Survivin WT is the predominant form found in blood and marrow samples from adult and pediatric patients with acute myeloid leukemia (AML) [97]. Splice variant expression did not correlate with stage or cytogenetic risk groups. In childhood AML, shorter overall survival was associated with higher expression of Survivin ΔEx3 and higher expression of Survivin 2B in adult AML showed a correlation with better event-free and overall survival. Analysis of the primers used in this study showed that they could also detect the recently identified Survivin ΔEx1/2, ΔptEx2/3 and ΔptEx1/2 variants [36]. The presence of these variants could alter the conclusions drawn, especially concerning Survivin WT and ΔEx3. Survivin WT was the predominant transcript detected in patients with acute (ALL) and chronic (CLL) lymphocytic leukemia with Survivin 2B and ΔEx3 expression higher in ALL compared to CLL [98]. No correlations with stage, age, etc., were found. The primers used were not specific to Survivin 2B and ΔEx3 and would also detect other variants.

Therapeutic Strategies

Targeted therapies that block Survivin function, destabilize Survivin protein or block Survivin transcription or translation are being pursued, as well as immunotherapeutic and gene therapy strategies. Preclinical studies show that Survivin knockdown results in appearance of multi-nucleated cells, increased caspase 3 activity, reduced tumor growth, apoptosis and increased sensitivity to chemotherapeutic drugs, suggesting that targeting Survivin in cancer is likely to be beneficial.

The interaction of Survivin WT with Hsp90 [99] requires residues ^{79}K- ^{90}K [62,63]. A peptide mimic of this interaction domain termed “Shepherdin” is being pursued clinically [63]. This domain is present in the Survivin 2B and 3B variants but not in the ΔEx3 or 2α variants. Although Hsp90 stabilizes Survivin WT, it is unknown if Hsp90 is required for stabilizing any of the alternatively spliced variants. Since ΔEx3 and 3B Survivin variants along with Survivin 2B are expressed in cancers and ΔEx3 and 3B have been associated with poor prognosis, Hsp90/Survivin destabilization by might not be effective in all tumors. A similar approach using the Hsp90 inhibitor geldenamycin is also in early clinical stages. However, since Hsp90 interacts with other proteins not just Survivin, the effect on blocking these Hsp90 interactions is unknown.

Survivin WT is stabilized during spindle checkpoint activation through the phosphorylation of T34 by cyclin B/cdk1 [100] and expression of a kinase dead dominant-negative mutant of cdk1 increases apoptosis following taxol treatment as a consequence of decreased Survivin stability [101]. These studies have led to a clinical trial based on inhibiting cdk1. However, even though T34 is conserved in all the Survivin splice variants, except ΔEx1/2 and ΔEx1/2G/T, the role of T34 phosphorylation in stabilization of variants other than Survivin WT is unknown.

A second generation antisense oligonucleotide with O-methoxyethyl modification in the last 4 bases at either ends of the oligo can specifically knock down Survivin mRNA and reduce Survivin protein [43]. It also induces caspase-dependent apoptosis, multi-polar mitotic spindles and multinucleated polyploid cells. The oligonucleotide targets the 3' untranslated region (UTR) of Survivin WT as well as ΔEx3 and 2B, therefore, it could block all three variants provided that these transcripts are expressed in cancers with these large 3' UTRs. The Survivin 3B variant is also over expressed in cancer and associated with poor prognosis but it is unknown if this antisense oligonucleotide will inhibit this variant, since complete information on the 3'UTR sequence of the 3B variant is not yet available. Similar to the antisense oligonucleotide, ribozymes [102-104] and RNAi's [105] directed against Survivin have been tested in preclinical cancer models. One ribozyme being developed should block all splice variants since they share the sequence in Exon 1 being targeted [103]. However, a second construct targets a sequence in exon III [102,104] that is not present in the anti-apoptotic ΔEx3 variant. Survivin directed RNAi's enhance growth inhibitory/apoptosis promoting effects of cytotoxic compounds [106] [59,107] and induce apoptosis in a number of cancer cell line models [108-110]. Analyses of the RNAi sequences indicate they would target multiple splice variants; however to date they have only been tested in cell lines in vitro. Intratumor injection of a cocktail of shRNAs directed towards exon III and I, which would be expected to block all the variants, showed a 70% reduction in tumor growth in a mouse xenograft model [105]. Although a good proof of concept study, route of delivery and susceptibility to nucleases in human plasma need to be addressed.

A phase I clinical trial in advanced or recurrent colon cancer using a Survivin 2B antigenic peptide has been initiated based on the rationale that Survivin 2B is expressed in various cancer cell lines and cancer tissues and the observation that CD8+ cytotoxic T lymphocytes recognize AYACNTSTL, a part of exon 2B of Survivin. Increasing dose of the antigen was administered in patients with only mild toxicity including anemia, fever and malaise observed in 3 patients, however no preliminary dose dependant efficacy was observed even at the highest dose tested [111]. The Survivin 2B/3B variant also contains this peptide sequence but expression of this variant in cancer is currently unclear. In addition, expression of Survivin 2B correlates with better prognosis in colon cancer [82], so stimulating antigenic response against this variant might not be a sound therapeutic strategy for all cancers. An oral DNA vaccine developed using Survivin WT and co-expressing CCL21 to facilitate development of cytotoxic T lymphocytes (CTL) shows promise in suppressing metastasis and blocking angiogenesis [112], however, its effect on normal hematopoietic stem and progenitor cells (HSPC) has not been evaluated. This is critical since

HSPC and endothelial progenitor cells both derive from the same cell, the hemangioblast. Furthermore, since CTL were induced by Survivin WT, it is unknown which epitope induced the CTL, therefore the effect of this vaccine on Survivin splice variants is unknown.

Gene Therapy strategies utilizing DNA constructs with a cytotoxic gene [113-115] or a gene that can convert a pro-drug to an active drug driven by a Survivin promoter [116] have been developed. In Survivin expressing cells, these constructs become activated and the cytotoxic gene(s) is expressed and the tumor cells die. T34A Survivin delivered as a plasmid or by an adenovirus can induce cell death by destabilizing Survivin [117,118]. However, although T34 is conserved in all the Survivin splice variants except ΔEx1-2 and ΔEx1-2G/T, it is currently unknown if the variants are all stabilized through phosphorylation by cdk1 and would be susceptible to destabilization by dominant-negative T34A. A final concern is how best to target constructs into tumors in order to circumvent possible systemic toxicity, in particular to the hematopoietic system (reviewed in [78]).

Summary/Potential Alternate Strategies to Target Survivin and Its Splice Variants

It is becoming increasingly clear that analysis of Survivin splice variants in various cancers although studied in some detail, is complicated by the lack of studies using specific primers/probes that distinguish all the now known splice variants reported to date and a lack of understanding of the biology of these variants. Considerable work in this area will be required to define the expression and function of these variants. Although much work is needed, some progress has been made. Since mice do not have a splice variant that is similar to human Survivin ΔEx3, one can assume that this variant might not play a normal physiological role and hence it can be specifically targeted in cancer cells. Also, this is the only variant where cancer specific expression is undisputable. Since the C-terminus of this protein is present in one other newly described variant, a large molecule drug targeted against Survivin ΔEx3 would be of potential value. However, since the Survivin ΔEx3 variant maintains its anti-apoptotic potential in spite of not having an intact BIR domain, additional work is required to identify proteins that interact with this variant and then target these interactions specifically.

A final point to consider is that since Survivin is also expressed in several normal stem cell populations [78], systemic treatment with any drug that targets Survivin could potentially lead to toxicities. Targeted knock down of Survivin in HSPC induces apoptosis and reduces the frequency and total number of HSPC and impairs hematopoietic reconstitution [119]. Conditional Survivin knock down *in vivo* in mice leads to death of all the animals due to hematopoietic complications [120]. Therefore a critical look at the potential side effects as well as the effects on splice variants is extremely important.

References

[1] Ambrosini, G., Adida, C., and Altieri, D.C. (1997). A novel anti-apoptosis gene, survivin, expressed in cancer and lymphoma. *Nat Med*, **3**, 917-921.

[2] Fukuda, S. and Pelus, L.M. (2002). Elevation of Survivin levels by hematopoietic growth factors occurs in quiescent CD34+ hematopoietic stem and progenitor cells before cell cycle entry. *Cell Cycle*, **1**, 322-326.

[3] Fukuda, S., Foster, R.G., Porter, S.B., and Pelus, L.M. (2002). The antiapoptosis protein survivin is associated with cell cycle entry of normal cord blood CD34(+) cells and modulates cell cycle and proliferation of mouse hematopoietic progenitor cells. *Blood*, **100**, 2463-2471.

[4] Fukuda, S. and Pelus, L.M. (2001). Regulation of the inhibitor-of-apoptosis family member survivin in normal cord blood and bone marrow CD34(+) cells by hematopoietic growth factors: implication of survivin expression in normal hematopoiesis. *Blood*, **98**, 2091-2100.

[5] Mesri, M., Morales-Ruiz, M., Ackermann, E.J., Bennett, C.F., Pober, J.S., Sessa, W.C. et al. (2001). Suppression of vascular endothelial growth factor-mediated endothelial cell protection by survivin targeting. *Am J Pathol*, **158**, 1757-1765.

[6] Conway, E.M., Zwerts, F., Van, E., V, DeVriese, A., Nagai, N., Luo, W. et al. (2003). Survivin-dependent angiogenesis in ischemic brain: molecular mechanisms of hypoxia-induced up-regulation. *Am J Pathol*, **163**, 935-946.

[7] Marconi, A., Dallaglio, K., Lotti, R., Vaschieri, C., Truzzi, F., Fantini, F. et al. (2007). Survivin identifies keratinocyte stem cells and is downregulated by anti-beta1 integrin during anoikis. *Stem Cells*, **25**, 149-155.

[8] Altura, R.A., Olshefski, R.S., Jiang, Y., and Boue, D.R. (2003). Nuclear expression of Survivin in paediatric ependymomas and choroid plexus tumours correlates with morphologic tumour grade. *Br J Cancer*, **89**, 1743-1749.

[9] Zwerts, F., Lupu, F., De, V.A., Pollefeyt, S., Moons, L., Altura, R.A. et al. (2007). Lack of endothelial cell survivin causes embryonic defects in angiogenesis, cardiogenesis, and neural tube closure. *Blood*, **109**, 4742-4752.

[10] Altznauer, F., Martinelli, S., Yousefi, S., Thurig, C., Schmid, I., Conway, E.M. et al. (2004). Inflammation-associated cell cycle-independent block of apoptosis by survivin in terminally differentiated neutrophils. *J Exp Med*, **199**, 1343-1354.

[11] Xing, Z., Conway, E.M., Kang, C., and Winoto, A. (2004). Essential role of survivin, an inhibitor of apoptosis protein, in T cell development, maturation, and homeostasis. *J Exp Med*, **199**, 69-80.

[12] Martinelli, S., Kostylina, G., Niggli, V., Baumann, C., Fey, M.F., Wendel, H.G. et al. (2006). Targeting survivin via PI3K but not c-akt/PKB by anticancer drugs in immature neutrophils. *Oncogene*, **25**, 6915-6923.

[13] Zhang, Y., Nagata, Y., Yu, G., Nguyen, H.G., Jones, M.R., Toselli, P. et al. (2004). Aberrant quantity and localization of Aurora-B/AIM-1 and survivin during megakaryocyte polyploidization and the consequences of Aurora-B/AIM-1-deregulated expression. *Blood*, **103**, 3717-3726.

[14] Kornacker, M., Verneris, M.R., Kornacker, B., Scheffold, C., and Negrin, R.S. (2001). Survivin expression correlates with apoptosis resistance after lymphocyte activation and is found preferentially in memory T cells. *Immunol Lett*, **76**, 169-173.

[15] Gurbuxani, S., Xu, Y., Keerthivasan, G., Wickrema, A., and Crispino, J.D. (2005). Differential requirements for survivin in hematopoietic cell development. *Proc Natl Acad Sci U S A*, **102**, 11480-11485.

[16] Sampath, J. and Pelus, L.M. (2007). Alternative splice variants of survivin as potential targets in cancer. *Curr Drug Discov Technol*, **4**, 174-191.

[17] Li, F. and Altieri, D.C. (1999). Transcriptional analysis of human survivin gene expression. *Biochem J,* **344** Pt 2, 305-311.

[18] Xu, R., Zhang, P., Huang, J., Ge, S., Lu, J., and Qian, G. (2007). Spl and Sp3 regulate basal transcription of the survivin gene. Biochem Biophys Res Commun, 356, 286-292.

[19] Jiang, Y., Saavedra, H.I., Holloway, M.P., Leone, G., and Altura, R.A. (2004). Aberrant regulation of survivin by the RB/E2F family of proteins. *J Biol Chem*, **279**, 40511-40520.

[20] Mirza, A., McGuirk, M., Hockenberry, T.N., Wu, Q., Ashar, H., Black, S. et al. (2002). Human survivin is negatively regulated by wild-type p53 and participates in p53-dependent apoptotic pathway. *Oncogene*, **21**, 2613-2622.

[21] Hoffman, W.H., Biade, S., Zilfou, J.T., Chen, J., and Murphy, M. (2002). Transcriptional repression of the anti-apoptotic survivin gene by wild type p53. *J Biol Chem*, **277**, 3247-3257.

[22] Wang, Z., Sampath, J., Fukuda, S., and Pelus, L.M. (2005). Disruption of the inhibitor of apoptosis protein survivin sensitizes Bcr-abl-positive cells to STI571-induced apoptosis. *Cancer Res*, **65**, 8224-8232.

[23] Fukuda, S. and Pelus, L.M. (2004). Activated H-Ras regulates hematopoietic cell survival by modulating Survivin. *Biochem Biophys Res Commun*, **323**, 636-644.

[24] Sayeed, A., Konduri, S.D., Liu, W., Bansal, S., Li, F., and Das, G.M. (2007). Estrogen receptor alpha inhibits p53-mediated transcriptional repression: implications for the regulation of apoptosis. *Cancer Res,* **67**, 7746-7755.

[25] Mitsiades, C.S., Mitsiades, N., Poulaki, V., Schlossman, R., Akiyama, M., Chauhan, D. et al. (2002). Activation of NF-kappaB and upregulation of intracellular anti-apoptotic proteins via the IGF-1/Akt signaling in human multiple myeloma cells: therapeutic implications. *Oncogene*, **21**, 5673-5683.

[26] Kawakami, H., Tomita, M., Matsuda, T., Ohta, T., Tanaka, Y., Fujii, M. et al. (2005). Transcriptional activation of survivin through the NF-kappaB pathway by human T-cell leukemia virus type I tax. *Int J Cancer*, **115**, 967-974.

[27] Cosgrave, N., Hill, A.D., and Young, L.S. (2006). Growth factor-dependent regulation of survivin by c-myc in human breast cancer. *J Mol Endocrinol*, **37**, 377-390.

[28] Li, Y., Xie, M., Yang, J., Yang, D., Deng, R., Wan, Y. et al. (2006). The expression of antiapoptotic protein survivin is transcriptionally upregulated by DEC1 primarily through multiple sp1 binding sites in the proximal promoter. *Oncogene*, **25**, 3296-3306.

[29] Gritsko, T., Williams, A., Turkson, J., Kaneko, S., Bowman, T., Huang, M. et al. (2006). Persistent activation of stat3 signaling induces survivin gene expression and confers resistance to apoptosis in human breast cancer cells. *Clin Cancer Res*, **12**,11-19.

[30] Chen, C.L., Hsieh, F.C., Lieblein, J.C., Brown, J., Chan, C., Wallace, J.A. et al. (2007). Stat3 activation in human endometrial and cervical cancers. Br J Cancer, 96, 591-599.

[31] Scheper, M.A., Nikitakis, N.G., and Sauk, J.J. (2007). Survivin is a downstream target and effector of sulindac-sensitive oncogenic Stat3 signalling in head and neck cancer. *Int J Oral Maxillofac Surg*, **36**, 632-639.

[32] Kelly, L.M., Liu, Q., Kutok, J.L., Williams, I.R., Boulton, C.L., and Gilliland, D.G. (2002). FLT3 internal tandem duplication mutations associated with human acute myeloid leukemias induce myeloproliferative disease in a murine bone marrow transplant model. *Blood*, **99**, 310-318.

[33] Fukuda, S., Bian, H.-M., Conway, E.M., and Pelus, L.M. (2005). The Survivin/p21WAF1/Cip1 axis and the CXC chemokine SDF1 regulate growth factor independent survival of primary hematopoietic progenitor cells induced by Internal Tandem Duplication mutations of Flt3. *Blood*, **106**, 461a.

[34] Fukuda, S., Singh, P., Conway, E.M., Yamaguchi, S., and Pelus, L.M. (2007). Survivin regulates aberrant proliferation of hematopoietic progenitor cells with self renewal capability and development of acute leukemia induced by Internal-Tandem duplication of Flt3. *Blood*, **110**, 184a.

[35] Vaira, V., Lee, C.W., Goel, H.L., Bosari, S., Languino, L.R., and Altieri, D.C. (2007). Regulation of survivin expression by IGF-1/mTOR signaling. *Oncogene*, **26**, 2678-2684.

[36] Li, F. and Ling, X. (2006). Survivin study: an update of "what is the next wave"? *J Cell Physiol*, **208**, 476-486.

[37] Mola, G., Vela, E., Fernandez-Figueras, M.T., Isamat, M., and Munoz-Marmol, A.M. (2007). Exonization of Alu-generated splice variants in the survivin gene of human and non-human primates. *J Mol Biol*, **366**, 1055-1063.

[38] Thompson, J.D., Higgins, D.G., and Gibson, T.J. (1994). CLUSTAL W: improving the sensitivity of progressive multiple sequence alignment through sequence weighting, position-specific gap penalties and weight matrix choice. *Nucleic Acids Res,* **22**, 4673-4680.

[39] Sun, C., Nettesheim, D., Liu, Z., and Olejniczak, E.T. (2005). Solution structure of human survivin and its binding interface with Smac/Diablo. *Biochemistry*, **44**, 11-17.

[40] Chantalat, L., Skoufias, D.A., Kleman, J.P., Jung, B., Dideberg, O., and Margolis, R.L. (2000). Crystal structure of human survivin reveals a bow tie-shaped dimer with two unusual alpha-helical extensions. *Mol Cell,* **6**, 183-189.

[41] Verdecia, M.A., Huang, H., Dutil, E., Kaiser, D.A., Hunter, T., and Noel, J.P. (2000). Structure of the human anti-apoptotic protein survivin reveals a dimeric arrangement. *Nat Struct Biol*, **7**, 602-608.

[42] Cuff, J.A., Clamp, M.E., Siddiqui, A.S., Finlay, M., and Barton, G.J. (1998). JPred: a consensus secondary structure prediction server. *Bioinformatics*, **14**, 892-893.

[43] Li, F., Ackermann, E.J., Bennett, C.F., Rothermel, A.L., Plescia, J., Tognin, S. et al. (1999). Pleiotropic cell-division defects and apoptosis induced by interference with survivin function. *Nat Cell Biol*, **1**, 461-466.

[44] Li, F., Ambrosini, G., Chu, E.Y., Plescia, J., Tognin, S., Marchisio, P.C. et al. (1998). Control of apoptosis and mitotic spindle checkpoint by survivin. *Nature*, **396**, 580-584.

[45] Caldas, H., Jiang, Y., Holloway, M.P., Fangusaro, J., Mahotka, C., Conway, E.M. et al. (2005). Survivin splice variants regulate the balance between proliferation and cell death. *Oncogene*, **24**, 1994-2007.

[46] Mahotka, C., Wenzel, M., Springer, E., Gabbert, H.E., and Gerharz, C.D. (1999). Survivin-deltaEx3 and survivin-2B: two novel splice variants of the apoptosis inhibitor survivin with different antiapoptotic properties. *Cancer Res*, **59**, 6097-6102.

[47] Colnaghi, R., Connell, C.M., Barrett, R.M., and Wheatley, S.P. (2006). Separating the anti-apoptotic and mitotic roles of survivin. *J Biol Chem,* **281**, 33450-33456.

[48] Knauer, S.K., Bier, C., Schlag, P., Fritzmann, J., Dietmaier, W., Rodel, F. et al. (2007). The survivin isoform survivin-3B is cytoprotective and can function as a chromosomal passenger complex protein. *Cell Cycle*, **6**, 1502-1509.

[49] Noton, E.A., Colnaghi, R., Tate, S., Starck, C., Carvalho, A., Ko, F.P. et al. (2006). Molecular analysis of survivin isoforms: evidence that alternatively spliced variants do not play a role in mitosis. *J Biol Chem*, **281**, 1286-1295.

[50] Caldas, H., Honsey, L.E., and Altura, R.A. (2005). Survivin 2alpha: a novel Survivin splice variant expressed in human malignancies. *Mol Cancer*, **4**, 11-19.

[51] Tamm, I., Wang, Y., Sausville, E., Scudiero, D.A., Vigna, N., Oltersdorf, T. et al. (1998). IAP-family protein survivin inhibits caspase activity and apoptosis induced by Fas (CD95), Bax, caspases, and anticancer drugs. *Cancer Res*, **58**, 5315-5320.

[52] Zaffaroni, N. and Daidone, M.G. (2002). Survivin expression and resistance to anticancer treatments: perspectives for new therapeutic interventions. *Drug Resist Updat*, **5**, 65-72.

[53] Jiang, X., Wilford, C., Duensing, S., Munger, K., Jones, G., and Jones, D. (2001). Participation of Survivin in mitotic and apoptotic activities of normal and tumor-derived cells. *J Cell Biochem*, **83**, 342-354.

[54] Malcles, M.H., Wang, H.W., Koumi, A., Tsai, Y.H., Yu, M., Godfrey, A. et al. (2007). Characterisation of the anti-apoptotic function of survivin-DeltaEx3 during TNFalpha-mediated cell death. *Br J Cancer*, **96**, 1659-1666.

[55] Azuhata, T., Scott, D., Takamizawa, S., Wen, J., Davidoff, A., Fukuzawa, M. et al. (2001). The inhibitor of apoptosis protein survivin is associated with high-risk behavior of neuroblastoma. *J Pediatr Surg,* **36**, 1785-1791.

[56] Conway, E.M., Pollefeyt, S., Steiner-Mosonyi, M., Luo, W., DeVriese, A., Lupu, F. et al. (2002). Deficiency of survivin in transgenic mice exacerbates Fas-induced apoptosis via mitochondrial pathways. *Gastroenterology*, **123**, 619-631.

[57] Azuhata, T., Scott, D., Griffith, T.S., Miller, M., and Sandler, A.D. (2006). Survivin inhibits apoptosis induced by TRAIL, and the ratio between survivin and TRAIL receptors is predictive of recurrent disease in neuroblastoma. *J Pediatr Surg*, **41**, 1431-1440.

[58] Suzuki, A., Ito, T., Kawano, H., Hayashida, M., Hayasaki, Y., Tsutomi, Y. et al. (2000). Survivin initiates procaspase 3/p21 complex formation as a result of interaction with Cdk4 to resist Fas-mediated cell death. *Oncogene*, **19**, 1346-1353.

[59] Nakao, K., Hamasaki, K., Ichikawa, T., Arima, K., Eguchi, K., and Ishii, N. (2006). Survivin downregulation by siRNA sensitizes human hepatoma cells to TRAIL-induced apoptosis. *Oncol Rep*, **16**, 389-392.

[60] Shin, S., Sung, B.J., Cho, Y.S., Kim, H.J., Ha, N.C., Hwang, J.I. et al. (2001). An anti-apoptotic protein human survivin is a direct inhibitor of caspase-3 and -7. *Biochemistry*, **40**, 1117-1123.

[61] Dohi, T., Beltrami, E., Wall, N.R., Plescia, J., and Altieri, D.C. (2004). Mitochondrial survivin inhibits apoptosis and promotes tumorigenesis. *J Clin Invest*, **114**, 1117-1127.

[62] Gyurkocza, B., Plescia, J., Raskett, C.M., Garlick, D.S., Lowry, P.A., Carter, B.Z. et al. (2006). Antileukemic activity of shepherdin and molecular diversity of hsp90 inhibitors. *J Natl Cancer Inst*, **98**, 1068-1077.

[63] Plescia, J., Salz, W., Xia, F., Pennati, M., Zaffaroni, N., Daidone, M.G. et al. (2005). Rational design of shepherdin, a novel anticancer agent. *Cancer Cell*, **7**, 457-468.

[64] Fortugno, P., Wall, N.R., Giodini, A., O'Connor, D.S., Plescia, J., Padgett, K.M. et al. (2002). Survivin exists in immunochemically distinct subcellular pools and is involved in spindle microtubule function. *J Cell Sci*, **115**, 575-585.

[65] Kang, B.H. and Altieri, D.C. (2006). Regulation of survivin stability by the aryl hydrocarbon receptor-interacting protein. *J Biol Chem*, **281**, 24721-24727.

[66] Skoufias, D.A., Mollinari, C., Lacroix, F.B., and Margolis, R.L. (2000). Human survivin is a kinetochore-associated passenger protein. *J Cell Biol*, **151**, 1575-1582.

[67] Uren, A.G., Wong, L., Pakusch, M., Fowler, K.J., Burrows, F.J., Vaux, D.L. et al. (2000). Survivin and the inner centromere protein INCENP show similar cell-cycle localization and gene knockout phenotype. *Curr Biol*, **10**, 1319-1328.

[68] Wheatley, S.P., Carvalho, A., Vagnarelli, P., and Earnshaw, W.C. (2001). INCENP is required for proper targeting of Survivin to the centromeres and the anaphase spindle during mitosis. *Curr Biol*, **11**, 886-890.

[69] Chang, J.L., Chen, T.H., Wang, C.F., Chiang, Y.H., Huang, Y.L., Wong, F.H. et al. (2006). Borealin/Dasra B is a cell cycle-regulated chromosomal passenger protein and its nuclear accumulation is linked to poor prognosis for human gastric cancer. *Exp Cell Res*, **312**, 962-973.

[70] Gassmann, R., Carvalho, A., Henzing, A.J., Ruchaud, S., Hudson, D.F., Honda, R. et al. (2004). Borealin: a novel chromosomal passenger required for stability of the bipolar mitotic spindle. *J Cell Biol*, **166**, 179-191.

[71] Bolton, M.A., Lan, W., Powers, S.E., McCleland, M.L., Kuang, J., and Stukenberg, P.T. (2002). Aurora B kinase exists in a complex with survivin and INCENP and its kinase activity is stimulated by survivin binding and phosphorylation. *Mol Biol Cell*, **13**, 3064-3077.

[72] Adams, R.R., Carmena, M., and Earnshaw, W.C. (2001). Chromosomal passengers and the (aurora) ABCs of mitosis. *Trends Cell Biol,* **11**, 49-54.

[73] Wheatley, S.P., Henzing, A.J., Dodson, H., Khaled, W., and Earnshaw, W.C. (2004). Aurora-B phosphorylation in vitro identifies a residue of survivin that is essential for its localization and binding to inner centromere protein (INCENP) in vivo. *J Biol Chem,* **279**, 5655-5660.

[74] Cao, L., Yan, X., Wu, Y., Hu, H., Li, Q., Zhou, T. et al. (2006). Survivin mutant (Surv-DD70, 71AA) disrupts the interaction of Survivin with Aurora B and causes multinucleation in HeLa cells. *Biochem Biophys Res Commun*, **346**, 400-407.

[75] Knauer, S.K., Bier, C., Habtemichael, N., and Stauber, R.H. (2006). The Survivin-Crm1 interaction is essential for chromosomal passenger complex localization and function. *EMBO Rep*, **7,** 1259-1265.

[76] Mahotka, C., Liebmann, J., Wenzel, M., Suschek, C.V., Schmitt, M., Gabbert, H.E. et al. (2002). Differential subcellular localization of functionally divergent survivin splice variants. *Cell Death Differ*, **9,** 1334-1342.

[77] Song, Z. and Wu, M. (2005). Identification of a novel nucleolar localization signal and a degradation signal in Survivin-deltaEx3: a potential link between nucleolus and protein degradation. *Oncogene*, **24**, 2723-2734.

[78] Fukuda, S. and Pelus, L.M. (2006). Survivin, a cancer target with an emerging role in normal adult tissues. *Mol Cancer Ther*, **5**, 1087-1098.

[79] Kappler, M., Kohler, T., Kampf, C., Diestelkotter, P., Wurl, P., Schmitz, M. et al. (2001). Increased survivin transcript levels: an independent negative predictor of survival in soft tissue sarcoma patients. *Int J Cancer*, **95**, 360-363.

[80] Taubert, H., Kappler, M., Bache, M., Bartel, F., Kohler, T., Lautenschlager, C. et al. (2005). Elevated expression of survivin-splice variants predicts a poor outcome for soft-tissue sarcomas patients. *Oncogene*, **24**, 5258-5261.

[81] Yamada, Y., Kuroiwa, T., Nakagawa, T., Kajimoto, Y., Dohi, T., Azuma, H. et al. (2003). Transcriptional expression of survivin and its splice variants in brain tumors in humans. *J Neurosurg*, **99**, 738-745.

[82] Suga, K., Yamamoto, T., Yamada, Y., Miyatake, S., Nakagawa, T., and Tanigawa, N. (2005). Correlation between transcriptional expression of survivin isoforms and clinicopathological findings in human colorectal carcinomas. *Oncol Rep*, **13**, 891-897.

[83] Krieg, A., Mahotka, C., Krieg, T., Grabsch, H., Muller, W., Takeno, S. et al. (2002). Expression of different survivin variants in gastric carcinomas: first clues to a role of survivin-2B in tumour progression. *Br J Cancer*, **86**, 737-743.

[84] Meng, H., Lu, C., Mabuchi, H., and Tanigawa, N. (2004). Prognostic significance and different properties of survivin splicing variants in gastric cancer. *Cancer Lett*, **216**, 147-155.

[85] Takashima, H., Nakajima, T., Moriguchi, M., Sekoguchi, S., Nishikawa, T., Watanabe, T. et al. (2005). In vivo expression patterns of survivin and its splicing variants in chronic liver disease and hepatocellular carcinoma. *Liver Int*, **25**, 77-84.

[86] Mahotka, C., Krieg, T., Krieg, A., Wenzel, M., Suschek, C.V., Heydthausen, M. et al. (2002). Distinct in vivo expression patterns of survivin splice variants in renal cell carcinomas. *Int J Cancer*, **100**, 30-36.

[87] Espinosa, M., Cantu, D., Herrera, N., Lopez, C.M., De la Garza, J.G., Maldonado, V. et al. (2006). Inhibitors of apoptosis proteins in human cervical cancer. *BMC Cancer*, **6,** 45.

[88] Fangusaro, J.R., Jiang, Y., Holloway, M.P., Caldas, H., Singh, V., Boue, D.R. et al. (2005). Survivin, Survivin-2B, and Survivin-deltaEx3 expression in medulloblastoma: biologic markers of tumour morphology and clinical outcome. *Br J Cancer*, **92**, 359-365.

[89] Li, X.N., Shu, Q., Su, J.M., Adesina, A.M., Wong, K.K., Perlaky, L. et al. (2007). Differential expression of survivin splice isoforms in medulloblastomas. *Neuropathol Appl Neurobiol*, **33**, 67-76.

[90] O'Driscoll, L., Linehan, R., Kennedy, M., Cronin, D., Purcell, R., Glynn, S. et al. (2003). Lack of prognostic significance of survivin, survivin-deltaEx3, survivin-2B, galectin-3, bag-1, bax-alpha and MRP-1 mRNAs in breast cancer. *Cancer Lett*, **201**, 225-236.

[91] Ryan, B., O'Donovan, N., Browne, B., O'Shea, C., Crown, J., Hill, A.D. et al. (2005). Expression of survivin and its splice variants survivin-2B and survivin-DeltaEx3 in breast cancer. *Br J Cancer*, **92**, 120-124.

[92] Vegran, F., Boidot, R., Oudin, C., Riedinger, J.M., and Lizard-Nacol, S. (2005). Distinct expression of Survivin splice variants in breast carcinomas. *Int J Oncol*, **27**, 1151-1157.

[93] Span, P.N., Tjan-Heijnen, V.C., Heuvel, J.J., de Kok, J.B., Foekens, J.A., and Sweep, F.C. (2006). Do the survivin (BIRC5) splice variants modulate or add to the prognostic value of total survivin in breast cancer? *Clin Chem*, **52**, 1693-1700.

[94] Vegran, F., Boidot, R., Oudin, C., Defrain, C., Rebucci, M., and Lizard-Nacol, S. (2007). Association of p53 gene alterations with the expression of antiapoptotic survivin splice variants in breast cancer. *Oncogene*, **26**, 290-297.

[95] Royds, J.A. and Iacopetta, B. (2006). p53 and disease: when the guardian angel fails. *Cell Death Differ*, **13**, 1017-1026.

[96] Savouret, J.F., Rauch, M., Redeuilh, G., Sar, S., Chauchereau, A., Woodruff, K. et al. (1994). Interplay between estrogens, progestins, retinoic acid and AP-1 on a single regulatory site in the progesterone receptor gene. *J Biol Chem*, **269**, 28955-28962.

[97] Wagner, M., Schmelz, K., Wuchter, C., Ludwig, W.D., Dorken, B., and Tamm, I. (2006). In vivo expression of survivin and its splice variant survivin-2B: impact on clinical outcome in acute myeloid leukemia. *Int J Cancer,* **119**, 1291-1297.

[98] Nakagawa, Y., Yamaguchi, S., Hasegawa, M., Nemoto, T., Inoue, M., Suzuki, K. et al. (2004). Differential expression of survivin in bone marrow cells from patients with acute lymphocytic leukemia and chronic lymphocytic leukemia. *Leuk Res*, **28**, 487-494.

[99] Fortugno, P., Beltrami, E., Plescia, J., Fontana, J., Pradhan, D., Marchisio, P.C. et al. (2003). Regulation of survivin function by Hsp90. *Proc Natl Acad Sci U S A,* **100**, 13791-13796.

[100] O'Connor, D.S., Grossman, D., Plescia, J., Li, F., Zhang, H., Villa, A. et al. (2000). Regulation of apoptosis at cell division by p34cdc2 phosphorylation of survivin. *Proc Natl Acad Sci U S A,* **97**, 13103-13107.

[101] Shapiro, G.I. (2004). Preclinical and clinical development of the cyclin-dependent kinase inhibitor flavopiridol. *Clin Cancer Res*, **10**, 4270s-4275s.

[102] Pennati, M., Binda, M., De, C.M., Pratesi, G., Folini, M., Citti, L. et al. (2004). Ribozyme-mediated down-regulation of survivin expression sensitizes human melanoma cells to topotecan in vitro and in vivo. *Carcinogenesis*, **25**, 1129-1136.

[103] Pennati, M., Binda, M., Colella, G., Zoppe', M., Folini, M., Vignati, S. et al. (2004). Ribozyme-mediated inhibition of survivin expression increases spontaneous and drug-induced apoptosis and decreases the tumorigenic potential of human prostate cancer cells. *Oncogene*, **23**, 386-394.

[104] Choi, K.S., Lee, T.H., and Jung, M.H. (2003). Ribozyme-mediated cleavage of the human survivin mRNA and inhibition of antiapoptotic function of survivin in MCF-7 cells. *Cancer Gene Ther*, **10**, 87-95.

[105] Caldas, H., Holloway, M.P., Hall, B.M., Qualman, S.J., and Altura, R.A. (2006). Survivin-directed RNA interference cocktail is a potent suppressor of tumour growth in vivo. *J Med Genet,* **43**, 119-128.

[106] Paduano, F., Villa, R., Pennati, M., Folini, M., Binda, M., Daidone, M.G. et al. (2006). Silencing of survivin gene by small interfering RNAs produces supra-additive growth suppression in combination with 17-allylamino-17-demethoxygeldanamycin in human prostate cancer cells. *Mol Cancer Ther*, **5**, 179-186.

[107] Yonesaka, K., Tamura, K., Kurata, T., Satoh, T., Ikeda, M., Fukuoka, M. et al. (2006). Small interfering RNA targeting survivin sensitizes lung cancer cell with mutant p53 to adriamycin. *Int J Cancer*, **118**, 812-820.

[108] Miao, G.Y., Lu, Q.M., and Zhang, X.L. (2007). Downregulation of survivin by RNAi inhibits growth of human gastric carcinoma cells. *World J Gastroenterol*, **13**, 1170-1174.

[109] Guan, H.T., Xue, X.H., Dai, Z.J., Wang, X.J., Li, A., and Qin, Z.Y. (2006). Down-regulation of survivin expression by small interfering RNA induces pancreatic cancer cell apoptosis and enhances its radiosensitivity. *World J Gastroenterol*, **12**, 2901-2907.

[110] Congmin, G., Mu, Z., Yihui, M., and Hanliang, L. (2006). Survivin--an attractive target for RNAi in non-Hodgkin's lymphoma, Daudi cell line as a model. *Leuk Lymphoma*, **47**, 1941-1948.

[111] Tsuruma, T., Hata, F., Torigoe, T., Furuhata, T., Idenoue, S., Kurotaki, T. et al. (2004). Phase I clinical study of anti-apoptosis protein, survivin-derived peptide vaccine therapy for patients with advanced or recurrent colorectal cancer. *J Transl Med*, **2**, 19.

[112] Xiang, R., Mizutani, N., Luo, Y., Chiodoni, C., Zhou, H., Mizutani, M. et al. (2005). A DNA vaccine targeting survivin combines apoptosis with suppression of angiogenesis in lung tumor eradication. *Cancer Res*, **65**, 553-561.

[113] Zhu, Z.B., Makhija, S.K., Lu, B., Wang, M., Kaliberova, L., Liu, B. et al. (2004). Transcriptional targeting of tumors with a novel tumor-specific survivin promoter. *Cancer Gene Ther*, **11**, 256-262.

[114] Caldas, H., Jaynes, F.O., Boyer, M.W., Hammond, S., and Altura, R.A. (2006). Survivin and Granzyme B-induced apoptosis, a novel anticancer therapy. *Mol Cancer Ther*, **5**, 693-703.

[115] Chen, J.S., Liu, J.C., Shen, L., Rau, K.M., Kuo, H.P., Li, Y.M. et al. (2004). Cancer-specific activation of the survivin promoter and its potential use in gene therapy. *Cancer Gene Ther*, **11**, 740-747.

[116] Kawamura, K., Yu, L., Tomizawa, M., Shimozato, O., Ma, G., Li, Q. et al. (2007). Transcriptional regulatory regions of the survivin gene activate an exogenous suicide gene in human tumors and enhance the sensitivity to a prodrug. *Anticancer Res,* **27**, 89-93.

[117] Mesri, M., Wall, N.R., Li, J., Kim, R.W., and Altieri, D.C. (2001). Cancer gene therapy using a survivin mutant adenovirus. *J Clin Invest*, **108**, 981-990.

[118] Zhang, M., Mukherjee, N., Bermudez, R.S., Latham, D.E., Delaney, M.A., Zietman, A.L. et al. (2005). Adenovirus-mediated inhibition of survivin expression sensitizes human prostate cancer cells to paclitaxel in vitro and in vivo. *Prostate*, **64**, 293-302.

[119] Fukuda, S., Conway, E.M., and Pelus, L.M. (2006). Survivin Regulates Proliferation of Normal Hematopoietic Stem and Progenitor Cells In Vivo. *Blood*, **108**, 258a.

[120] Leung, C.G., Xu, Y., Mularski, B., Liu, H., Gurbuxani, S., and Crispino, J.D. (2007). Requirements for survivin in terminal differentiation of erythroid cells and maintenance of hematopoietic stem and progenitor cells. *J Exp Med*, **204**, 1603-1611.

In: Oncogene Proteins: Structure, Functions and Analyses
Editors: P.B. Murphy and J.R. Clarke, pp. 161-190
ISBN: 978-1-60456-259-0

Chapter VI

Toy Story: Small-Molecule Inhibitors and Small Peptides of Bcl-2 Family Proteins

Yannis Guillemin, Juan Garcia Valero, Clara Locher, Jérôme Kucharczak, Germain Gillet and Abdel Aouacheria[*]
Apoptosis and Oncogenesis Laboratory, IBCP, Institut de Biologie et Chimie des Protéines, 7 passage du Vercors, Lyon, F-69367, France; CNRS, UMR 5086; Université de Lyon, France; université Lyon 1, Lyon, F-69007, France; IFR 128, Lyon, F-69007, France

Abstract

In metazoans including human, the mitochondrial pathway of apoptosis depends on orchestrated interactions between Bcl-2 and a number of closely related proteins. The Bcl-2 family of proteins is composed of pro- and anti-apoptotic members that regulate mitochondrial outer membrane permeabilization, a pivotal step in programmed cell death. Death inducers (e.g. Bax, Bak, BH3-only proteins) promote cytochrome c release from mitochondria, leading to the activation of intracellular cysteine proteases termed caspases that mediate cell demise. Conversely, pro-survival members such as Bcl-2 or Bcl-xL decrease cell susceptibility to apoptosis by preventing the activation of Bax-like or BH3-only proteins. Overexpression of anti-apoptotic proteins commonly occur in human tumors, and can contribute not only to disease development and progression but also to resistance to chemotherapy. Therefore, pro-survival Bcl-2 family members represent attractive targets for the development of novel anticancer drugs. On the other hand, alternative strategies are also emerging that use cytoprotective agents targeting pro-apoptotic Bcl-2 family proteins to treat disorders associated with excess apoptosis (e.g. stroke or neurodegenerative diseases). To date, three-dimensional structures of about a dozen Bcl-2 family members have been determined, contributing to a better understanding of Bcl-2 family proteins and to the design of selective inhibitors. After a brief overview of the apoptosis pathways, we review the recent small-

[*] E-mail address: a.aouacheria@ibcp.fr. Tel: +33-472-72-26-11; Fax: +33-472-72-26-01; IBCP, CNRS UMR 5086, 7 passage du Vercors, 69367 LYON Cedex France. (Principal author. To whom correspondence should be addressed)

molecule drugs targeting Bcl-2 family proteins with an emphasis on those discovered through fragment-based approaches. We also examine the literature on potential therapeutic use of non-BH3, biologically active fragments derived from Bcl-2 family proteins, which has begun to emerge in addition to BH3 mimetics.

Introduction

Apoptosis (programmed cell death) is an essential process permitting the destruction of damaged, abnormal or superfluous cells in a wide variety of physiological contexts. Disturbance of apoptosis (either too little or too much) has been linked to many human diseases including cancer, autoimmune disorders as well as neurodegenerative diseases, AIDS and ischemia (Hipfner and Cohen, 2004). Cancer cells most often exhibit a lack of apoptosis, which contributes not only to tumor progression but also to resistance to chemotherapy and tumor recurrence. Along with sustained proliferation, the ability of cancer cells to evade apoptosis (through expression of antiapoptotic proteins or inactivation of proapoptotic proteins) is considered a "hallmark of cancer" (Hanahan and Weinberg, 2000). There is current research interest in developing novel apoptosis-based anticancer drugs that would selectively induce cell death in cancer cells without damaging normal cells. Conversely, basic apoptosis research lays the foundation for developing novel therapeutic strategies to combat diseases hallmarked by an excessive amount of apoptosis, such as brain injury and Alzheimer's disease.

The pathway to cellular suicide is controlled by a number of intracellular checkpoints of which the Bcl-2 family plays a prominent role. The Bcl-2 family of proteins comprises both antagonists of apoptosis (e.g. Bcl-2 or Bcl-xL) and agonists of apoptosis (e.g. Bax or Bak), which function at least in part through protein-protein interactions. In the simplest possible model, the ratio of pro- versus anti-apoptotic members dictates the response to an apoptotic signal. However, recent findings have expanded and complicated this simple paradigm (Adams and Cory, 2007) (part I and II). Here, we discuss two (complementary) aspects of current biomedical research related to human Bcl-2 family proteins (part III-1). Nonpeptidic, small-molecule inhibitors of Bcl-2-like proteins developed through 'fragment assembly' are described (part III-2). Only compounds generated by covalently bridging chemical entities that bind to separate regions on Bcl-2 proteins will be considered. Then, the therapeutic potential of short peptide fragments from selected anti- or pro-apoptotic Bcl-2 family members is presented (part IV). After a brief synthesis of the literature on BH3 peptides (part IV-1), the last section (part IV-2) will focus on non-BH3, Bcl-2-like derived peptides, which are potential novel candidates for therapeutic exploration.

I. Apoptosis Signaling Pathways

In mammalian cells, apoptosis is mainly mediated through two pathways: the (intrinsic) mitochondrial pathway and the (extrinsic) death receptor pathway (Danial and Korsmeyer, 2004) (Figure 1). The death receptor pathway is initiated at the cell surface by the ligation of death receptors of the TNF receptor superfamily (e.g. Fas or TRAIL-R1) by their cognate ligands. In the intrinsic pathway, mitochondria play a critical role by releasing cytochrome *c* from intermembrane space to the cytosol, where, in association with adaptor protein Apaf-1 (apoptotic protease-activating factor-1) and dATP, it forms the apoptosome complex, activating caspase-9, which in turn activates caspase-3 (Jiang and Wang, 2004). Once activated, the effector caspases (e.g. caspase-3 and -7) cleave structural proteins, signaling molecules, DNA-repairing enzymes and proteins important for cell survival, resulting in cell demise. Permeabilization of the mitochondrial outer membrane, with subsequent release of mitochondrial intermembrane space proteins and caspase activation, is therefore considered a key event in apoptosis induction.

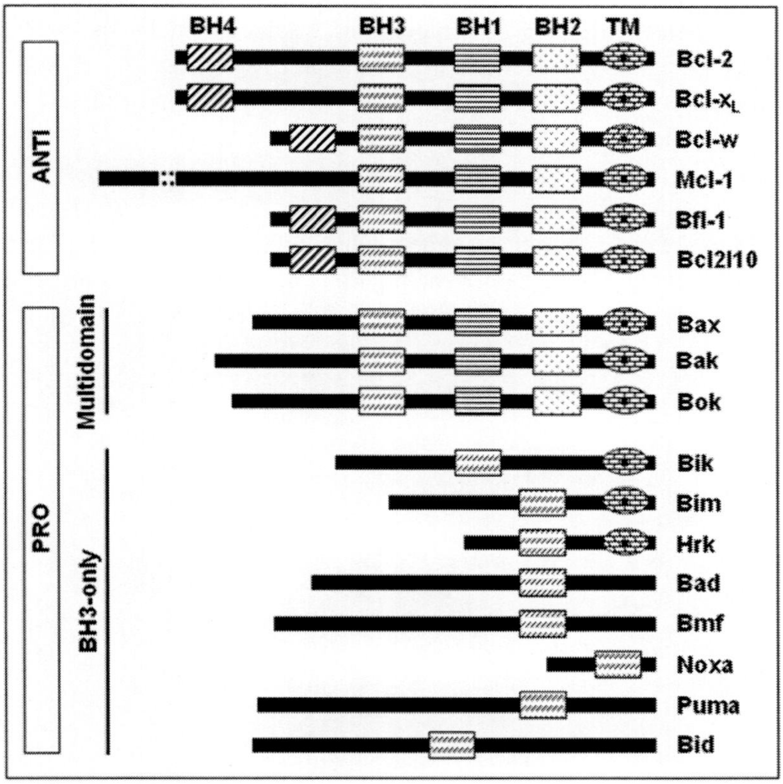

Figure 1. Extrinsic and intrinsic cell death signaling pathways. There are two major apoptotic pathways involved in programmed cell death: the mitochondrial (intrinsic) pathway and the death receptor (extrinsic) pathway. In the mitochondrial pathway, apoptotic signals are relayed by pro-apoptotic members of the Bcl-2 family ('Bax/Bak') to mitochondria, where release of apoptogenic factors such as cytochrome c and Smac/Diablo is induced. Anti-apoptotic members ('Bcl-2/xL') inhibit mitochondrial outer membrane permeabilization and cytochrome c release. For the majority of cell death stimuli, the intrinsic death signal is communicated to the mitochondrion through BH3-only proteins. Once released, cytochrome c interacts with Apaf-1 in the presence of dATP to form the apoptosome, which binds and activates procaspase-9. Caspase-9 in turn activates downstream caspases such as caspase-3, -6 and -7, which cleave a variety of substrates and the cell undergoes apoptosis. In contrast, the death receptor apoptotic pathway is initiated by the binding of death ligands (i.e. TNF, FasL, TRAIL) to their cognate transmembrane death receptors. Upon interaction, caspase-8 is activated and, similar to caspase-9 in the intrinsic pathway, processes downstream caspases. Cross-talk can occur between the extrinsic and intrinsic pathways *via* the proteolysis of Bid by caspase-8. Caspase-8 cleaves cytosolic Bid generating a truncated Bid (tBid) that translocates to the mitochondria and promotes the release of cytochrome c in concert with Bax/Bak.

II-Bcl-2 Family Proteins: Key Regulators of the Mitochondrial Apoptotic Pathway

II.1. Structure and Function

Proteins of the Bcl-2 family are considered critical arbiters of cell survival and are functionally categorized as either pro-apoptotic or anti-apoptotic based on their ability to promote or inhibit mitochondrial outer membrane permeabilization. At the structural level, this evolutionary conserved group of proteins has been defined by the presence of Bcl-2 homology (BH) domains (Aouacheria et al., 2005; Cory et al., 2003). Prosurvival multidomain members generally share sequence similarity in four conserved BH domains (BH1–4), while pro-death multidomain members contain only two or three BH domains (BH1-3) (Figure 2). To date, the human repertoire of multidomain members comprises six or seven prosurvival members (Bcl-2, Bcl-xL, Bcl-w, Mcl-1, Bcl2l10, Bfl-1 and also possibly Bcl2l12) and up to six pro-apoptotic members (Bax, Bak, Bok, Bcl-G, Bcl-rambo and Bfk). Among the death-inducing Bcl-2 family members is a group of death proteins collectively termed the ''BH3-only'' subfamily (including Bid, Bad, Bik, Bim, Bmf, Noxa, Puma and Hrk

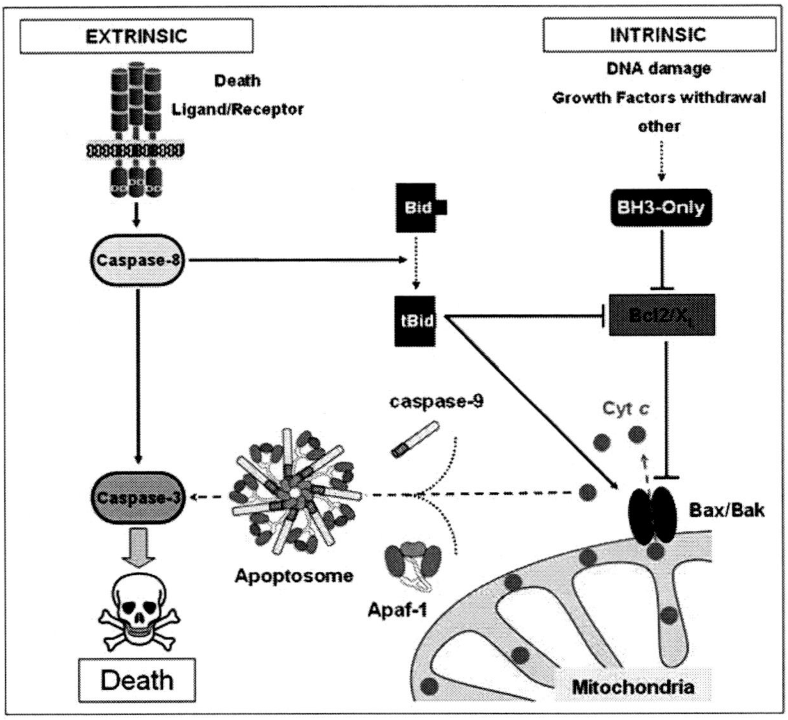

Figure 2. Shematic representation of Bcl-2 family members. Pro- and anti-apoptotic members differ in the number of their BH domains (Bcl-2 homology) (legend at the top) and in the presence or absence of a transmembrane domain (TM). Multidomain members can be divided into antiapoptotic proteins that possess Bcl-2 homology (BH) domains BH1, BH2, BH3 and BH4, and proapoptotic proteins with three homology domains (BH1-BH3). Pro-apoptotic members can be further subdivided into pro-apoptotic multidomain members and BH3-only proteins. BH3-only members display sequence conservation only in the third BH domain.

in humans) that exhibit sequence similarity only in the BH3 domain. Most multidomain members and several BH3-only proteins contain a hydrophobic segment at their C-terminal end thought to confer membrane anchorage. This transmembrane domain (TM) facilitates the interaction with the endoplasmic reticulum, the nuclear envelope and the mitochondrial outer membrane, where these proteins reside or translocate during apoptosis. Dynamic control over the different subcellular localizations can occur through interaction with a binding protein (e.g. dynein in the case of Bim), heterodimerization, conformational change (e.g. Bax) or other post-translation mechanisms such as phosphorylation (e.g. Bad) and proteolysis (e.g. Bid) (Gross et al., 1998; Kang et al., 2005; Kaufmann et al., 2004; Puthalakath et al., 1999; Shirane and Nakayama, 2003; Wang et al., 1999).

Experimentally-determined three-dimensional (3D) structures and secondary structure predictions suggest that the core fold is conserved between multidomain proteins of the Bcl-2 family (including viral members), and is composed of a globular bundle of amphipathic α-helices surrounding one or two central hydrophobic α-helices (see Table 1 for an up-to-date listing of the available 3D structures). This helical bundle fold has structural analogy with the pore-forming subunits of bacterial toxins (Petros et al., 2004). Based on this resemblance, at least four members of the Bcl-2 family have been shown to produce ion-conducting pores *in vitro* in synthetic lipid membranes (Antonsson et al., 1997; Minn et al., 1997; Schendel et al., 1999; Schlesinger et al., 1997). A hydrophobic groove formed on the surface of Bcl-2-like members by their BH1, BH2 and BH3 domains is required for binding to BH3-only proteins (and BH3-only domains) and for prosurvival function. Structure of Bax revealed that the C-terminal hydrophobic helix occupies the BH3-binding pocket (Suzuki et al., 2000). Displacement of the C-terminal tail from the hydrophobic groove is therefore a key event involved in the activation and targeting of Bax to the mitochondrial membrane upon apoptosis induction. Intriguingly, in members such as Bcl-w, Bcl-xL and Mcl-1, hydrophobic residues from the C-terminal region also obstruct the hydrophobic BH3 binding groove (Day et al., 2005; Denisov et al., 2003; Hinds et al., 2003; Jeong et al., 2004), suggesting that conformational changes of helix-bundled Bcl-2 family members may be a general mechanism of regulation, functioning aside or in addition to functional homo- and heterodimerizations (Dlugosz et al., 2006; Zhang et al., 2004). Most BH3-only proteins have unrelated predicted structures (with the exception of Bid), and some members of this clan (Bim, Bmf and Bad) were assigned to the class of intrinsically unstructured proteins (Hinds et al., 2007).Despite wide differences in amino acid sequence, the BH3-only protein Bid exhibits a conserved structure very similar to that of Bcl-2, Bcl-xL and Bax, i.e. multidomain members of the Bcl-2 family (Chou et al., 1999; McDonnell et al., 1999). Different from other proapoptotic Bcl-2 family members, Bid has evolved unique functions in interconnecting the death receptor pathway to the mitochondrial intrinsic pathway (Chou et al., 1999; McDonnell et al., 1999) and, although controversial (Kaufmann et al., 2007a; Kaufmann et al., 2007b), in controlling cell cycle progression (Kamer et al., 2005; Zinkel et al., 2005).

Table 1. Structure of Bcl-2 family proteins

PDB code	Method	Resolution (Å)	R-factor	Protein name	Notes	Ligands	Organism	Reference
1af3	X-ray	2.50Å	0.225	Bcl-XL A	1-196		Rat	Aritomi et al. 1997
1maz	X-ray	2.20Å	0.192	Bcl-XL A	-		Human	Muchmore et al. 1996
1pq1	X-ray	1.65Å	0.219	Bcl-XL A	1-196	mBim-BH3	Mouse	Liu et al. 2003
1pq0	X-ray	2,2	0,217	Bcl-XL A	1-196		Mouse	Liu et al. 2003
2bzw	X-ray	2.30 Å	0.219	Bcl-XL A	1-211	mBad	Mouse	Lee et al. 2004
2b48	X-ray	3.45 Å	0.261	Bcl-XL A	1-211	dimer	Human	O'Neill et al. 2006
1ysg	NMR			Bcl-XL A	-	4FC + TN1	Human	Oltersdorf et al. 2005
1ysi	NMR			Bcl-XL A	-	N3B	Human	Oltersdorf et al. 2005
1ysn	NMR			Bcl-XL A	-	43B	Human	Oltersdorf et al. 2005
1bxl	NMR			Bcl-XL A	-	Pep Bak	Human	Sattler et al. 1997
1g5j	NMR			Bcl-XL A	1-209	Pep BH3-Bad	Human	Petros et al. 2000
1lxl	NMR			Bcl-XL A	-		Human	Muchmore et al. 1996
1r2d	X-ray	1.95 Å	0.211	Bcl-XL A	-		Human	Manion et al. 2004
2o2m	NMR			Bcl-XL A	Bcl-2/Bcl-X		Human	Bruncko, M et al. 2007
2o2n	NMR			Bcl-XL A	Chimera Bcl-2/Bcl-X	LIW	Human	Bruncko, M et al. 2007
2oly	NMR			Bcl-XL A	Chimera Bcl-2/Bcl-X	43B	Human	Bruncko, M et al. 2007

Table 1. Continued

PDB code	Method	Resolution (Å)	R-factor	Protein name	Notes	Ligands	Organism	Reference
2yxj	X-ray	2,2	0,201	Bcl-XL A,B	-	ABT-737	Human	Lee et al. To be published
2pon	NMR			Bcl-XL B	Δ(45-84+197-233)	hBeclin-1	Huamn	Feng et al. 2007
2p1l	X-ray	2,5	0,225	Bcl-XL	Δ27-82	hBH3-Beclin-1	Human	Oberstein et al. 2007
1g5m	NMR			Bcl-2 isoform 1	Chimera Bcl-2/Bcl-X		Human	Petros et al. 2001
1gjh	NMR			Bcl-2 isoform 2	Chimera Bcl-2/Bcl-X		Human	Petros et al. 2001
2o22	NMR			Bcl-2 A	Chimera Bcl-2/Bcl-X	LIU	Human	Bruncko et al., 2007
2o21	NMR			Bcl-2 A	Chimera Bcl-2/Bcl-X	43B	Human	Bruncko et al. 2007
2o2f	NMR			Bcl-2 A	Chimera Bcl-2/Bcl-X	LIO	Human	Bruncko et al. 2007
1mk3	NMR			Bcl-W	P116V		Human	Denisov et al. 2003
1o0l	NMR			Bcl-W	A128E		Human	Hinds et al. 2003
1zy3	NMR			Bcl-W	2-171	BH3-hBid	Human	Denisov et al. 2006
1wsx	NMR			Mcl-1 A	152-308		Mouse	Day et al. 2005
2jm6	NMR			Mcl-1 B	152-308	mNoxaB-BH3	Mouse	Czabotar et al. 2007
2nI9	X-ray	1,55	0,174	Mcl-1	171-327 (Chimera)	BH3-hBim	Rat/human	Czabotar et al. 2007

Table 1. Continued

PDB code	Method	Resolution (Å)	R-factor	Protein name	Notes	Ligands	Organism	Reference
2nla	X-ray	2,8	0,211	Mcl-1	171-327 (Chimera)	BH3-mNoxaB	Rat/human	Czabotar et al. 2007
2pqk	X-ray	2	0,196	Mcl-1	171-327	Pep BH3-Bim	Human	Bare et al. To be published 2007
1ohu	X-ray	2.03	0.212	Ced-9	-		C. elegans	Jeong et al. 2003
1ty4	X-ray	2.20	0.220	Ced-9	68-237	Egl-1	C. elegans	Yan et al. 2004
2a5y	X-ray	2.6	0.249	Ced-9	48-251 (C107/135/164S)	Ced-4	C.elegans	Yan et al. 2005
1k3kA	NMR			vBcl-2			HHV8 (Viral)	Huang et al. 2001
1q59A	NMR			BHRF1			EBV (Viral)	Huang et al. 2003
2o42	X-ray	2,91	0.251	M11L			Myxoma virus (viral)	Douglas et al. 2007
2jby	X-ray	2.41	0.288	M11L	1-132	Bak-BH3	Myxoma virus (viral)	Kvansakul et al. 2007
2jbx	Xray	2.73	0.218	M11L	1-132		Myxoma virus (viral)	Kvansakul et al. 2007
2 abo	NMR			γHV68 v-Bcl-2			Murid herpesvirus 4	Loh et al. 2005
2ims	X-ray	1,48	0,181	Bak	Δ187-211	homodimer	Human	Moldoveanu et al. 2006
2imt	X-ray	1,49	0,194	Bak	Δ187-211	homodimer	Human	Moldoveanu et al.. 2006

Table 1. Continued

PDB code	Method	Resolution (Å)	R-factor	Protein name	Notes	Ligands	Organism	Reference
2jcn	X-ray	1.80	0.189	Bak			Human	Moche et al. To be published, 2007
2bid	NMR			Bid		GS-Bid	Human	Chou et al. 1999
1ddb	NMR			Bid	2-171	P116V	Mouse	Mcdonnell et al. 1999
1f16	NMR			Bax A			Human	Suzuki et al. 2000

Footnotes: m: mouse ; h : human ; Pep :Peptide ; GS-Bid : Glycine-Serine before Bid sequence. 4FC : 4'-FLUORO-1,1'-BIPHENYL-4-CARBOXYLIC ACID ($C_{13}H_9FO_2$);TN1 : 5,6,7,8-TETRAHYDRONAPHTHALEN-1-OL ($C_{10}H_{12}O$); N3B : N-[(4'-FLUORO-1,1'-BIPHENYL-4-YL)CARBONYL]- 3-NITRO-4-{[2-(PHENYLSULFANYL)ETHYL]AMINO}BENZENESULFONAMIDE ($C_{27}H_{22}FN_3\ O_5S_2$);43B : 3-NITRO-N-{4-[2-(2-PHENYLETHYL)-1,3-BENZOTHIAZOL- 5-YL]BENZOYL}-4-{[2-(PHENYLSULFANYL)ETHYL]AMINO}BENZENESULFONAMIDE ($C_{36}\ H_{30}\ N_4\ O_5\ S_3$);LIW : 4-[4-(BIPHENYL-2-YLMETHYL) PIPERAZIN-1-YL]- N-[(4-{[1,1-DIMETHYL-2-(PHENYLTHIO)ETHYL]AMINO}- 3-NITROPHENYL)SULFONYL]BENZAMIDE($C_{40}H_{41}N_5O_5S_2$);LIU: N-[(4-{[1,1-dimethyl-2-(phenylthio)ethyl]amino}- 3-nitrophenyl)sulfonyl]-4-(4,4-dimethylpiperidin- 1-yl)benzamide ($C_{30}\ H_{36}N_4O_5S_2$) ;LIO : 4-(4-BENZYL-4-METHOXYPIPERIDIN-1-YL)-N-[(4- {[1,1-DIMETHYL-2-(PHENYLTHIO)ETHYL]AMINO}- 3-NITROPHENYL)SULFONYL]BENZAMIDE($C_{36}H_{40}N_4O_6S_2$).

II.2. Mechanisms of Apoptosis Regulation by the Bcl-2 Family

Extensive evidence demonstrates that the relative balance of opposing Bcl-2 family members and their mutual interactions is crucial for regulating the mitochondrial cell death pathway. However, molecular details as to how the three subgroups of Bcl-2 family members (multidomain antiapoptotic, multidomain proapoptotic and BH3-only proteins) control apoptosis are still unclear. Cells isolated from Bax/Bak double knock-out mice exhibit a dramatic resistance to a wide range of noxious stimuli and do not undergo mitochondrial outer membrane permeabilization in response to apoptosis induction (Lum et al., 2005; Wei et al., 2001). Mechanistically, these critical apoptosis inducers are thought to homo-oligomerize and form pores in the mitochondrial outer membrane, thus causing (either alone or in cooperation with other factors) the efflux of cytochrome c and other proteins sequestered in the mitochondrial intermembrane space (Adams and Cory, 2007). The proapoptotic action of Bax/Bak is antagonized by anti-apoptotic Bcl-2 family proteins, which probably carry out their prosurvival function by acting on the cytosolic sides of the mitochondrial and ER/perinuclear membranes. BH3-only proteins have progressively emerged as crucial intermediates connecting multiple cytotoxic signals to the core apoptotic pathway. These death factors cannot induce apoptosis in cells lacking both Bax and Bak and therefore function upstream of the multidomain proteins (Chen et al., 2005; Zong et al., 2001). Different apoptotic stimuli can trigger apoptosis through activation of distinct BH3-only proteins, and this activation can be achieved either at the transcriptional (e.g. Puma and Noxa) or post-translational level (e.g. Bid or Bad) (Labi et al., 2006). Interestingly, BH3-only members display distinct preferences of association to their antiapoptotic Bcl-2-like relatives, with certain BH3-only proteins able to bind to all known prosurvival proteins (e.g. tBid, Bim and Puma) and others exhibiting marked selectivity (e.g. Bad and Noxa) (Figure 3A) (Certo et al., 2006; Chen et al., 2005; Kuwana et al., 2005). The characterization of the molecular mode of action of BH3-only proteins is currently undergoing intense investigation (Figure 3B). In one model, the BH3-only proteins Bid and Bim can directly promote Bax and Bak activation and are therefore called “activators” (Kuwana et al., 2005; Letai et al., 2002; Walensky, 2006). Other BH3-only proteins such as Bad, Puma and Noxa (termed “sensitizers”) may instead induce apoptosis by releasing Bid and Bim from prosurvival members (e.g. Bcl-2 or Bcl-xL). In another model, BH3-only proteins act by binding to and inhibiting prosurvival Bcl-2 proteins, this neutralization being sufficient to promote Bax activation and permeabilization of the mitochondrial outer membrane (Chen et al., 2005; Willis et al., 2005; Willis et al., 2007). The picture is likely to be more complex because different cells may express variable levels of Bcl-2-like prosurvival proteins, Bax/Bak effectors and Bax/Bak activators, i.e. BH3-only proteins. This view has recently received a striking illustration from Del Gaizo Moore et al. and Deng et al. who applied a technique called ‘BH3 profiling’ to examine the response of cancer cells to the Bcl-2-like antagonist ABT-737 (Del Gaizo Moore et al., 2007; Deng et al., 2007). Their observations suggest that not only expression levels of anti-apoptotic Bcl-2 proteins is critical for cancer cell survival, but that their binding status to death activators (e.g. Bim) determine if a cell is ‘primed’ for death or not (i.e. undergo apoptosis in response to a Bcl-2 antagonist).

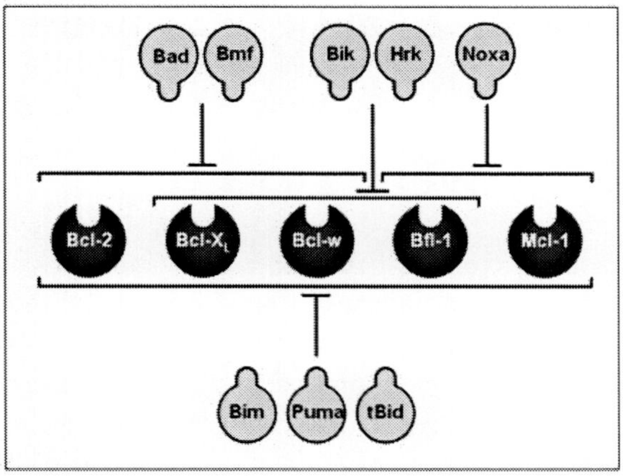

(a)

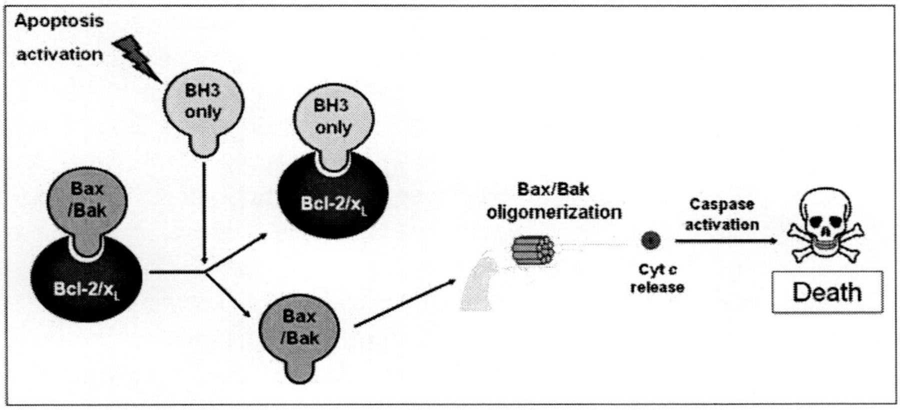

(b)

Figure 3. Models of Bcl-2 family protein regulation in apoptosis. (A) Different BH3 only proteins bind to and inhibit specific antiapoptotic Bcl-2 proteins. Bim Puma and tBid neutralize Bcl2, Bcl-xL, Bcl-w, Mcl-1 and Bfl-1 while other BH3-only proteins are more selective. (B) Anti-apoptotic members such as Bcl-2 and Bcl-xL inhibit Bax/Bak pro-apoptotic members and prevent them from inducing mitochondrial outer membrane permeabilization. BH3-only proteins act as upstream sentinels for cellular stress or damage and either activate Bax/Bak ("activators") or inactivate Bcl-2/xL ("sensitizers").

A last point, worthy of note, is that proteins from outside the Bcl-2 family have been reported to bind to and modulate the function of all three subgroups of Bcl-2 proteins (Cheng et al., 2003; Puthalakath et al., 1999; Puthalakath et al., 2001; Shibasaki et al., 1997). In addition, post-translational modifications of Bcl-2 family members by phosphorylation (Blagosklonny, 2001; del Peso et al., 1997) or proteolysis (Clem et al., 1998; Fadeel et al., 1999) can significantly affect their activity, while alternative splicing produces many Bcl-2 family protein variants with different functions (Akgul et al., 2004).

III. Targeting BCL-2 Family Members

III.1. Current Approaches in Bcl-2 Family Protein Inhibition

III.1.1. Anti-apoptotic Members

The *bcl-2* gene, isolated from follicular B-cell lymphoma cells bearing the t14;18 translocation, was one of the first oncogenes to be identified and the first found to enhance cell survival rather than cell proliferation (Tsujimoto et al., 1985; Vaux et al., 1988). Since this initial discovery, antiapoptotic Bcl-2 family members have been shown to be expressed in many human cancers and are thought to play a significant role in carcinogenesis, tumor progression and resistance to chemotherapy (Adams and Cory, 2007). Significantly, data from The National Cancer Institute Anticancer Drug Screen (NCI-ACDS) revealed a strong correlation between Bcl-xL expression and resistance to chemotherapy in a panel of human tumor cell lines (Amundson et al., 2000). It is therefore believed that an effective therapy for cancer may be provided by blocking Bcl-2-like prosurvival members. A number of studies exemplified the interest in targeting anti-apoptotic Bcl-2 family genes at the level of gene expression, e.g. through antisense oligonucleotides (Kim et al., 2007) or RNA interference (Bai et al., 2005). These genetic approaches have several disadvantages including possible off-target effects, insufficient delivery or stability, and the need of prolonged suppression of mRNA accumulation for proteins with slow turnover rates (which proved to be the case for *bcl-2*). The most common *bcl-2* antisense oligonucleotide is known as Genasense® or oblimersen sodium (formerly G3139), which has shown promising activity in some, but not all trials. Drugs targeting and directly inhibiting anti-apoptotic Bcl-2 family proteins are the alternative that is currently being extensively evaluated. An attractive approach for achieving functional inhibition of prosurvival Bcl-2 proteins is to mimic the activity of pro-apoptotic members, and in particular BH3-only proteins. Cell-permeable peptides, designed to mimic the BH3 domain of Bcl-2 family death members, were shown to bind to prosurvival Bcl-2-like proteins and induce apoptosis in Bcl-2 or Bcl-xL-expressing cells (Holinger et al., 1999; Petros et al., 2000; Wang et al., 2000b) (see part V-I). However, a number of obstacles are associated with the development of peptides as therapeutics including predicted high inhibitory concentrations *in vivo* and high costs. Synthetic small-molecules inhibitor drugs that bind the BH3 pocket of Bcl-2 proteins represent another implementation of the BH3-mimetic screening strategy. A number of such antagonists are currently at various stages of development, such as GX15-070 (Trudel et al., 2007), gossypol derivatives (Becattini et al., 2004a; Kitada et al., 2003), antimycin analogs (Tzung et al., 2001), HA14-1 (Wang et al., 2000a), TW-37 (Mohammad et al., 2007), BH3I-1 and BH3I-2 (Degterev et al., 2001) (reviewed in (Reed and Pellecchia, 2005)). While most of these compounds have less affinity for the BH3 binding groove of prosurvival Bcl-2 members than BH3 peptides (Baell and Huang, 2002; O'Neill et al., 2004), a number of small molecule ligands (e.g. ABT-737, pyrogallol derivatives, terephthalamides and terphenyls) were recently reported that bind tightly to Bcl-xL with Kd values in the nanomolar range (Oltersdorf et al., 2005; Tang et al., 2007; Yin et al., 2005a; Yin et al., 2005b).

III.1.2. Pro-apoptotic Members

Because excessive apoptosis plays a role in degenerative diseases, modulation of pro-apoptotic members is becoming therapeutically relevant and potent compounds directly targeting Bax/Bak or BH3-only proteins could emerge as promising drug targets. However, most of the published strategies have investigated Bax translocation to mitochondria during apoptosis, i.e. a post-translational event. Bax is largely localized within the cytosol of healthy cells and, in response to apoptosis induction, undergoes a conformational change and translocates to mitochondrial membranes. Several proteins including Ku70 (Cohen et al., 2004; Sawada et al., 2003b) and Humanin (Guo et al., 2003) have been shown to bind Bax and to block its activation. Under normal growth conditions, Ku70 in its unacetylated form sequesters Bax in the cytosol, and cell damage causes Ku70 acetylation (by CBP or PCAF) and Bax release. Ku70 suppression of the Bax-mediated cell death pathway offers new perspectives for the control of inappropriate apoptosis, either in the form of cell-permeable peptides based on the Bax-inhibiting domain of Ku70 or chemical modulators of Ku70 acetylation (Sawada et al., 2003a; Sawada et al., 2003b). Combining these agents with CBP or PCAF inhibitors is an attractive strategy for reducing apoptosis, although selective inhibitors have yet to be discovered for both enzymes. Humanin is a 24-aminoacid secreted peptide that has also been shown to inhibit Bax translocation to mitochondria and cell death. In addition, Humanin may also suppress apoptosis by interacting with Bid or BimEL (a Bim isoform) and preventing Bax activation induced by these BH3-only proteins (Luciano et al., 2007; Zhai et al., 2005). Screening programs have been initiated to identify compounds that may modulate the interaction of Humanin with Bax (or Bid and Bim). As Humanin was reported to block neuronal cell death in Alzheimer disease (Hashimoto et al., 2001), β-amyloid toxicity in human cerebrovascular smooth muscle cells (Jung and Van Nostrand, 2003) and prion-peptide toxicity in rat cortical neurons (Sponne et al., 2004), Humanin derivatives may be of great interest for the treatment of neurodegenerative diseases. However, the risk of developing cancer should be carefully assessed because up-regulation of the mitochondrial 16S mRNA (which encodes Humanin) was suggested in some malignancies (Maximov et al., 2002). Last, Bax carries a number of regulatory sequences controlling the exposition of its membrane-targeting domains, such as the N-terminal ART motif (for apoptosis-regulating targeting sequence) that prevents its insertion into the mitochondrial outer membrane in cells that are not committed to die (Cartron et al., 2005; Goping et al., 1998). Clearly, these regulatory sequences represent attractive targets for selective cytoprotection, and compounds that specifically bind to the ART domain and inhibit Bax insertion into the mitochondrial membrane may have cell-protective activity. Although promising, it remains to be determined which, if any, of these strategies may be successful in preventing Bax activation *in vivo*. Perhaps the most exciting aspect of this research is that it paves the ground for the development of agents that would inhibit post-translational events involving other Bcl-2-like death factors, such as Bid cleavage, Bad phosphorylation or Bim translocation from microtubules to mitochondria.

III.2. Fragment-Based drug Discovery of Bcl-2 Family Protein Inhibitors

Structure-based drug design combined to fragment-based screening represent a powerful method for identifying novel, high affinity, leads in modern biomedical research (Erlanson, 2006; Hajduk and Greer, 2007). Chemical libraries may be small (from dozens to hundreds of compounds) and focus on a limited range of structures, or may be large (e.g., thousands of compounds) and encompass a wide variety of structural variants. Low affinity fragments (with K_D in the millimolar range) can be converted to high affinity compounds (with K_D values in the nano- or micromolar range) upon chemical bonding. In this part, we focus on two key examples of small molecule drugs that were discovered through fragment-based approaches and that inhibit Bcl-2 family proteins with opposite function (Figure 4).

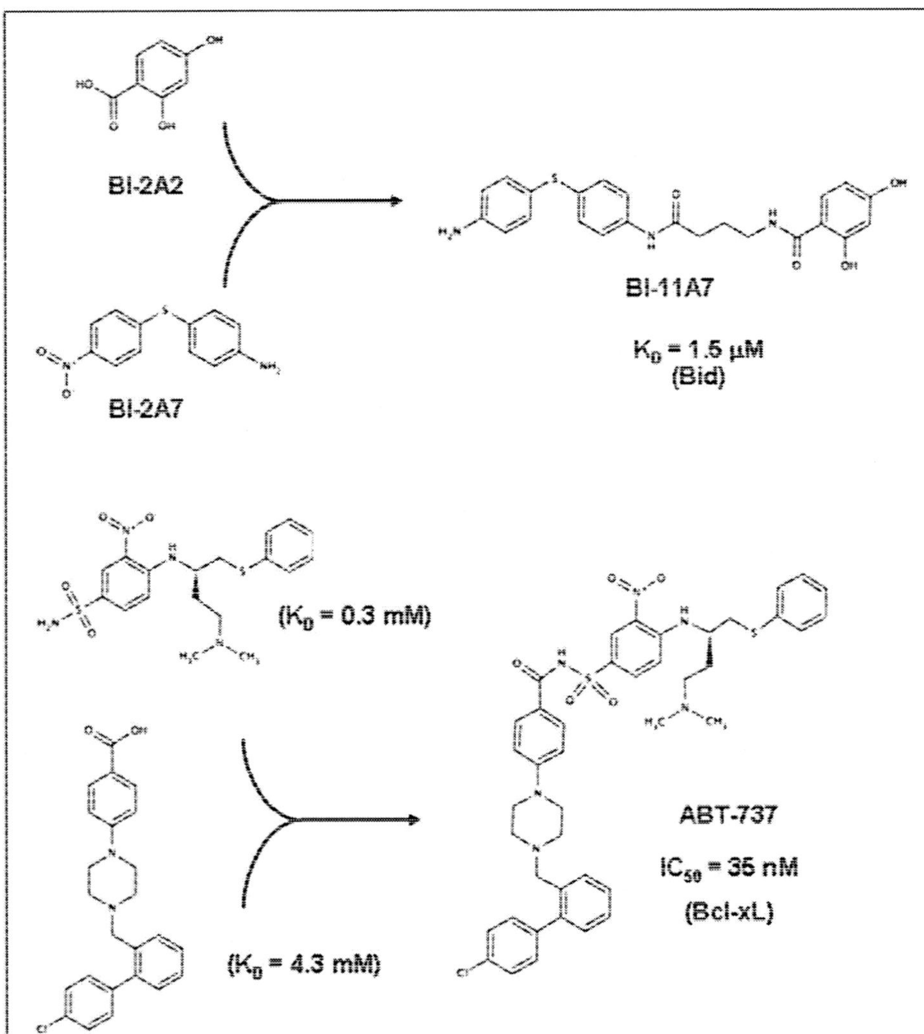

(a)

Figure 4. Continued on next page.

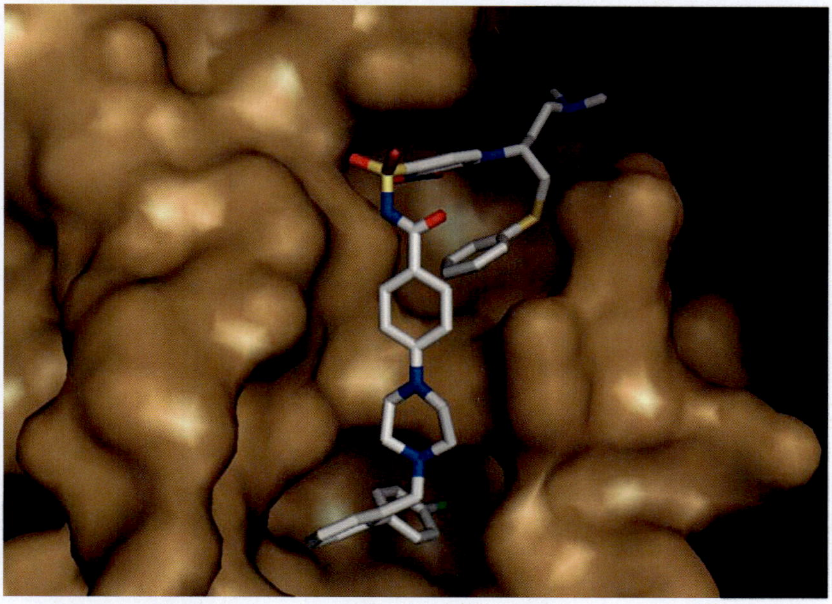

(b)

Figure 4. Fragment based drug discovery on Bcl-2 family proteins. (A) Chemical structures of ABT-737 and BI-11A7. The molecule ABT-737, which binds and inhibits Bcl-2, Bcl-xL and Bcl-w, showed potent anti-tumour effects *in vivo*. BI-11A7 is a dual ligand engineered through 'SAR-by-NMR' that occupies a deep pocket on the surface of Bid. The drug inhibited Bid-induced cell death *in vitro*. The chemical structures of the initial fragments and of the optimized small-molecule inhibitors are drawn. (B) Crystal structure of Bcl-xL in complex with ABT-737 (PDB code 2YXJ).

III.2.1. Inhibitors of Prosurvival Bcl-2-Family Proteins: ABT-737

ABT-737 is a BH3-mimic developed at Abbott Laboratories through a combination of structure-based design and medicinal chemistry. The discovery of this drug is the result of a nuclear magnetic resonance–based high-throughput screen (called 'SAR-by-NMR') for compounds that specifically binds to the BH3-binding cleft of Bcl-xL (Oltersdorf et al., 2005). The project made extensive use of NMR-based structural data pertaining to Bcl-xL alone or complexed to a Bak BH3 peptide. A pair of small molecules that bind simultaneously to the hydrophobic groove of Bcl-xL were isolated (from libraries containing 10,000 and 3,500 fragments) and chemically linked together (Figure 4A). The resulting compound was then optimized to reduce binding to human serum albumin. The final product, ABT-737, was found to fit with high affinity (Ki ≤ 1nM) and specificity into the BH3-binding groove of Bcl-xL. Further investigations indicated that, like Bad, ABT-737 only targeted the highly homologous Bcl-2, Bcl-xL and Bcl-w proteins. As expected from the sequence divergence in their BH3 docking region, ABT-737 does not bind tightly to the less homologous proteins Mcl-1, Bcl2l10, and A1. Interestingly, recent crystal structure determination of ABT-737 bound to Bcl-xL revealed that the recognition sites for the drug and for the Bak BH3 peptide (used in the initial discovery process) are slightly different. As the key binding interactions do not involve identical residues, ABT-737 and BH3 peptides should therefore be considered as functional but not structural analogs (Lee et al., 2007).

ABT-737 triggered apoptosis of cancer cells lines and primary tumor cells from patients (e.g. chronic lymphocytic leukemia cells) (Kohl et al., 2007; Konopleva et al., 2006; Oltersdorf et al., 2005). Moreover, it has demonstrated anti-cancer activity in tumor animal models (including small-cell lung cancer xenografts) and showed synergy with conventional cytotoxics both *in vitro* and *in vivo* (Kang et al., 2007; Oltersdorf et al., 2005). However, because it cannot neutralize Mcl-1, Bfl1/A1 and Bcl2l10, ABT-737 was not expected to be cytotoxic as a single-agent in cancer cells that overexpress these anti-apoptotic proteins. Consistently, ABT-737 exhibited reduced killing effects in tumor cells with high levels of Mcl-1 (Konopleva et al., 2006; van Delft et al., 2006) and resistant cells became sensitized to ABT-737-induced apoptosis when Mcl-1 was eliminated by RNAi (Lin et al., 2007) or antagonized by its selective ligand, Noxa (Tahir et al., 2007). These results suggest that most cell types are protected by more than one of the prosurvival proteins, and that combining the Bcl-2/xL inhibitor ABT-737 with antagonists of Mcl-1, Bcl2l10 or Bfl-1/A1 might be a useful strategy against cancer cells. The orphan steroid receptor Nur77 has been reported to bind to Bcl-2, Bcl2l10 and Bfl-1/A1 (but not to Bcl-xL, Mcl-1, or Bcl-w) and a Nur-77 derived synthetic peptide was shown to antagonize Bcl2l10 prosurvival function (Luciano et al., 2007), providing a strategy for inhibition of these prosurvival members. Combining ABT-737 or Nur77 peptidomimetics with agents that antagonize Mcl-1 may prove to be a potent apoptosis-inducing approach in cancer therapy. Advanced technologies such as RNA interference, enhancement of protein turnover boosted by kinase inhibitors (Breitschopf et al., 2000; Kucharczak et al., 2005; Zhong et al., 2005), and peptide aptamers (Nouvion et al., 2007) provide alternate routes for inhibition of Mcl-1, Bfl-1/A1 and Bcl2l10.

III.2.2. Inhibitors of Pro-apoptotic Bcl-2-Family Members: BI-11A7

Sustained cell death causes the loss of neurons and neurological deficits following stroke or neurodegenerative diseases. Bid is a key pro-apoptotic member of the Bcl-2 family which acts upstream in the control of the apoptotic program, leading to neuronal cell death in numerous experimental models relevant to neurological disorders (Bermpohl et al., 2006; Culmsee and Plesnila, 2006; Culmsee et al., 2005; Franz et al., 2002; Henshall et al., 2001; Plesnila, 2004; Plesnila et al., 2002; Plesnila et al., 2001; Yamamoto et al., 2006; Yin et al., 2002). In particular, Bid is a critical mediator of ischemic cell death within the central nervous system, as shown by investigations in Bid-/- knock out cells and animals. Bid therefore constitutes an attractive target for strategies directed towards the development of neuroprotective therapeutics. Upon apoptosis induction, Bid is activated by proteolytic cleavage, moves to mitochondria and permeabilizes the outer mitochondrial membrane, thereby promoting cytochrome c release into the cytosol and cell death (see Figure 1 and (Yin, 2006) for a review). Pellechia's group recently employed an approach called 'SAR by ILOES'(structure-activity relationships by interligand nuclear Overhauser effect) to screen a library of about 300 chemical fragments to identify potential Bid small-molecule inhibitors (Becattini et al., 2006; Becattini et al., 2004b). Two weak affinity, low molecular weight compounds interacting at proximal sites within a deep hydrophobic crevice on the surface of Bid (near to the BH3 region) were isolated. Using the NMR three-dimensional solution structure of Bid as a guide for computer-based docking, chemical linkers of various lengths were added between the two small molecules (Figure 4B). One of the resulting compounds (BI-11A7), containing a three-carbon chain linker, bound with enhanced affinity to Bid.

Intriguingly, this bidentate compound was reported to bind with higher affinity to tBid, suggesting that proteolytic cleavage of Bid elicits only minor structural alterations in the remaining fragment (tBid). Although the small molecule was not able to bind Bcl-xL, the specificity this pharmacologic inhibitor requires further demonstration.

BI-11A7 prevented Bid-induced release of Smac/Diablo (an intermembrane space mitochondrial protein) from purified mitochondria *in vitro* and reduced significantly Bid-mediated apoptosis in cultured cell lines. In addition, the compound conferred significant protection to rat primary neurons exposed to glutamate (Becattini et al., 2006). Unfortunately, this small molecule Bid inhibitor was not examined in animal models and concerns about potential toxicity remain to be resolved. In any event, nonpeptide chemical inhibitors of Bid (and of its active form tBid) represent attractive tools for elucidating the role of this death protein in neurodegenerative disorders. Moreover, small-molecule compounds selectively targeting Bid might serve as lead compounds for future drug development in the treatment of cerebrovascular accidents (as pharmacologic adjuncts to current anti-thrombolytic therapy).

IV. Manipulating BCL-2 Family Members

IV.1. BH3 Peptides

Since pro-apoptotic Bcl-2 family proteins contact the hydrophobic groove of prosurvival members *via* their BH3 domain, it has been hypothesized that BH3 mimetics, discovered through rational design or functional screening, could induce cancer cell apoptosis by targeting anti-apoptotic members. Based on the finding that peptides comprising the BH3 domain of pro-apoptotic Bcl-2 family members could indeed act as specific death inducers, agonistic BH3 peptides from selected BH3-only proteins (e.g. Bid and Bad) were first developed. Second-generation BH3 peptides with enhanced pharmacological properties such as protease-resistance, improved α-helicity, cell-permeability and increased binding potency to Bcl-2 proteins were recently described (Dharap and Minko, 2003; Goldsmith et al., 2006; Walensky et al., 2004; Yang et al., 2004). In one of the reports, the authors envisioned utilizing olefin metathesis strategy for BH3 peptide synthesis and generated a Bid BH3 α-helix stabilized by an all-hydrocarbon staple. This Bid-derived stapled peptide activated apoptosis in leukaemia cell lines and inhibited the growth of human leukaemia xenografts in mice (Walensky et al., 2004). In another study, cyclic lactam bridged BH3 peptide analogues with increased helical structure were synthesized (Yang et al., 2004). Cell-permeable peptides were developed that fuse a D-isomer arginine homopolymer (termed r8) to the BH3 domains of Bid or Bad, and these r8-BH3 peptides induced apoptosis of neuroblastoma cells *in vitro* and *in vivo* in a mouse xenograft model (Goldsmith et al., 2006). Peptides were also engineered that conjugate the BH3 motif to moieties that target specific cell surface molecules on tumor cells, in order to minimize unwanted side effects. For instance, tumor cells overexpressing the luteinizing hormone-releasing hormone (LHRH) receptor were killed by a Bak-derived BH3 peptide conjugated to a synthetic LHRH peptide (Dharap and Minko, 2003). Current drug discovery programs continue to focus their efforts on converting biologically active BH3 peptides into clinically and commercially viable agents and assessing their efficacy in animal models. As previously mentioned, BH3 peptides from pro-apoptotic Bcl-2 family proteins vary from one another in their binding profiles among prosurvival Bcl-2 family proteins. Therefore, a next step could be the design of peptidomimetics with improved specificity (i.e. that can discriminate against the BH3-binding clefts of pro-apoptotic members) and selectivity (i.e. that can bind to the BH3-binding pocket of one prosurvival member relative to others) (Gemperli et al., 2005). Foldamer-based ligands were recently designed that display some selectivity in binding to Bcl-xL and Bcl-w relative to Bcl-2 and Mcl-1 (Sadowsky et al., 2007).

IV.2. Non-BH3 Active Peptides Derived from Bcl-2 Family Proteins

While Bcl-2 proteins are long recognized as targets for drug-discovery to treat apoptosis-related human diseases, only few attempts have been made to use of non-BH3, (poly) peptide fragments of these proteins as tools for biotechnological applications (Table 2).

IV.2.1. Pro-apoptotic Peptides

Following the demonstration that Bax-derived fragments could induce cell death when overpexressed in E. coli and cultured cell lines (Ishibashi et al., 1998), an adenoviral construct encoding a truncated form of Bax under the control of the Cre/LoxP system was shown to induce apoptosis both *in vitro* and *in vivo* in the A549 subcutaneous xenograft lung tumor model (Usui et al., 2003). The N-terminally truncated form of Bax (ΔN Bax) used in the study lacks BH3 and the first helix but retains part of the channel-forming domain and a complete hydrophobic tail. Data indicated that there were marked differences in the mechanism of cytotoxicity following full-length Bax or ΔN Bax expression. Most striking was the observation that the cell death induced by ΔN Bax could not be prevented by a pan-caspase inhibitor whereas it significantly reduced the cell-death-inducing activity of wild-type Bax. Furthermore, deletion of the N-terminal portion of Bax significantly enhanced the pro-apoptotic activity when compared to the full-length protein, suggesting that the N-terminal region of pro-apoptotic multidomain members exerts regulatory function over the C-terminal moiety. Interestingly, cleavage releasing the C-terminal part can convert an anti-apoptotic Bcl-2 family member into a pro-apoptotic one (e.g. cleavage of Bcl-2, Bcl-xL and Bfl-1) (Basanez et al., 2001; Cheng et al., 1997; Kirsch et al., 1999; Kucharczak et al., 2005; Liang et al., 2002). These observations raise the possibility that the C-terminal region (including the putative pore-forming helices and TM domain) perform a toxic function both in pro-apoptotic and anti-apoptotic members (Aouacheria et al., 2005). In this regard, the C-terminal domain of Bfl-1 was reported to be a potent cell death inducer when fused to GFP or other label (Ko et al., 2003; Yang et al., 2005). A synthetic peptide corresponding to this tail-anchor domain induced cytochrome c release from purified mitochondria. Moreover, the Bfl-1 tail-anchor domain was found to exhibit mitochondrial-targeting and amphipathic properties, providing evidence that this domain likely kill cells by compromising the integrity of the mitochondrial membrane. Consistent with this, the cell death induced by the tail-anchor region was shown to be independent of Bax and Bak and could not be rescued by overexpression of prosurvival proteins such as Bcl-xL.

Table 2. Non-BH3 (poly)peptides derived from Bcl-2 family proteins

Name	Reference	Sequence	Product
PRO-APOPTOTIC			
HCCS-1 Tail-Anchoring	Ko et al., 2007	KFLEPKSGWMTFLEVTGKICEMLFCPEAILL TRKDTPYC (34-71)	GFP-Fusion Protein
Bfl1 Tail-Anchoring	Ko et al., 2007 Yang et al., 2005	KFEPKSGWMTFLEVTGKICEMLSLLKQYC (147-175)	GFP-Fusion Protein Synthetic Peptide
Calpain-cleaved Bax	Gao et al., 2000	Bax 18KD MW	GFP-Fusion Protein
ΔN Bax	Usui et al. 2003	(112-192)	Adenoviral
C-Terminal Noxa (BH3+MTD)	Seo et al., 2003	MPGKKARKNAQPSPARAPAELEVECATQL RRFGDKLNFRQKLLNLISKLFCSGT (1-54)	GFP-Fusion Protein
ANTI-APOPTOTIC			
Bcl-X_L – BH4	Shimizu et al., 2000	SNRELVVDFLSYKSQKGYS (4-23)	Synthetic Peptide
Bcl-2-BH4	Shimizu et al., 2000	TGYDNREIVMKYIHYKLSQRGYEW (7-30)	Synthetic Peptide
TAT-BH4 (Bcl-xL)	Sugioka Et al., 2003	Ac-CRKKRRQRRR-β-A-SNRELVVDFLSYKLSQKGYS-COOH	Synthetic Peptide
TAT-BH4 (Bcl-xL)	Hotchkiss et al., 2006	Ac-RKKRROrn-RRR-β-A-(*l*)-SNRELVVDFLSYKLSQKGYS-COOH	TAT-Fusion Protein Synthetic Peptide
TAT-Bcl-xL	Hotchkiss et al., 2006	Bcl-xL Complete	TAT-Fusion Protein

The cell death-inducing mechanisms utilized by fragments derived from Bax, Bfl-1 and possibly Noxa (Seo et al., 2003) appear novel as they might bypass the Bcl-2/Bax checkpoint and thus function irrespective of the anti-apoptotic Bcl-2 protein status in cancer cells. Hence, active peptidic fragments derived from Bcl-2 family proteins are certainly deserving of further evaluation for their efficacy as potential anti-cancer agents, when introduced into cancer cells, e.g. by adenoviral gene therapy, or directly delivered to tumors through coupling to specific ligands.

IV.2.2. Anti-apoptotic Peptides

The BH4 domain is unique to anti-apoptotic members of the Bcl-2 family, and interacts principally with proteins outside of the Bcl-2 family such as Raf-1, calcineurin and paxillin (Denis et al., 2003; Huang et al., 1998; Reed et al., 1996). Bcl-2 or its prosurvival homologues (e.g. Bcl-xL and Bcl2l10) deleted of their BH4 motif are no longer anti-apoptotic (Huang et al., 1998; Zhang et al., 2001), and proteolytic cleavage of Bcl-2 by caspase-3 or calpain produces a fragment deleted of the BH4 region which shows pro-apoptotic activity (Grandgirard et al., 1998; Vance et al., 1996). Interestingly, Mcl-1 lacks any recognizable BH4 domain (Aouacheria et al., 2005). The BH4 domain of Bcl-xL was found to close VDAC, an outer mitochondrial channel, and HIV Tat-conjugated synthetic BH4 peptides protected HeLa cells from etoposide-induced apoptosis (Shimizu et al., 2000). Moreover, Tat-BH4 peptides were shown to be readily internalized in a variety of cells, to prevent apoptosis *in vitro* and *in vivo* in several animal models including ischemia/reperfusion, DNA damage and sepsis-induced T-cell apoptosis (Hotchkiss et al., 2006; Sugioka et al., 2003). Thus, although less extensively studied than the BH3 motif, the BH4 domain has demonstrated important functions in the regulation of antiapoptotic Bcl-2 proteins, and may prove useful in treating diseases associated with excessive apoptosis. However, the BH4 signature is evasive compared to the other BH domains, which renders the design and engineering of BH4-like peptides more difficult. In addition, lack of knowledge about the structural determinants for interaction with regulatory proteins outside of the Bcl-2 family may hinder future development of BH4 mimetics.

Conclusion

While the precise *modus operandi* of Bcl-2 family proteins remains unclear, pharmacological manipulation of anti-apoptotic members has attracted much attention for cancer therapy. In contrast, only a few approaches have been reported for prolonging the life of cells by targeting key pro-apoptotic proteins (e.g. Bax, Bak or Bid). Although pan-inhibitors may prove useful in some circumstances, future work should aim at developing agents targeting selected members of the Bcl-2 family, as recently initiated for prosurvival Bcl-2 proteins (Chen et al., 2005; Gemperli et al., 2005; Zhai et al., 2006) or the pro-apoptotic Bid protein (Becattini et al., 2006; Becattini et al., 2004b). The ideal small-molecule Bcl-2 inhibitor should displays 'druggable' characteristics, can be administered orally, exhibits adequate distribution kinetics and has potent and specific activity at low concentrations in circulation. Extensive pre-clinical and clinical trials are needed for each new compound to satisfy such requirements. Because membrane association and/or integration are critical for the function of Bcl-2 family multidomain members, the development of innovative drugs may be delayed by the lack of structural information about their conformation at the membrane surface or upon membrane insertion. The design of drugs able to interfere with the activity of Bcl-2 proteins at the membrane level, by inhibiting membrane translocation, integration or oligomerization, is worthy of further attention, because it could give fine control over mitochondrial membrane permeabilization. Last, peptides and peptidomimetics derived from natural Bcl-2 family protein templates (e.g. BH3 peptides and Bax or Bfl-1 membrane-destabilizing fragments) offer promising new possibilities for cancer treatment. To finish up by coming back to the title of this review, inhibiting protein-protein interactions with small molecules sounds like assembling Lego blocks of different shapes into appropriate surface areas. If this game metaphor is to be employed, subverting the activity of Bcl-2 family proteins through the use of truncated fragments is then much more like destroying the toy to play with its small pieces.

Acknowledgements

YG is a recipient of. MRT (Ministère de la Recherche des Sciences et Techniques) PhD. Fellowship. JGV is a recipient of a doctoral fellowship from Region Rhône Alpes. We are grateful to Dr. Julien Thibaut for help with the illustrations and to Mrs. Aurélie Cornut for assistance. We also thank Mrs. Chouki Socks for critical reading of the paper. This work was supported by Ligue Contre le Cancer, Comité de la Drôme and by Fondation Thérèse et René Planiol.

References

Adams, J.M. and Cory, S. (2007) Bcl-2-regulated apoptosis: mechanism and therapeutic potential. *Curr Opin Immunol*, **19**, 488-496.

Akgul, C., Moulding, D.A. and Edwards, S.W. (2004) Alternative splicing of Bcl-2-related genes: functional consequences and potential therapeutic applications. *Cell Mol Life Sci*, **61**, 2189-2199.

Amundson, S.A., Myers, T.G., Scudiero, D., Kitada, S., Reed, J.C. and Fornace, A.J., Jr. (2000) An informatics approach identifying markers of chemosensitivity in human cancer cell lines. *Cancer Res*, **60**, 6101-6110.

Antonsson, B., Conti, F., Ciavatta, A., Montessuit, S., Lewis, S., Martinou, I., Bernasconi, L., Bernard, A., Mermod, J.J., Mazzei, G., Maundrell, K., Gambale, F., Sadoul, R. and Martinou, J.C. (1997) Inhibition of Bax channel-forming activity by Bcl-2. *Science*, **277**, 370-372.

Aouacheria, A., Brunet, F. and Gouy, M. (2005) Phylogenomics of life-or-death switches in multicellular animals: Bcl-2, BH3-Only, and BNip families of apoptotic regulators. *Mol Biol Evol*, **22**, 2395-2416.

Baell, J.B. and Huang, D.C. (2002) Prospects for targeting the Bcl-2 family of proteins to develop novel cytotoxic drugs. *Biochem Pharmacol*, **64**, 851-863.

Bai, J., Sui, J., Demirjian, A., Vollmer, C.M., Jr., Marasco, W. and Callery, M.P. (2005) Predominant Bcl-XL knockdown disables antiapoptotic mechanisms: tumor necrosis factor-related apoptosis-inducing ligand-based triple chemotherapy overcomes chemoresistance in pancreatic cancer cells in vitro. *Cancer Res*, **65**, 2344-2352.

Basanez, G., Zhang, J., Chau, B.N., Maksaev, G.I., Frolov, V.A., Brandt, T.A., Burch, J., Hardwick, J.M. and Zimmerberg, J. (2001) Pro-apoptotic cleavage products of Bcl-xL form cytochrome c-conducting pores in pure lipid membranes. *J Biol Chem*, **276**, 31083-31091.

Becattini, B., Culmsee, C., Leone, M., Zhai, D., Zhang, X., Crowell, K.J., Rega, M.F., Landshamer, S., Reed, J.C., Plesnila, N. and Pellecchia, M. (2006) Structure-activity relationships by interligand NOE-based design and synthesis of antiapoptotic compounds targeting Bid. *Proc Natl Acad Sci U S A*, **103**, 12602-12606.

Becattini, B., Kitada, S., Leone, M., Monosov, E., Chandler, S., Zhai, D., Kipps, T.J., Reed, J.C. and Pellecchia, M. (2004a) Rational design and real time, in-cell detection of the proapoptotic activity of a novel compound targeting Bcl-X(L). *Chem Biol*, **11**, 389-395.

Becattini, B., Sareth, S., Zhai, D., Crowell, K.J., Leone, M., Reed, J.C. and Pellecchia, M. (2004b) Targeting apoptosis via chemical design: inhibition of bid-induced cell death by small organic molecules. *Chem Biol*, **11**, 1107-1117.

Bermpohl, D., You, Z., Korsmeyer, S.J., Moskowitz, M.A. and Whalen, M.J. (2006) Traumatic brain injury in mice deficient in Bid: effects on histopathology and functional outcome. *J Cereb Blood Flow Metab*, **26**, 625-633.

Blagosklonny, M.V. (2001) Unwinding the loop of Bcl-2 phosphorylation. *Leukemia*, **15**, 869-874.

Breitschopf, K., Haendeler, J., Malchow, P., Zeiher, A.M. and Dimmeler, S. (2000) Posttranslational modification of Bcl-2 facilitates its proteasome-dependent degradation: molecular characterization of the involved signaling pathway. *Mol Cell Biol*, **20**, 1886-1896.

Cartron, P.F., Arokium, H., Oliver, L., Meflah, K., Manon, S. and Vallette, F.M. (2005) Distinct domains control the addressing and the insertion of Bax into mitochondria. *J Biol Chem*, **280**, 10587-10598.

Certo, M., Del Gaizo Moore, V., Nishino, M., Wei, G., Korsmeyer, S., Armstrong, S.A. and Letai, A. (2006) Mitochondria primed by death signals determine cellular addiction to antiapoptotic BCL-2 family members. *Cancer Cell*, **9**, 351-365.

Chen, L., Willis, S.N., Wei, A., Smith, B.J., Fletcher, J.I., Hinds, M.G., Colman, P.M., Day, C.L., Adams, J.M. and Huang, D.C. (2005) Differential targeting of prosurvival Bcl-2 proteins by their BH3-only ligands allows complementary apoptotic function. *Mol Cell*, **17**, 393-403.

Cheng, E.H., Kirsch, D.G., Clem, R.J., Ravi, R., Kastan, M.B., Bedi, A., Ueno, K. and Hardwick, J.M. (1997) Conversion of Bcl-2 to a Bax-like death effector by caspases. *Science*, **278**, 1966-1968.

Cheng, E.H., Sheiko, T.V., Fisher, J.K., Craigen, W.J. and Korsmeyer, S.J. (2003) VDAC2 inhibits BAK activation and mitochondrial apoptosis. *Science*, **301**, 513-517.

Chou, J.J., Li, H., Salvesen, G.S., Yuan, J. and Wagner, G. (1999) Solution structure of BID, an intracellular amplifier of apoptotic signaling. *Cell*, **96**, 615-624.

Clem, R.J., Cheng, E.H., Karp, C.L., Kirsch, D.G., Ueno, K., Takahashi, A., Kastan, M.B., Griffin, D.E., Earnshaw, W.C., Veliuona, M.A. and Hardwick, J.M. (1998) Modulation of cell death by Bcl-XL through caspase interaction. *Proc Natl Acad Sci U S A*, **95**, 554-559.

Cohen, H.Y., Lavu, S., Bitterman, K.J., Hekking, B., Imahiyerobo, T.A., Miller, C., Frye, R., Ploegh, H., Kessler, B.M. and Sinclair, D.A. (2004) Acetylation of the C terminus of Ku70 by CBP and PCAF controls Bax-mediated apoptosis. *Mol Cell*, **13**, 627-638.

Cory, S., Huang, D.C. and Adams, J.M. (2003) The Bcl-2 family: roles in cell survival and oncogenesis. *Oncogene*, **22**, 8590-8607.

Culmsee, C. and Plesnila, N. (2006) Targeting Bid to prevent programmed cell death in neurons. *Biochem Soc Trans*, **34**, 1334-1340.

Culmsee, C., Zhu, C., Landshamer, S., Becattini, B., Wagner, E., Pellecchia, M., Blomgren, K. and Plesnila, N. (2005) Apoptosis-inducing factor triggered by poly(ADP-ribose) polymerase and Bid mediates neuronal cell death after oxygen-glucose deprivation and focal cerebral ischemia. *J Neurosci*, **25**, 10262-10272.

Danial, N.N. and Korsmeyer, S.J. (2004) Cell death: critical control points. *Cell*, **116**, 205-219.

Day, C.L., Chen, L., Richardson, S.J., Harrison, P.J., Huang, D.C. and Hinds, M.G. (2005) Solution structure of prosurvival Mcl-1 and characterization of its binding by proapoptotic BH3-only ligands. *J Biol Chem*, **280**, 4738-4744.

Degterev, A., Lugovskoy, A., Cardone, M., Mulley, B., Wagner, G., Mitchison, T. and Yuan, J. (2001) Identification of small-molecule inhibitors of interaction between the BH3 domain and Bcl-xL. *Nat Cell Biol*, **3,** 173-182.

Del Gaizo Moore, V., Brown, J.R., Certo, M., Love, T.M., Novina, C.D. and Letai, A. (2007) Chronic lymphocytic leukemia requires BCL2 to sequester prodeath BIM, explaining sensitivity to BCL2 antagonist ABT-737. *J Clin Invest*, **117**, 112-121.

del Peso, L., Gonzalez-Garcia, M., Page, C., Herrera, R. and Nunez, G. (1997) Interleukin-3-induced phosphorylation of BAD through the protein kinase Akt. *Science*, **278**, 687-689.

Deng, J., Carlson, N., Takeyama, K., Dal Cin, P., Shipp, M. and Letai, A. (2007) BH3 profiling identifies three distinct classes of apoptotic blocks to predict response to ABT-737 and conventional chemotherapeutic agents. *Cancer Cell*, **12**, 171-185.

Denis, G.V., Yu, Q., Ma, P., Deeds, L., Faller, D.V. and Chen, C.Y. (2003) Bcl-2, via its BH4 domain, blocks apoptotic signaling mediated by mitochondrial Ras. *J Biol Chem*, **278**, 5775-5785.

Denisov, A.Y., Madiraju, M.S., Chen, G., Khadir, A., Beauparlant, P., Attardo, G., Shore, G.C. and Gehring, K. (2003) Solution structure of human BCL-w: modulation of ligand binding by the C-terminal helix. *J Biol Chem*, **278**, 21124-21128.

Dharap, S.S. and Minko, T. (2003) Targeted proapoptotic LHRH-BH3 peptide. *Pharm Res*, **20**, 889-896.

Dlugosz, P.J., Billen, L.P., Annis, M.G., Zhu, W., Zhang, Z., Lin, J., Leber, B. and Andrews, D.W. (2006) Bcl-2 changes conformation to inhibit Bax oligomerization. *Embo J*, **25**, 2287-2296.

Erlanson, D.A. (2006) Fragment-based lead discovery: a chemical update. *Curr Opin Biotechnol*, **17**, 643-652.

Fadeel, B., Hassan, Z., Hellstrom-Lindberg, E., Henter, J.I., Orrenius, S. and Zhivotovsky, B. (1999) Cleavage of Bcl-2 is an early event in chemotherapy-induced apoptosis of human myeloid leukemia cells. *Leukemia*, **13**, 719-728.

Franz, G., Beer, R., Intemann, D., Krajewski, S., Reed, J.C., Engelhardt, K., Pike, B.R., Hayes, R.L., Wang, K.K., Schmutzhard, E. and Kampfl, A. (2002) Temporal and spatial profile of Bid cleavage after experimental traumatic brain injury. *J Cereb Blood Flow Metab*, **22**, 951-958.

Gemperli, A.C., Rutledge, S.E., Maranda, A. and Schepartz, A. (2005) Paralog-selective ligands for bcl-2 proteins. *J Am Chem Soc*, **127**, 1596-1597.

Goldsmith, K.C., Liu, X., Dam, V., Morgan, B.T., Shabbout, M., Cnaan, A., Letai, A., Korsmeyer, S.J. and Hogarty, M.D. (2006) BH3 peptidomimetics potently activate apoptosis and demonstrate single agent efficacy in neuroblastoma. *Oncogene*, **25**, 4525-4533.

Goping, I.S., Gross, A., Lavoie, J.N., Nguyen, M., Jemmerson, R., Roth, K., Korsmeyer, S.J. and Shore, G.C. (1998) Regulated targeting of BAX to mitochondria. *J Cell Biol*, **143**, 207-215.

Grandgirard, D., Studer, E., Monney, L., Belser, T., Fellay, I., Borner, C. and Michel, M.R. (1998) Alphaviruses induce apoptosis in Bcl-2-overexpressing cells: evidence for a caspase-mediated, proteolytic inactivation of Bcl-2. *Embo J*, **17**, 1268-1278.

Gross, A., Jockel, J., Wei, M.C. and Korsmeyer, S.J. (1998) Enforced dimerization of BAX results in its translocation, mitochondrial dysfunction and apoptosis. *Embo J*, **17**, 3878-3885.

Guo, B., Zhai, D., Cabezas, E., Welsh, K., Nouraini, S., Satterthwait, A.C. and Reed, J.C. (2003) Humanin peptide suppresses apoptosis by interfering with Bax activation. *Nature*, **423**, 456-461.

Hajduk, P.J. and Greer, J. (2007) A decade of fragment-based drug design: strategic advances and lessons learned. *Nat Rev Drug Discov*, **6,** 211-219.

Hanahan, D. and Weinberg, R.A. (2000) The hallmarks of cancer. *Cell*, **100**, 57-70.

Hashimoto, Y., Niikura, T., Tajima, H., Yasukawa, T., Sudo, H., Ito, Y., Kita, Y., Kawasumi, M., Kouyama, K., Doyu, M., Sobue, G., Koide, T., Tsuji, S., Lang, J., Kurokawa, K. and Nishimoto, I. (2001) A rescue factor abolishing neuronal cell death by a wide spectrum of familial Alzheimer's disease genes and Abeta. *Proc Natl Acad Sci U S A*, **98**, 6336-6341.

Henshall, D.C., Bonislawski, D.P., Skradski, S.L., Lan, J.Q., Meller, R. and Simon, R.P. (2001) Cleavage of bid may amplify caspase-8-induced neuronal death following focally evoked limbic seizures. *Neurobiol Dis*, **8**, 568-580.

Hinds, M.G., Lackmann, M., Skea, G.L., Harrison, P.J., Huang, D.C. and Day, C.L. (2003) The structure of Bcl-w reveals a role for the C-terminal residues in modulating biological activity. *Embo J*, **22**, 1497-1507.

Hinds, M.G., Smits, C., Fredericks-Short, R., Risk, J.M., Bailey, M., Huang, D.C. and Day, C.L. (2007) Bim, Bad and Bmf: intrinsically unstructured BH3-only proteins that undergo a localized conformational change upon binding to prosurvival Bcl-2 targets. *Cell Death Differ*, **14**, 128-136.

Hipfner, D.R. and Cohen, S.M. (2004) Connecting proliferation and apoptosis in development and disease. *Nat Rev Mol Cell Biol*, **5,** 805-815.

Holinger, E.P., Chittenden, T. and Lutz, R.J. (1999) Bak BH3 peptides antagonize Bcl-xL function and induce apoptosis through cytochrome c-independent activation of caspases. *J Biol Chem*, 274, 13298-13304.

Hotchkiss, R.S., McConnell, K.W., Bullok, K., Davis, C.G., Chang, K.C., Schwulst, S.J., Dunne, J.C., Dietz, G.P., Bahr, M., McDunn, J.E., Karl, I.E., Wagner, T.H., Cobb, J.P., Coopersmith, C.M. and Piwnica-Worms, D. (2006) TAT-BH4 and TAT-Bcl-xL peptides protect against sepsis-induced lymphocyte apoptosis in vivo. *J Immunol*, **176**, 5471-5477.

Huang, D.C., Adams, J.M. and Cory, S. (1998) The conserved N-terminal BH4 domain of Bcl-2 homologues is essential for inhibition of apoptosis and interaction with CED-4. *Embo J*, **17**, 1029-1039.

Ishibashi, Y., Nishimaki, K., Asoh, S., Nanbu-Wakao, R., Yamada, T. and Ohta, S. (1998) Pore formation domain of human pro-apoptotic Bax induces mammalian apoptosis as well as bacterial death without antagonizing anti-apoptotic factors. *Biochem Biophys Res Commun*, **243**, 609-616.

Jeong, S.Y., Gaume, B., Lee, Y.J., Hsu, Y.T., Ryu, S.W., Yoon, S.H. and Youle, R.J. (2004) Bcl-x(L) sequesters its C-terminal membrane anchor in soluble, cytosolic homodimers. *Embo J*, **23**, 2146-2155.

Jiang, X. and Wang, X. (2004) Cytochrome C-mediated apoptosis. *Annu Rev Biochem*, **73**, 87-106.

Jung, S.S. and Van Nostrand, W.E. (2003) Humanin rescues human cerebrovascular smooth muscle cells from Abeta-induced toxicity. *J Neurochem*, **84**, 266-272.

Kamer, I., Sarig, R., Zaltsman, Y., Niv, H., Oberkovitz, G., Regev, L., Haimovich, G., Lerenthal, Y., Marcellus, R.C. and Gross, A. (2005) Proapoptotic BID is an ATM effector in the DNA-damage response. *Cell*, **122**, 593-603.

Kang, C.B., Tai, J., Chia, J. and Yoon, H.S. (2005) The flexible loop of Bcl-2 is required for molecular interaction with immunosuppressant FK-506 binding protein 38 (FKBP38). *FEBS Lett*, **579**, 1469-1476.

Kang, M.H., Kang, Y.H., Szymanska, B., Wilczynska-Kalak, U., Sheard, M.A., Harned, T.M., Lock, R.B. and Reynolds, C.P. (2007) Activity of vincristine, L-ASP, and dexamethasone against acute lymphoblastic leukemia is enhanced by the BH3-mimetic ABT-737 in vitro and in vivo. *Blood*, **110**, 2057-2066.

Kaufmann, T., Schinzel, A. and Borner, C. (2004) Bcl-w(edding) with mitochondria. *Trends Cell Biol*, **14**, 8-12.

Kaufmann, T., Tai, L., Ekert, P.G., Huang, D.C., Norris, F., Lindemann, R.K., Johnstone, R.W., Dixit, V.M., Arnoult, D. and Strasser, A. (2007a) Response: does bid play a role in the DNA damage response? *Cell*, **130**, 10-11.

Kaufmann, T., Tai, L., Ekert, P.G., Huang, D.C., Norris, F., Lindemann, R.K., Johnstone, R.W., Dixit, V.M. and Strasser, A. (2007b) The BH3-only protein bid is dispensable for DNA damage- and replicative stress-induced apoptosis or cell-cycle arrest. *Cell*, **129**, 423-433.

Kim, R., Emi, M., Matsuura, K. and Tanabe, K. (2007) Antisense and nonantisense effects of antisense Bcl-2 on multiple roles of Bcl-2 as a chemosensitizer in cancer therapy. *Cancer Gene Ther*, **14**, 1-11.

Kirsch, D.G., Doseff, A., Chau, B.N., Lim, D.S., de Souza-Pinto, N.C., Hansford, R., Kastan, M.B., Lazebnik, Y.A. and Hardwick, J.M. (1999) Caspase-3-dependent cleavage of Bcl-2 promotes release of cytochrome c. *J Biol Chem*, **274**, 21155-21161.

Kitada, S., Leone, M., Sareth, S., Zhai, D., Reed, J.C. and Pellecchia, M. (2003) Discovery, characterization, and structure-activity relationships studies of proapoptotic polyphenols targeting B-cell lymphocyte/leukemia-2 proteins. *J Med Chem*, **46**, 4259-4264.

Ko, J.K., Choi, K.H., Kim, H.J., Choi, H.Y., Yeo, D.J., Park, S.O., Yang, W.S., Kim, Y.N. and Kim, C.W. (2003) Conversion of Bfl-1, an anti-apoptotic Bcl-2 family protein, to a potent pro-apoptotic protein by fusion with green fluorescent protein (GFP). *FEBS Lett*, **551**, 29-36.

Kohl, T.M., Hellinger, C., Ahmed, F., Buske, C., Hiddemann, W., Bohlander, S.K. and Spiekermann, K. (2007) BH3 mimetic ABT-737 neutralizes resistance to FLT3 inhibitor treatment mediated by FLT3-independent expression of BCL2 in primary AML blasts. *Leukemia*, **21**, 1763-1772.

Konopleva, M., Contractor, R., Tsao, T., Samudio, I., Ruvolo, P.P., Kitada, S., Deng, X., Zhai, D., Shi, Y.X., Sneed, T., Verhaegen, M., Soengas, M., Ruvolo, V.R., McQueen, T., Schober, W.D., Watt, J.C., Jiffar, T., Ling, X., Marini, F.C., Harris, D., Dietrich, M., Estrov, Z., McCubrey, J., May, W.S., Reed, J.C. and Andreeff, M. (2006) Mechanisms of apoptosis sensitivity and resistance to the BH3 mimetic ABT-737 in acute myeloid leukemia. *Cancer Cell*, **10**, 375-388.

Kucharczak, J.F., Simmons, M.J., Duckett, C.S. and Gelinas, C. (2005) Constitutive proteasome-mediated turnover of Bfl-1/A1 and its processing in response to TNF receptor activation in FL5.12 pro-B cells convert it into a prodeath factor. *Cell Death Differ*, **12**, 1225-1239.

Kuwana, T., Bouchier-Hayes, L., Chipuk, J.E., Bonzon, C., Sullivan, B.A., Green, D.R. and Newmeyer, D.D. (2005) BH3 domains of BH3-only proteins differentially regulate Bax-mediated mitochondrial membrane permeabilization both directly and indirectly. *Mol Cell*, **17**, 525-535.

Labi, V., Erlacher, M., Kiessling, S. and Villunger, A. (2006) BH3-only proteins in cell death initiation, malignant disease and anticancer therapy. *Cell Death Differ*, **13**, 1325-1338.

Lee, E.F., Czabotar, P.E., Smith, B.J., Deshayes, K., Zobel, K., Colman, P.M. and Fairlie, W.D. (2007) Crystal structure of ABT-737 complexed with Bcl-xL: implications for selectivity of antagonists of the Bcl-2 family. *Cell Death Differ*, **14**, 1711-1713.

Letai, A., Bassik, M.C., Walensky, L.D., Sorcinelli, M.D., Weiler, S. and Korsmeyer, S.J. (2002) Distinct BH3 domains either sensitize or activate mitochondrial apoptosis, serving as prototype cancer therapeutics. *Cancer Cell*, **2**, 183-192.

Liang, Y., Nylander, K.D., Yan, C. and Schor, N.F. (2002) Role of caspase 3-dependent Bcl-2 cleavage in potentiation of apoptosis by Bcl-2. *Mol Pharmacol*, **61**, 142-149.

Lin, X., Morgan-Lappe, S., Huang, X., Li, L., Zakula, D.M., Vernetti, L.A., Fesik, S.W. and Shen, Y. (2007) 'Seed' analysis of off-target siRNAs reveals an essential role of Mcl-1 in resistance to the small-molecule Bcl-2/Bcl-XL inhibitor ABT-737. *Oncogene*, **26**, 3972-3979.

Luciano, F., Krajewska, M., Ortiz-Rubio, P., Krajewski, S., Zhai, D., Faustin, B., Bruey, J.M., Bailly-Maitre, B., Lichtenstein, A., Kolluri, S.K., Satterthwait, A.C., Zhang, X.K. and Reed, J.C. (2007) Nur77 converts phenotype of Bcl-B, an antiapoptotic protein expressed in plasma cells and myeloma. *Blood*, **109**, 3849-3855.

Lum, J.J., Bauer, D.E., Kong, M., Harris, M.H., Li, C., Lindsten, T. and Thompson, C.B. (2005) Growth factor regulation of autophagy and cell survival in the absence of apoptosis. *Cell*, **120**, 237-248.

Maximov, V., Martynenko, A., Hunsmann, G. and Tarantul, V. (2002) Mitochondrial 16S rRNA gene encodes a functional peptide, a potential drug for Alzheimer's disease and target for cancer therapy. *Med Hypotheses*, **59**, 670-673.

McDonnell, J.M., Fushman, D., Milliman, C.L., Korsmeyer, S.J. and Cowburn, D. (1999) Solution structure of the proapoptotic molecule BID: a structural basis for apoptotic agonists and antagonists. *Cell*, **96**, 625-634.

Minn, A.J., Velez, P., Schendel, S.L., Liang, H., Muchmore, S.W., Fesik, S.W., Fill, M. and Thompson, C.B. (1997) Bcl-x(L) forms an ion channel in synthetic lipid membranes. *Nature*, **385**, 353-357.

Mohammad, R.M., Goustin, A.S., Aboukameel, A., Chen, B., Banerjee, S., Wang, G., Nikolovska-Coleska, Z., Wang, S. and Al-Katib, A. (2007) Preclinical studies of TW-37, a new nonpeptidic small-molecule inhibitor of Bcl-2, in diffuse large cell lymphoma xenograft model reveal drug action on both Bcl-2 and Mcl-1. *Clin Cancer Res*, **13**, 2226-2235.

Nouvion, A.L., Thibaut, J., Lohez, O.D., Venet, S., Colas, P., Gillet, G. and Lalle, P. (2007) Modulation of Nr-13 antideath activity by peptide aptamers. *Oncogene*, **26**, 701-710.

Oltersdorf, T., Elmore, S.W., Shoemaker, A.R., Armstrong, R.C., Augeri, D.J., Belli, B.A., Bruncko, M., Deckwerth, T.L., Dinges, J., Hajduk, P.J., Joseph, M.K., Kitada, S., Korsmeyer, S.J., Kunzer, A.R., Letai, A., Li, C., Mitten, M.J., Nettesheim, D.G., Ng, S., Nimmer, P.M., O'Connor, J.M., Oleksijew, A., Petros, A.M., Reed, J.C., Shen, W., Tahir, S.K., Thompson, C.B., Tomaselli, K.J., Wang, B., Wendt, M.D., Zhang, H., Fesik, S.W. and Rosenberg, S.H. (2005) An inhibitor of Bcl-2 family proteins induces regression of solid tumours. *Nature*, **435**, 677-681.

O'Neill, J., Manion, M., Schwartz, P. and Hockenbery, D.M. (2004) Promises and challenges of targeting Bcl-2 anti-apoptotic proteins for cancer therapy. *Biochim Biophys Acta*, **1705**, 43-51.

Petros, A.M., Nettesheim, D.G., Wang, Y., Olejniczak, E.T., Meadows, R.P., Mack, J., Swift, K., Matayoshi, E.D., Zhang, H., Thompson, C.B. and Fesik, S.W. (2000) Rationale for Bcl-xL/Bad peptide complex formation from structure, mutagenesis, and biophysical studies. *Protein Sci*, **9**, 2528-2534.

Petros, A.M., Olejniczak, E.T. and Fesik, S.W. (2004) Structural biology of the Bcl-2 family of proteins. *Biochim Biophys Acta*, **1644**, 83-94.

Plesnila, N. (2004) Role of mitochondrial proteins for neuronal cell death after focal cerebral ischemia. *Acta Neurochir Suppl*, **89**, 15-19.

Plesnila, N., Zinkel, S., Amin-Hanjani, S., Qiu, J., Korsmeyer, S.J. and Moskowitz, M.A. (2002) Function of BID -- a molecule of the bcl-2 family -- in ischemic cell death in the brain. *Eur Surg Res*, **34**, 37-41.

Plesnila, N., Zinkel, S., Le, D.A., Amin-Hanjani, S., Wu, Y., Qiu, J., Chiarugi, A., Thomas, S.S., Kohane, D.S., Korsmeyer, S.J. and Moskowitz, M.A. (2001) BID mediates neuronal cell death after oxygen/ glucose deprivation and focal cerebral ischemia. *Proc Natl Acad Sci U S A*, **98**, 15318-15323.

Puthalakath, H., Huang, D.C., O'Reilly, L.A., King, S.M. and Strasser, A. (1999) The proapoptotic activity of the Bcl-2 family member Bim is regulated by interaction with the dynein motor complex. *Mol Cell*, **3**, 287-296.

Puthalakath, H., Villunger, A., O'Reilly, L.A., Beaumont, J.G., Coultas, L., Cheney, R.E., Huang, D.C. and Strasser, A. (2001) Bmf: a proapoptotic BH3-only protein regulated by interaction with the myosin V actin motor complex, activated by anoikis. *Science*, **293**, 1829-1832.

Reed, J.C. and Pellecchia, M. (2005) Apoptosis-based therapies for hematologic malignancies. *Blood*, **106**, 408-418.

Reed, J.C., Zha, H., Aime-Sempe, C., Takayama, S. and Wang, H.G. (1996) Structure-function analysis of Bcl-2 family proteins. Regulators of programmed cell death. *Adv Exp Med Biol*, **406**, 99-112.

Sadowsky, J.D., Fairlie, W.D., Hadley, E.B., Lee, H.S., Umezawa, N., Nikolovska-Coleska, Z., Wang, S., Huang, D.C., Tomita, Y. and Gellman, S.H. (2007) (alpha/beta+alpha)-peptide antagonists of BH3 domain/Bcl-x(L) recognition: toward general strategies for foldamer-based inhibition of protein-protein interactions. *J Am Chem Soc*, **129**, 139-154.

Sawada, M., Hayes, P. and Matsuyama, S. (2003a) Cytoprotective membrane-permeable peptides designed from the Bax-binding domain of Ku70. *Nat Cell Biol*, **5**, 352-357.

Sawada, M., Sun, W., Hayes, P., Leskov, K., Boothman, D.A. and Matsuyama, S. (2003b) Ku70 suppresses the apoptotic translocation of Bax to mitochondria. *Nat Cell Biol*, **5**, 320-329.

Schendel, S.L., Azimov, R., Pawlowski, K., Godzik, A., Kagan, B.L. and Reed, J.C. (1999) Ion channel activity of the BH3 only Bcl-2 family member, BID. *J Biol Chem*, **274**, 21932-21936.

Schlesinger, P.H., Gross, A., Yin, X.M., Yamamoto, K., Saito, M., Waksman, G. and Korsmeyer, S.J. (1997) Comparison of the ion channel characteristics of proapoptotic BAX and antiapoptotic BCL-2. *Proc Natl Acad Sci U S A*, **94**, 11357-11362.

Seo, Y.W., Shin, J.N., Ko, K.H., Cha, J.H., Park, J.Y., Lee, B.R., Yun, C.W., Kim, Y.M., Seol, D.W., Kim, D.W., Yin, X.M. and Kim, T.H. (2003) The molecular mechanism of Noxa-induced mitochondrial dysfunction in p53-mediated cell death. *J Biol Chem*, **278**, 48292-48299.

Shibasaki, F., Kondo, E., Akagi, T. and McKeon, F. (1997) Suppression of signalling through transcription factor NF-AT by interactions between calcineurin and Bcl-2. *Nature*, **386**, 728-731.

Shimizu, S., Konishi, A., Kodama, T. and Tsujimoto, Y. (2000) BH4 domain of antiapoptotic Bcl-2 family members closes voltage-dependent anion channel and inhibits apoptotic mitochondrial changes and cell death. *Proc Natl Acad Sci U S A*, **97**, 3100-3105.

Shirane, M. and Nakayama, K.I. (2003) Inherent calcineurin inhibitor FKBP38 targets Bcl-2 to mitochondria and inhibits apoptosis. *Nat Cell Biol*, 5, 28-37.

Sponne, I., Fifre, A., Koziel, V., Kriem, B., Oster, T. and Pillot, T. (2004) Humanin rescues cortical neurons from prion-peptide-induced apoptosis. *Mol Cell Neurosci*, 25, 95-102.

Sugioka, R., Shimizu, S., Funatsu, T., Tamagawa, H., Sawa, Y., Kawakami, T. and Tsujimoto, Y. (2003) BH4-domain peptide from Bcl-xL exerts anti-apoptotic activity in vivo. *Oncogene*, **22**, 8432-8440.

Suzuki, M., Youle, R.J. and Tjandra, N. (2000) Structure of Bax: coregulation of dimer formation and intracellular localization. *Cell*, **103**, 645-654.

Tahir, S.K., Yang, X., Anderson, M.G., Morgan-Lappe, S.E., Sarthy, A.V., Chen, J., Warner, R.B., Ng, S.C., Fesik, S.W., Elmore, S.W., Rosenberg, S.H. and Tse, C. (2007) Influence of Bcl-2 family members on the cellular response of small-cell lung cancer cell lines to ABT-737. *Cancer Res*, **67**, 1176-1183.

Tang, G., Yang, C.Y., Nikolovska-Coleska, Z., Guo, J., Qiu, S., Wang, R., Gao, W., Wang, G., Stuckey, J., Krajewski, K., Jiang, S., Roller, P.P. and Wang, S. (2007) Pyrogallol-based molecules as potent inhibitors of the antiapoptotic Bcl-2 proteins. *J Med Chem*, **50**, 1723-1726.

Trudel, S., Li, Z.H., Rauw, J., Tiedemann, R.E., Wen, X.Y. and Stewart, A.K. (2007) Preclinical studies of the pan-Bcl inhibitor obatoclax (GX015-070) in multiple myeloma. *Blood*, **109**, 5430-5438.

Tsujimoto, Y., Cossman, J., Jaffe, E. and Croce, C.M. (1985) Involvement of the bcl-2 gene in human follicular lymphoma. *Science*, **228**, 1440-1443.

Tzung, S.P., Kim, K.M., Basanez, G., Giedt, C.D., Simon, J., Zimmerberg, J., Zhang, K.Y. and Hockenbery, D.M. (2001) Antimycin A mimics a cell-death-inducing Bcl-2 homology domain 3. *Nat Cell Biol*, **3**, 183-191.

Usui, K., Saijo, Y., Narumi, K., Koyama, S., Maemondo, M., Kikuchi, T., Tazawa, R., Hagiwara, K., Ishibashi, Y., Ohta, S. and Nukiwa, T. (2003) N-terminal deletion augments the cell-death-inducing activity of BAX in adenoviral gene delivery to nonsmall cell lung cancers. *Oncogene*, **22**, 2655-2663.

van Delft, M.F., Wei, A.H., Mason, K.D., Vandenberg, C.J., Chen, L., Czabotar, P.E., Willis, S.N., Scott, C.L., Day, C.L., Cory, S., Adams, J.M., Roberts, A.W. and Huang, D.C. (2006) The BH3 mimetic ABT-737 targets selective Bcl-2 proteins and efficiently induces apoptosis via Bak/Bax if Mcl-1 is neutralized. *Cancer Cell*, **10**, 389-399.

Vance, B.A., Zacharchuk, C.M. and Segal, D.M. (1996) Recombinant mouse Bcl-2(1-203). Two domains connected by a long protease-sensitive linker. *J Biol Chem*, **271**, 30811-30815.

Vaux, D.L., Cory, S. and Adams, J.M. (1988) Bcl-2 gene promotes haemopoietic cell survival and cooperates with c-myc to immortalize pre-B cells. *Nature*, **335**, 440-442.

Walensky, L.D. (2006) BCL-2 in the crosshairs: tipping the balance of life and death. *Cell Death Differ*, **13**, 1339-1350.

Walensky, L.D., Kung, A.L., Escher, I., Malia, T.J., Barbuto, S., Wright, R.D., Wagner, G., Verdine, G.L. and Korsmeyer, S.J. (2004) Activation of apoptosis in vivo by a hydrocarbon-stapled BH3 helix. *Science*, **305**, 1466-1470.

Wang, H.G., Pathan, N., Ethell, I.M., Krajewski, S., Yamaguchi, Y., Shibasaki, F., McKeon, F., Bobo, T., Franke, T.F. and Reed, J.C. (1999) Ca2+-induced apoptosis through calcineurin dephosphorylation of BAD. *Science*, **284**, 339-343.

Wang, J.L., Liu, D., Zhang, Z.J., Shan, S., Han, X., Srinivasula, S.M., Croce, C.M., Alnemri, E.S. and Huang, Z. (2000a) Structure-based discovery of an organic compound that binds Bcl-2 protein and induces apoptosis of tumor cells. *Proc Natl Acad Sci U S A*, **97**, 7124-7129.

Wang, J.L., Zhang, Z.J., Choksi, S., Shan, S., Lu, Z., Croce, C.M., Alnemri, E.S., Korngold, R. and Huang, Z. (2000b) Cell permeable Bcl-2 binding peptides: a chemical approach to apoptosis induction in tumor cells. *Cancer Res*, **60**, 1498-1502.

Wei, M.C., Zong, W.X., Cheng, E.H., Lindsten, T., Panoutsakopoulou, V., Ross, A.J., Roth, K.A., MacGregor, G.R., Thompson, C.B. and Korsmeyer, S.J. (2001) Proapoptotic BAX and BAK: a requisite gateway to mitochondrial dysfunction and death. *Science*, **292**, 727-730.

Willis, S.N., Chen, L., Dewson, G., Wei, A., Naik, E., Fletcher, J.I., Adams, J.M. and Huang, D.C. (2005) Proapoptotic Bak is sequestered by Mcl-1 and Bcl-xL, but not Bcl-2, until displaced by BH3-only proteins. *Genes Dev*, **19**, 1294-1305.

Willis, S.N., Fletcher, J.I., Kaufmann, T., van Delft, M.F., Chen, L., Czabotar, P.E., Ierino, H., Lee, E.F., Fairlie, W.D., Bouillet, P., Strasser, A., Kluck, R.M., Adams, J.M. and

Huang, D.C. (2007) Apoptosis initiated when BH3 ligands engage multiple Bcl-2 homologs, not Bax or Bak. *Science*, **315**, 856-859.

Yamamoto, A., Murphy, N., Schindler, C.K., So, N.K., Stohr, S., Taki, W., Prehn, J.H. and Henshall, D.C. (2006) Endoplasmic reticulum stress and apoptosis signaling in human temporal lobe epilepsy. *J Neuropathol Exp Neurol*, **65**, 217-225.

Yang, B., Liu, D. and Huang, Z. (2004) Synthesis and helical structure of lactam bridged BH3 peptides derived from pro-apoptotic Bcl-2 family proteins. *Bioorg Med Chem Lett*, **14**, 1403-1406.

Yang, W.S., Ko, J.K., Park, S.O., Choi, H.Y., Kim, Y.N. and Kim, C.W. (2005) C-terminal region of Bfl-1 induces cell death that accompanies caspase activation when fused with GFP. *J Cell Biochem*, **94**, 1234-1247.

Yin, H., Lee, G.I., Sedey, K.A., Kutzki, O., Park, H.S., Orner, B.P., Ernst, J.T., Wang, H.G., Sebti, S.M. and Hamilton, A.D. (2005a) Terphenyl-Based Bak BH3 alpha-helical proteomimetics as low-molecular-weight antagonists of Bcl-xL. *J Am Chem Soc*, **127**, 10191-10196.

Yin, H., Lee, G.I., Sedey, K.A., Rodriguez, J.M., Wang, H.G., Sebti, S.M. and Hamilton, A.D. (2005b) Terephthalamide derivatives as mimetics of helical peptides: disruption of the Bcl-x(L)/Bak interaction. *J Am Chem Soc*, **127**, 5463-5468.

Yin, X.M. (2006) Bid, a BH3-only multi-functional molecule, is at the cross road of life and death. *Gene*, **369**, 7-19.

Yin, X.M., Luo, Y., Cao, G., Bai, L., Pei, W., Kuharsky, D.K. and Chen, J. (2002) Bid-mediated mitochondrial pathway is critical to ischemic neuronal apoptosis and focal cerebral ischemia. *J Biol Chem*, **277**, 42074-42081.

Zhai, D., Jin, C., Satterthwait, A.C. and Reed, J.C. (2006) Comparison of chemical inhibitors of antiapoptotic Bcl-2-family proteins. *Cell Death Differ*, **13**, 1419-1421.

Zhai, D., Luciano, F., Zhu, X., Guo, B., Satterthwait, A.C. and Reed, J.C. (2005) Humanin binds and nullifies Bid activity by blocking its activation of Bax and Bak. *J Biol Chem*, **280**, 15815-15824.

Zhang, H., Holzgreve, W. and De Geyter, C. (2001) Bcl2-L-10, a novel anti-apoptotic member of the Bcl-2 family, blocks apoptosis in the mitochondria death pathway but not in the death receptor pathway. *Hum Mol Genet*, **10**, 2329-2339.

Zhang, Z., Lapolla, S.M., Annis, M.G., Truscott, M., Roberts, G.J., Miao, Y., Shao, Y., Tan, C., Peng, J., Johnson, A.E., Zhang, X.C., Andrews, D.W. and Lin, J. (2004) Bcl-2 homodimerization involves two distinct binding surfaces, a topographic arrangement that provides an effective mechanism for Bcl-2 to capture activated Bax. *J Biol Chem*, **279**, 43920-43928.

Zhong, Q., Gao, W., Du, F. and Wang, X. (2005) Mule/ARF-BP1, a BH3-only E3 ubiquitin ligase, catalyzes the polyubiquitination of Mcl-1 and regulates apoptosis. *Cell*, **121**, 1085-1095.

Zinkel, S.S., Hurov, K.E., Ong, C., Abtahi, F.M., Gross, A. and Korsmeyer, S.J. (2005) A role for proapoptotic BID in the DNA-damage response. *Cell*, **122**, 579-591.

Zong, W.X., Lindsten, T., Ross, A.J., MacGregor, G.R. and Thompson, C.B. (2001) BH3-only proteins that bind pro-survival Bcl-2 family members fail to induce apoptosis in the absence of Bax and Bak. *Genes Dev*, **15**, 1481-1486.

In: Oncogene Proteins: Structure, Functions and Analyses ISBN: 978-1-60456-259-0
Editors: P.B. Murphy and J.R. Clarke, pp. 191-208

Chapter VII

New Aspects of AKT Signaling: Nucleus, Calcium, and Hormone

Takeo Minaguchi
Department of Obstetrics and Gynecology, Toranomon Hospital, Tokyo, Japan

Abstract

AKT/PKB is a serine/threonine kinase which represents a central node of the PI3K signaling, one of critical pathways contributing to pathogenesis of diverse cancers and diseases. A variety of extracellular stimuli such as growth factors, hormones, and cytokines activate PI3K, resulting in phosphorylation and activation of AKT. AKT in turn phosphorylates downstream substrates, *e.g.* p21, p27, MDM2, GSK3, BAD, Caspase-9, FOXO, mTOR, and TSC1/2, thereby exerting various cellular processes including cell growth, proliferation, survival, angiogenesis, migration and metabolism. In a past decade, enormous number of improvements regarding this oncogenic pathway have been achieved including the identification of direct AKT substrates and their downstream components as well as the development of targeted therapeutic small molecules. Nonetheless, novel intriguing aspects of the AKT pathway have been unraveled more recently. Accumulative evidence now suggest that subcellular compartmentalization of AKT is involved in carcinogenesis through nuclear-cytoplasmic specific networks. In endocrine-related cancers, antitumor action by hormonal agents and acquired hormone resistance are both likely to be mediated by the AKT pathway. The crosstalk between Ca^{2+} signaling and the AKT pathway has been demonstrated in cardiomyocytes and other cell types. This review highlights and discusses these lines of emerging evidence regarding the AKT signaling and their therapeutic significance.

Introduction

AKT/PKB is a serine/threonine kinase and exists as three isoforms, α, β, and γ (Akt1, Akt2, and Akt3, respectively). AKT is a central node of the phosphatidylinositol 3-kinase (PI3K) signaling which is one of the pivotal pathways contributing to pathogenesis of diverse cancers and diseases. A variety of extracellular stimuli such as growth factors, hormones, and cytokines activate PI3K, leading to phosphorylation and activation of AKT. AKT in turn phosphorylates downstream substrates, *e.g.* p21, p27, MDM2, GSK3, BAD, Caspase-9, FOXO, mTOR, and TSC1/2, thus exerting diverse cellular processes including cell growth, proliferation, survival, angiogenesis, migration and metabolism. In a past decade, tremendous progress regarding this oncogenic pathway have been made including the identification of direct AKT substrates and their downstream components as well as the development of targeted therapeutic small molecules such as receptor tyrosine kinase inhibitors. Nonetheless, novel intriguing aspects of the AKT pathway have been uncovered more recently, including the role of subcellular compartmentalization of AKT in carcinogenesis, the hormonal regulation of the AKT pathway in endocrine-related cancers, and the crosstalk between Ca^{2+} signaling and the AKT pathway. This review highlights and discusses these lines of burgeoning evidence on AKT signaling and their therapeutic significance.

Nuclear AKT Network

Not only an independent nuclear phospholipid cycle has been found to exist, but in the past decade it has been demonstrated that numerous proteins participating in the phosphoinositide signaling cascade are present also in the nucleus, including even diverse cell surface receptors, implicating that an additional mode of signaling network inside the nucleus may be complementing the cytoplasmic pathways (Deleris et al., 2006; Gillooly et al., 2000; Irvine, 2003; Lian and Di Cristofano, 2005; Martelli et al., 2003). PI3K, the first enzyme in this cascade, generates phosphatidylinositol (3,4,5)-triphosphate ($PI(3,4,5)P_3$). All three known classes of PI3Ks (class I, II, and III) have been found in the nucleus either shuttling into the nucleus upon specific mitogenic (Kim, 1998; Martelli et al., 2000; Metjian et al., 1999) and differentiation (Neri et al., 1999; Neri et al., 1994) stimuli, or constitutively co-localized with the nuclear speckles (Didichenko and Thelen, 2001; Neri et al., 2002). Cytoplasmic PI3K is activated by receptor tyrosine kinases and membrane-associated GTPases, but the latter proteins are not localized in the nucleus. However, recent studies have demonstrated that, in response to NGF stimulation, the nuclear GTPase, PIKE, can activate nuclear PI3K, thus playing a similar role to that of Ras at the plasma membrane (Rong et al., 2003; Ye et al., 2000). PIKE is activated by NGF through the nuclear translocation of PLCγ1, which functions as guanine nucleotide exchange factor for PIKE (Ye et al., 2002). The generation of $PI(3,4,5)P_3$ at the plasma membrane results in the recruitment and activation of signaling proteins containing a PH domain, including phosphoinositide-dependent kinase 1 (PDK1) and AKT. The recent evidence that both these proteins also exist in the nucleus further support the hypothesis that the entire PI3K/PDK/AKT circuitry is at work in this compartment. Indeed, recent studies have shown that NGF elicits nuclear $PI(3,4,5)P_3$-dependent anti-apoptotic signaling in PC12 cells, where NGF induces the translocation of PI3K and its activation by nuclear PIKE. PI3K generates $PI(3,4,5)P_3$, thus leading to the activation of nuclear Akt (Ahn et al., 2004). However, nuclear Akt is necessary, but not sufficient to mediate the anti-apoptotic action of NGF. In fact, the B23 protein, a major nucleolar phosphoprotein, interacts with $PI(3,4,5)P_3$, and the nuclear B23-$PI(3,4,5)P_3$ complex associates with caspase-activated DNase, thereby inhibiting DNA fragmentation (Ahn et al., 2005).

AKT requires a double phosphorylation to be fully activated: $PI(3,4,5)P_3$ drives the phosphorylation on Thr^{308} in AKT1 (309 in AKT2 and 305 in AKT3) through PDK1, whereas

Ser^{473} phosphorylation (474 in AKT2 and 472 in AKT3) by an unknown kinase depends on $PI(3,4)P_2$ (Bellacosa et al., 2005; Scheid et al., 2002). Recently, the PI3K-related DNA-dependent protein kinase has been identified as a prominent Ser^{473} kinase in membrane extracts from HEK293 cells (Feng et al., 2004). AKT was thought to translocate to the nucleus after these phosphorylation steps. Recent studies demonstrated, however, that mitogens, including insulin (Lim et al., 2003), IGF-1 and PDGF (Scheid et al., 2005) stimulated the nuclear shuttling of PDK1 in various cell lines. Moreover, PDK1 nuclear localization is increased in PTEN deficient cells and in cells treated with PI3K inhibitors, LY294002 or wortmannin, indicating that PI3K pathway is involved in PDK1 nuclear trafficking (Lim et al., 2003). It was also shown that Ser^{396} phosphorylation on PDK1 is necessary for its nuclear translocation (Scheid et al., 2005). Ser^{396} is a residue adjacent to the functional nuclear export signal (NES) (residues 382-391) of PDK1, which is responsible for chromosome region maintenance 1 (CRM1)-mediated nuclear export. The NES-deleted mutant of PDK1, which accumulates in the nucleus, is as active as wt-PDK1 (Lim et al., 2003). Moreover, co-transfection experiments in MCF-7 cells revealed that nuclear PDK1 colocalizes with AKT, and inhibits the transcriptional activity and the nuclear localization of the transcription factor FOXO (Forkhead box, class O) 3 (Scheid et al., 2005). Forkhead transcription factors of FOXO subfamily are important downstream targets of nuclear AKT. Direct phosphorylation of FKHR/FOXO1, FOXO2, FKHRL1/FOXO3, and AFX/FOXO4 transcription factors by AKT results in their nuclear export and inhibition of transcription. FOXO proteins play an essential role in many of the effects of AKT on cell proliferation and survival. FOXO proteins control the transcription of pro-apoptotic genes, such as Fas ligand or Bim (Bcl-2 interacting mediator) and anti-apoptotic genes, such as the transcriptional repressor Bcl-6 (Birkenkamp and Coffer, 2003; Song et al., 2005). Phosphorylation of cAMP-response element-binding protein (CREB), another nuclear transcription factor, by AKT results in increasing affinity of CREB to its co-activator CRE and transcriptional activity of CREB. CREB mediates AKT-induced expression of some anti-apoptotic genes, such as Bcl-2 and Mcl-1 (Song et al., 2005).

A large body of literature indicates the presence of activated AKT in the nucleus of tumor cells. Like PDK1, AKT has a CRM1-binding NES that promotes its nuclear export (Saji et al., 2005). However, the mechanism of nuclear entry still remains to be identified. The presence of nuclear, activated AKT has been detected in lung and breast primary tumors (Lee et al., 2002; Nicholson et al., 2003). In the prostate, nuclear localization of activated AKT increases during the progression from normal tissue to low-grade prostatic intraepithelial neoplasia (PIN), high-grade PIN, and tumor (Van de Sande et al., 2005). Moreover, in the tumors, the extent of nuclear localization strongly correlates with the Gleason score. In primary follicular thyroid tumors, the localization of activated AKT inversely correlates to the presence of PTEN in the nucleus: while nuclear PTEN levels diminish during the progression from normal tissue to adenoma to carcinoma, nuclear phosphorylated AKT levels increase (Vasko et al., 2004). An ovarian steroid, estrogen, increases nuclear AKT activation and localization in murine and human cardiomyocytes, and young women have higher levels of nuclear, active AKT than men or postmenopausal women (Camper-Kirby et al., 2001). The effects of estrogen on the activation and nuclear localization of AKT are well described by the endometrial model system, revealing a strong inverse correlation between the presence of PTEN and active AKT in the nucleus. The levels of active nuclear AKT in the glandular epithelium are low in the early proliferative phase, when the levels of nuclear PTEN are very high and selected cell populations undergo apoptosis (Kokawa et al., 1996) and then increase to peak in the late proliferative phase, when nuclear PTEN starts declining and proliferation is at its peak (Guzeloglu-Kayisli et al., 2003). In the secretory phase, when cells do not proliferate and prepare for apoptosis, nuclear AKT levels are very low, as are those of PTEN. Stromal cells display a similar pattern, but the levels of phosphorylated, nuclear AKT drop to a minimum during decidualization, when nuclear PTEN is very high. In primary hepatocytes, AKT is not activated and is localized to the cytoplasm. Upon insulin stimulation, AKT becomes phosphorylated at the cell membrane and then seems to move to the nuclear membrane. Conversely, in the transformed HepG2 cells, AKT is mostly phosphorylated and localized to the nucleus. Insulin stimulation further increases AKT activation and nuclear localization (Syed et al., 2002). Immunofluorescence experiments in 293T and REF-52 cells have shown that, upon growth factor stimulation, AKT is immediately recruited at the membrane, but 15 min later it is found mostly in the nucleus (Andjelkovic et al., 1997; Meier et al., 1997). Hence, the common notion was that AKT translocates into the nucleus following activation at the cell membrane. However, accumulative evidence now suggests that AKT activation occurs, at least in part, in the nuclear compartment. It has been recently shown that phosphorylation of AKT is not required for nuclear translocation, even though activation of PI3K enhances AKT nuclear import and PI3K inhibitors prevent it (Saji et al., 2005). These results are consistent with previous data demonstrating that Akt nuclear translocation in primary mouse endothelial cells depends on PI3K activation and the presence of RhoB, a small GTPase functioning in intracellular vesicle trafficking, but does not require Akt phosphorylation (Adini et al., 2003). Additionally, IGF-I and PDGF stimulations increase nuclear, active Akt in MC3T3-E1 mouse calvaria fibroblasts with a slower kinetic than PI3K migrates into the nucleus (Borgatti et al., 2000; Martelli et al., 2000), suggesting that PI3K migration precedes nuclear accumulation of phosphorylated AKT. Lastly, enforced Akt expression in the nucleus of primary cardiomyocytes or in the heart of transgenic mice results in the accumulation of phosphorylated Akt, indicating that Akt can undergo phosphorylation inside the nucleus (Shiraishi et al., 2004).

Trotman et al. have recently explored the nuclear Akt-mediated mechanism of tumorigenesis in a murine model (Trotman et al., 2006). The authors showed that the Pml tumor suppressor prevents cancers by inactivating phosphorylated Akt (pAkt) inside the nucleus. PML-nuclear body, a subnuclear macromolecule structure, is involved in apoptosis, proliferation, and senescence. They found that *Pml* loss markedly accelerates tumor onset, incidence and progression in *Pten*-heterozygous mutants, and leads to female sterility with features that recapitulate the phenotype of *Foxo3a* knockout mice. They showed that *Pml* deficiency on its own leads to tumorigenesis in the prostate, a tissue that is exquisitely sensitive to pAkt levels, and further demonstrate that Pml specifically recruits the pAkt phosphatase PP2a as well as pAkt into Pml nuclear bodies. Notably, they found that *Pml*-null cells are impaired in PP2a phosphatase activity towards Akt, and thus accumulate nuclear pAkt. Consequently, the progressive reduction in *Pml* dose leads to inactivation of Foxo3a-mediated transcription of proapoptotic *Bim* and the cell cycle inhibitor $p27^{kip1}$. Together, these results highlight the importance of nuclear AKT network in human cancer pathogenesis and treatment.

The Crosstalk between Ca^{2+} Signaling and the AKT Pathway

The interactions between Ca^{2+} signaling and the AKT pathway have been demonstrated in many cell types including cardiomyocytes and breast cancer cells.

Previous reports showed that cardiac-specific overexpression of active mutant Akt induces hypertrophy *in vivo* in transgenic mice (Condorelli et al., 2002; Cook et al., 2002; Shioi et al., 2002). Akt activation also exerts a cardioprotective effect by inhibiting cardiomyocyte death during ischemia/reperfusion (Fujio et al., 2000; Matsui et al., 1999; Matsui et al., 2001; Miao et al., 2000; Mockridge et al., 2000). Besides, transgenic mice with cardiac-specific overexpression of constitutively active Akt increases base-line contractility (Condorelli et al., 2002). Kim et al. have investigated the molecular mechanism responsible for increased myocardial function by Akt activation (Kim et al., 2003). Heart/body weight, LV ejection fraction, contractile function in isolated ventricular myocyte, the rate of relaxation, Fura-2 Ca^{2+} transients, L-type Ca^{2+} channel current (I_{Ca}) were all increased in transgenic mice with cardiac-specific overexpression of active Akt (TG) compared with wild type (WT). Interestingly, sarcoplasmic reticulum Ca^{2+} ATPase 2a (SERCA2a) protein levels were also increased in TG, which could be recapitulated *in vitro* by adenovirus-mediated Akt overexpression in cultured adult ventricular myocytes. Conversely, pharmacological inhibition of SERCA2a with either ryanodine or thapsigargin attenuated myocyte contraction and relaxation and Ca^{2+} channel kinetics more in TG than in WT. To summarize, it has been demonstrated that overexpressed Akt in cardiomyocytes induces enhanced contractility and relaxation, Ca^{2+} transients, and I_{Ca} via increased SERCA2a expression (Kim et al., 2003).

Class IA PI3K (PI3Kα, β and δ) can be activated by receptor tyrosine kinase/cytokine receptor activation, and class IB PI3K (PI3Kγ) by G protein-coupled receptors (GPCR). Under the negative control by the 3'-lipid phosphatase PTEN, class IA and class IB PI3Ks are known to mediate distinct phenotypes in the heart (Kamp and Chiamvimonvat, 2006). Accumulated evidence has shown that PI3Kα, γ, and PTEN are expressed in cardiomyocytes, fibroblasts, endothelial cells, and vascular smooth muscle cells, where they affect cell survival, hypertrophy, and contractility. In recent years, several transgenic and knockout mouse models have demonstrated a fundamental role of PI3K/PTEN signaling in the regulation of myocardial hypertrophy and contractility. Genetic ablation of Pten caused both hypertrophy and impaired contractility in the knockout mice (Crackower et al., 2002). Notably, the cardiac hypertrophy effect of Pten knockout was specifically attributable to the PI3Kα-mediated pathways, whereas the decrease in contractility was directed by PI3Kγ, suggesting a possible dichotomy in cardiac PI3K signaling based on the distinct isoforms. However, the downstream mediators responsible for the different phenotypic responses have not been fully defined. Sun and colleagues, utilizing combinations of genetically altered mice, have recently provided new evidence linking the regulation of cardiac L-type Ca^{2+} channels by IGF-1 and Pten to PI3Kα-mediated hypertrophy response, but not the PI3Kγ-mediated impaired contractility (Sun et al., 2006). The authors have shown that IGF-1 or inactivation of Pten enhances Ca^{2+} influx through the L-type Ca^{2+} channel via PI3Kα-dependent Akt activation. This observation is consistent with the previous notion that IGF-1 receptor tyrosine kinase stimulation activates Akt via PI3Kα (McMullen et al., 2004), as well as the recent finding described above that constitutively active overexpressed Akt increases I_{Ca} in cardiomyocytes (Kim et al., 2003). On the other hand, specific regulation of L-type Ca^{2+} channels by PI3K isoforms have previously been shown in portal vein myocytes where platelet derived growth factor (PDGF) increases I_{Ca} via the class IA isoform, p110β (Macrez et al., 2001), while angiotensin II–mediated increases in I_{Ca} are dependent on the class IB isoform, p110γ (Quignard et al., 2001). Sun et al. (Sun et al., 2006) further documented that basal Akt activity was insufficient to up-regulate cardiac I_{Ca} when Pten phosphatase activity was normal; when Pten activity is abolished, the PI3Kα activity was sufficient to activate Akt and increase I_{Ca}. In contrast, knockout of PI3Kγ in Pten-null mice did not affect basal or IGF-1–stimulated I_{Ca}. However, the identification of specific downstream components of the PI3K pathway responsible for the different cardiac responses still awaits further mechanistic works.

In rat myocardiac H9c2 cells, it was previously shown that cell survival signaling of the Akt pathway was influenced by a change in the cytoplasmic free calcium concentration ($[Ca^{2+}]i$) via the regulation of a Ser/Thr phosphatase, protein phosphatase 2Ac (PP2Ac): Akt was down-regulated when $[Ca^{2+}]i$ was elevated by thapsigargin, an inhibitor of the endoplasmic reticulum Ca^{2+}-ATPase, but was up-regulated when it was suppressed by 1,2-bis(o-aminophenoxy)ethane-N,N,N',N'-tetraacetic acid tetra(acetoxymethyl)ester (BAPTA-AM), a cell permeable Ca^{2+} chelator (Kageyama et al., 2002). The inactivation of Akt was well correlated with the susceptibility to oxidant-induced apoptosis in H9c2 cells (Kageyama et al., 2002). Yasuoka et al. have further examined the transcriptional regulation of PP2Acα in H9c2 cells with Ca^{2+} modulators (Yasuoka et al., 2004). Transcription of the PP2Acα gene was increased by thapsigargin but decreased by BAPTA-AM. cAMP response element (CRE) was found responsible for the Ca^{2+}-dependent regulation of PP2Acα. Taken together, a long term change of $[Ca^{2+}]i$ regulates PP2Acα transcription via CRE, resulting in a change in the activation status of Akt leading to an altered susceptibility to apoptosis.

c-Myc-overexpressing mammary epithelial cells are proapoptotic: their survival is strongly promoted by epidermal growth factor (EGF) (Ramljak et al., 2003). Deb et al. have demonstrated that EGF-induced Akt activation and survival in c-Myc transgenic mouse mammary carcinoma cells are both calcium/calmodulin (Ca^{2+}/CaM)-dependent (Deb et al., 2004). Akt activation is abolished by the phospholipase C-γ inhibitor U-73122, BAPTA-AM, and the specific CaM antagonist W-7, implicating Ca^{2+}/CaM-mediated activation of Akt in these cells. Additionally, Akt activation by serum and insulin is also inhibited by W-7. EGF-induced and Ca^{2+}/CaM-mediated Akt activation occurs in both tumorigenic and non-tumorigenic mouse and human mammary epithelial cells, independent of c-Myc overexpression. These results suggest that Ca^{2+}/CaM may be common regulators of AKT activation, irrespective of upstream receptor activator, mammalian species, and transformation status in mammary epithelial cells. However, only c-Myc-overexpressing mouse mammary carcinoma cells, but not normal mouse mammary epithelial cells, undergo apoptosis in the presence of W-7, pointing to the selective role of CaM for survival of these cells. Ca^{2+}/CaM-regulated Akt activation is mediated directly by neither CaM kinases nor PI3K: pharmacological inhibitors of CaM kinase kinase and CaM kinases II and III do not inhibit EGF-induced Akt activation, and W-7 does not inhibit phosphotyrosine-associated PI3K activation. However, Akt is co-immunoprecipitated with CaM in an EGF-dependent manner, which is inhibited by W-7. The authors speculate that Ca^{2+}-bound CaM transports Akt to the plasma membrane, where Akt binds to PI3K products and is subsequently phosphorylated. Further efforts are necessary to clarify the exact mechanism of Ca^{2+}/CaM-regulated AKT activation. Additionally, it will be quite interesting to look into the effect of Ca^{2+}/CaM on the subcellular compartmentalization of AKT and PTEN, and its association with AKT activation, as we have recently shown that Ca^{2+} appears to regulate nuclear-cytoplasmic localization of PTEN in breast cancer MCF-7 cells (Minaguchi et al., 2006).

The detailed mechanism of AKT-mediated I_{Ca} regulation remains elusive: *e.g.* the involvement of direct phosphorylation by AKT of $Ca_v1.2$ (the α_{1C} subunit of L-type Ca^{2+} channels) or other subunits of the channel including the various $Ca_v\beta$ subunit isoforms, or alternatively, other intermediate substrate for AKT such as SERCA2a. It should be also addressed whether the differential effects of PI3Kα and PI3Kγ in the heart are attributable to differences in compartmentalization of these isoforms. The importance of precise cellular signaling in this PI3Kα/AKT/$Ca_v1.2$ pathway is demonstrated by the essential role of PI3Kα signaling in physiological hypertrophy characterized by normal organization of cardiac structure and normal or enhanced cardiac function in response to exercise training (McMullen et al., 2003), whereas signaling via PI3Kγ has been implicated in pathological hypertrophy associated with pressure or volume overload in disease (Kamp and Chiamvimonvat, 2006; McMullen and Jennings, 2007; McMullen et al., 2003). The precise integration of these isoform-specific signaling cascades in given cellular compartments is likely crucial for the appropriate cardiomyocyte responses.

Alogether, in cardiomyocytes, AKT upregulates Ca^{2+} current through L-type channels via SERCA2a, thereby modulating cardiac function. A long-term increase of cytoplasmic Ca^{2+} in turn downregulates AKT via PP2Acα transcription and induces apoptosis. These findings suggest the presence of feedback mechanism in the Ca^{2+}/AKT-mediated regulation of cardiac functions, further providing implications for the pathogenesis of heart failure. By contrast, in breast tumor cells, Ca^{2+} seems to upregulate AKT, leading to attenuation of apoptosis. Hence, the mode of crosstalk between Ca^{2+} and AKT signaling appears to be cell-type specific. Previously Chung et al. have demonstrated that nuclear-cytoplasmic partitioning of PTEN differentially regulates the cell cycle and apoptosis: nuclear PTEN down-regulates phosphorylation of p44/42 and cyclin D1 and induces G_1 cell cycle arrest, whereas cytoplasmic PTEN down-regulates phosphorylation of AKT, up-regulates p27, and induces apoptosis (Chung and Eng, 2005). Accordingly, it is plausible to assume that Ca^{2+} regulates apoptosis and cell cycle through modulating nuclear-cytoplasmic compartmentalization of PTEN and their downstream signaling cascades in breast tumor cells. Indeed, in breast cancer cell line MCF-7, we have recently shown that Ca^{2+} ionophore A23187 increases nuclear localization of PTEN and decreases phosphorylation of nuclear p44/42, possibly through affecting the interaction between PTEN and the hypothetical carrier molecule for PTEN nuclear import, MVP (Minaguchi et al., 2006).

Given that AKT protects cardiomyocytes from apoptosis in response to pathologic stimuli (Fujio et al., 2000), AKT activation may be an ideal method of enhancing the cardiac contractility and improving Ca^{2+} handling in response to pressure overload. Indeed, it has been recently found that exercise and enhanced cardiac PI3K activity improve survival by 15–20% in the model of dilated cardiomyopathy, whereas depressed PI3K activity shortens life span by almost 50% (McMullen et al., 2007). McMullen and colleagues have also demonstrated that supranormal cardiac PI3K activity resulted in functional improvement in mice subjected to pressure overload (aortic banding), whereas reduced cardiac PI3K activity markedly accelerated the progression of heart failure in the same model (McMullen et al.,

2007; McMullen et al., 2003). Recent findings further suggest that the IGF1/PI3K/AKT axis may have a beneficial effect, at least in part, by inhibiting pathological growth and signaling molecules downstream of GPCR (DeBosch et al., 2006; McMullen et al., 2007). PI3K (p110α) signaling negatively regulates GPCR-stimulated ERK activation in isolated cardiomyocytes (McMullen et al., 2007). Recently the possible use of isoform-specific inhibitors of PI3K have generated great interest in the cancer field (Fan et al., 2006; Foukas et al., 2006). However, potential adverse effects include diabetes and heart disease: recent findings suggest that such inhibitors may not cause unmanageable metabolic disturbances (Foukas et al., 2006). Although mice with reduced cardiac PI3K (p110α) activity had normal cardiac function under basal conditions (Shioi et al., 2000), they exhibited accelerated heart failure in a setting of dilated cardiomyopathy or pressure overload (McMullen et al., 2007). Therapeutically targeting PI3K isoforms may still require drug delivery systems that specifically target particular cell types or organs.

Hormonal Regulation of the AKT Pathway in Endocrine-Related Cancers

Reciprocal Regulation of the PI3K/AKT Pathway by Ovarian Steroids in Endometrial Carcinogenesis

Loss of PTEN function characterizes at least 20% of endometrial hyperplasias, precancerous lesions, and more than 50% of endometrial endometrioid adenocarcinomas. PTEN inactivation is the earliest step in endometrial carcinogenesis, as normal premenopausal naturally cycling women have small numbers of immunohistochemically detected PTEN-deficient endometrial glands, which have been shown to harbor acquired mutations and deletions of *PTEN* when microdissected (Mutter et al., 2001). *PTEN* mutation is highly associated with endometrial endometrioid adenocarcinoma, estrogen-dependent type (type I) endometrial cancer (Risinger et al., 1998). Endogenous or exogenous estrogen unopposed with progesterone is thought to contribute to the pathogenesis of this subtype of endometrial tumor (Hecht and Mutter, 2006). Estrogen receptor (ER) is known to exert non-genomic, extranuclear action as well as genomic, intranuclear function as a transcription factor: ERα binds in a ligand-dependent manner to p85α, the regulatory subunit of PI3K, leading to AKT activation (Simoncini et al., 2000). Accordingly, loss of PTEN function is thought to accelerate the signaling of estrogen/PI3K/AKT pathway in tumors. However, the precise mechanism of opposing effect by progesterone remains unclear. It has been reported that the level of phosphorylated AKT is reduced in endometrial stromal cells undergoing progesterone-induced *in vitro* decidualization (Yoshino et al., 2003). Marino et al. have reported that estradiol-induced AKT phosphorylation is modulated by PTEN via MAPK pathway (Marino et al., 2003). Published findings including these suggest that anti-tumor action of progesterone on endometrial tumor might be mediated by up-regulation of PTEN, which subsequently down-regulates AKT phosphorylation.

Normal endometrium is hormonally regulated in a cyclic manner and undergoes dynamic changes, which are carried out by ovarian steroids-regulated mechanisms of proliferation, differentiation, and apoptosis. The tumor suppressor PTEN regulates cell growth, apoptosis, and proliferation, and phosphorylation in PTEN tail is known to decrease both its activity and degradation. Based on the hypotheses that PTEN expression in human normal endometrium varies throughout the menstrual cycle, regulated by ovarian steroids, Guzeloglu-Kayisli et al. have examined a direct regulation of PTEN expression by ovarian steroids (Guzeloglu-Kayisli et al., 2003). Estradiol increased PTEN phosphorylation at 5-15 min, while progesterone induced a significant increase in PTEN protein levels after 24-h treatment. Consistent with the previous data (Mutter et al., 2000b), endometrial PTEN expression revealed temporal and spatial changes throughout the menstrual cycle. Interestingly enough, phosphorylated PTEN is mostly localized in the nucleus, whereas unphosphorylated PTEN is mainly in the cytoplasm. Furthermore, estradiol treatment increased nuclear PTEN immunoreactivity, and decreased that of cytoplasmic PTEN. The authors propose that ovarian steroids may regulate the endometrial PTEN pool, thus affecting endometrial cell proliferation and/or apoptosis. Given that nuclear-cytoplasmic partitioning of PTEN has been recently shown to differentially regulate the cell cycle and apoptosis in breast cancer cells (Chung and Eng, 2005), it would be reasonable to postulate that estrogen and progesterone reciprocally regulate endometrial cell cycle and apoptosis through phosphorylation-dependent compartmentalization of PTEN. More effort is clearly warranted to resolve this issue.

Systemic progestins are known to be effective in the treatment of precancerous endometrial lesions (Gal et al., 1983; Heikkinen et al., 2000; Wentz, 1974). Zheng et al. have examined the effect of this hormonal therapy on clearance of immunohistochemically detected PTEN-null endometrial glands (Zheng et al., 2004). PTEN immunohistochemistry of pre and postprogestin-treated endometria was performed on 17 women presenting with biopsy-diagnosed hyperplasia, precancerous endometrial lesion. The persistence of PTEN-null glands in progestin-treated patients was compared to that in normal proliferative endometrium of endogenously cycling premenopausal women. PTEN-null glands in 90% (9/10) of prehormonal patients disappeared after progestin therapy in contrast with only 17% (2/12) disappearance in normal cycling women (P=0.002) (Zheng et al., 2004). The same group has also tested the efficacy of locally administered progestin via intrauterine device (IUD) on PTEN-null endometrial glands, the presence of which by immunohistochemistry represent latent precancers (Orbo et al., 2006). Perimenopausal and postmenopausal women had an endometrial biopsy and underwent hormonal therapy with progestin-impregnated IUD (n = 21), cyclic oral low-dose progestins (n = 28), or surveillance only (n = 22) with follow-up biopsies. The IUD group showed the highest rate of regression, 62% pre-therapy and 5% post-therapy rate of PTEN-null endometrial glands, contrasting to nonsignificant changes for the oral progestin and untreated control groups, suggesting that IUD-mediated local delivery of high-dose progestins leads to ablation of PTEN-inactivated endometrial latent precancers. The observed disappearances of PTEN-null glands following systemic/local progestin therapy cannot be explained by reactivation of a silent *PTEN* gene, given that structurally damaged *PTEN* alleles (deletions, mutations) comprise the majority of PTEN-inactivating events in endometrial neoplasia (Mutter et al., 2000a) and histologically unremarkable PTEN-null glands (Mutter et al., 2001). Accordingly, progestin may inhibit proliferation through an effect downstream of PTEN, possibly at or below the level of AKT. Indeed, by the immunohistochemical analysis in medroxyprogesterone acetate (MPA)-treated 31 cases with endometrial precancerous lesion or early-stage adenocarcinoma, we have recently observed significant decrease of phosphorylated AKT expression by MPA treatment (p=0.002), which suggests that anti-tumor action of MPA is mediated by dephosphorylation of AKT

(Minaguchi et al., 2007). Our observation *in vivo* keeps in line also with the recent finding by Chen et al. that progesterone inhibits the estradiol-stimulated AKT/GSK3β/cyclin D1/pRb pathway to block uterine epithelial cell proliferation (Chen et al., 2005). Besides, progesterone has been reported to up-regulate p27 in normal and cancer endometrial cells (Kawaguchi et al., 2006; Shiozawa et al., 2001; Shiozawa et al., 1998), and the significance of p27 expression as a predictive marker for the effectiveness of progestin therapy in endometrial carcinoma has been recently presented (Watanabe et al., 2006). The progesterone/AKT signaling cascade is likely to play a pivotal role for this endocrine therapy in endometrial carcinoma.

Utilizing a murine model system, where Pten(+/-) mice develop endometrial neoplastic lesions with full penetrance, Vilgelm et al. have recently provided *in vivo* evidence that loss of Pten and subsequent Akt activation result in the activation of ERα-dependent pathways (Vilgelm et al., 2006). The investigators have shown that loss of Pten in the mouse endometrium activates Akt and results in increased phosphorylation of ERα on Ser^{167}. ERα phosphorylation in turn leads to its increased transcription activity, even in the absence of ligand. Moreover, reduced endometrial ERα levels and activity dramatically decreases the neoplastic effect of Pten loss in the endometrium, in contrast to complete estrogen depletion. These results provide evidence for the novel crosstalk mechanism between genomic and non-genomic signaling pathways of ER, and may account for the reason why PTEN inactivation is highly associated with estrogen-dependent type I endometrial carcinoma.

Role of the AKT Pathway in Endocrine Resistance of Breast Cancer—from Bench to Bedside

The PI3K/AKT pathway is a cell survival pathway that is also activated in hormone-resistant breast cancer. Among downstream effectors of AKT, mTOR activates p70S6 kinase and 4E binding protein-1, which in turn regulate the translation of pro- and anti-apoptotic mRNAs. Approaches to targeting these pathways include either specific PI3K inhibitors such as LY294002, or rapamycin analogs such as temsirolimus (CCI-779) and everolimus (RAD-001). Breast cancer cell lines with activated AKT have been shown to be especially sensitive to mTOR inhibition (Yu et al., 2001). More recently, evidence has emerged to support the combination of mTOR antagonists with a hormonal agent, either tamoxifen or an aromatase inhibitor in preclinical models of ER-positive hormone-sensitive and hormone-resistant breast cancer (Boulay et al., 2005; deGraffenried et al., 2004). The estrogen-dependent growth of both wild-type MCF-7 and aromatase-expressing breast cancer cells (MCF-7/Aro) was suppressed in a dose-dependent manner by everolimus (RAD-001), suggesting that mTOR signaling is required for the estrogen-dependent proliferation of these cells. Subsequently, the combination of the aromatase inhibitor letrozole and everolimus have produced maximal growth suppression of MCF-7/Aro cells, indicating additive/synergistic effects (Boulay et al., 2005). Evidence has emerged that activation of AKT and the downstream mTOR pathway can cause resistance to tamoxifen (Clark et al., 2002). MCF-7 cells expressing constitutively active AKT were able to proliferate under reduced estrogen conditions and were resistant to the growth inhibitory effects of tamoxifen in xenograft models (deGraffenried et al., 2004). Cotreatment with the mTOR antagonist temsirolimus inhibited mTOR activity and restored sensitivity to tamoxifen through induction of apoptosis, thus suggesting that AKT-induced tamoxifen resistance may be partially mediated by the mTOR pathway.

Single-agent activity with partial response of 11% and stable disease of 33% has been reported for temsirolimus in 109 heavily pretreated patients with locally advanced or metastatic breast cancer (Chan et al., 2005). Clinical trials examining combined therapy have been initiated, based on the encouraging preclinical data for possible synergistic effects by combined mTOR antagonists and endocrine therapy (Boulay et al., 2005; deGraffenried et al., 2004). A three-arm randomized phase II study compared two different schedules of oral temsirolimus combined with letrozole to letrozole alone: a nonsignificant enhanced clinical benefit rate of 80% for the combination compared with 69% for the letrozole alone has been reported (Johnston, 2006). A large-scale phase III trial with the oral temsirolimus in combination with letrozole compared with letrozole alone has started for postmenopausal patients with ER-positive, locally advanced or metastatic breast cancer. Similar randomized phase III studies assessing letrozole with or without everolimus are planned both in the neoadjuvant and advanced breast cancer settings (Johnston, 2006).

Conclusion

New observations reviewed here indicate cell-type specific interactions among the three signaling cascades, nuclear, calcium, and hormone pathways, and AKT serves as a nodal point for the crosstalk. In cardiomyocytes, AKT is regulating cellular Ca^{2+}, which in turn regulates AKT activity, thereby affecting cardiac functions and contributing to pathogenesis of the heart. In breast cancer cells, intracellular Ca^{2+} seems to affect both the activity of AKT and subcellular compartmentalization of PTEN, thus regulating cell cycle and/or apoptosis. In the endometrium, ovarian steroids, estrogen and progesterone, appear to reciprocally regulate phosphorylation and compartmentalization of PTEN, which correlate to phosphorylation of nuclear AKT. These accumulating lines of evidence further underline the importance of the AKT signaling network as a target for therapeutic intervention in endocrine-related cancers and heart diseases. On the other hand, because of the involvement in diverse signaling cascades, it is suggested that targeting the AKT pathways may require drug delivery systems that specifically target particular cell types or organs to circumvent potential adverse toxicities. Emerging evidence described above now produce a number of additional intriguing questions, answering which will provide significant mechanistic and clinical insights into the AKT pathways of potential benefit to many cancer and heart disease patients.

References

Adini, I., Rabinovitz, I., Sun, J. F., Prendergast, G. C., And Benjamin, L. E. (2003). Rhob Controls Akt Trafficking And Stage-Specific Survival Of Endothelial Cells During Vascular Development. *Genes Dev* **17**, 2721-2732.

Ahn, J. Y., Liu, X., Cheng, D., Peng, J., Chan, P. K., Wade, P. A., And Ye, K. (2005). Nucleophosmin/B23, A Nuclear Pi(3,4,5)P(3) Receptor, Mediates The Antiapoptotic Actions Of Ngf By Inhibiting Cad. *Mol Cell* **18**, 435-445.

Ahn, J. Y., Rong, R., Liu, X., And Ye, K. (2004). Pike/Nuclear Pi 3-Kinase Signaling Mediates The Antiapoptotic Actions Of Ngf In The Nucleus. *Embo J* **23**, 3995-4006.

Andjelkovic, M., Alessi, D. R., Meier, R., Fernandez, A., Lamb, N. J., Frech, M., Cron, P., Cohen, P., Lucocq, J. M., And Hemmings, B. A. (1997). Role Of Translocation In The Activation And Function Of Protein Kinase B. *J Biol Chem* **272**, 31515-31524.

Bellacosa, A., Kumar, C. C., Di Cristofano, A., And Testa, J. R. (2005). Activation Of Akt Kinases In Cancer: Implications For Therapeutic Targeting. *Adv Cancer Res* **94**, 29-86.

Birkenkamp, K. U., And Coffer, P. J. (2003). Regulation Of Cell Survival And Proliferation By The Foxo (Forkhead Box, Class O) Subfamily Of Forkhead Transcription Factors. *Biochem Soc Trans* **31**, 292-297.

Borgatti, P., Martelli, A. M., Bellacosa, A., Casto, R., Massari, L., Capitani, S., And Neri, L. M. (2000). Translocation Of Akt/Pkb To The Nucleus Of Osteoblast-Like Mc3t3-E1 Cells Exposed To Proliferative Growth Factors. *Febs Lett* **477**, 27-32.

Boulay, A., Rudloff, J., Ye, J., Zumstein-Mecker, S., O'reilly, T., Evans, D. B., Chen, S., And Lane, H. A. (2005). Dual Inhibition Of Mtor And Estrogen Receptor Signaling In Vitro Induces Cell Death In Models Of Breast Cancer. *Clin Cancer Res* **11**, 5319-5328.

Camper-Kirby, D., Welch, S., Walker, A., Shiraishi, I., Setchell, K. D., Schaefer, E., Kajstura, J., Anversa, P., And Sussman, M. A. (2001). Myocardial Akt Activation And Gender: Increased Nuclear Activity In Females Versus Males. *Circ Res* **88**, 1020-1027.

Chan, S., Scheulen, M. E., Johnston, S., Mross, K., Cardoso, F., Dittrich, C., Eiermann, W., Hess, D., Morant, R., Semiglazov, V., *Et Al.* (2005). Phase Ii Study Of Temsirolimus (Cci-779), A Novel Inhibitor Of Mtor, In Heavily Pretreated Patients With Locally Advanced Or Metastatic Breast Cancer. *J Clin Oncol* **23**, 5314-5322.

Chen, B., Pan, H., Zhu, L., Deng, Y., And Pollard, J. W. (2005). Progesterone Inhibits The Estrogen-Induced Phosphoinositide 3-Kinase-->Akt-->Gsk-3beta-->Cyclin D1-->Prb Pathway To Block Uterine Epithelial Cell Proliferation. *Mol Endocrinol* **19**, 1978-1990.

Chung, J. H., And Eng, C. (2005). Nuclear-Cytoplasmic Partitioning Of Phosphatase And Tensin Homologue Deleted On Chromosome 10 (Pten) Differentially Regulates The Cell Cycle And Apoptosis. *Cancer Res* **65**, 8096-8100.

Clark, A. S., West, K., Streicher, S., And Dennis, P. A. (2002). Constitutive And Inducible Akt Activity Promotes Resistance To Chemotherapy, Trastuzumab, Or Tamoxifen In Breast Cancer Cells. *Mol Cancer Ther* **1,** 707-717.

Condorelli, G., Drusco, A., Stassi, G., Bellacosa, A., Roncarati, R., Iaccarino, G., Russo, M. A., Gu, Y., Dalton, N., Chung, C., *Et Al.* (2002). Akt Induces Enhanced Myocardial Contractility And Cell Size In Vivo In Transgenic Mice. *Proc Natl Acad Sci U S A* **99**, 12333-12338.

Cook, S. A., Matsui, T., Li, L., And Rosenzweig, A. (2002). Transcriptional Effects Of Chronic Akt Activation In The Heart. *J Biol Chem* **277**, 22528-22533.

Crackower, M. A., Oudit, G. Y., Kozieradzki, I., Sarao, R., Sun, H., Sasaki, T., Hirsch, E., Suzuki, A., Shioi, T., Irie-Sasaki, J., *Et Al.* (2002). Regulation Of Myocardial Contractility And Cell Size By Distinct Pi3k-Pten Signaling Pathways. *Cell* **110**,737-749.

Deb, T. B., Coticchia, C. M., And Dickson, R. B. (2004). Calmodulin-Mediated Activation Of Akt Regulates Survival Of C-Myc-Overexpressing Mouse Mammary Carcinoma Cells. *J Biol Chem* **279**, 38903-38911.

Debosch, B., Treskov, I., Lupu, T. S., Weinheimer, C., Kovacs, A., Courtois, M., And Muslin, A. J. (2006). Akt1 Is Required For Physiological Cardiac Growth. *Circulation* **113**, 2097-2104.

Degraffenried, L. A., Friedrichs, W. E., Russell, D. H., Donzis, E. J., Middleton, A. K., Silva, J. M., Roth, R. A., And Hidalgo, M. (2004). Inhibition Of Mtor Activity Restores Tamoxifen Response In Breast Cancer Cells With Aberrant Akt Activity. *Clin Cancer Res* **10**, 8059-8067.

Deleris, P., Gayral, S., And Breton-Douillon, M. (2006). Nuclear Ptdlns(3,4,5)P3 Signaling: An Ongoing Story. *J Cell Biochem* **98**, 469-485.

Didichenko, S. A., And Thelen, M. (2001). Phosphatidylinositol 3-Kinase C2alpha Contains A Nuclear Localization Sequence And Associates With Nuclear Speckles. *J Biol Chem* **276**, 48135-48142.

Fan, Q. W., Knight, Z. A., Goldenberg, D. D., Yu, W., Mostov, K. E., Stokoe, D., Shokat, K. M., And Weiss, W. A. (2006). A Dual Pi3 Kinase/Mtor Inhibitor Reveals Emergent Efficacy In Glioma. *Cancer Cell* **9,** 341-349.

Feng, J., Park, J., Cron, P., Hess, D., And Hemmings, B. A. (2004). Identification Of A Pkb/Akt Hydrophobic Motif Ser-473 Kinase As Dna-Dependent Protein Kinase. *J Biol Chem* **279**, 41189-41196.

Foukas, L. C., Claret, M., Pearce, W., Okkenhaug, K., Meek, S., Peskett, E., Sancho, S., Smith, A. J., Withers, D. J., And Vanhaesebroeck, B. (2006). Critical Role For The P110alpha Phosphoinositide-3-Oh Kinase In Growth And Metabolic Regulation. *Nature* **441**, 366-370.

Fujio, Y., Nguyen, T., Wencker, D., Kitsis, R. N., And Walsh, K. (2000). Akt Promotes Survival Of Cardiomyocytes In Vitro And Protects Against Ischemia-Reperfusion Injury In Mouse Heart. *Circulation* **101**, 660-667.

Gal, D., Edman, C. D., Vellios, F., And Forney, J. P. (1983). Long-Term Effect Of Megestrol Acetate In The Treatment Of Endometrial Hyperplasia. *Am J Obstet Gynecol* **146**, 316-322.

Gillooly, D. J., Morrow, I. C., Lindsay, M., Gould, R., Bryant, N. J., Gaullier, J. M., Parton, R. G., And Stenmark, H. (2000). Localization Of Phosphatidylinositol 3-Phosphate In Yeast And Mammalian Cells. *Embo J* **19**, 4577-4588.

Guzeloglu-Kayisli, O., Kayisli, U. A., Al-Rejjal, R., Zheng, W., Luleci, G., And Arici, A. (2003). Regulation Of Pten (Phosphatase And Tensin Homolog Deleted On Chromosome 10) Expression By Estradiol And Progesterone In Human Endometrium. *J Clin Endocrinol Metab* **88**, 5017-5026.

Hecht, J. L., And Mutter, G. L. (2006). Molecular And Pathologic Aspects Of Endometrial Carcinogenesis. *J Clin Oncol* **24**, 4783-4791.

Heikkinen, J., Vaheri, R., Kainulainen, P., And Timonen, U. (2000). Long-Term Continuous Combined Hormone Replacement Therapy In The Prevention Of Postmenopausal Bone Loss: A Comparison Of High- And Low-Dose Estrogen-Progestin Regimens. *Osteoporos Int* **11**, 929-937.

Irvine, R. F. (2003). Nuclear Lipid Signalling. *Nat Rev Mol Cell Biol* **4,** 349-360.

Johnston, S. R. (2006). Clinical Efforts To Combine Endocrine Agents With Targeted Therapies Against Epidermal Growth Factor Receptor/Human Epidermal Growth Factor Receptor 2 And Mammalian Target Of Rapamycin In Breast Cancer. *Clin Cancer Res* **12**, 1061s-1068s.

Kageyama, K., Ihara, Y., Goto, S., Urata, Y., Toda, G., Yano, K., And Kondo, T. (2002). Overexpression Of Calreticulin Modulates Protein Kinase B/Akt Signaling To Promote Apoptosis During Cardiac Differentiation Of Cardiomyoblast H9c2 Cells. *J Biol Chem* **277**, 19255-19264.

Kamp, T. J., And Chiamvimonvat, N. (2006). Mission Impossible: Igf-1 And Pten Specifically "Akt"Ing On Cardiac L-Type Ca2+ Channels. *Circ Res* **98**, 1349-1351.

Kawaguchi, M., Watanabe, J., Hamano, M., Kamata, Y., Arai, T., Nishimura, Y., Obokata, A., Jobo, T., And Kuramoto, H. (2006). Medroxyprogesterone Acetate Stimulates Cdk Inhibitors, P21 And P27, In Endometrial Carcinoma Cells Transfected With Progesterone Receptor-B Cdna. *Eur J Gynaecol Oncol* **27**, 33-38.

Kim, S. J. (1998). Insulin Rapidly Induces Nuclear Translocation Of Pi3-Kinase In Hepg2 Cells. *Biochem Mol Biol Int* **46**, 187-196.

Kim, Y. K., Kim, S. J., Yatani, A., Huang, Y., Castelli, G., Vatner, D. E., Liu, J., Zhang, Q., Diaz, G., Zieba, R., *Et Al.* (2003). Mechanism Of Enhanced Cardiac Function In Mice With Hypertrophy Induced By Overexpressed Akt. *J Biol Chem* **278**, 47622-47628.

Kokawa, K., Shikone, T., And Nakano, R. (1996). Apoptosis In The Human Uterine Endometrium During The Menstrual Cycle. *J Clin Endocrinol Metab* **81**, 4144-4147.

Lee, S. H., Kim, H. S., Park, W. S., Kim, S. Y., Lee, K. Y., Kim, S. H., Lee, J. Y., And Yoo, N. J. (2002). Non-Small Cell Lung Cancers Frequently Express Phosphorylated Akt; An Immunohistochemical Study. *Apmis* **110**, 587-592.

Lian, Z., And Di Cristofano, A. (2005). Class Reunion: Pten Joins The Nuclear Crew. *Oncogene* **24**, 7394-7400.

Lim, M. A., Kikani, C. K., Wick, M. J., And Dong, L. Q. (2003). Nuclear Translocation Of 3'-Phosphoinositide-Dependent Protein Kinase 1 (Pdk-1): A Potential Regulatory Mechanism For Pdk-1 Function. *Proc Natl Acad Sci U S A* **100**, 14006-14011.

Macrez, N., Mironneau, C., Carricaburu, V., Quignard, J. F., Babich, A., Czupalla, C., Nurnberg, B., And Mironneau, J. (2001). Phosphoinositide 3-Kinase Isoforms Selectively Couple Receptors To Vascular L-Type Ca(2+) Channels. *Circ Res* **89**, 692-699.

Marino, M., Acconcia, F., And Trentalance, A. (2003). Biphasic Estradiol-Induced Akt Phosphorylation Is Modulated By Pten Via Map Kinase In Hepg2 Cells. *Mol Biol Cell* **14**, 2583-2591.

Martelli, A. M., Borgatti, P., Bortul, R., Manfredini, M., Massari, L., Capitani, S., And Neri, L. M. (2000). Phosphatidylinositol 3-Kinase Translocates To The Nucleus Of Osteoblast-Like Mc3t3-E1 Cells In Response To Insulin-Like Growth Factor I And Platelet-Derived Growth Factor But Not To The Proapoptotic Cytokine Tumor Necrosis Factor Alpha. *J Bone Miner Res* **15**, 1716-1730.

Martelli, A. M., Tabellini, G., Borgatti, P., Bortul, R., Capitani, S., And Neri, L. M. (2003). Nuclear Lipids: New Functions For Old Molecules? *J Cell Biochem* **88**, 455-461.

Matsui, T., Li, L., Del Monte, F., Fukui, Y., Franke, T. F., Hajjar, R. J., And Rosenzweig, A. (1999). Adenoviral Gene Transfer Of Activated Phosphatidylinositol 3'-Kinase And Akt Inhibits Apoptosis Of Hypoxic Cardiomyocytes In Vitro. *Circulation* **100**, 2373-2379.

Matsui, T., Tao, J., Del Monte, F., Lee, K. H., Li, L., Picard, M., Force, T. L., Franke, T. F., Hajjar, R. J., And Rosenzweig, A. (2001). Akt Activation Preserves Cardiac Function And Prevents Injury After Transient Cardiac Ischemia In Vivo. *Circulation* **104**, 330-335.

Mcmullen, J. R., Amirahmadi, F., Woodcock, E. A., Schinke-Braun, M., Bouwman, R. D., Hewitt, K. A., Mollica, J. P., Zhang, L., Zhang, Y., Shioi, T., *Et Al.* (2007). Protective

Effects Of Exercise And Phosphoinositide 3-Kinase(P110alpha) Signaling In Dilated And Hypertrophic Cardiomyopathy. *Proc Natl Acad Sci U S A* **104**, 612-617.

Mcmullen, J. R., And Jennings, G. L. (2007). Differences Between Pathological And Physiological Cardiac Hypertrophy: Novel Therapeutic Strategies To Treat Heart Failure. *Clin Exp Pharmacol Physiol* **34**, 255-262.

Mcmullen, J. R., Shioi, T., Huang, W. Y., Zhang, L., Tarnavski, O., Bisping, E., Schinke, M., Kong, S., Sherwood, M. C., Brown, J., *Et Al.* (2004). The Insulin-Like Growth Factor 1 Receptor Induces Physiological Heart Growth Via The Phosphoinositide 3-Kinase(P110alpha) Pathway. *J Biol Chem* **279**, 4782-4793.

Mcmullen, J. R., Shioi, T., Zhang, L., Tarnavski, O., Sherwood, M. C., Kang, P. M., And Izumo, S. (2003). Phosphoinositide 3-Kinase(P110alpha) Plays A Critical Role For The Induction Of Physiological, But Not Pathological, Cardiac Hypertrophy. *Proc Natl Acad Sci U S A* **100**, 12355-12360.

Meier, R., Alessi, D. R., Cron, P., Andjelkovic, M., And Hemmings, B. A. (1997). Mitogenic Activation, Phosphorylation, And Nuclear Translocation Of Protein Kinase Bbeta. *J Biol Chem* **272**, 30491-30497.

Metjian, A., Roll, R. L., Ma, A. D., And Abrams, C. S. (1999). Agonists Cause Nuclear Translocation Of Phosphatidylinositol 3-Kinase Gamma. A Gbetagamma-Dependent Pathway That Requires The P110gamma Amino Terminus. *J Biol Chem* **274**, 27943-27947.

Miao, W., Luo, Z., Kitsis, R. N., And Walsh, K. (2000). Intracoronary, Adenovirus-Mediated Akt Gene Transfer In Heart Limits Infarct Size Following Ischemia-Reperfusion Injury In Vivo. *J Mol Cell Cardiol* **32**, 2397-2402.

Minaguchi, T., Nakagawa, S., Takazawa, Y., Nei, T., Horie, K., Fujiwara, T., Osuga, Y., Yasugi, T., Kugu, K., Yano, T., *Et Al.* (2007). Combined Phospho-Akt And Pten Expressions Associated With Post-Treatment Hysterectomy After Conservative Progestin Therapy In Complex Atypical Hyperplasia And Stage Ia, G1 Adenocarcinoma Of The Endometrium. *Cancer Lett* **248**, 112-122.

Minaguchi, T., Waite, K. A., And Eng, C. (2006). Nuclear Localization Of Pten Is Regulated By Ca(2+) Through A Tyrosil Phosphorylation-Independent Conformational Modification In Major Vault Protein. *Cancer Res* **66**, 11677-11682.

Mockridge, J. W., Marber, M. S., And Heads, R. J. (2000). Activation Of Akt During Simulated Ischemia/Reperfusion In Cardiac Myocytes. *Biochem Biophys Res Commun* **270**, 947-952.

Mutter, G. L., Ince, T. A., Baak, J. P., Kust, G. A., Zhou, X. P., And Eng, C. (2001). Molecular Identification Of Latent Precancers In Histologically Normal Endometrium. *Cancer Res* **61**, 4311-4314.

Mutter, G. L., Lin, M. C., Fitzgerald, J. T., Kum, J. B., Baak, J. P., Lees, J. A., Weng, L. P., And Eng, C. (2000a). Altered Pten Expression As A Diagnostic Marker For The Earliest Endometrial Precancers. *J Natl Cancer Inst* **92**, 924-930.

Mutter, G. L., Lin, M. C., Fitzgerald, J. T., Kum, J. B., And Eng, C. (2000b). Changes In Endometrial Pten Expression Throughout The Human Menstrual Cycle. *J Clin Endocrinol Metab* **85**, 2334-2338.

Neri, L. M., Borgatti, P., Capitani, S., And Martelli, A. M. (2002). The Nuclear Phosphoinositide 3-Kinase/Akt Pathway: A New Second Messenger System. *Biochim Biophys Acta* **1584**, 73-80.

Neri, L. M., Marchisio, M., Colamussi, M. L., And Bertagnolo, V. (1999). Monocytic Differentiation Of Hl-60 Cells Is Characterized By The Nuclear Translocation Of Phosphatidylinositol 3-Kinase And Of Definite Phosphatidylinositol-Specific Phospholipase C Isoforms. *Biochem Biophys Res Commun* **259**, 314-320.

Neri, L. M., Milani, D., Bertolaso, L., Stroscio, M., Bertagnolo, V., And Capitani, S. (1994). Nuclear Translocation Of Phosphatidylinositol 3-Kinase In Rat Pheochromocytoma Pc 12 Cells After Treatment With Nerve Growth Factor. *Cell Mol Biol (Noisy-Le-Grand)* **40**, 619-626.

Nicholson, K. M., Streuli, C. H., And Anderson, N. G. (2003). Autocrine Signalling Through Erbb Receptors Promotes Constitutive Activation Of Protein Kinase B/Akt In Breast Cancer Cell Lines. *Breast Cancer Res Treat* **81**, 117-128.

Orbo, A., Rise, C. E., And Mutter, G. L. (2006). Regression Of Latent Endometrial Precancers By Progestin Infiltrated Intrauterine Device. *Cancer Res* **66**, 5613-5617.

Quignard, J. F., Mironneau, J., Carricaburu, V., Fournier, B., Babich, A., Nurnberg, B., Mironneau, C., And Macrez, N. (2001). Phosphoinositide 3-Kinase Gamma Mediates Angiotensin Ii-Induced Stimulation Of L-Type Calcium Channels In Vascular Myocytes. *J Biol Chem* **276**, 32545-32551.

Ramljak, D., Coticchia, C. M., Nishanian, T. G., Saji, M., Ringel, M. D., Conzen, S. D., And Dickson, R. B. (2003). Epidermal Growth Factor Inhibition Of C-Myc-Mediated Apoptosis Through Akt And Erk Involves Bcl-Xl Upregulation In Mammary Epithelial Cells. *Exp Cell Res* **287**, 397-410.

Risinger, J. I., Hayes, K., Maxwell, G. L., Carney, M. E., Dodge, R. K., Barrett, J. C., And Berchuck, A. (1998). Pten Mutation In Endometrial Cancers Is Associated With Favorable Clinical And Pathologic Characteristics. *Clin Cancer Res* **4,** 3005-3010.

Rong, R., Ahn, J. Y., Huang, H., Nagata, E., Kalman, D., Kapp, J. A., Tu, J., Worley, P. F., Snyder, S. H., And Ye, K. (2003). Pi3 Kinase Enhancer-Homer Complex Couples Mgluri To Pi3 Kinase, Preventing Neuronal Apoptosis. *Nat Neurosci* **6,** 1153-1161.

Saji, M., Vasko, V., Kada, F., Allbritton, E. H., Burman, K. D., And Ringel, M. D. (2005). Akt1 Contains A Functional Leucine-Rich Nuclear Export Sequence. Biochem Biophys Res Commun **332**, 167-173.

Scheid, M. P., Huber, M., Damen, J. E., Hughes, M., Kang, V., Neilsen, P., Prestwich, G. D., Krystal, G., And Duronio, V. (2002). Phosphatidylinositol (3,4,5)P3 Is Essential But Not Sufficient For Protein Kinase B (Pkb) Activation; Phosphatidylinositol (3,4)P2 Is Required For Pkb Phosphorylation At Ser-473: Studies Using Cells From Sh2-Containing Inositol-5-Phosphatase Knockout Mice. *J Biol Chem* **277**, 9027-9035.

Scheid, M. P., Parsons, M., And Woodgett, J. R. (2005). Phosphoinositide-Dependent Phosphorylation Of Pdk1 Regulates Nuclear Translocation. Mol Cell Biol *25*, 2347-2363.

Shioi, T., Kang, P. M., Douglas, P. S., Hampe, J., Yballe, C. M., Lawitts, J., Cantley, L. C., And Izumo, S. (2000). The Conserved Phosphoinositide 3-Kinase Pathway Determines Heart Size In Mice. *Embo J* **19**, 2537-2548.

Shioi, T., Mcmullen, J. R., Kang, P. M., Douglas, P. S., Obata, T., Franke, T. F., Cantley, L. C., And Izumo, S. (2002). Akt/Protein Kinase B Promotes Organ Growth In Transgenic Mice. *Mol Cell Biol* **22**, 2799-2809.

Shiozawa, T., Horiuchi, A., Kato, K., Obinata, M., Konishi, I., Fujii, S., And Nikaido, T. (2001). Up-Regulation Of P27kip1 By Progestins Is Involved In The Growth Suppression

Of The Normal And Malignant Human Endometrial Glandular Cells. *Endocrinology* **142**, 4182-4188.

Shiozawa, T., Nikaido, T., Nakayama, K., Lu, X., And Fujii, S. (1998). Involvement Of Cyclin-Dependent Kinase Inhibitor P27kip1 In Growth Inhibition Of Endometrium In The Secretory Phase And Of Hyperplastic Endometrium Treated With Progesterone. *Mol Hum Reprod* **4,** 899-905.

Shiraishi, I., Melendez, J., Ahn, Y., Skavdahl, M., Murphy, E., Welch, S., Schaefer, E., Walsh, K., Rosenzweig, A., Torella, D., *Et Al.* (2004). Nuclear Targeting Of Akt Enhances Kinase Activity And Survival Of Cardiomyocytes. *Circ Res* **94**, 884-891.

Simoncini, T., Hafezi-Moghadam, A., Brazil, D. P., Ley, K., Chin, W. W., And Liao, J. K. (2000). Interaction Of Oestrogen Receptor With The Regulatory Subunit Of Phosphatidylinositol-3-Oh Kinase. *Nature* **407**, 538-541.

Song, G., Ouyang, G., And Bao, S. (2005). The Activation Of Akt/Pkb Signaling Pathway And Cell Survival. *J Cell Mol Med* **9,** 59-71.

Sun, H., Kerfant, B. G., Zhao, D., Trivieri, M. G., Oudit, G. Y., Penninger, J. M., And Backx, P. H. (2006). Insulin-Like Growth Factor-1 And Pten Deletion Enhance Cardiac L-Type Ca2+ Currents Via Increased Pi3kalpha/Pkb Signaling. *Circ Res* **98**, 1390-1397.

Syed, N. A., Horner, K. N., Misra, V., And Khandelwal, R. L. (2002). Different Cellular Localization, Translocation, And Insulin-Induced Phosphorylation Of Pkbalpha In Hepg2 Cells And Hepatocytes. *J Cell Biochem* **86**, 118-127.

Trotman, L. C., Alimonti, A., Scaglioni, P. P., Koutcher, J. A., Cordon-Cardo, C., And Pandolfi, P. P. (2006). Identification Of A Tumour Suppressor Network Opposing Nuclear Akt Function. *Nature* **441**, 523-527.

Van De Sande, T., Roskams, T., Lerut, E., Joniau, S., Van Poppel, H., Verhoeven, G., And Swinnen, J. V. (2005). High-Level Expression Of Fatty Acid Synthase In Human Prostate Cancer Tissues Is Linked To Activation And Nuclear Localization Of Akt/Pkb. *J Pathol* **206**, 214-219.

Vasko, V., Saji, M., Hardy, E., Kruhlak, M., Larin, A., Savchenko, V., Miyakawa, M., Isozaki, O., Murakami, H., Tsushima, T., *Et Al.* (2004). Akt Activation And Localisation Correlate With Tumour Invasion And Oncogene Expression In Thyroid Cancer. *J Med Genet* **41**, 161-170.

Vilgelm, A., Lian, Z., Wang, H., Beauparlant, S. L., Klein-Szanto, A., Ellenson, L. H., And Di Cristofano, A. (2006). Akt-Mediated Phosphorylation And Activation Of Estrogen Receptor Alpha Is Required For Endometrial Neoplastic Transformation In Pten+/- Mice. *Cancer Res* **66**, 3375-3380.

Watanabe, J., Watanabe, K., Jobo, T., Kamata, Y., Kawaguchi, M., Imai, M., Okayasu, I., And Kuramoto, H. (2006). Significance Of P27 As A Predicting Marker For Medroxyprogesterone Acetate Therapy Against Endometrial Endometrioid Adenocarcinoma. *Int J Gynecol Cancer* **16** Suppl 1, 452-457.

Wentz, W. B. (1974). Progestin Therapy In Endometrial Hyperplasia. *Gynecol Oncol* **2,** 362-367.

Yasuoka, C., Ihara, Y., Ikeda, S., Miyahara, Y., Kondo, T., And Kohno, S. (2004). Antiapoptotic Activity Of Akt Is Down-Regulated By Ca2+ In Myocardiac H9c2 Cells. Evidence Of Ca(2+)-Dependent Regulation Of Protein Phosphatase 2ac. *J Biol Chem* **279**, 51182-51192.

Ye, K., Aghdasi, B., Luo, H. R., Moriarity, J. L., Wu, F. Y., Hong, J. J., Hurt, K. J., Bae, S. S., Suh, P. G., And Snyder, S. H. (2002). Phospholipase C Gamma 1 Is A Physiological Guanine Nucleotide Exchange Factor For The Nuclear Gtpase Pike. *Nature* **415**,541-544.

Ye, K., Hurt, K. J., Wu, F. Y., Fang, M., Luo, H. R., Hong, J. J., Blackshaw, S., Ferris, C. D., And Snyder, S. H. (2000). Pike. A Nuclear Gtpase That Enhances Pi3kinase Activity And Is Regulated By Protein 4.1n. *Cell* **103**, 919-930.

Yoshino, O., Osuga, Y., Hirota, Y., Koga, K., Yano, T., Tsutsumi, O., And Taketani, Y. (2003). Akt As A Possible Intracellular Mediator For Decidualization In Human Endometrial Stromal Cells. *Mol Hum Reprod* **9,** 265-269.

Yu, K., Toral-Barza, L., Discafani, C., Zhang, W. G., Skotnicki, J., Frost, P., And Gibbons, J. J. (2001). Mtor, A Novel Target In Breast Cancer: The Effect Of Cci-779, An Mtor Inhibitor, In Preclinical Models Of Breast Cancer. *Endocr Relat Cancer* **8**, 249-258.

Zheng, W., Baker, H. E., And Mutter, G. L. (2004). Involution Of Pten-Null Endometrial Glands With Progestin Therapy. *Gynecol Oncol* **92**, 1008-1013.

In: Oncogene Proteins: Structure, Functions and Analyses ISBN: 978-1-60456-259-0
Editors: P.B. Murphy and J.R. Clarke, pp. 209-221

Chapter VIII

Non-receptor Protein Tyrosine Kinase Fes as a Candidate for Anticancer Molecular Targeting Therapy

***Shigeru Kanda*[1,3*], *Yasuyoshi Miyata*[2] *and Hiroshi Kanetake*[2]**

[1]Department of Molecular Microbiology and Immunology,
Division of Endothelial Cell Biology

[2]Department of Urology, Nagasaki University Graduate School of Biomedical Sciences,
1-7-1 Sakamoto, Nagasaki 852-8501, Japan

[3]Department of Experimental and Clinical Laboratory Medicine, National Hospital Organization, Nagasaki Hospital, 41-6 Sakuragi-machi, Nagasaki 851-0251, Japan

Abstract

Fes (also known as Fps), together with Fer, form the nonreceptor protein tyrosine kinase subfamily. The expression of Fes in normal tissues has been shown to be limited to monocytic hematopoietic cells, endothelial cells, and some neuronal and epithelial cells. Studies using mice expressing activated mutant Fes or cultured endothelial cells indicated that Fes might be involved in pathological angiogenesis, suggesting it could be a potential target for anti-angiogenic therapy. Fes was initially identified as a protooncogene product. However, recent studies have indicated that Fes may act also as a tumor suppressor protein in some carcinomas, such as colon cancer, based on the identification of inactivating mutated Fes in colon cancer tissues and downregulated expression of Fes in some cancer cells. Thus, during treatment of these cancers, inhibition of Fes activity may stimulate progression of the pathological process. In this short communication, we will review the recent reports on the roles of Fes in tumor progression and discuss the problems that must be solved before targeting of this kinase in any anti-cancer therapy.

* E-mail address: skanda-jua@umin.net. Phone: +81 95 823 2261. Fax: +81 95 828 2616. Address all correspondence to: Shigeru Kanda, M.D., Ph.D., National Hospital Organization, Nagasaki Hospital, 41-6 Sakuragi-machi, Nagasaki 851-0251, Japan

Introduction

Fes (also known as Fps) and Fer define a unique subfamily of non-receptor protein-tyrosine kinases encoded by a protooncogene, *c-fes/fps* [1, 2]. Transforming oncogenes, *v-fps* (from Fujinami and PRC-type chicken sarcoma virus) and *v-fes* (from feline sarcoma virus) encode proteins consisting of N-terminal viral *gag* sequence fused to *c-fes-* or *c-fps*-derived kinase sequence. The gag-fusion forms exhibit constitutive activation of tyrosine kinase that is essential for their transforming potential. Unlike the ubiquitously distributed Fer, Fes is only expressed on myeloid hematopoietic cells, vascular endothelial cells, some neuronal and epithelial cells [3]. Structurally, Fes consists of a long N-terminal unique region followed by an SH2 domain and a protein-tyrosine kinase domain with two autophosphorylation sites (Y713 and Y811) [1, 4]. Y713 is a key regulatory residue for Fes kinase activity [5]. The unique N-terminal region contains two coiled-coil homology domains (CCDs), which are involved in homophilic intermolecular oligomerization, followed by autophosphorylation and subsequent activation of tyrosine kinase [6]. Studies using mutant Fes proteins suggested that the first CCD negatively regulates kinase activity while the second CCD has a positive regulatory role [7]. Regulation of the protein-tyrosine kinase activity by coiled-coil domains is a unique defining feature of the Fes/Fer kinase family.

Mice with targeted deletion of *fes gene* or mutant mice expressing kinase-inactive Fes are viable and fertile, and showed no significant defect during development [8, 9]. Mice expressing kinase-inactive Fes showed markedly reduced tyrosine phosphorylation of Stat3 and Stat5A in bone marrow-derived macrophages in response to granulocyte-macrophage colony-stimulating factor (GM-CSF) but not to interleukin (IL)-3 or IL-6 [8]. This suggests a distinct nonredundant role for Fps/Fes in signaling from the GM-CSF receptor that does not extend to the closely related IL-3 receptor. Mice lacking the expression of Fes had reduced numbers of bone marrow myeloid progenitors and circulating mature myeloid cells, and these cells were more sensitive to lipopolysaccharide than those from wild type mice [9]. These results suggest that the bulk of Fes function *in vivo* could be dispensable for other protein tyrosine kinases. The critical point is that mice positive for the expression of activated mutant Fes or v-Fps exhibited increased vascularity or tumorigenicity, indicating that Fes may play important roles in tumor development and progression [10, 11].

Signal Transduction Pathways Upstream or Downstream of Fes (Tables 1 and 2)

Fes is activated by a number of external stimuli. In cells of hematopoietic origin, IL-3, IL-4, IL-6, GM-CSF, and erythropoietin activate Fes and Fes is also activated downstream of c-kit and IgE receptor FcεRI [12-18]. Furthermore, fibroblast growth factor-2 (FGF-2), vascular endothelial growth factor-A (VEGF-A), angiopoietins (Ang), stromal cell-derived factor-1α (SDF-1α), and sonic hedgehog (Shh) activate Fes in endothelial cells [19-26]. Nerve growth factor (NGF) also activates Fes in PC12 cells [27]. These results indicate that growth factor receptor tyrosine kinases and activated cytokine receptors or G-protein-coupled receptors are involved at least in part as upstream activators of Fes. Activation of Fes requires oligomerization. The common beta subunit of the IL-3 receptor is shared by the receptors for IL-5 and GM-CSF. GM-CSF induces physical association between Fes and the common beta subunit of the IL-3 receptor [12]. Receptors for IL-6, oncostatin M, and leukemia inhibitory factor, transduce signal through a common β-chain called gp130. Fes associates with gp130 independently of ligand stimulation [13]. IL-4 induces the association of Fes with the IL-4

Table 1. Upstream activators of Fes.

Ligand	Possible receptors and/or components	Cell type	References
IL-3	Common β-subunit of IL-3R	Hematopoietic cells	[12]
GM-CSF	Common β-subunit of IL-3R	Hematopoietic cells	[12]
IL-6	gp130	Hematopoietic cells	[13]
IL-4	α chain of IL-4R	Hematopoietic cells	[14]
EPO	EPOR	Hematopoietic cells	[15]
	c-kit	Hematopoietic cells	[16,17]
IgE	FcεRI	Hematopoietic cells	[18]
FGF-2	FGFR-1	Endothelial cells	[19,20]
VEGF-A	VEGFR-2	Endothelial cells	[21,22]
Ang1, 2	Tie 2	Endothelial cells	[23,24]
SDF-1α	CXCR4	Endothelial cells	[25]
Shh	Smoothened	Endothelial cells	[26]
PEDF	unknown	Endothelial cells	[35]
NGF	NGFR	PC-12 cells	[27]
No	Src	COS-7 cells	[34]
No	Bcr-Abl	Hematopoietic and 293T cells	[60]
No	KAP-1	293T cells	[40]

IL; interleukin, GM-CSF; granulocyte-macrophage colony stimulating factor, EPO; erythropoietin, EPOR; EPO receptor, IgE; immunoglobulin E, FGF; fibroblast growth factor, FGFR; FGF receptor, VEGF; vascular endothelial growth factor, VEGFR; VEGF receptor, Ang; angiopoietin, PDGF; platelet-derived growth factor, PDGFR; PDGF receptor, SDF-1α; stromal cell-derived factor-1α, Shh; sonic hedgehog, PEDF; pigment epithelium-derived factor, NGF; nerve growth factor, KAP-1; Kruppel-associated box-associated protein-1.

Table 2. Downstream signaling molecules of Fes.

Signaling molecule	Function	References
Bcr	Binding to Fes and recruitment of Grb2-Sos complex	[29]
	Activation of Rac1 and Cdc42, leading to neurite outgrowth of PC-12 cells	[31]
PI3-kinase	Differentiation of ECs treated by SDF-1α, Shh, and VEGF-A	[22, 25, 26]
	Migration of ECs treated by Ang2 and VEGF-A	[22, 23]
	Proliferation of ECs treated by Ang1	[24]
	Neurite outgrowth of PC12 cells treated by NGF	[27]
Fyn	Inhibition of differentiation of PEDF-treated ECs by inactivation	[35]
Src	Focal adhesion disassembly and migration of ECs treated by FGF-2	[20]
STAT1 and 3	Tyrosine phosphorylation and activation	[15, 37]
HSH2	Binding to Fes followed by tyrosine phosphorylation	[39]

EC; endothelial cells, SDF-1α; stromal cell-derived factor-1α, Shh; sonic hedgehog, VEGF; vascular endothelial growth factor, Ang; angiopoietin, NGF; nerve growth factor, PEDF; pigment epithelium-derived factor, FGF; fibroblast growth factor, STAT; signal transducers and activators of transcription, HSH2; hematopoietic SH2 protein.

receptor alpha chain [14]. These cytokine receptor subunits aggregate by ligand binding, which is subsequently followed by autophosphorylation and activation of non-receptor protein tyrosine kinases, such as JAK [28]. Therefore, the binding of Fes to aggregated receptor subunits may help the oligomerization of Fes, followed by its activation. However, neither direct association of Fes with receptor tyrosine kinases, such as FGF receptor-1, Tie 2, VEGF receptor-2, and NGF receptors, nor binding of Fes to particular adaptor proteins downstream of these receptor tyrosine kinases to promote Fes oligomerization have been demonstrated to date.

A panel of signaling molecules act downstream of activated Fes. The breakpoint-cluster region protein (Bcr) is composed of N-terminal serine/threonine kinase domain, a central region homologous to the Dbl guanine nucleotide exchange factor, and a C-terminal GTPase activating protein domain for Rho family small GTPases. Fes associates with Bcr between the N-terminal unique region and SH2 domain of Fes and N-terminal kinase domain of Bcr [29]. Fes-tyrosine phosphorylated Bcr associates with Grb2-Sos complex, an activator of Ras [29]. This observation implies the potential role of Fes in Ras/mitogen-activated protein kinase (MAPK) signaling. Indeed, constitutively activated MAPK/extracellular signal-regulated kinase kinase (MEK) and its target ERK are essential for the proliferation of v-Fes-overexpressing murine macrophages [30]. Other targets of Fes/Bcr pathway are Rho family GTPases, Rac1 and Cdc42. Fes-induced neurite extension in PC12 cells is inhibited by the expression of dominant negative Rac1 or Cdc42, and a fragment of Bcr containing Fes binding and tyrosine phosphorylation sites [31]. Therefore, Fes-induced neurite extension seems to require Bcr-driven Rac1 and Cdc42 activities. On the other hand, Ras, Rac, and Cdc42 and their downstream molecules, ERK and c-Jun N-terminal kinase, have been reported to be involved in v-Fps-induced fibroblast transformation [32]. These results suggest that Fes-regulated Bcr and its downstream signaling diverges either proliferation or differentiation in a cell context-dependent manner.

Another important downstream signaling molecule is phosphoinositide 3-kinase (PI3-kinase). One of the mechanisms involved in the activation of class I PI3-kinase is the binding to tyrosine phosphorylated proteins [33]. In mouse T-cells, Fes is tyrosine phosphorylated upon IL-4-stimulation, followed by association with PI3-kinase [14]. In endothelial cells, Ang1 and 2 induce autophosphorylation of Fes and activation of PI3-kinase [23, 24]. Ang-activated PI3-kinase activity was co-immunoprecipitated with wild type Fes, but not with kinase-inactive Fes [23]. In the yeast two hybrid system, wild type, but not kinase-inactive Fes associated with p85 regulatory subunit of PI3-kinase [23], as has been shown in IL-4-treated mouse T-cells [14]. Ang1- and 2-induced activation of PI3-kinase is dependent on Fes activity, because expression of kinase-inactive Fes (dominant negative for endogenous Fes) abolished Ang-induced PI3-kinase activation [23, 24]. Similar results were obtained in endothelial cells treated with SDF-1α and Shh [25, 26], suggesting that Fes is an activator of PI3-kinase in endothelial cells. PI3-kinase activity was involved in Abg2-directed chemotaxis [23] and morphological differentiation of endothelial cells promoted by SDF-1α and Shh [25, 26]. In VEGF-A-treated endothelial cells, activated PI3-kinase was associated with not only activated Fes, but also activated VEGF receptor-2, Src, and tyrosine phosphorylated insulin receptor substrate 1 [21]. Therefore, expression of kinase-inactive Fes does not exert a dominant negative effect on PI3-kinase-sensitive chemotaxis and capillary morphogenesis induced by VEGF-A in endothelial cells. However, our recent study have shown that downregulation of Fes by small interfering RNA (siRNA) significantly inhibited VEGF-A-promoted chemotaxis and capillary morphogenesis [22], suggesting that Fes may act as not only a protein tyrosine kinase, but also a scaffold protein for proper activation of PI3-kinase in VEGF-A-treated endothelial cells. In PC12 cells, PI3-kinase was activated and was required for neurite extension induced by NGF [27].

Interaction between Fes and Src family kinases is cell-context dependent. Co-expression with wild-type Src and Fyn resulted in potent tyrosine phosphorylation of kinase-inactive Fes in Sf-9 cells [19]. In this system, whether tyrosine phosphorylation of Fes results in its activation remains elusive. However, coexpression of Fes and active Src in COS-7 cells resulted in Fes activation, leading to colocalization of Fes with microtubules [34]. Thus, Src acts as an activator of Fes in transfected COS-7 cells. Fyn is inactivated by tyrosine phosphorylation of its C-terminal tail by Fes. This inactivation occurred within focal adhesions of endothelial cells treated with pigment epithelium-derived factor, which is linked to the inhibition of FGF-2-promoted capillary morphogenesis of endothelial cells [35]. In FGF-2-treated endothelial cells, activation of Fes enhanced FAK-dependent activation of Src within focal adhesions, leading to disassembly of focal adhesion and subsequent migration toward FGF-2 [20].

Signal transducers and activators of transcription (STATs) are SH2 domain containing transcription factors. They are tyrosine phosphorylated followed by their dimerization upon stimulation by a variety of cytokines/growth factors and subsequently enter into nuclei for transcriptional activation [36]. STAT3 is tyrosine phosphorylated and activated by Fes in human 293T cells or insect Sf-9 cells co-transfected with STAT3 and wild type Fes, but not with kinase-inactive Fes [37]. Unlike STAT3, activation of STAT5 was not observed in these systems [37]. Fes was involved in EPO-induced activation of STAT1 and STAT3 in UT-7/EPO cells and 293T cells [15]. Thus, STAT1 and STAT3 act as downstream effectors of Fes.

The adapter protein hematopoietic SH2 protein (HSH2) is involved in the survival and differentiation of B-cells [38]. It has been shown that Fes and activated Cdc42-associated tyrosine kinase bind to and tyrosine phosphorylate HSH2 [39]. Fes would thus be a possible regulator of B-cell survival and differentiation.

The transcriptional co-repressor Kruppel-associated box-associated protein (KAP)-1 was found to interact with Fes second coiled-coil domain [40]. KAP-1 forms a complex with heterochromatin proteins and histone deacetylases to repress transcriptions [41-43]. Co-expression of Fes and KAP-1 in 293T cells induced Fes autophosphorylation and subsequent tyrosine phosphorylation of KAP-1. Thus, KAP-1 acts as both an upstream and downstream signaling molecule of Fes [40].

Expression of Fes in Human Malignant Cells and Neoplastic Tissues

Although Fes is expressed in only a few types of normal cells, tumor cells from various origins have been found to express Fes gene and/or protein. Since early studies demonstrated that Fes was expressed in hematopoietic cells of myeloid origin [44], expression of Fes was preferentially observed in myelogenous leukemia cells rather than lymphoid leukemia cells [45, 46]. However, the expression has been also detected in some B-cell lymphomas and cell lines derived from patients with Hodgkin's disease [47, 48]. In solid tumors, Fes protein is expressed in lung cancers (51.2% of adenocarcinoma, 26.3% of squamous cell carcinoma, 35.7% of large cell carcinoma, and 15.4% of small cell carcinoma) [49]. Copy number gain of *fes gene* has been observed in diffuse gastric cancers [50], oral squamous cell carcinomas [51], and glioblastomas [52]. Fes transcript is increased in ovarian adenocarcinomas, mixed-Mullerian tumors [53], and renal cell carcinomas [54]. In contrast, expression of Fes gene is reported to be downregulated in melanomas and anaplastic thyroid carcinomas when compared with normal melanocytes or thyroid tissue, respectively [55, 56].

It is not clear at this stage whether the expression of Fes gene and protein correlates with tumor progression and prognosis because of the lack of large clinicopathological studies. Furthermore, whether the expression of Fes benefits the proliferation, survival, and invasion of tumor cells needs further analysis.

Is Fes a Tumor Suppressor Protein?

Fusion proteins, Bcr-Abl (p210 and p185), produced by chromosomal translocation, are associated with leukemogenesis (p210 in chronic myelogenous leukemia, CML and p185 in acute lymphocytic leukemia) [57, 58]. Fusion to the N-terminal Bcr sequence induces constitutive activation of Abl protein tyrosine kinase. Introduction of Fes into the CML cell line K-562 induced growth arrest and terminal differentiation [59], suggesting that Fes may inhibit the progression of CML through interaction with Bcr-Abl. Coexpression of Fes with Bcr-Abl in 293T cells promoted reciprocal transphosphorylation, resulting in the activation of Fes [60]. This result suggests that Bcr-Abl may act as an upstream activator of Fes. Interestingly, Bcr-Abl-induced cytokine-independent outgrowth of myeloid leukemia cells, DAGM cells, was markedly inhibited by the presence of wild type Fes, but not by kinase-inactive Fes [60]. Thus, it seems likely that Fes exerts antiproliferative effect on leukemia cells expressing Bcr-Abl and that inhibition of Fes activity may favor the progression of Bcr-Abl-driven leukemia.

Analysis of 147 colorectal cancer tissues by Bardelli and coworkers [61] found 4 somatic mutations of the kinase domain of Fes, including M704V, R706Q, V743M, and S759F. These mutations were not found in Korean or Japanese colorectal cancer patients [62, 63]. Functional studies by two independent groups have shown that M704V, V743M, and S759F, or M704V, R706Q, and V743M mutations downregulated the kinase activity of Fes [64, 65]. Both groups expressed the mutant Fes in 293T cells. It is not clear why either R706Q or S759F mutations did not commonly exhibit reduced Fes kinase activity. Sangrar et al. [64] also showed that mice targeted with either null or kinase-inactive *fes* developed early-onset breast cancer in transgenic ice with mammary epithelial-specific expression of polyoma middle T antigen. Furthermore, Delfino and colleagues [65] demonstrated that introduction of wild type Fes into colorectal cancer cells, HT-29 and HCT 116 cells, inhibited anchorage-independent growth of these cells in soft agar. These results suggest that Fes may have tumor suppressor functions in epithelial malignancies. However, the downstream signaling molecules that are involved in these tumor suppressive functions have not been identified yet.

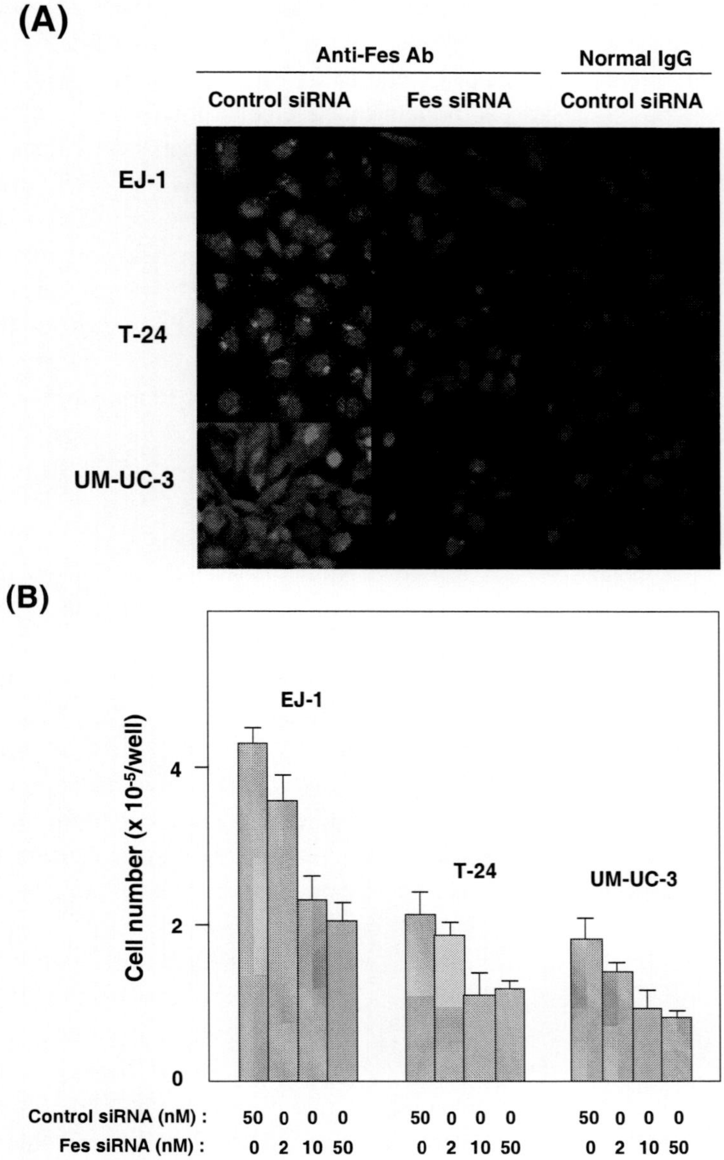

Figure 1. (A) Treatment of bladder cancer cells (EJ-1, T-24, and UM-UC-3 cells) with Fes siRNA (50 μM) downregulates the expression of endogenous Fes. EJ-1 and T-24 cells were obtained from Japanese Cancer Research Resources Bank and UM-UC-3 cells were kindly provided by Dr. Hiro-omi Kanayama at the Department of Urology, The University of Tokushima Graduate School Institute of Health Bioscience. Cells were seeded onto coverslips and then treated with either negative control siRNA or Fes siRNA (Qiagen, Ref. [22]). Three days later, cells were fixed and Fes protein was visualized with anti-Fes (H-65) antibody (Santa Cruz Biotechnologies) by indirect immunofluorescent staining. Endogenous Fes distributed mainly in the nuclei and perinuclear area in EJ-1 and T-24 cells, whereas Fes was observed throughout the cytoplasm in UM-UC-3 cells. (B) Treatment with Fes siRNA dose-dependently inhibits the proliferation of bladder cancer cells. Cells wee seeded into wells of 24-well plates and then treated with siRNA at indicated concentrations. Three days later, cells were detached by trypsin and their number was counted. Data are expressed as means ± SD for triplicate wells.

To our knowledge, there are no reports describing Fes expression in bladder carcinoma cells. We examined the expression and the effect of Fes downregulation by siRNA in three human bladder carcinoma cell lines. As shown in Fig. 1, treatment with siRNA for Fes downregulated the expression of Fes protein in these cells. Interestingly, downregulation of Fes inhibited the proliferation of three bladder carcinoma cell lines. Furthermore, downregulation of Fes expression suppresses its kinase activity and scaffold function of Fes protein. These results raise the possibility that targeting Fes by a specific siRNA may inhibit angiogenesis as well as proliferation of human bladder cancer cells in clinics. Although the absence of Fes protein accelerated breast carcinogenesis in mutant mice [64], downregulation of Fes decreased the proliferation of bladder cancer cells. The reason for the discrepancy between the results of mutant mice and ours is unknown at present. One possibility is that breast carcinogenesis involves initiation and progression steps to transform normal cells into cancer cells, whereas cancer cells do not require further progression steps. Thus, cancer cells may use Fes for proliferation, and expression of Fes protein may affect the transformation of normal cells during development.

Conclusion

Fes is involved in a variety of angiogenic cellular responses by endothelial cells. Approaches based on Fes targeting may be a potentially useful therapeutic strategy in cancer patients through inhibition of tumor-related angiogenesis. Before this strategy could be ascertained, it is important to examine whether inhibition or downregulation of Fes inhibits angiogenesis *in vivo*. Also, Fes may act as a tumor suppressor in certain malignancies, and thus inhibition of Fes kinase activity in any treatment of human cancers may favor the progression of the disease. Hence, it is important to examine the role of Fes in proliferation, survival and motility of a wide variety of tumor cells.

Because of the lack of small molecular weight synthetic kinase inhibitors for Fes [4], downregulation of Fes (*e.g.,* treatment with specific siRNA) or introduction of kinase-inactive Fes into tumor cells may be helpful [66, 67]. Clinically applicable strategy for *in vivo* gene delivery of Fes siRNA to tumor tissue can be achieved by transcatheter intra-arterial administration of liposomes or vectors containing siRNA via feeding arteries of tumors. This approach can be used for the treatment of advanced renal cell carcinomas, primary and secondary hepatic tumors, and bladder cancers.

Progress in proteome analyses and microarray studies may reveal a relationship between Fes expression and progression of human cancers. If overexpression of Fes is associated with high tumor stage and poor prognosis in certain cancers, downregulation of Fes may be effective in halting cancer progression. Correlation analysis of clinicopathological features and Fes expression is therefore important to conjecture whether Fes plays a role in tumor progression and suppression.

Finally, it is also important to identify the signaling pathways involved in Fes-mediated tumor suppressive function. Many signaling molecules have diverse functions, such as induction of differentiation, growth arrest and stimulation of cell survival. In many cases, the mechanisms how signaling molecules decide the cell fate (*e.g.,* proliferation or differentiation) in response to external stimuli are not well understood. However, if the effecter molecule that mediates Fes-induced tumor suppression is identified, simultaneous inhibition of this molecule and Fes kinase activity could be beneficial in the control of advanced cancers.

References

[1] Smithgall, T. E., Rogers, J. A., Peters, K. L., Li, J., Briggs, S. D., Lionberger, J. M., Cheng, H., Shibata, A., Scholtz, B., Schreiner, S., Dunham, N. (1998) The c-Fes family of protein-tyrosine kinases. *Crit Rev Oncog.* **9**, 43-62.

[2] Greer, P. (2002) Closing in on the biological functions of Fps/Fes and Fer. *Nat Rev Mol Cell Biol.* **3**, 278-289.

[3] Haigh, J., McVeigh, J., Greer, P. (1996) The fps/fes tyrosine kinase is expressed in myeloid, vascular endothelial, epithelial, and neuronal cells and is localized in the trans-Golgi network. *Cell Growth Differ.* **7**, 931-944.

[4] Kanda, S., Miyata, Y., Kanetake, H., Smithgall, T. E. (2007) Non-receptor protein-tyrosine kinases as molecular targets for anti-angiogenic therapy. *Int J Mol Med.* **20**, 113-121.

[5] Rogers, J. A., Read, R. D., Li, J., Peters, K. L., Smithgall, T. E. (1996) Autophosphorylation of the Fes tyrosine kinase. Evidence for an intermolecular mechanism involving two kinase domain tyrosine residues. *J Biol Chem.* **271**, 17519-17525.

[6] Read, R. D., Lionberger, J. M., Smithgall, T. E. (1997) Oligomerization of the Fes tyrosine kinase. Evidence for a coiled-coil domain in the unique N-terminal region. *J Biol Chem.* **272**, 18498-18503.

[7] Cheng, H. Y., Schiavone, A. P., Smithgall, T. E. (2001) A point mutation in the N-terminal coiled-coil domain releases c-Fes tyrosine kinase activity and survival signaling in myeloid leukemia cells. *Mol Cell Biol.* **21**, 6170-6180, 2001.

[8] Senis, Y., Zirngibl, R., McVeigh, J., Haman, A., Hoang, T., Greer, P. A. (1999) Targeted disruption of the murine fps/fes proto-oncogene reveals that Fps/Fes kinase activity is dispensable for hematopoiesis. *Mol Cell Biol.* **19**, 7436-7446.

[9] Zirngibl, R. A., Senis, Y., Greer, P. A. (2002) Enhanced endotoxin sensitivity in fps/fes-null mice with minimal defects in hematopoietic homeostasis. *Mol Cell Biol.* **22**, 2472-2486.

[10] Greer, P., Haigh, J., Mbamalu, G., Khoo, W., Bernstein, A., Pawson, T. (1994) The Fps/Fes protein-tyrosine kinase promotes angiogenesis in transgenic mice. *Mol Cell Biol.* **14**, 6755-6763.

[11] Yee, S. P., Mock, D., Greer, P., Maltby, V., Rossant, J., Bernstein, A., Pawson, T. (1989) Lymphoid and mesenchymal tumors in transgenic mice expressing the v-fps protein-tyrosine kinase. *Mol Cell Biol.* **9**, 5491-5499.

[12] Hanazono, Y., Chiba, S., Sasaki, K., Mano, H., Miyajima, A., Arai, K., Yazaki, Y., Hirai, H. (1993) c-fps/fes protein-tyrosine kinase is implicated in a signaling pathway triggered by granulocyte-macrophage colony-stimulating factor and interleukin-3. *EMBO J.* **12**, 1641-1616.

[13] Matsuda, T., Fukada, T., Takahashi-Tezuka, M., Okuyama, Y., Fujitani, Y., Hanazono, Y., Hirai, H., Hirano, T. (1995) Activation of Fes tyrosine kinase by gp130, an

interleukin-6 family cytokine signal transducer, and their association. *J Biol Chem.* **270**, 11037-11039.

[14] Izuhara, K., Feldman, R. A., Greer, P., Harada, N. (1996) Interleukin-4 induces association of the c-fes proto-oncogene product with phosphatidylinositol-3 kinase. *Blood.* **88**, 3910-3918.

[15] Kirito, K., Nakajima, K., Watanabe, T., Uchida, M., Tanaka, M., Ozawa, K., Komatsu, N. (2002) Identification of the human erythropoietin receptor region required for Stat1 and Stat3 activation. *Blood.* **99**, 102-110.

[16] Petti, F., Thelemann, A., Kahler, J., McCormack, S., Castaldo, L., Hunt, T., Nuwaysir, L., Zeiske, L., Haack, H., Sullivan, L., Garton, A., Haley, J. D. (2005) Temporal quantitation of mutant Kit tyrosine kinase signaling attenuated by a novel thiophene kinase inhibitor OSI-930. *Mol Cancer Ther.* **4,** 1186-1197.

[17] Voisset, E., Lopez, S., Dubreuil, P., De Sepulveda, P. (2007) The tyrosine kinase FES is an essential effector of KITD816V proliferation signal. *Blood.* **110**, 2593-2599.

[18] Udell, C. M., Samayawardhena, L. A., Kawakami, Y., Kawakami, T., Craig, A. W. (2006) Fer and Fps/Fes participate in a Lyn-dependent pathway from FcepsilonRI to platelet-endothelial cell adhesion molecule 1 to limit mast cell activation. *J Biol Chem.* **281**, 20949-20957.

[19] Kanda, S., Lerner, E. C., Tsuda, S., Shono, T., Kanetake, H., Smithgall, T. E. (2000) The nonreceptor protein-tyrosine kinase c-Fes is involved in fibroblast growth factor-2-induced chemotaxis of murine brain capillary endothelial cells. *J Biol Chem.* **275**, 10105-10111.

[20] Kanda, S., Miyata Y., Kanetake, H., Smithgall, T. E. (2006) Fibroblast growth factor-2 induces the activation of Src through Fes, which regulates focal adhesion disassembly. *Exp Cell Res.* **312**, 3015-3022.

[21] Kanda, S., Mochizuki, Y., Miyata, Y., Kanetake, H. (2003) The role of c-Fes in vascular endothelial growth factor-A-mediated signaling by endothelial cells. *Biochem Biophys Res Commun.* **306**, 1056-1063, 2003.

[22] Kanda, S., Kanetake, H., Miyata, Y. (2007) Downregulation of Fes inhibits VEGF-A-induced chemotaxis and capillary-like morphogenesis by cultured endothelial cells. *J Cell Mol Med.* **11**, 495-501.

[23] Mochizuki, Y., Nakamura, T., Kanetake, H., Kanda, S. (2002) Angiopoietin 2 stimulates migration and tube-like structure formation of murine brain capillary endothelial cells through c-Fes and c-Fyn. *J Cell Sci.* **115**, 175-183.

[24] Kanda, S., Miyata, Y., Mochizuki, Y., Kanetake, H. (2005) Angiopoietin 1 is mitogenic for cultured endothelial cells. *Cancer Res.* **65**, 6820-6827.

[25] Kanda, S., Mochizuki, Y., Kanetake, H. (2003) Stromal cell-derived factor-1α induces tube-like structure formation of endothelial cells through phosphoinositide 3-kinase. *J Biol Chem.* **278**, 257-262.

[26] Kanda, S., Mochizuki, Y., Suematsu, T., Miyata, Y., Nomata, K., Kanetake, H. (2003) Sonic hedgehog induces capillary morphogenesis by endothelial cells through phosphoinositide 3-kinase. *J Biol Chem.* **278**, 8244-8249.

[27] Shibata, A., Laurent, C. E., Smithgall, T. E. (2003) The c-Fes protein-tyrosine kinase accelerates NGF-induced differentiation of PC12 cells through a PI3K-dependent mechanism. *Cell Signal.* **15**, 279-288.

[28] Ihle, J. N., Kerr, I. M. (1995) Jaks and Stats in signaling by the cytokine receptor superfamily. *Trends Genet.* **11**, 69-74.

[29] Maru, Y., Peters, K. L., Afar, D. E., Shibuya, M., Witte, O. N., Smithgall, T. E. (1995) Tyrosine phosphorylation of BCR by FPS/FES protein-tyrosine kinases induces association of BCR with GRB-2/SOS. *Mol Cell Biol.* **15**, 835-842.

[30] Rovida, E., Marra, F., Baccarini, M., Dello Sbarba, P. (2000) Constitutive activation of the MAPK pathway mediates v-fes-induced mitogenesis in murine macrophages. *Blood.* **95**, 3959-63.

[31] Laurent, C. E., Smithgall, T. E. (2004) The c-Fes tyrosine kinase cooperates with the breakpoint cluster region protein (Bcr) to induce neurite extension in a Rac- and Cdc42-dependent manner. *Exp Cell Res.* **299**, 188-198.

[32] Li, J., Smithgall, T. E. (1998) Fibroblast transformation by Fps/Fes tyrosine kinases requires Ras, Rac, and Cdc42 and induces extracellular signal-regulated and c-Jun N-terminal kinase activation. *J Biol Chem.* **273**, 13828-13834.

[33] Fruman, D. A, Meyers, R. E. Cantley, L. C. (1998). Phosphoinositide kinases. *Annu Rev Biochem.* **67**, 481-507.

[34] Laurent, C. E., Delfino, F. J., Cheng, H. Y., Smithgall, T. E. (2004) The human c-Fes tyrosine kinase binds tubulin and microtubules through separate domains and promotes microtubule assembly. *Mol Cell Biol.* **24**, 9351-9358.

[35] Kanda, S., Mochizuki, Y., Nakamura, T., Miyata, Y., Kanetake, H. (2005) Pigment epithelium-derived factor inhibits fibroblast growth factor-2-induced capillary morphogenesis of endothelial cells through c-Fyn. *J Cell Sci.* **118**, 961-970.

[36] Finbloom, D. S., Larner, A. C. (1995) Regulation of the Jak/STAT signalling pathway. *Cell Signal.* **7**, 739-745.

[37] Nelson, K. L, Rogers, J. A., Bowman, T. L., Jove, R., Smithgall, T. E. (1998) Activation of STAT3 by the c-Fes protein-tyrosine kinase. *J Biol Chem.* **273**, 7072-7077.

[38] Herrin, B. R., Justement, L. B. (2006) Expression of the adaptor protein hematopoietic Src homology 2 is up-regulated in response to stimuli that promote survival and differentiation of B cells. *J Immunol.* **176**, 4163-4172.

[39] Oda, T., Muramatsu, M. A., Isogai, T., Masuho, Y., Asano, S., Yamashita, T. (2001) HSH2: a novel SH2 domain-containing adapter protein involved in tyrosine kinase signaling in hematopoietic cells. *Biochem Biophys Res Commun.* **288**, 1078-1086.

[40] Delfino, F. J., Shaffer, J. M., Smithgall, T. E. (2006) The KRAB-associated co-repressor KAP-1 is a coiled-coil binding partner, substrate and activator of the c-Fes protein tyrosine kinase. *Biochem J.* **399**, 141-150.

[41] Lechner, M. S., Begg, G. E., Speicher, D. W., Rauscher, F. J. 3rd. (2000) Molecular determinants for targeting heterochromatin protein 1-mediated gene silencing: direct chromoshadow domain-KAP-1 corepressor interaction is essential. *Mol Cell Biol.* **20**, 6449-6465.

[42] Matsuda, E., Agata, Y., Sugai, M., Katakai, T., Gonda, H., Shimizu, A. (2001) Targeting of Kruppel-associated box-containing zinc finger proteins to centromeric heterochromatin. Implication for the gene silencing mechanisms. *J Biol Chem.* **276**, 14222-14229.

[43] Schultz, D. C., Friedman, J. R., Rauscher, F. J. 3rd. (2001) Targeting histone deacetylase complexes via KRAB-zinc finger proteins: the PHD and bromodomains of KAP-1 form

a cooperative unit that recruits a novel isoform of the Mi-2alpha subunit of NuRD. *Genes Dev.* **15**, 428-443.

[44] Samarut, J., Mathey-Prevot, B., Hanafusa, H. (1985) Preferential expression of the *c-fps* protein in chicken macrophages and granulocytic cells. *Mol Cell Biol.* **5**, 1067-1072.

[45] MacDonald, I., Levy, J., Pawson, T. (1985) Expression of the mammalian c-fes protein in hematopoietic cells and identification of a distinct fes-related protein. *Mol Cell Biol.* **5**, 2543-2551.

[46] Mavilio, F., Sposi, N. M., Petrini, M., Bottero, L., Marinucci, M., De Rossi, G., Amadori, S., Mandelli, F., Peschle, C. (1986) Expression of cellular oncogenes in primary cells from human acute leukemias. *Proc Natl Acad Sci U S A.* **83**, 4394-4398.

[47] Lee, J., Talpaz, M., Walters, R., Gutterman, J. U., Blick, M. (1988) Aberrant expression of the K-ras proto-oncogene in B-cell malignancies. *Hematol Pathol.* **2,** 229-237.

[48] Jucker, M., Schaadt, M., Diehl, V., Poppema, S., Jones, D., Tesch, H. (1990) Heterogeneous expression of proto-oncogenes in Hodgkin's disease derived cell lines. *Hematol Oncol.* **8**, 191-204.

[49] Nishio, H., Nakamura, S., Horai, T., Ikegami, H., Matsuda, M. (1992) Clinical and histopathologic evaluation of the expression of Ha-ras and fes oncogene products in lung cancer. *Cancer.* **69**, 1130-1136.

[50] Peng, D. F., Sugihara, H., Mukaisho, K., Tsubosa, Y., Hattori, T. (2003) Alterations of chromosomal copy number during progression of diffuse-type gastric carcinomas: metaphase- and array-based comparative genomic hybridization analyses of multiple samples from individual tumours. *J Pathol.* **201**, 439-450.

[51] Liu, C. J., Lin, S. C., Chen, Y. J., Chang, K. M., Chang, K. W. (2006) Array-comparative genomic hybridization to detect genome-wide changes in microdissected primary and metastatic oral squamous cell carcinomas. *Mol Carcinog.* **45**, 721-731.

[52] Nakahara, Y., Shiraishi, T., Okamoto, H., Mineta, T., Oishi, T., Sasaki, K., Tabuchi, K. (2004) Detrended fluctuation analysis of genome-wide copy number profiles of glioblastomas using array-based comparative genomic hybridization. *Neuro Oncol.* **6,** 281-289.

[53] Kacinski, B. M., Carter, D., Kohorn, E. I., Mittal, K., Bloodgood, R. S., Donahue, J., Kramer, C. A., Fischer, D., Edwards, R., Chambers, S. K., et al. (1989) Oncogene expression in vivo by ovarian adenocarcinomas and mixed-mullerian tumors. *Yale J Biol Med.* **62**, 379-392.

[54] Karthaus, H. F., Schalken, J. A., Feitz, W. F., Debruyne, F. M., de Haan, P. T., Bloemers, H. P., Van de Ven, W. J. (1986) Expression of the human fes cellular oncogene in renal cell tumors. *Urol Res.* **14**, 123-127.

[55] Onda, M., Emi, M., Yoshida, A., Miyamoto, S., Akaishi, J., Asaka, S., Mizutani, K., Shimizu, K., Nagahama, M., Ito, K., Tanaka, T., Tsunoda, T. (2004) Comprehensive gene expression profiling of anaplastic thyroid cancers with cDNA microarray of 25 344 genes. *Endocr Relat Cancer.* **11**, 843-854.

[56] Easty, D. J., Herlyn, M., Bennett, D. C. (1995) Abnormal protein tyrosine kinase gene expression during melanoma progression and metastasis. *Int J Cancer. 60*, 129-136.

[57] Ben-Neriah, Y., Daley, G. Q., Mes-Masson, A. M., Witte, O. N., Baltimore, D. (1986) The chronic myelogenous leukemia-specific p210 protein is the product of the *bcr/abl* hybrid gene. *Science.* **233**, 212-214.

[58] Hermans, A., Heisterkamp, N., von Lindern, M., van Baal, S., Meijer, D., van der Plas, D., Wiedemann, L. M., Groffen, J., Bootsma, D., Grosveld, G. (1987) Unique fusion of *bcr* and c-*abl* genes in Philadelphia chromosome-positive acute lymphoblastic leukemia. *Cell.* **51**, 33-40.

[59] Yu, G,, Smithgall, T. E., Glazer, R. I. (1989) K562 leukemia cells transfected with the human *c-fes* gene acquire the ability to undergo myeloid differentiation. *J Biol Chem.* **264**, 10276-10281.

[60] Lionberger, J. M., Smithgall, T. E. (2000) The c-Fes protein-tyrosine kinase suppresses cytokine-independent outgrowth of myeloid leukemia cells induced by Bcr-Abl. *Cancer Res.* **60**, 1097-1103.

[61] Bardelli, A., Parsons, D. W., Silliman, N., Ptak, J., Szabo, S., Saha, S., Markowitz, S., Willson, J. K., Parmigiani, G., Kinzler, K. W., Vogelstein, B., Velculescu, V. E. (2003) Mutational analysis of the tyrosine kinase in colorectal cancers. *Science.* **300**, 949.

[62] Kim, H. S., Kim, S. Y., Nam, S. W., Park, W. S., Lee, J. Y., Yoo, N. J., Lee, S. H. (2006) Absence of FES exon 16 and 17 mutation in the colorectal carcinomas in Korean patients. *Dig Liver Dis.* **38**, 213-214.

[63] Shao, R. X., Kato, N., Lin, L. J., Murayama, R., Moriyama, M., Ikenoue, T., Watabe, H., Otsuka, M., Guleng, B., Ohta, M., Tanaka, Y., Kondo, S., Dharel, N., Chang, J. H., Yoshida, H., Kawabe, T., Omata, M. (2007) Absence of tyrosine kinase mutations in Japanese colorectal cancer patients. *Oncogene.* **26**, 2133-2135.

[64] Sangrar, W., Zirgnibl, R. A., Gao, Y., Muller, W. J., Jia, Z., Greer, P. A. (2005) An identity crisis for fps/fes: oncogene or tumor suppressor? *Cancer Res.* **65**, 3518-3522.

[65] Delfino, F. J., Stevenson, H., Smithgall, T. E. (2006) A growth-suppressive function for the c-fes protein-tyrosine kinase in colorectal cancer. *J Biol Chem.* **281**, 8829-8835.

[66] Putral, L. N., Gu, W., McMillan, N. A. (2006) RNA interference for the treatment of cancer. *Drug News Perspect.* **19**, 317-324.

[67] Kim, D. H., Rossi, J. J. (2007) Strategies for silencing human disease using RNA interference. *Nat Rev Genet.* **8**, 173-184.

In: Oncogene Proteins: Structure, Functions and Analyses ISBN: 978-1-60456-259-0
Editors: P.B. Murphy and J.R. Clarke, pp. 223-248

Chapter IX

Molecular and Cellular Alterations Produced by HER2 Overexpression and Modified by HER2 Inhibition in Cancer Cells

Xiao-Feng Le and Robert C. Bast, Jr.
Department of Experimental Therapeutics, The University of Texas M. D. Anderson Cancer Center, Houston, Texas 77030, USA

Abstract

The human epidermal growth factor receptor 2 (HER2, also known as c-neu or erbB-2) is an important member of the epidermal growth factor receptor (EGFR) family. As a preferred heterodimer partner for other members of the EGFR family, HER2 plays a critical role in signaling that generates a wide variety of cellular responses in both normal and malignant cells. HER2 is amplified/overexpressed in a number of cancers, including breast, ovarian, gastric, colon, and non-small cell lung carcinomas. HER2 amplification/overexpression is associated with poor prognosis largely due to aggressive tumor growth, early metastasis and resistance to chemotherapy. HER2 is not only a prognostic marker for patients with cancers that overexpress HER2, but is also a predictive marker for response to treatment. A humanized monoclonal antibody trastuzumab (Herceptin®) has been developed from the murine anti-HER2 monoclonal antibody 4D5 and has become the first-line therapy for patients with breast cancers that overexpress HER2. In this review, we have updated information regarding molecules and pathways that are influenced by HER2 signaling. Structure-function analysis indicates that HER2 is constitutively activated and prepared to heterodimerize with HER3 and EGFR. MUC4 levels can modulate binding of trastuzumab and susceptibility to treatment. Affymetrix microarray analysis has identified a number of new candidates for targets regulated by HER2. A number of HER2-regulated genes have also been implicated in G1 cell cycle arrest (p27Kip1, Jab1, Mirk, hKIS, AKT, CDK2, cyclin D1 and cyclin G2), non-receptor kinase signaling (BRK), estrogen responsiveness (ERRa), apoptosis (Bim), motility (alpha$_6$beta$_4$ integrin, MEMO, S100A4), invasion (CXCR4, RECK) and angiogenesis (VEGF, IL-8, TSP1). Independent activation of one or more of these targets may contribute to resistance to trastuzumab. Targeting both HER2 and one or more of these HER2 regulated molecules might provide greater anti-tumor activity in cancers that overexpress the receptor.

Introduction

The human epidermal growth factor receptor 2 (HER2, also known as c-neu or ErbB-2) is a member of the type I receptor tyrosine kinase family, which consists of four closely related family members: HER2, epidermal growth factor receptor (EGFR; ErbB1 or EGFR), HER3 (ErbB3) and HER4 (ErbB4) (1-4). The *HER2* gene is located on human chromosome 17q12 and encodes a 185-kDa member of the HER family of cell surface receptor tyrosine kinases (5). Among the HER family members, the HER2 receptor appears unique in that it lacks a known ligand, functioning rather as a preferred co-receptor in the dimerization and activation of the other HER receptors (6,7).

Amplification of the *HER2* gene and/or overexpression of the HER2 protein are detected in approximately 25-30% of human breast cancers (8-10). *HER2* gene amplification can result in dramatically increased number (average 50-fold) of surface HER-2 receptors on cancer cells compared to the normal mammary epithelium (11,12). Abnormal HER2 amplification or overexpression is predictive of poor clinical outcome (9-10).

Aberrant HER2 protein expression or gene amplification is believed to play a direct role in malignant transformation and progression. In preclinical models, a 10-fold increase in HER2 expression with an exogenous promoter can transform NIH3T3 cells (13,14). Activated HER2 can uniquely promote a ductal carcinoma *in situ* (DCIS)-like phenotype in *in vitro* models of mammary acini (15). The oncogenic potential of HER2 in mammary tissues has been further documented in a transgenic mouse model. Transgenic mice carrying the activated c-neu oncogene under the control of the mouse mammary tumor virus (MMTV) promoter/enhancer uniformly develop mammary adenocarcinomas (16). Thus, expression of an activated *c-neu* oncogene appears to be sufficient to induce malignant transformation in the mammary tissue in a single step. Additional studies showed that mice bearing the rat *c-neu* oncogene under transcriptional control of an MMTV promoter stochastically develop focal mammary tumors and pulmonary metastases after a long latency (17,18). Clinical studies also indicate that *HER2* amplification and/or overexpression is a relatively early event in the clinical pathogenesis of human breast cancer based on its frequent occurrence in pre-invasive DCIS (19-21). The frequency of *HER2* amplification detected in high grade DCIS has been reported to be as high as 77% (22). All of these observations are consistent with the hypothesis that aberrant HER2 activity plays an important role in tumor initiation as well as in the emergence of aggressive cellular behavior associated with progressive disease.

The importance of HER2 overexpression for initiating and maintaining the malignant phenotype, expression of the receptor on the cell membrane, and the differential expression of HER2 expression on cancer cells all make HER2 an ideal target for tumor therapy. Trastuzumab, a humanized anti-HER2 monoclonal antibody (Herceptin®) developed specifically against HER2 receptor from the murine monoclonal antibody 4D5 (23), has been remarkably effective in treating patients with metastatic breast cancers that overexpress HER2. Trastuzumab treatment can produce a response rate of 10% ~ 15% as a single agent in heavily pre-treated patients (24). As a first-line, single-agent, trastuzumab produced an objective response rate of 26% in women with HER2-overexpressing metastatic breast cancer

(25). More importantly, trastuzumab has been shown to significantly enhance the effectiveness of chemotherapy in patients whose tumors over-express HER2. The combination of chemotherapy plus trastuzumab has a much higher rate of response than chemotherapy alone (50% *vs* 32%) (26). The combination of trastuzumab and chemotherapy also improves the time to disease progression (7.4 *vs* 4.6 months) and the median response duration (9.1 *vs* 6.1 months) when compared to chemotherapy alone (26). Addition of trastuzumab to paclitaxel treatment after the standard regimen of doxorubicin and cyclophosphamide was associated with a 33 percent reduction in the risk of death among women with surgically resected breast cancers that overexpressed HER2 (27). A clinical trial further showed that one-year of treatment with trastuzumab after completion of surgery and chemotherapy significantly improved both disease-free survival and overall survival among women with HER2-positive breast cancer (28,29).

Several other approaches have been used to target HER2 in pre-clinical models and early clinical trials. Pertuzumab, also known as 2C4 or Omnitarg, is a humanized monoclonal antibody that binds to an epitope on HER2 distinct from that recognized by Herceptin (30). Unlike trastuzumab, pertuzumab can inhibit growth of tumors that express low levels of HER2. Structural analysis indicates pertuzumab blocks the dimerization of the HER receptors. Transcriptional repression of HER2 with E1A-liposomes has been shown to effectively suppress breast and ovarian cancers in xenograft models (31). Recently, lapatinib (Tykerb®, GW572016), a small molecule inhibitor of EGFR and HER2, has been approved by the U.S. Food and Drug Administration for use in combination with capecitabine to treat patients with metastatic breast cancer that overexpress HER2 (32). The addition of lapatinib to capecitabine resulted in significantly prolonged time to progression (8.4 months in combination therapy *vs* 4.4 months with capecitabine alone) (32). Large randomized trials using lapatinib following chemotherapy and surgery are ongoing for early stage breast cancers that overexpress HER2.

A number of review articles summarize the function of HER2 and the downstream signaling pathways that it regulates (33-44). Treatment of cancer cells that overexpress HER2 with anti-HER2 antibodies such as trastuzumab produces G1 cell cycle arrest, inhibits motility and angiogenesis, and enhances susceptibility to cytotoxic drugs including doxorubicin and paclitaxel. Trastuzumab downregulates HER2 and inhibits dimerization of HER2 with HER3, decreasing signaling through the PI3 kinase pathway. This chapter focuses on more recent reports of the molecules and pathways that are modified or induced by HER2 signaling.

2. Structure-Function Analysis of HER Family Members

The solution of structures from X-ray diffraction images for crystals of the four HER family proteins (EGFR, HER2, HER3, and HER4) has provided new insight into HER2 function, its interaction with other proteins, and its role as a target for trastuzumab. The extracellular regions of all four HER family members contain four domains (I, II, III, IV) which have high structural homology to the extracellular domains found in the type I insulin-like growth factor receptor (45). Ligands have been identified for EGFR, HER3 and HER4, but not for HER2. In the absence of specific ligand binding, EGFR, HER3, and HER4 receptors exhibit a closed C-shape arrangement of all extracellular domains (I, II, III, or IV) (45-48). This closed structure constrains dimerization of the receptors. Once a ligand binds and docks between domains I and III or IV of the respective receptors (e. g. EGF to EGFR, heregulin to HER3 and HER4), the closed C shape structure is extended and becomes active (45-48). Once activated, homo- or hetero-dimerization can occur between HER family members, initiating downstream signaling.

The crystal structure of HER2 differs significantly from that of EGFR, HER3, or HER4. HER2 exhibits a ligand-activated conformation, suggesting that HER2 is constitutively able to dimerize with other HER receptors in the absence of direct ligand binding (49,50). This unique structure provides a possible explanation for HER2 acting as the preferred partner for other HER receptors. This structure also suggests why HER2 fails to bind known ligands.

Structural analysis of the trastuzumab-HER2 interaction indicates that the antibody binds to extracellular domain IV in the juxtamembrane region of HER2 (49). By contrast, structural analysis of the HER2-pertuzumab complex has demonstrated that pertuzumab binds to the center of HER2's extracellular domain II, sterically blocking receptor dimerization (51). Mutagenesis and biochemical analyses have identified a number of residues in domain II that are essential for HER2-HER3 heterodimerization and subsequent signal transduction (51). Mutagenesis experiments have also identified a requirement for domain IV in HER2-HER3 heterodimerization (52). Thus, pertuzumab completely blocks dimerization of HER2 with other HER family receptors, whereas trastuzumab partially blocks dimerization and may also promote cleavage of the extracellular structure of HER2. Analysis of crystal structure and amino acid sequence also suggest that critical residues in domain II of the different HER family members may determine the interaction between monomers and may reflect the different preferences for homo- and heterodimerization between the HER receptors (53).

3. Mucin 4 (MUC4) and HER2

Mucins are membrane-associated, highly glycosylated macromolecules that are implicated in protection and lubrication of epithelial surfaces (54). There are three structurally and functionally distinct classes of mucins: secreted gel-forming mucins (MUC2, MUC5AC, MUC5B, and MUC6), transmembrane mucins (MUC1, MUC3A, MUC3B, MUC4, MUC12, MUC16 and MUC17), and miscellaneous mucins (MUC7, MUC8, MUC9, MUC13, MUC15) (54). Recent data suggest that mucins can also function as signaling modulators in response to the stimulation or stress of epithelia. For example, the cytoplasmic tail of MUC1 (recognized by the CA15-3 or CA27.29 assays) can bind beta-catenin, a key component of adhesion and Wnt signaling (55). The MUC16 gene encodes the human CA125 protein that is a widely used as an ovarian cancer biomarker (56). MUC4 has been identified as a novel intramembrane ligand for the HER2 receptor, which can regulate HER2 signaling and sensitivity to trastuzumab-based therapy.

MUC4 contains MUC4α, a heavily *O*-glycosylated subunit, and MUC4β, a 90-kDa transmembrane subunit containing EGF-like domains (57). The molecular weight of MUC4α varies between 500 and 900 kDa, and the whole molecule of MUC4 extends 1 to 2 μm from the cell surface (58). The MUC4β subunit interacts with the HER2 receptor. Inhibition of expression of MUC4 with an antisense RNA against MUC4 reduces expression of HER2 and slows growth of pancreatic cancer cells (59). Inhibition of expression of MUC4 with benzyl-GalNac, an inhibitor of mucin secretion, can block phosphorylation of HER2 and HER3 (60). MUC4 overexpression can induce HER2 phosphorylation and can alter the membrane localization of HER2 (and possibly HER3) (61, 62). The relocation of HER2 decreases the ligand-induced receptor internalization and enhances HER2 signaling (62). These observations indicate that MUC4 can positively modulate HER2 signaling. On the other hand, overexpression of rat MUC4 reduces binding of trastuzumab to HER2-expressing tumor cells due to MUC4-masking the epitope of HER2 that binds antibody (63). This observation is further supported by an observation that a human trastuzumab-resistant breast cancer cell line has fewer HER2 binding sites, associated with overexpression of human MUC4 (64). Knockdown of MUC4 expression by siRNA increases the binding of trastuzumab to this cell line (64). Therefore, MUC4 is structurally and functionally linked to the HER2 receptor and its associated signaling pathways.

4. Regulation of Gene Expression by Anti-HER2 Monoclonal Antibodies

Since human trastuzumab or murine 4D5 binds specifically to the HER2 receptor, these antibodies can be used as probes to identify downstream targets of HER2. Signaling through HER2 can regulate expression of multiple genes. We have evaluated the impact of anti-HER2 antibody on gene expression in two breast cancer cell lines that overexpress HER2 (SKBr3 and BT474). SKBr3 cells were treated with 4D5 or control antibody MOPC21. BT474 cells were treated trastuzumab or control human immunoglobin G (hIgG). Total RNA was isolated to permit synthesis of cRNA for Affymetrix array hybridization. The U95Av2 gene chip was used to measure expression of 12,000 known human genes as previously described (65).

Two methods were used to analyze differential gene expression. A standard Affymetrix MAS 5.0 statistical analysis tool package was used for each probe set to measure fold change (FC) on a $\log_2$ scale, a 95% confidence interval (CI) for FC on the $\log_2$ scale, a p-value for detection used to make a presence or absence calls for each gene and a p-value for detecting change or differential expression between samples. The default standard Affymetrix change calls were not used. A second method, designated the Position Dependent Nearest Neighbor (PDNN) model, was developed by Dr. Li Zhang in the Department of Biostatistics of the M.D. Anderson Cancer Center. The PDNN model, which is available at web site: http://odin.mdacc.tmc.edu/~zhangli/PerfectMatch/, relies on a relationship between DNA base pair stacking energies and probe binding efficiencies. Only a perfect match is used to estimate the intensity of a probe set. Cross hybridization is accounted for in model estimation. Remarkable reproducibility between replicate samples has been attained at our Gene Microarray Core Facility. The candidate gene lists were generated if the log ratio exceeded 1 (corresponding to at least a 2-fold change) shown with both analytic methods.

Based on the two methods of array analysis describe above, anti-HER2 antibody downregulated expression of 26 genes in SKBr3 cells (Table 1). Sixteen of these 26 genes listed in Table 1 have been reported previously (65).

Table 1. Genes that were downregulated by anti-HER2 antibody in SKBr3 cells

Number	Gene symbol	Unigene #	Gene full name
1	RFC4	Hs.35120	Replication factor C (activator 1) 4 (37kD)
2	UBE2C	Hs.93002	Ubiquitin-conjugating enzyme E2C
3	PCNA	Hs.78996	Proliferating cell nuclear antigen
4	CCNA2	Hs.85137	Cyclin A2
5	TSSC3	Hs.154036	Tumor suppressing subtransferable candidate 3
6	DEEPEST	Hs.16244	Mitotic spindle coiled-coil related protein
7	NRM	Hs.519993	Nurim (nuclear envelope membrane protein)
8	TUBB2	Hs.251653	Tubulin, beta, 2
9	STK15	Hs.250822	Serine/threonine kinase 15
10	BUB1B	Hs.36708	Budding uninhibited by benzimidazoles 1
11	ZWINT	Hs.42650	ZW10 interactor
12	TUBA1	Hs.75318	Tubulin, alpha 1 (testis specific)
13	AMD1	Hs.262476	S-adenosylmethionine decarboxylase 1
14	KNSL5	Hs.270845	Kinesin-like 5 (mitotic kinesin-like protein 1)
15	EZH2	Hs.77256	Enhancer of zeste (Drosophila) homolog 2
16	TYMS	Hs.82962	Thymidylate synthetase
17	HMG2	Hs.80684	High-mobility group protein 2
18	TPX2	Hs.244580	TPX2, microtubule-associated, homolog
19	GINS1	Hs.658464	GINS complex subunit 1 (Psf1 homolog)
20	H4FG	Hs.46423	H4 histone family, member G
21	HEC	Hs.58169	Highly expressed in cancer
22	MCM6	Hs.155462	Minichromosome maintenance deficient 6
23	TOP2A	Hs.156346	Topoisomerase (DNA) II alpha (170kD)
24	CKS2	Hs.83758	CDC28 protein kinase 2
25	KNSL1	Hs.8878	Kinesin-like 1
26	FEN1	Hs.4756	Flap structure-specific endonuclease 1

The 26 genes repressed by anti-HER2 antibody can be sorted into four functional areas: cell cycle regulation (G2 phase), DNA replication/repair, and chromatin structure, and microtubule structure (Table 1). As 4D5 antibody inhibits HER2 function (23), these data suggest that overexpression of HER2 in breast cancer cells activates cell cycle progression through G2, increases DNA replication, and modifies chromatin and microtubule structures.

As shown in Table 2, seventeen genes were upregulated by 4D5 in SKBr3 cells. Among the seventeen genes, activation of a number of tumor suppressors (VDUP1, cyclin G2, NUPR1, GPNMB, HBP1, and EMP1) is noticeable (Table 2).

Trastuzumab altered the expression of a greater number of genes in BT474 cells than did 4D5 in SKBr3 cells. Anti-HER2 antibody downregulated 76 genes in BT474 cells (Table 3). Changes in the expression for 24 of the 76 genes correlated with changes in SKBr3 cells. Although alterations were more frequent in BT474 cells, most of the changes still fell within four functional areas: cell cycle regulation (G2 phase), DNA replication/repair, chromatin structure, and microtubule structure (Table 3). Therefore, results from BT474 cells confirmed that overexpression of HER2 in breast cancer cells may drive progression of the cell cycle through G2, increase DNA replication, and modify chromatin and microtubule structures that contribute to the malignant phenotype.

Table 2. Genes that were upregulated by anti-HER2 antibody in SKBr3 cells

Number	Gene symbol	Unigene #	Gene full name
1	ERBB2	Hs.323910	v-erb-b2 avian erythroblastic leukemia viral oncogene homolog 2
2	CCNG2	Hs.79069	cyclin G2
3	ERBB3	Hs.199067	v-erb-b2 avian erythroblastic leukemia viral oncogene homolog 3
4	PRKAR1A	Hs.183037	protein kinase, cAMP-dependent, regulatory, type I, alpha
5	VDUP1	Hs.179526	Upregulated by 1,25-dihydroxyvitamin D-3
6	DST	Hs.631992	Dystonin
7	SEPP1	Hs.275775	Selenoprotein P, plasma, 1
8	NUPR1	Hs.513463	Nuclear protein 1
9	LITAF	Hs.459940	Lipopolysaccharide-induced TNF factor
10	PCP4	Hs.80296	Purkinje cell protein 4
11	EMP1	Hs.696241	Epithelial membrane protein 1
12	GPNMB	Hs.190495	Glycoprotein (transmembrane) nmb
13	EIF4B	Hs.648394	Eukaryotic translation initiation factor 4B
14	HBP1	Hs.10882	HMG-box containing protein 1
15	ATP5A1	Hs.298280	ATP synthase, mitochondrial F1 complex, alpha, isoform 1
16	USP9X	Hs.77578	Ubiquitin specific protease 9, X chromosome
17	GAPD	Hs.169476	Glyceraldehyde-3-phosphate dehydrogenase

Table 3. Genes that were downregulated by anti-HER2 antibody in BT474 cells

Number	Gene symbol	Unigene #	Gene full name
1	H4FG	Hs.46423	H4 histone family, member G
2	HEC	Hs.58169	highly expressed in cancer, rich in leucine heptad repeats
3	PAF	Hs.81892	PCNA-associated factor KIAA0101/p15(PAF) or KIAA0101
4	STK15	Hs.250822	serine/threonine kinase 15
5	CDKN3	Hs.84113	cyclin-dependent kinase inhibitor 3
6	ZWINT	Hs.42650	ZW10 interactor
7	TYMS	Hs.82962	thymidylate synthetase
8	CCNB1	Hs.23960	cyclin B1
9	TOP2A	Hs.156346	Topoisomerase (DNA) II alpha 170kDa
10	BUB1B	Hs.36708	budding uninhibited by benzimidazoles 1, beta
11	TTK	Hs.169840	TTK protein kinase
12	RFC4	Hs.35120	replication factor C (activator 1) 4 (37kD)
13	PCNA	Hs.78996	proliferating cell nuclear antigen
14	MCM6	Hs.155462	Minichromosome maintenance deficient (mis5, S. pombe) 6
15	FEN1	Hs.4756	flap structure-specific endonuclease 1
16	UBE2C	Hs.93002	ubiquitin-conjugating enzyme E2C

Table 3. Continued

Number	Gene symbol	Unigene #	Gene full name
17	TPX2	Hs.244580	TPX2, microtubule-associated, homolog (Xenopus laevis)
18	MGC1780	Hs.77550	hypothetical protein MGC1780
19	CCNE2	Hs.30464	cyclin E2
20	CENPF	Hs.77204	centromere protein F (350/400kD, mitosin)
21	APOBEC3B	Hs.226307	Apolipoprotein B mRNA editing enzyme, catalytic 3B
22	CCNB2	Hs.194698	cyclin B2
23	DEEPEST	Hs.16244	mitotic spindle coiled-coil related protein
24	GGH	Hs.78619	gamma-glutamyl hydrolase
25	TRIP13	Hs.6566	thyroid hormone receptor interactor 13
26	CCNB1	Hs.23960	cyclin B1
27	KNSL1	Hs.8878	kinesin-like 1
28	CDC2	Hs.184572	cell division cycle 2, G1 to S and G2 to M
29	TYMS	Hs.82962	thymidylate synthetase
30	MELK	Hs.184339	Maternal embryonic leucine zipper kinase
31	MAD2L1	Hs.79078	MAD2 (mitotic arrest deficient, yeast, homolog)-like 1
32	HMG2	Hs.80684	high-mobility group (nonhistone chromosomal) protein 2
33	CKS2	Hs.83758	CDC28 protein kinase 2
34	GINS1	Hs.658464	GINS complex subunit 1 (Psf1 homolog)
35	PTTG1	Hs.252587	pituitary tumor-transforming 1
36	DUT	Hs.82113	dUTP pyrophosphatase
37	MAD2L1	Hs.79078	MAD2 (mitotic arrest deficient, yeast, homolog)-like 1
38	RRM1	Hs.2934	ribonucleotide reductase M1 polypeptide
39	BUB1	Hs.98658	budding uninhibited by benzimidazoles 1 (yeast homolog)
40	DLG7	Hs.77695	Discs, large homolog 7 (Drosophila)
41	MCM2	Hs.57101	Minichromosome maintenance deficient 2 (mitotin)
42	NUP205	Hs.632299	Nucleoporin 205kDa
43	TK1	Hs.105097	thymidine kinase 1, soluble
44	H2AFX	Hs.147097	H2A histone family, member X
45	TSSC3	Hs.154036	tumor suppressing subtransferable candidate 3
46	CDC7L1	Hs.28853	CDC7 (cell division cycle 7, S. cerevisiae, homolog)-like 1
47	FANCI	Hs.513126	Fanconi anemia, complementation group I
48	MTHFD1	Hs.172665	methylenetetrahydrofolate dehydrogenase (NADP+)
49	K-ALPHA-1	Hs.334842	tubulin, alpha, ubiquitous
50	ODC1	Hs.75212	ornithine decarboxylase 1
51	FKBP1A	Hs.179661	FK506-binding protein 1A (12kD)
52	LDHB	Hs.234489	lactate dehydrogenase B
53	CIT	Hs.15767	citron (rho-interacting, serine/threonine kinase 21)

Table 3. Continued

Number	Gene symbol	Unigene #	Gene full name
54	KIAA0165	Hs.153479	extra spindle poles, S. cerevisiae, homolog of
55	MCM3	Hs.179565	Minichromosome maintenance deficient (S. cerevisiae) 3
56	H2AFZ	Hs.119192	H2A histone family, member Z
57	CENPA	Hs.1594	centromere protein A (17kD)
58	CDC20	Hs.82906	CDC20 (cell division cycle 20, S. cerevisiae, homolog)
59	BARD1	Hs.54089	BRCA1 associated RING domain 1
60	TUBA2	Hs.98102	tubulin, alpha 2
61	KNSL6	Hs.69360	kinesin-like 6 (mitotic centromere-associated kinesin)
62	SLC4A1AP	Hs.306000	solute carrier family 4 (anion), member 1, adapter protein
63	PIM3	Hs.530381	Pim-3 oncogene
64	ASK	Hs.152759	activator of S phase kinase
65	KIF14	Hs.3104	Kinesin family member 14
66	TUBG1	Hs.21635	tubulin, gamma 1
67	EZH2	Hs.77256	enhancer of zeste (Drosophila) homolog 2
68	PRIM1	Hs.82741	primase, polypeptide 1 (49kD)
69	KPNA2	Hs.159557	karyopherin alpha 2 (RAG cohort 1, importin alpha 1)
70	FOXM1	Hs.239	forkhead box M1
71	NME1	Hs.118638	non-metastatic cells 1, protein (NM23A) expressed in
72	TIMELESS	Hs.118631	timeless (Drosophila) homolog
73	MTHFD2	Hs.154672	methylene tetrahydrofolate dehydrogenase (NAD+)
74	STK12	Hs.180655	serine/threonine kinase 12
75	RPA3	Hs.1608	replication protein A3 (14kD)
76	MSH2	Hs.78934	mutS (E. coli) homolog 2 (colon cancer, nonpolyposis 1)

In contrast to the correlation between cell lines observed for downregulated genes, only four genes were upregulated by anti-HER2 antibodies in both BT474 and SKBr3 cells (ERBB3, ATP5A1, SEPP1, and GAPD) (Table 4). ErbB3 or HER3 modulates PI3 kinase signaling. SEPP1 functions as an antioxidant in the extracellular space and is associated with endothelial cells. In contrast to the effect of anti-HER2 antibody on SKBr3 cells, only one tumor suppressor (RBL2) was significantly upregulated by trastuzumab in BT474 cells. Among the 34 genes upregulated by inhibition of HER2, elevation of enzymes that metabolize cellular constituents and xenobiotics such as CYP1A1, GNS, OGT, ATP5A1, GAPD, SAT, and AMY2B) were noticeable. Surprisingly, some genes involved in cytoskeleton such as ACTB, SRI, and TUBA3 were stimulated by treatment with trastuzumab in BT474 cells.

Table 4. Genes that were upregulated by anti-HER2 antibody in BT474 cells

Number	Gene symbol	Unigene #	Gene full name
1	CST1	Hs.123114	cystatin SN
2	CBX6	Hs.107374	chromobox homolog 6
3	CYP1A1	Hs.72912	cytochrome P450, subfamily 1polypeptide 1
4	GNS	Hs.164036	glucosamine (N-acetyl)-6-sulfatase
5	BRD3	Hs.522472	Bromodomain containing 3
6	JAK1	Hs.50651	Janus kinase 1 (a protein tyrosine kinase)
7	RBL2	Hs.79362	retinoblastoma-like 2 (p130)
8	TNFSF10	Hs.83429	tumor necrosis factor (ligand) superfamily, member 10
9	ERBB3	Hs.199067	v-erb-b2 avian erythroblastic leukemia viral oncogene homolog 3
10	TEGT	Hs.74637	testis enhanced gene transcript (BAX inhibitor 1)
11	RAB6A	Hs.5636	RAB6A, member RAS oncogene family
12	TFRC	Hs.77356	transferrin receptor (p90, CD71)
13	DSG2	Hs.94560	desmoglein 2
14	CALCOCO1	Hs.156667	Calcium binding and coiled-coil domain 1
15	ACTB	Hs.288061	actin, beta
16	H2BFC	Hs.137594	H2B histone family, member C
17	OGT	Hs.100293	O-linked N-acetylglucosamine (GlcNAc) transferase
18	SRI	Hs.489040	Sorcin
19	ATP5A1	Hs.155101	ATP synthase, mitochondrial F1 complex, alpha, isoform 1
20	PIP	Hs.99949	prolactin-induced protein
21	GAPD	Hs.169476	glyceraldehyde-3-phosphate dehydrogenase
22	TUBA3	Hs.272897	Tubulin, alpha, brain-specific
23	MXI1	Hs.118630	MAX-interacting protein 1
24	VIM	Hs.297753	Vimentin
25	H2AFL	Hs.28777	H2A histone family, member L
26	5T4	Hs.82128	5T4 oncofetal trophoblast glycoprotein
27	INADL	Hs.321197	PDZ domain protein (Drosophila inaD-like)
28	ITM2B	Hs.239625	integral membrane protein 2B
29	SEPP1	Hs.3314	selenoprotein P, plasma, 1
30	ANXA3	Hs.1378	annexin A3
31	SAT	Hs.28491	Spermidine/spermine N1-acetyltransferase
32	AMY2B	Hs.335493	amylase, alpha 2B; pancreatic
33	GTF2I	Hs.278589	general transcription factor II, i
34	SATB1	Hs.74592	special AT-rich sequence binding protein 1

5. Regulation of the Cell Cycle by HER2

Arrest of the cell cycle is one of the major mechanisms by which trastuzumab inhibits HER2 signaling and tumor cell growth (66,67). In breast cancer cells that overexpress HER2, treatment with anti-HER2 antibody upregulates the $p27^{Kip1}$ protein, one of the key negative regulators of the G1 phase of the cell cycle (68). As shown in Figure 1, treatment of cells that overexpress HER2 with trastuzumab alters phosphorylation of $p27^{Kip1}$, stabilizing the protein and favoring its nuclear localization.

Six signaling targets and pathways are modulated by trastuzumab. By inhibiting CDK2 expression and activity and decreasing Thr187 phosphorylation of $p27^{Kip1}$ protein, trastuzumab abrogates the targeting of SCF-ubiquitin E3 ligase and minimizes the proteasome degradation of $p27^{Kip1}$ protein. By inhibiting the expression and activity of AKT and hKIS, trastuzumab blocks Thr157, Thr198-, and Ser10 phosphorylation, which causes $p27^{Kip1}$ to translocate from the cytosol to the nucleus, which stabilizes $p27^{Kip1}$ protein and increases the inhibitory effect of $p27^{Kip1}$ on CDK2. By inhibiting the expression of Jab1 and increasing nuclear retention of $p27^{Kip1}$, trastuzumab increases the inhibitory action of $p27^{Kip1}$ on CDK2 and stabilizes $p27^{Kip1}$. By inhibiting the expression and activity of cyclin D and c-Myc, trastuzumab releases the sequestrated $p27^{Kip1}$ protein from cyclin D-CDK4/6 complexes and so increases its effect on CDK2-cyclin E complexes. By stimulating the expression and activity of MIRK, trastuzumab stabilizes $p27^{Kip1}$ in the nucleus, which increases inhibitory action of $p27^{Kip1}$ on CDK2 and stabilizes $p27^{Kip1}$ protein. These six signaling targets and pathways induced by trastuzumab work in concert to maximize the expression and inhibitory action of $p27^{Kip1}$ protein, contributing to G1 cell cycle arrest and growth inhibition.

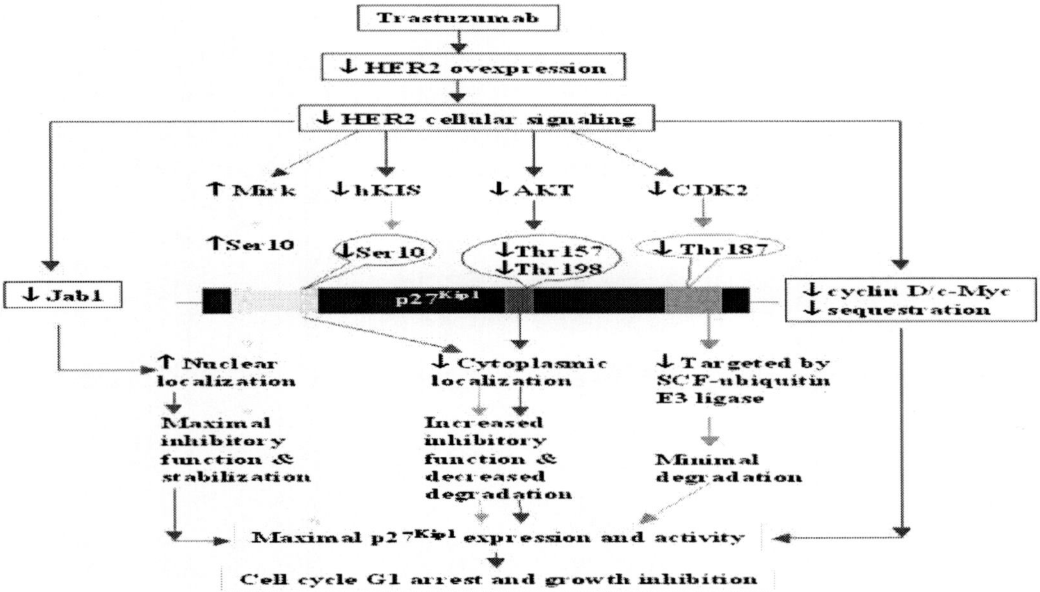

Figure 1. HER2 signaling modulates the activity and levels of $p27^{Kip1}$ protein.

A number of other cell cycle regulators are influenced by HER2 signaling. Trastuzumab downregulates phosphatidylinositol 3 (PI3) -kinase signaling, inhibiting the nuclear protein mapped to the ATM locus (NPAT or p220) and histone H4 that are required for DNA and histone synthesis during progression from G1 to S phase of the cell cycle (69). Trastuzumab also upregulates an unconventional cyclin, cyclin G2 (Table 2), that produces cell cycle arrest and inhibits cell proliferation without altering $p27^{Kip1}$ levels (70). Two recent studies have identified cyclin D1 as a critical downstream target of HER2 in prostate epithelium (71) and breast cancer cells (72). Regulation of cyclin D1 by HER2 is believed to be through the

(PI3K)-AKT signaling pathway (72). CDK5 is reported to directly phosphorylate HER2 receptor at site of Ser-1176 and HER3 receptor at sites of Thr-871 and Ser-1120 (73). The phosphorylation of HER2 and HER3 receptors by CDK5 is associated with activation of PI3K-AKT signaling (73).

6. Estrogen-Related Receptor Alpha (ERRa) and HER2

ERRa, also called NR3B1, has recently been proposed as a target of HER2 signaling (74-76). ERRa is a nuclear receptor that structurally relates to the estrogen receptor but does not bind naturally occurring estrogens. ERRa is widely expressed in normal tissues and is constitutively active (77). The constitutive ERRa can interact with and be modulated by members of the steroid receptor coactivator (SRC) and the peroxisome proliferator-activated receptor (PPAR) gamma co-activator 1 (PGC-1) families of coactivators. The identified function of ERRa is to modulate estrogen signaling and estrogen resistance, and to regulate mitochondrial biogenesis and energy metabolism (78).

Evidence linking ERRa to HER2 signaling includes: 1) overexpression of an activated HER2 oncogene converts ERRa from a repressor to an activator of estrogen-response element (ERE)-regulated transcription (74); 2) disruption of HER2 signaling in BT-474 cells with trastuzumab converts ERRa from an activator to a repressor of ERE-regulated transcription (74); 3) ERRa mRNA levels correlate in breast cancers with those of HER2 (75); and 4) ERRa is phosphorylated by epidermal growth factor (EGF)-EGF receptor signaling pathway (76), which is associated with HER2 signaling. The link between HER2 and ERRa is important as it links HER2 to estrogen response signaling and to the modulation of metabolism.

7. Bcl-2-Interacting Mediator of Cell Death (Bim) and HER2

Bim is a BH3-only protein, a pro-apoptotic member of the Bcl-2 protein family (79). There are three isoforms of Bim protein in mammalian cells: short (Bim-S), long (Bim-L) and extra long protein (Bim-EL) (79). Recent studies have demonstrated that the apoptotic activity of Bim-EL, the most abundant Bim variant, can be regulated through post-translational mechanisms that involve both ERK1/2- and PI3K-AKT-dependent phosphorylation (80-82). Phosphorylation of Bim-EL by ERK1/2 on serine 69 (80) and by AKT on serine 87 (81) leads to dissociation of Bim-EL from the pro-survival protein Bcl-2, promoting its subsequent ubiquitination and proteasomal degradation.

Our group has previously shown that HER2 signaling activates the ERK1/2- and PI3K-AKT signaling pathways (65,80). Indeed, when HER2 is overexpressed, Bim is constitutively suppressed through a MAPK-dependent mechanism (83). During mammary acinar morphogenesis, Bim is found to be specifically and negatively regulated by the HER2 oncogene, but not by the HPV16 E2 oncogene (84). Induction of Bim protein by gefitinib

(Iressa) is more prominent in HER2-overpressing breast cancer cells (85). These data suggest that Bim is downregulated by HER2 signaling.

8. Regulation of the Cell Motility/Migration/Invasion by HER2

HER2 signaling enhances cell motility, adhesion, migration, invasion, and tumor metastasis (18). Several new targets of HER2 signaling have been identified during the past several years. The $alpha_6beta_4$ integrin is a heterodimeric transmembrane molecule that functions primarily as an adhesion receptor required for migration of normal epithelial and carcinoma cells (86). A physical interaction between the $alpha_6beta_4$ integrin and HER2 oncoprotein has been observed (87). Interestingly, $alpha_6beta_4$ integrin regulates HER2 through translational control. Depletion of $alpha_6beta_4$ integrin by small interfering RNA (siRNA) reduces HER2 protein levels without affecting HER2 mRNA level, whereas reexpression of $alpha_6beta_4$ increases the HER2 protein level (88). The $alpha_6beta_4$ integrin interaction with HER2 stimulates proliferation and invasiveness of breast cancer cells (89). A complex formed between $alpha_6beta_4$ integrin and HER2 enhances activation of the signal transducer and activator of transcription 3 (STAT3) and c-Jun (89). Deletion of $alpha_6beta_4$ integrin in a mouse model suppresses mammary tumor onset and invasive growth (89). STAT3 signaling enhances proliferation, increases motility and contributes to disruption of epithelial adhesion and polarity, whereas c-Jun signaling is required for increased cell proliferation (89). This study suggests that $alpha_6beta_4$ integrin is a signaling partner of HER2. These observations are also consistent with the possibility that targeting $alpha_6beta_4$ integrin signaling could enhance the efficacy of anti-HER2 therapy.

The spatial and temporal assembly of actin filaments controls cell motility including the formation of lamellipodia and filopodia (90). A novel molecule, MEMO (Mediator of ErbB2-driven cell MOtility, also known as CGI-27; C2orf4; NS5ATP7) has been implicated in HER2-dependent cell motility and metastasis (90). MEMO is a 33-kDa cytoplasmic protein that does not interact directly with HER2 (91), but rather binds to a Shc adaptor protein that, in turn, interacts with the HER2 receptor at Tyr 1227 (91). Our group has previously shown that both anti-HER2 antibody and heregulin increase the association of Shc adaptor protein with HER2 (92). However, anti-HER2 antibody preferentially induces binding of the 46-kDa isoform of Shc to HER2, whereas heregulin preferentially induces binding of 52-kDa isoform of Shc to HER3 (92). MEMO does not affect early cytoskeletal actin reorganization, since a reduction in MEMO levels does not affect heregulin-induced lamellipodia formation (91). MEMO is able to stimulate cell migration by relaying extracellular signals to the microtubule cytoskeleton through Shc (91).

S100A4 (also known as Mts1, metastasin, p9Ka, pEL98, CAPL, calvasculin, Fsp-1, placental calcium-binding protein) belongs to the family of EF-hand calcium-binding proteins, whose expression is elevated in cancer cells and contributes to tumor cell motility and metastatic potential (93). S100A4 is a direct target of HER2 signaling through PI3K and extracellular signal-regulated kinase 1/2 (94). Microarray expression profiling has shown that HER2 upregulates the expression of prometastatic genes including S100A4. Levels of HER2 and S100A4 are tightly correlated in samples of primary medulloblastoma (94).

The chemokines and their receptors have implicated in tumor cell migration and metastasis. CXC chemokine receptors 4, 7, and 10 (CXCR4, CCR7, CCR10) have been shown to play a critical role in determining the sites of metastasis for cancer cells (95). Certain organs such as lung, bone, and liver exhibit high levels of chemokine stromal cell-derived factor-1α (SDF-1α, also known as CXC chemokine ligand 12, CXCL12), which is the natural ligand for CXCR4 (95). Human breast cancer cells, particularly HER2-overexpressing breast cancer cells, express high levels of CXCR4 (96). *HER2/neu* expression enhances the expression of *CXCR4* and CXCR4 is required for HER2-mediated invasion *in vitro* and lung metastasis *in vivo* (96). Not only HER2 can enhance CXCR4 expression, but also SDF-1α-CXCR4 signaling can enhance HER2 expression through the Src kinase (97). This positive feedback may contribute to the highly metastatic potential of HER2-overexpressing breast cancer cells.

The HER-2 oncogene inhibits the expression of the matrix metalloproteinase (MMP) inhibitor reversion-inducing-cysteine-rich protein with Kazal motifs (RECK) to promote cell invasion (98). The *RECK* gene encodes a membrane-anchored glycoprotein of about 110 kDa with multiple epidermal growth factor-like repeats and serine-protease inhibitor-like domains (99). The RECK protein negatively regulates at least three members of the matrix metalloproteinase (MMP) family, MMP-2, MMP-9, and MT1-MMP (99). HER-2 induces the binding of Sp proteins and HDAC1 to the RECK promoter to transcriptionally inhibit RECK expression and to promote cell invasion. Restoration of RECK may provide a novel strategy for the inhibition of HER-2-induced metastasis.

9. Breast Tumor Kinase (BRK) and HER2

Breast tumor kinase (BRK, also called protein tyrosine kinase 6 or PTK6) is a potential target for HER2 signaling. BRK, originally identified from a human metastatic breast tumor, is a cytoplasmic nonreceptor tyrosine kinase containing SH3 and SH2 domains (100). Although BRK is not expressed in the normal mammary gland, it is induced and overexpressed in a majority of breast cancers (101, 102). Structurally, BRK has a 46% amino acid identity to Src (103, 104), which implies that the functions of BRK are related to those of the Src family of tyrosine kinases. Indeed, BRK stimulates the proliferation of breast cancer cells (105). BRK also directly activates Rac GTPase and phosphorylates paxillin, which promotes cell migration and invasion (106, 107).

The transcriptional factors STAT3 and STAT5b have been recently identified as the physiological targets of BRK (108, 109). Therefore, in response to the stimulation of cytokines and growth factors, BRK may possess cellular functions in addition to those of cell proliferation and cell migration. BRK is known to enhance the response of mammary epithelial cells to EGF stimulation (110). BRK can phosphoylate the HER3 receptor, as well as AKT kinase (111,112). A recent report showed that the HER3/HER4 ligand, heregulin, activates BRK kinase activity and knockdown of BRK by siRNA in breast cancer cells decreases proliferation (107). Expression of BR and HER2 predict poor prognosis, independently and in combination (113). All of these data suggest that BRK is a potential target of HER2 signaling.

10. Regulation of Angiogenesis by HER2

HER2 signaling has been linked to expression of the pro-angiogenic vascular endothelial growth factor (VEGF). In HER2-overexpressing breast cancer cells, VEGF expression is increased (114-119). Our own data demonstrated that increasing HER2 expression upregulates VEGF levels, whereas inhibition of HER2 signaling with trastuzumab and siRNA against HER2 decrease VEGF levels through the PI3K pathway (119). In agreement with these preclinical data, there is a strong and positive correlation of VEGF levels with HER2 expression at a tissue level in primary breast cancers (120,121). Patients with high levels of VEGF tend to have a worse clinical outcome (120,121). These preclinical and clinical data indicate that VEGF expression is linked to the HER2 signaling pathway in breast cancer cells and that high levels of VEGF may contribute to the aggressive properties of breast cancers with HER2 amplification and overexpression.

In addition to the angiogenic factor VEGF, we have identified interleukin-8 (IL-8) and thrombspondin-1 (TSP-1) as downstream targets of HER2 signaling (119). Reexpression of the *HER2* oncogene in MCF-7 and T47D cells that express low levels of HER2 results in elevated expression of IL-8, decreased expression of TSP-1, but no significant change in bFGF levels on RT-PCR, ELISA, or Western blot analysis. Reexpression of *HER2* is associated with elevated PI3K-AKT activity. Inhibition of HER2 function in BT474 cells that overexpress HER2 with its specific antibody trastuzumab leads to a decrease in IL-8 expression and to an increase in TSP-1 expression. Inhibition of *HER2* with a specific siRNA results in similar effects to trastuzumab. Inhibition of HER2 with trastuzumab or HER2 siRNA does not affect bFGF expression (119).

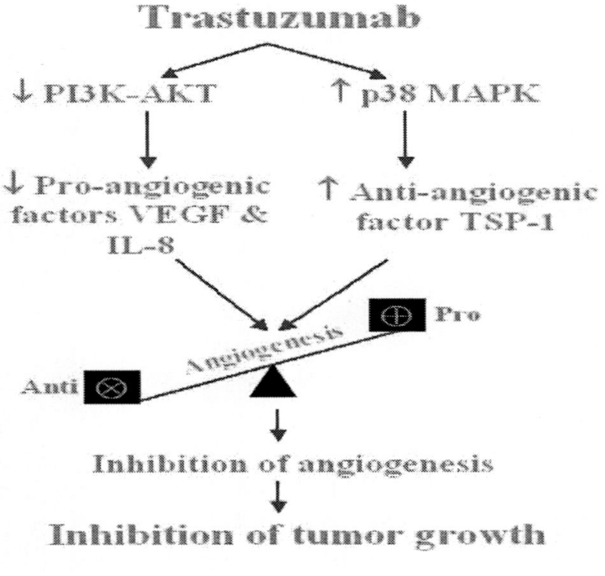

Figure 2. Effects of HER2 signaling modulated by trastuzumab on angiogeneic factors VEGF, IL8 and TSP1.

What are the signaling pathways by which HER2 signaling regulates the expression of VEGF, IL-8, and TSP-1? We have illustrated that inhibition of the PI3K-AKT pathway with

LY294002 downregulates IL-8 and VEGF expression. Inhibition of HER2 with trastuzumab and siRNA inhibits the PI3K-AKT activity. However, LY294002 has no effect on TSP-1 expression, suggesting TSP-1 production, unlike IL-8 and VEGF, does not depend on the PI3K-AKT pathway. Instead, the p38 MAPK specific inhibitor SB203580 inhibits TSP-1 expression. Conversely, overexpression of p38 MAPK increases TSP-1 expression. Trastuzumab transiently activates p38 MAPK and is able to reverse the TSP-1 inhibition induced by SB203580 (119). We conclude that HER2 signaling is able to upregulate pro-angiogenic factors VEGF and IL-8 via the PI3K-AKT pathway and to downregulate the anti-angiogenic factor TSP-1 via the p38 MAPK pathway (Fig. 2).

The anti-HER2 antibody trastuzumab is a potent inhibitor of angiogenesis in breast cancer, simultaneously affecting different pro- and anti- angiogenic factors. The regulation of VEGF by HER2 may also involve signaling events via the extracellular signal-regulated kinase and mammalian target of rapamycin (mTOR)- p70S6K (122).

Breast cancers that overexpress HER2 are highly aggressive (4,19-22). Overexpression of HER-2 is associated with younger patient age, premenopausal status, lack of estrogen or progresterone receptor expression and high nuclear grade in invasive breast cancers (22). Significant association is also observed between HER-2 and increasing number of involved lymph nodes (123, 124). Furthermore, overexpression of HER-2 is associated with more chest effusion formation in breast cancer (125). As lymph node metastasis is typical of breast cancer, it is of interest that HER2 signaling can affect lymphangiogenesis. Indeed, expression of two well-characterized lymphangiogenic factors, VEGF-C and VEGF-D, has been linked to HER2 signaling (120,126,127). Blockade of lymphatics by HER2 positive cancers may contribute to the formation of ascites and pleural effusions (126,127). In addition, VEGF increases vascular permeability and may also contribute to the accumulation of effusion fluid.

11. Conclusion

The HER2 oncoprotein plays an important role in the transformation and progression of several human cancers. Constitutive activation of the HER2 receptor, favors dimerization with HER3 or EGFR. Sensitivity to trastuzumab can be modulated by levels of the bulky MUC4 molecule that can hinder antibody binding. Molecules that regulate the cell cycle, apoptosis, cell migration/motility/invasion, angiogenesis and tumor metastasis are among the targets modulated by the HER2 oncoprotein at a transcriptional or post-transcriptional level. Independent activation of one or more of these targets may contribute to resistance to trastuzumab. Targeting both HER2 and one or more of these HER2 regulated molecules might provide greater anti-tumor activity in cancers that overexpress the receptor. Many of the functional changes induced by HER2 oncoprotein result from activation of the PI3K-AKT pathway following formation of HER2-HER3 heterodimers. Other signaling pathways can also participate including p38 MAP Kinase that could be activated by HER2-EGFR heterodimerization. Some novel drugs such as lapatinib inhibit both HER2-HER3 and HER2-EGFR signaling. Further attention should be given to aberrant activation of HER2-regulated targets that persists after inhibition of both pathways.

Acknowledgements

This work was supported by the Anne and Henry Zarrow Foundation and by a grant CA39930 from the National Cancer Institute.

References

[1] Schechter AL, Stern DF, Vaidyanathan L, Decker SJ, Drebin JA, Greene MI, Weinberg RA. The neu oncogene: an erb-B-related gene encoding a 185,000-Mr tumour antigen. *Nature*. 1984; 312(5994): 513-516.

[2] King CR, Kraus MH, Aaronson SA. Amplification of a novel v-erbB-related gene in a human mammary carcinoma. *Science*. 1985; 229: 974–976.

[3] Coussens L, Yang-Feng TL, Liao YC, Chen E, Gray A, McGrath J, Seeburg PH, Libermann TA, Schlessinger J, Francke U, et al. Tyrosine kinase receptor with extensive homology to EGF receptor shares chromosomal location with neu oncogene. *Science*. 1985; 230:1132–1139.

[4] Di Fiore PP, Kraus MH. Mechanisms involving an expanding erbB/EGF receptor family of tyrosine kinases in human neoplasia. *Cancer Treat Res*. 1992; 61:139-160.

[5] Yamamoto T, Ikawa S, Akiyama T, Semba K, Nomura N, Miyajima N, Saito T, Toyoshima K. Similarity of protein encoded by the human c-erb-B-2 gene to epidermal growth factor receptor. *Nature*. 1986; 319: 230–234.

[6] Tzahar E, Waterman H, Chen X, Levkowitz G, Karunagaran D, Lavi S, Ratzkin BJ, Yarden Y. A hierarchical network of interreceptor interactions determines signal transduction by Neu differentiation factor/neuregulin and epidermal growth factor. *Mol Cell Biol*. 1996; 16:5276–5287.

[7] Graus-Porta D, Beerli RR, Daly JM, Hynes NE. ErbB-2, the preferred heterodimerization partner of all ErbB receptors, is a mediator of lateral signaling. *EMBO J*. 1997; 16: 1647–1655.

[8] Slamon DJ, Clark GM, Wong SG, Levin WJ, Ullrich A, McGuire WL. Human breast cancer: correlation of relapse and survival with amplification of the HER-2/neu oncogene. *Science*. 1987; 235: 177–182.

[9] Slamon DJ, Godolphin W, Jones LA, Holt JA, Wong SG, Keith DE, Levin WJ, Stuart SG, Udove J, Ullrich A, et al. Studies of the HER-2/neu proto-oncogene in human breast and ovarian cancer. *Science*. 1989; 244: 707–712.

[10] Toikkanen S, Helin H, Isola J, Joensuu H. Prognostic significance of HER-2 oncoprotein expression in breast cancer: a 30-year follow-up. *J Clin Oncol*. 1992; 10: 1044–1048.

[11] Chazin VR, Kaleko M, Miller AD, Slamon DJ. Transformation mediated by the human HER-2 gene independent of the epidermal growth factor receptor. *Oncogene*. 1992; 7: 1859–1866.

[12] Press MF, Pike MC, Chazin VR, Hung G, Udove JA, Markowicz M, Danyluk J, Godolphin W, Sliwkowski M, Akita R, et al. Her-2/neu expression in node-negative breast cancer: direct tissue quantitation by computerized image analysis and association of overexpression with increased risk of recurrent disease. *Cancer Res*. 1993; 53: 4960–4970.

[13] Di Fiore PP, Pierce JH, Kraus MH, Segatto O, King CR, Aaronson SA. erbB-2 is a potent oncogene when overexpressed in NIH/3T3 cells. *Science*. 1987; 237(4811): 178-182.
[14] Hudziak RM, Schlessinger J, Ullrich A. Increased expression of the putative growth factor receptor p185HER2 causes transformation and tumorigenesis of NIH 3T3 cells. *Proc Natl Acad Sci U S A*. 1987; 84(20):7159-63.
[15] Muthuswamy SK, Li D, Lelievre S, Bissell MJ, Brugge JS. ErbB2, but not ErbB1, reinitiates proliferation and induces luminal repopulation in epithelial acini. *Nat Cell Biol*. 2001; 3:785–792.
[16] Muller WJ, Sinn E, Pattengale PK, Wallace R, Leder P. Single-step induction of mammary adenocarcinoma in transgenic mice bearing the activated c-neu oncogene. *Cell*. 1988; 54(1): 105-115.
[17] Bouchard L, Lamarre L, Tremblay PJ, Jolicoeur P. Stochastic appearance of mammary tumors in transgenic mice carrying the MMTV/c-neu oncogene. *Cell*. 1989; 57(6): 931-936.
[18] Guy CT, Webster MA, Schaller M, Parsons TJ, Cardiff RD, Muller WJ. Expression of the neu protooncogene in the mammary epithelium of transgenic mice induces metastatic disease. *Proc Natl Acad Sci USA* 1992; 89:10578–10582.
[19] Gusterson BA, Machin LG, Gullick WJ, Gibbs NM, Powles TJ, Price P, McKinna A, Harrison S. Immunohistochemical distribution of c-erbB-2 in infiltrating and in situ breast cancer. *Int J Cancer* 1988; 42: 842–845.
[20] Ho GH, Calvano JE, Bisogna M, Borgen PI, Rosen PP, Tan LK, Van Zee KJ. In microdissected ductal carcinoma in situ, HER-2/neu amplification, but not p53 mutation, is associated with high nuclear grade and comedo histology. *Cancer* 2000; 89: 2153–2160.
[21] Somerville JE, Clarke LA, Biggart JD. c-erbB-2 overexpression and histological type of in situ and invasive breast carcinoma. *J Clin Pathol*. 1992; 45: 16–20.
[22] Allred DC, Clark GM, Molina R, Tandon AK, Schnitt SJ, Gilchrist KW, Osborne CK, Tormey DC, McGuire WL. Overexpression of HER-2/neu and its relationship with other prognostic factors change during the progression of in situ to invasive breast cancer. *Hum Pathol*. 1992; 23: 974–979.
[23] Hudziak RM, Lewis GD, Winget M, Fendly BM, Shepard HM, Ullrich A. p185HER2 monoclonal antibody has antiproliferative effects in vitro and sensitizes human breast tumor cells to tumor necrosis factor. *Mol Cell Biol*. 1989; 9(3): 1165-1172.
[24] Cobleigh MA, Vogel CL, Tripathy D, Robert NJ, Scholl S, Fehrenbacher L, et al. Multinational study of the efficacy and safety of humanized anti-HER2 monoclonal antibody in women who have HER2-overexpressing metastatic breast cancer that has progressed after chemotherapy for metastatic disease. *J Clin Oncol* 1999; 17:2639-2648.
[25] Vogel CL, Cobleigh MA, Tripathy D, Gutheil JC, Harris LN, Fehrenbacher L, Slamon DJ, Murphy M, Novotny WF, Burchmore M, Shak S, Stewart SJ, Press M. Efficacy and safety of trastuzumab as a single agent in first-line treatment of HER2-overexpressing metastatic breast cancer. *J Clin Oncol*. 2002; 20(3): 719-726.
[26] Slamon DJ, Leyland-Jones B, Shak S, Fuchs H, Paton V, Bajamonde A, Fleming T, Eiermann W, Wolter J, Pegram M, Baselga J, Norton L. Use of chemotherapy plus a

monoclonal antibody against HER2 for metastatic breast cancer that overexpresses HER2. *N Engl J Med.* 2001; 344(11): 783-792.

[27] Romond EH, Perez EA, Bryant J, Suman VJ, Geyer CE Jr, Davidson NE, Tan-Chiu E, Martino S, Paik S, Kaufman PA, Swain SM, Pisansky TM, Fehrenbacher L, Kutteh LA, Vogel VG, Visscher DW, Yothers G, Jenkins RB, Brown AM, Dakhil SR, Mamounas EP, Lingle WL, Klein PM, Ingle JN, Wolmark N. Trastuzumab plus adjuvant chemotherapy for operable HER2-positive breast cancer. *N Engl J Med.* 2005; 353(16): 1673-84.

[28] Piccart-Gebhart MJ, Procter M, Leyland-Jones B, Goldhirsch A, Untch M, Smith I, Gianni L, Baselga J, Bell R, Jackisch C, Cameron D, Dowsett M, Barrios CH, Steger G, Huang CS, Andersson M, Inbar M, Lichinitser M, Lang I, Nitz U, Iwata H, Thomssen C, Lohrisch C, Suter TM, Ruschoff J, Suto T, Greatorex V, Ward C, Straehle C, McFadden E, Dolci MS, Gelber RD; Herceptin Adjuvant (HERA) Trial Study Team. Trastuzumab after adjuvant chemotherapy in HER2-positive breast cancer. *N Engl J Med.* 2005; 353(16): 1659-1672.

[29] Smith I, Procter M, Gelber RD, Guillaume S, Feyereislova A, Dowsett M, Goldhirsch A, Untch M, Mariani G, Baselga J, Kaufmann M, Cameron D, Bell R, Bergh J, Coleman R, Wardley A, Harbeck N, Lopez RI, Mallmann P, Gelmon K, Wilcken N, Wist E, Sanchez Rovira P, Piccart-Gebhart MJ; HERA study team. 2-year follow-up of trastuzumab after adjuvant chemotherapy in HER2-positive breast cancer: a randomised controlled trial. *Lancet.* 2007; 369(9555): 29-36.

[30] Pal SK, Pegram M. Targeting HER2 Epitopes. *Semin Oncol.* 2006; 33(4):386-391.

[31] Wang SC, Zhang L, Hortobagyi GN, Hung MC. Targeting HER2: recent developments and future directions for breast cancer patients. *Semin Oncol.* 2001; 28(6 Suppl 18): 21-29.

[32] Burris HA 3rd, Hurwitz HI, Dees EC, Dowlati A, Blackwell KL, O'Neil B, Marcom PK, Ellis MJ, Overmoyer B, Jones SF, Harris JL, Smith DA, Koch KM, Stead A, Mangum S, Spector NL. Phase I safety, pharmacokinetics, and clinical activity study of lapatinib (GW572016), a reversible dual inhibitor of epidermal growth factor receptor tyrosine kinases, in heavily pretreated patients with metastatic carcinomas. *J Clin Oncol.* 2005; 23(23): 5305-5313.

[33] Bublil EM, Yarden Y. The EGF receptor family: spearheading a merger of signaling and therapeutics. *Curr Opin Cell Biol.* 2007; 19(2): 124-134.

[34] Pegram MD, Konecny G, Slamon DJ. The molecular and cellular biology of HER2/neu gene amplification/overexpression and the clinical development of Trastuzumab (trastuzumab) therapy for breast cancer. *Cancer Treat Res* 2000; 103:57-75.

[35] Mendelsohn J, Baselga J. The EGF receptor family as targets for cancer therapy. *Oncogene.* 2000; 19:6550-6565.

[36] Harari D, Yarden Y. Molecular mechanisms underlying ErbB2/HER2 action in breast cancer. *Oncogene.* 2000; 19(53): 6102-6114.

[37] Stern DF. Tyrosine kinase signalling in breast cancer: ErbB family receptor tyrosine kinases. *Breast Cancer Res.* 2000; 2(3): 176-83.

[38] Yarden Y. The EGFR family and its ligands in human cancer. signalling mechanisms and therapeutic opportunities. *Eur J Cancer.* 2001; 37 Suppl 4: S3-8.

[39] Stern DF. ErbBs in mammary development. Exp Cell Res. 2003; 284(1): 89-98.

[40] Zhou BP, Hung MC. Dysregulation of cellular signaling by HER2/neu in breast cancer. *Semin Oncol.* 2003; 30(5 Suppl 16): 38-48.

[41] Holbro T, Hynes NE. ErbB receptors: directing key signaling networks throughout life. *Annu Rev Pharmacol Toxicol.* 2004; 44: 195-217.

[42] Marmor MD, Skaria KB, Yarden Y. Signal transduction and oncogenesis by ErbB/HER receptors. *Int J Radiat Oncol Biol Phys.* 2004; 58(3):903-913.

[43] Citri A, Yarden Y. EGF-ERBB signalling: towards the systems level. *Nat Rev Mol Cell Biol.* 2006; 7(7): 505-16.

[44] Zhang H, Berezov A, Wang Q, Zhang G, Drebin J, Murali R, Greene MI. ErbB receptors: from oncogenes to targeted cancer therapies. *J Clin Invest.* 2007; 117(8): 2051-2058.

[45] Cho HS, Leahy DJ. Structure of the extracellular region of HER3 reveals an interdomain tether. *Science* 2002; 297(5585): 1330-1333.

[46] Ogiso H, Ishitani R, Nureki O, Fukai S, Yamanaka M, Kim JH, Saito K, Sakamoto A, Inoue M, Shirouzu M, Yokoyama S. Crystal structure of the complex of human epidermal growth factor and receptor extracellular domains. *Cell* 2002; 110(6):775-787.

[47] Ferguson KM, Berger MB, Mendrola JM, Cho HS, Leahy DJ, Lemmon MA: EGF activates its receptor by removing interactions that autoinhibit ectodomain dimerization. *Mol Cell* 2003, 11:507-517.

[48] Bouyain S, Longo PA, Li S, Ferguson KM, Leahy DJ: The extracellular region of ErbB4 adopts a tethered conformation in the absence of ligand. *Proc Natl Acad Sci USA* 2005, 102:15024-15029.

[49] Cho HS, Mason K, Ramyar KX, Stanley AM, Gabelli SB, Denney DW Jr, Leahy DJ. Structure of the extracellular region of HER2 alone and in complex with the Herceptin Fab. *Nature.* 2003; 421(6924): 756-760.

[50] Garrett TP, McKern NM, Lou M, Elleman TC, Adams TE, Lovrecz GO, Kofler M, Jorissen RN, Nice EC, Burgess AW, Ward CW. The crystal structure of a truncated ErbB2 ectodomain reveals an active conformation, poised to interact with other ErbB receptors. *Mol Cell.* 2003; 11(2): 495-505.

[51] Franklin MC, Carey KD, Vajdos FF, Leahy DJ, de Vos AM, Sliwkowski MX. Insights into ErbB signaling from the structure of the ErbB2-pertuzumab complex. *Cancer Cell.* 2004; 5(4): 317-328.

[52] Saxon ML, Lee DC. Mutagenesis reveals a role for epidermal growth factor receptor extracellular subdomain IV in ligand binding. *J Biol Chem.* 1999; 274(40): 28356-28362.

[53] Stein RA and Staros JV. *BMC Evolutionary Biology* 2006, 6:79-89.

[54] Byrd JC, Bresalier RS. Mucins and mucin binding proteins in colorectal cancer. *Cancer Metastasis Rev.* 2004; 23(1-2): 77-99.

[55] Carraway KL, Ramsauer VP, Haq B, Carothers Carraway CA. Cell signaling through membrane mucins. *Bioessays.* 2003; 5(1): 66-71.

[56] Yin BW, Lloyd KO. Molecular cloning of the CA125 ovarian cancer antigen: identification as a new mucin, MUC16. *J Biol Chem.* 2001; 276(29): 27371-27375.

[57] Carraway KL, Price-Schiavi SA, Komatsu M, Jepson S, Perez A, Carraway CA. Muc4/sialomucin complex in the mammary gland and breast cancer. *J Mammary Gland Biol Neoplasia* 2001; 6:323–337.

[58] Moniaux N, Nollet S, Porchet N, Degand P, Laine A, Aubert JP. Complete sequence of the human mucin MUC4: a putative cell membrane-associated mucin. *Biochem J* 1999; 338: 325–333.

[59] Singh AP, Moniaux N, Chauhan SC, Meza JL, Batra SK. Inhibition of MUC4 expression suppresses pancreatic tumor cell growth and metastasis. *Cancer Res*. 2004; 64: 622–630.

[60] Yokoyama A, Shi BH, Kawai T, Konishi H, Andoh R, Tachikawa H, Ihara S, Fukui Y. Muc4 is required for activation of ErbB2 in signet ring carcinoma cell lines. *Biochem Biophys Res Commun*. 2007; 355(1): 200-203.

[61] Funes M, Miller JK, Lai C, Carraway KL III, Sweeney C. The mucin Muc4 potentiates neuregulin signaling by increasing the cell-surface populations of ErbB2 and ErbB3. *J Biol Chem* 2006;281:19310–9.

[62] Ramsauer VP, Pino V, Farooq A, Carothers Carraway CA, Salas PJ, Carraway KL Muc4-ErbB2 complex formation and signaling in polarized CACO-2 epithelial cells indicate that Muc4 acts as an unorthodox ligand for ErbB2. *Mol Biol Cell*. 2006; 17(7): 2931-41

[63] Price-Schiavi SA, Jepson S, Li P, Arango M, Rudland PS, Yee L, Carraway KL. Rat Muc4 (sialomucin complex) reduces binding of anti-ErbB2 antibodies to tumor cell surfaces, a potential mechanism for Herceptin resistance. *Int J Cancer* 2002; 99: 783–791.

[64] Nagy P, Friedländer E, Tanner M, Kapanen AI, Carraway KL, Isola J, Jovin TM. Decreased accessibility and lack of activation of ErbB2 in JIMT-1, a Herceptin-resistant, MUC4-expressing breast cancer cell line. *Cancer Res* 2005; 65: 473–482.

[65] Le XF, Lammayot A, David Gold, Yiling Lu, Deepa Deshpande, Teresa Chang, Gordon B. Mills, Bast RC Jr. Genes Affecting Cell Cycle, Growth/Maintenance and Drug Sensitivity are Preferentially Regulated by Anti-HER2 Antibody through PI3K-AKT Signaling. *J Biol Chem*. 2005; 280(3): 2092-2104.

[66] Le XF, McWatters A, Wiener J, Mills GB, Bast RC, Jr. Anti-HER2 antibody and heregulin suppress growth of HER2-overexpressing human breast cancer cells through different mechanisms. *Clin Cancer Res*. 2000; 6: 260-270.

[67] Le XF, Claret FX, Lammayot A, Tian L, Deshpande D, LaPushin R, Tari AM, Bast RC Jr. The role of cyclin-dependent kinase inhibitor p27Kip1 in anti-HER2 antibody-induced G1 cell cycle arrest and tumor growth inhibition. *J Biol Chem*. 2003; 278:23441-23450.

[68] Le XF, Pruefer F, Bast RC Jr. HER2-targeting antibodies modulate the cyclin-dependent kinase inhibitor p27Kip1 via multiple signaling pathways. *Cell Cycle* 2005; 4: 87-95.

[69] Le XF, Bedrosian I, Mao W, Murray M, Lu Z, Keyomarsi K, Lee MH, Zhao J, Bast RC Jr. Anti-HER2 antibody trastuzumab inhibits CDK2-mediated NPAT and histone H4 expression via the PI3K pathway. *Cell Cycle*. 2006; 5(15): 1654-1661.

[70] Le XF, Arachchige-Don AS, W. Mao, MC. Horne, and RC. Bast, Jr. Roles of HER2, JNK, PI3K, and p70S6K pathways in regulation of cyclin G2 in human breast cancer cells. *Mol Cancer Ther*. 2007; 6 (11): 2843-2857.

[71] Casimiro M, Rodriguez O, Pootrakul L, Aventian M, Lushina N, Cromelin C, Ferzli G, Johnson K, Fricke S, Diba F, Kallakury B, Ohanyerenwa C, Chen M, Ostrowski M, Hung MC, Rabbani SA, Datar R, Cote R, Pestell R, Albanese C. ErbB-2 induces the

cyclin D1 gene in prostate epithelial cells in vitro and in vivo. *Cancer Res.* 2007; 67(9): 4364-4372.

[72] Koziczak M, Hynes NE. Cooperation between fibroblast growth factor receptor-4 and ErbB2 in regulation of cyclin D1 translation. *J Biol Chem.* 2004; 279(48): 50004-50011.

[73] Li BS, Ma W, Jaffe H, Zheng Y, Takahashi S, Zhang L, Kulkarni AB, Pant HC. Cyclin-dependent kinase-5 is involved in neuregulin-dependent activation of phosphatidylinositol 3-kinase and Akt activity mediating neuronal survival. *J Biol Chem.* 2003; 278(37): 35702-35709.

[74] Ariazi EA, Kraus RJ, Farrell ML, Jordan VC, Mertz JE. Estrogen-related receptor alpha1 transcriptional activities are regulated in part via the ErbB2/HER2 signaling pathway. *Mol Cancer Res.* 2007; 5(1): 71-85.

[75] Ariazi EA, Clark GM, Mertz JE. Estrogen-related receptor alpha and estrogenrelated receptor g associate with unfavorable and favorable biomarkers, respectively, in human breast cancer. *Cancer Res* 2002; 62: 6510-6518.

[76] Barry JB, Giguere V. Epidermal growth factor induced signaling in breast cancer cells results in selective target gene activation by orphan nuclear receptor estrogen-related receptor alpha. *Cancer Res* 2005; 65:6120–6129.

[77] Sladek R, Bader J-A, Giguère V. The orphan nuclear receptor estrogen-related receptor αis a transcriptional regulator of the human medium-chain acyl coenzyme A dehydrogenase gene. *Mol Cell Biol* 1997; 17: 5400–5409.

[78] Stein RA and McDonnell DP. Estrogen-related receptor alpha as a therapeutic target in cancer. *Endocrine-Related Cancer* 2006; 13: S25–S32.

[79] Adams JM, Cory S. The Bcl-2 apoptotic switch in cancer development and therapy. *Oncogene.* 2007; 26(9): 1324-1337.

[80] Luciano F, Jacquel A, Colosetti P, Herrant M, Cagnol S, Pages G, Auberger P. Phosphorylation of Bim-EL by Erk1/2 on serine 69 promotes its degradation via the proteasome pathway and regulates its proapoptotic function. *Oncogene.* 2003; 22(43): 6785-6793.

[81] Ewings KE, Hadfield-Moorhouse K, Wiggins CM, Wickenden JA, Balmanno K, Gilley R, Degenhardt K, White E, Cook SJ. ERK1/2-dependent phosphorylation of BimEL promotes its rapid dissociation from Mcl-1 and Bcl-xL. *EMBO J.* 2007; 26(12): 2856-2867.

[82] Qi XJ, Wildey GM, Howe PH. Evidence that Ser87 of BimEL is phosphorylated by Akt and regulates BimEL apoptotic function. *J Biol Chem.* 2006; 281(2): 813-823.

[83] Quadros MR, Connelly S, Kari C, Abrams MT, Wickstrom E, Rodeck U. EGFR-dependent downregulation of Bim in epithelial cells requires MAPK and PKC-δ activities. *Cancer Biol Ther* 2006; 5: 498–504.

[84] Reginato MJ, Mills KR, Becker EB, Lynch DK, Bonni A, Muthuswamy SK, Brugge JS. Bim regulation of lumen formation in cultured mammary epithelial acini is targeted by oncogenes. *Mol Cell Biol.* 2005; 25(11): 4591-601.

[85] Piechocki MP, Yoo GH, Dibbley SK, Lonardo F. Breast cancer expressing the activated HER2/neu is sensitive to gefitinib in vitro and in vivo and acquires resistance through a novel point mutation in the HER2/neu. *Cancer Res.* 2007; 67(14): 6825-6843.

[86] Lipscomb EA, Mercurio AM. Mobilization and activation of a signaling competent alpha6beta4integrin underlies its contribution to carcinoma progression. *Cancer Metastasis Rev*. 2005;24(3):413-423.

[87] Campiglio M, Tagliabue E, Srinivas U, Pellegrini R, Martignone S, Ménard S, Colnaghi MI, Lombardi L, Marchisio PC. Colocalization of the p185HER2 oncoprotein and integrin alpha 6 beta 4 in Calu-3 lung carcinoma cells. *J Cell Biochem*. 1994; 55(4): 409-418.

[88] Yoon SO, Shin S, Lipscomb EA. Novel mechanism for integrin-mediated ras activation in breast carcinoma cells: the alpha6beta4 integrin regulates ErbB2 translation and transactivates epidermal growth factor receptor/ErbB2 signaling. *Cancer Res*. 2006; 66(5): 2732-2739.

[89] Guo W, Pylayeva Y, Pepe A, Yoshioka T, Muller WJ, Inghirami G, Giancotti FG. Beta 4 integrin amplifies ErbB2 signaling to promote mammary tumorigenesis. *Cell*. 2006; 126(3): 489-502.

[90] Rafelski, S. M., and Theriot, J. A. Crawling toward a unified model of cell mobility: spatial and temporal regulation of actin dynamics. *Annu. Rev. Biochem*. 2004; 73: 209–239.

[91] Marone R, Hess D, Dankort D, Muller WJ, Hynes NE, Badache A. Memo mediates ErbB2-driven cell motility. *Nat Cell Biol*. 2004; 6(6): 515-522.

[92] Le XF, Vadlamudi R, McWatters A, Bae DS, Mills GB, Kumar R, Bast RC Jr. Differential signaling by an anti-p185(HER2) antibody and heregulin. *Cancer Res*. 2000; 60(13): 3522-3531.

[93] Helfman DM, Kim EJ, Lukanidin E, Grigorian M. The metastasis associated protein S100A4: role in tumour progression and metastasis. *Br J Cancer*. 2005; 92(11): 1955-1958.

[94] Hernan R, Fasheh R, Calabrese C, Frank AJ, Maclean KH, Allard D, Barraclough R, Gilbertson RJ. ERBB2 up-regulates S100A4 and several other prometastatic genes in medulloblastoma. *Cancer Res*. 2003; 63(1): 140-148.

[95] Müller A, Homey B, Soto H, Ge N, Catron D, Buchanan ME, McClanahan T, Murphy E, Yuan W, Wagner SN, Barrera JL, Mohar A, Verástegui E, Zlotnik A. Involvement of chemokine receptors in breast cancer metastasis. *Nature*. 2001; 410(6824): 50-56.

[96] Li YM, Pan Y, Wei Y, Cheng X, Zhou BP, Tan M, Zhou X, Xia W, Hortobagyi GN, Yu D, Hung MC. Upregulation of CXCR4 is essential for HER2-mediated tumor metastasis. *Cancer Cell* 2004; 6, 459–469.

[97] Cabioglu N, Summy J, Miller C, Parikh NU, Sahin AA, Tuzlali S, Pumiglia K, Gallick GE, Price JE. CXCL-12/stromal cell-derived factor-1alpha transactivates HER2-neu in breast cancer cells by a novel pathway involving Src kinase activation. *Cancer Res*. 2005; 65(15): 6493-7.

[98] Hsu MC, Chang HC, Hung WC. HER-2/neu represses the metastasis suppressor RECK via ERK and Sp transcription factors to promote cell invasion. *J Biol Chem*. 2006; 281(8): 4718-25.

[99] Oh J, Takahashi R, Kondo S, Mizoguchi A, Adachi E, Sasahara RM, Nishimura S, Imamura Y, Kitayama H, Alexander DB, Ide C, Horan TP, Arakawa T, Yoshida H, Nishikawa S, Itoh Y, Seiki M, Itohara S, Takahashi C, Noda M. The membrane-anchored MMP inhibitor RECK is a key regulator of extracellular matrix integrity and angiogenesis. *Cell*. 2001; 107(6): 789-800.

[100] Harvey AJ, Crompton MR. The Brk protein tyrosine kinase as a therapeutic target in cancer: opportunities and challenges. *Anticancer Drugs*. 2004; 15(2): 107-11.

[101] Mitchell PJ, Barker K, Martindale JE, Kamalati T, Lowe PN, Page MJ, Gusterson BA, Crompton M: Cloning and characterization of cDNAs encoding a novel non-receptor tyrosine kinase, brk, expressed in human breast tumors. *Oncogene* 1994, 9:2383-2390.

[102] Barker K, Jackson L, Crompton M: BRK tyrosine kinase expression in a high proportion of human breast carcinomas. *Oncogene* 1997, 15:799-805.

[103] Qiu H, Miller WT: Regulation of the nonreceptor tyrosine kinase Brk by auto phosphorylation and by autoinhibition. *Journal of Biological Chemistry* [Abbreviate] 2002, 277(37):34634-34641.

[104] Serfas M, Tyner AL: Brk, Srm, Frk, and Src42A from a distinct family of intracellular Src-like tyrosine kinases. *Oncology Research* 2002, 13:409-419.

[105] Harvey AJ, Crompton M: Use of RNA interference to validate Brk as a novel therapeutic target in breast cancer: Brk promotes breast cancer carcinoma cell proliferation. *Oncogene* 2003, 22:5006-5010.

[106] Chen HY, Shen CH, Tsai YT, Lin FC, Huang YP, Chen RH. Brk activates rac1 and promotes cell migration and invasion by phosphorylating paxillin. *Mol Cell Biol.* 2004; 24(24): 10558-72.

[107] Ostrander JH, Daniel AR, Lofgren K, Kleer CG, Lange CA. Breast tumor kinase (protein tyrosine kinase 6) regulates heregulin-induced activation of ERK5 and p38 MAP kinases in breast cancer cells. *Cancer Res.* 2007; 67(9): 4199-209.

[108] Liu L, Gao Y, Qiu H, Miller WT, Poli V, Reich NC. Identification of STAT3 as a specific substrate of breast tumor kinase. *Oncogene*. 2006; 25(35): 4904-12.

[109] Weaver AM, Silva CM. Signal transducer and activator of transcription 5b (STAT5b): A new target of breast tumor kinase/protein tyrosine kinase 6 (Brk/PTK6). *Breast Cancer Res*. 2007; 9(6):R79

[110] Kamalati T, Jolin HE, Mitchell PJ, Barker KT, Jackson LE, Dean CJ, Page MJ, Gusterson BA, Crompton MR. Brk, a breast tumor-derived non-receptor protein-tyrosine kinase, sensitizes mammary epithelial cells to epidermal growth factor.*J Biol Chem*. 1996; 271(48): 30956-63.

[111] Kamalati T, Jolin H, Fry M, Crompton M: Expression of BRK tyrosine kinase in mammary epithelial cells enhance the coupling of EGF signaling to PI-3 kinase and Akt, via erbB3 phosphorylation. *Oncogene* 2000, 19:5471-5476.

[112] Zhang P, Ostrander JH, Faivre EJ, Olsen A, Fitzsimmons D, Lange CA. Regulated association of protein kinase B/Akt with breast tumor kinase. *J Biol Chem*. 2005; 280(3): 1982-1991.

[113] Aubele M, Auer G, Walch AK, Munro A, Atkinson MJ, Braselmann H, Fornander T, Bartlett JM. PTK (protein tyrosine kinase)-6 and HER2 and 4, but not EGFR and 3 predict long-term survival in breast carcinomas. *Br J Cancer*. 2007;96(5):801-807.

[114] Klos KS, Wyszomierski SL, Sun M, Tan M, Zhou X, Li P, Yang W, Yin G, Hittelman WN, Yu D. ErbB2 increases vascular endothelial growth factor protein synthesis via activation of mammalian target of rapamycin/p70S6K leading to increased angiogenesis and spontaneous metastasis of human breast cancer cells. *Cancer Res*. 2006; 66(4):2028-2037.

[115] Bagheri-Yarmand R, Vadlamudi RK, Wang RA, Mendelsohn J, Kumar R. Vascular Endothelial Growth Factor Up-regulation via p21-activated Kinase-1 Signaling Regulates Heregulin-mediated Angiogenesis. *J Biol Chem* 2000; 275: 39451-39457.

[116] Yen L, You XL, Al Moustafa AE, Batist G, Hynes NE, Mader S, Meloche S, Alaoui-Jamali MA. Heregulin selectively upregulates vascular endothelial growth factor secretion in cancer cells and stimulates angiogenesis. *Oncogene*. 2000; 19(31): 3460-3469.

[117] Laughner E, Taghavi P, Chiles K, Mahon PC, Semenza GL. HER2 (neu) signaling increases the rate of hypoxia-inducible factor 1alpha (HIF-1alphA), synthesis: novel mechanism for HIF-1-mediated vascular endothelial growth factor expression. *Mol Cell Biol*. 2001; 21(12): 3995-4004.

[118] Loureiro RM, Maharaj AS, Dankort D, Muller WJ, D'Amore PA. ErbB2 overexpression in mammary cells upregulates VEGF through the core promoter. *Biochem Biophys Res Commun*. 2005; 326:455-465.

[119] Wen XF, Yang G, Mao W, Thornton A, Liu J, Bast RC Jr, Le XF. HER2 signaling modulates the equilibrium between pro- and antiangiogenic factors via distinct pathways: implications for HER2-targeted antibody therapy. *Oncogene*. 2006; 25(52): 6986-6996.

[120] Yang W, Klos K, Yang Y, Smith TL, Shi D, Yu D. ErbB2 overexpression correlates with increased expression of vascular endothelial growth factors A, C, and D in human breast carcinoma. *Cancer* 2002;94: 2855–2861.

[121] Konecny GE, Meng YG, Untch M, Wang HJ, Bauerfeind I, Epstein M, Stieber P, Vernes JM, Gutierrez J, Hong K, Beryt M, Hepp H, Slamon DJ, Pegram MD. Association between HER-2/neu and vascular endothelial growth factor expression predicts clinical outcome in primary breast cancer patients. *Clin Cancer Res*. 2004; 10(5): 1706-1716.

[122] Klos KS, Wyszomierski SL, Sun M, Tan M, Zhou X, Li P, Yang W, Yin G, Hittelman WN, Yu D. ErbB2 increases vascular endothelial growth factor protein synthesis via activation of mammalian target of rapamycin/p70S6K leading to increased angiogenesis and spontaneous metastasis of human breast cancer cells. *Cancer Res*. 2006; 66(4): 2028-2037.

[123] Têtu B, Brisson J. Prognostic significance of HER-2/neu oncoprotein expression in node-positive breast cancer. The influence of the pattern of immunostaining and adjuvant therapy. *Cancer* 1994; 73(9): 2359-2365.

[124] Tokatli F, Altaner S, Uzal C, Ture M, Kocak Z, Uygun K, Bilgi S. Association of HER-2/neu overexpression with the number of involved axillary lymph nodes in hormone receptor positive breast cancer patients. *Exp Oncol*. 2005; 27(2): 145-149.

[125] Davidson B, Konstantinovsky S, Nielsen S, Dong HP, Berner A, Vyberg M, Reich R. Altered expression of metastasis-associated and regulatory molecules in effusions from breast cancer patients: a novel model for tumor progression. *Clin Cancer Res*. 2004; 10(21): 7335-7346.

[126] Su JL, Shih JY, Yen ML, Jeng YM, Chang CC, Hsieh CY, Wei LH, Yang PC, Kuo ML. Cyclooxygenase-2 induces EP1- and HER-2/Neu-dependent vascular endothelial growth factor-C up-regulation: a novel mechanism of lymphangiogenesis in lung adenocarcinoma. *Cancer Res*. 2004; 64(2): 554-564.

[127] Hsieh CY, Chen CA, Chou CH, Lai KP, Jeng YM, Kuo ML, Wei LH. Overexpression of Her-2/NEU in epithelial ovarian carcinoma induces vascular endothelial growth factor C by activating NF-kappaB: Implications for malignant ascites formation and tumor lymphangiogenesis. *J Biomed Sci.* 2004; 11: 249–259.

In: Oncogene Proteins: Structure, Functions and Analyses ISBN: 978-1-60456-259-0
Editors: P.B. Murphy and J.R. Clarke, pp. 249-268

Chapter X

Aurora Kinases as Therapeutic Targets of Antioncogenic Drugs

Victor M. Bolanos-Garcia[*]
Department of Biochemistry, University of Cambridge;
80 Tennis Court Road CB2 1GA, Cambridge, UK.

Abstract

The human kinome constitutes one of the most promising targets for the development of novel chemotherapeutic drugs. Deregulation of many of the 518 members of the human kinome is associated with inflammation, neurodegenerative diseases and cancer. The several drugs that have reached clinical trials and other compounds, such as imatinib (Gleevec) and gefitinib (Iressa), that are already in clinical use as chemotherapeutic drugs, proof that the inhibition of malfunctioning tyrosine and serine-threonine kinases is a valid approach to the treatment of numerous diseases. Several small molecule inhibitors that selectively target the ATP binding site of Aurora kinases have been developed in recent years. Aurora A function is important for the proper progression through mitosis as it regulates mitotic checkpoint proteins. Aurora B plays an essential role in chromosome segregation and bipolar chromosome orientation and condensation while Aurora C cooperates with Aurora B to regulate mitotic chromosome dynamics in mammalian cells. This work describes the unique structural characteristics of Aurora kinases that have conducted to the development of novel antioncogenic agents. It also discusses the potentiality of dual-specificity inhibitors to interfere with the function of Aurora kinases through the disruption of specific protein-protein interactions.

Role in Homeostasis

Mammalian genomes encode for three Aurora kinases: Aurora A (also known as Aurora, Aurora-2, AIK, AIR-1, AIRK1, AYK1, BTAK, Eg2, MmIAK1 and STK15), Aurora B (also known as Aurora-1, AIM-1, AIK2, AIR-2, AIRK-2, ARK2, IAL-1 and STK12) and Aurora C (also known as AIK3) (Bolanos-Garcia, 2005). Phylogenetic tree analysis suggests that all

[*] E-mail address: victor@cryst.bioc.cam.ac.uk. Tel: 44-0-1223-766029. Fax: 44-0-1223-766002

three vertebrate Aurora kinases evolved from a single urochordate ancestor. Aurora A is an orthologus lineage in cold blooded vertebrates and mammals, while Auroras B and C seem to have evolved more recently in mammals from a duplication of an ancestral *aurora b/c* gene found in cold-blooded vertebrates (Brown et al., 2004). In other metazoans, including *X. laevis*, *D. melanogaster* and *C. elegans*, only Aurora A and B kinases have been identified (Bischoff et al., 1998) while in *S. cerevisiae* and *S. pombe* only one Aurora-like homolog is known (Fig. 1), suggesting that the functions of these kinases have diverged throughout evolution.

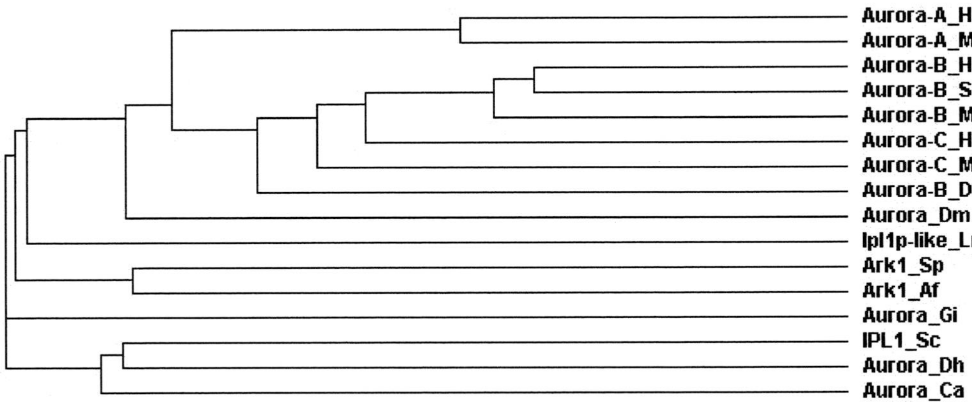

Figure 1. Phylogenetic tree showing the amino acid sequence divergence of the three Aurora A kinases A, B and C. Hs, *Homo sapiens*; Mm, *Mus musculus*; Ss, *Schizosaccharomyces pombe*; Dr, *Dario reno*; Dm, *Drosophila melanogaster*; Lm, *Leishmania major*; Af, *Aspergillus fumigatus*; Gi, *Giardia intestinalis*; Sc, *Saccharomyces cerevisiae*; Dh, *Debaryomyces hansenii*; Ca, *Candida albicans*.

Aurora A

Aurora kinase A localises to the duplicated centrosomes and to the spindle poles in mitosis (Kufer et al., 2003). This kinase has been implicated in centrosome maturation and separation and binds to the protein TPX2 (target protein for xenopus kinesin-like protein 2), which is required for spindle assembly (Tsai et al., 2003). The crystal structure of Aurora A complexed with TPX2 reveals that TPX2 binding induces a conformational change in Aurora A which hides the activating phosphoryl group from attack by protein phosphatase I (PPI) (Bayliss et al., 2003). Aurora kinase A is a key regulatory component of the p53 pathway (Katayama et al., 2004). Other potential substrates that are phosphorylated by activated Aurora A are transforming acidic coiled-coil protein (TACC), activator of histone deacetylase 1 (HDAC-1), centrosomin, and BRCA1, all of which localize to the centrosome.

Aurora A undergoes cell cycle dependent regulation: it is inactivated or degraded when the cell proceeds in G1 phase and its maximal expression occurs in G_2/M phase. At the end of the G_2/M phase Aurora A is degraded by the anaphase-promoting complex (APC) via the ubiquitin-proteosome pathway (Walter et al., 2000). Several methods to abrogate Aurora A activity have confirmed the assigned roles for this kinase: RNA interference (RNAi) delays mitotic entry (Hirota et al., 2003) while anti-Aurora A antibody microinjection into late G_2 cells prevents centriole pair separation at prophase and chromosome mal-alignment in the metaphase plate when injected after centriole separation (Marumoto et al., 2003). Also, studies on rat spermatocytes have shown that the inhibition of Aurora kinase activity perturbs chromosome alignment and spindle checkpoint signalling (Wang et al., 2006b). Although it seems that Aurora A activation depends on cdc2/cyclinB activity, the kinase that phosphorylates Aurora A remains to be identified.

Aurora B

Normal proliferating cells express Aurora B with levels peaking at the G_2/M phase of the cell cycle. Aurora B is a 'chromosomal passenger protein' that forms a quaternary complex with the inner centromere protein (INCENP) (Bolton et al., 2002), survivin (Honda et al., 2003) and borealin (Gassmann et al., 2004), all of which show functional interdependence and are required to localize the kinase to centromeres and the mitotic spindle. This complex first localizes along the chromosome during prophase, becomes concentrated at the inner centromere region during prometaphase and leaves the chromosome and localizes to the central spindle in anaphase. At the beginning of prophase, Aurora B is decorated along the length of the chromosomes, at metaphase it becomes concentrated at the inner centromere, at anaphase migrates to the central region of the mitotic spindle and finally in telophase to the site of the contractile ring (Meraldi et al., 2002). These features reveal that Aurora B is an important regulator of kinetochore function and is required for correct chromosomal alignment and segregation as well as cytokinetic functions and spindle checkpoint (Dewar et al., 2004). Although Aurora B plays a major role in the kinetochore assembly pathway in mammalian cells, only one Aurora B target, centromere protein-A, has been identified to date.

Additional substrates of Aurora B include histone H3 (a structural component of chromatin) (Hsu et al., 2000), topoisomerase IIα (a chromosomal scaffolding protein) (Morrison et al., 2002) and vimentin, which localizes to the cleavage furrow (Goto et al., 2003). During M-phase, Aurora B phosphorylates histone H3 on Serine 10 and this is thought to be important for chromosomal condensation and sister chromatid cohesion (Giet and Glover, 2001). An inactive Aurora B mutant causes multiple defects in the mitotic machinery, including the loss of kinetochore attachment to microtubules and the exit from mitosis without anaphase or cytokinesis (Murata-Hori and Wang, 2002). Similarly, abrogation of Aurora B by gene knockdown or RNAi seems to disrupt the multi-protein chromosomal passenger complex, hence affecting the normal localization and function of binding partners. However, normal cells treated with a small molecule inhibitor that affects only the catalytic function of Aurora B show mal-oriented chromosomes at anaphase, a defective spindle checkpoint and the inhibition of cytokinesis (Ditchfield et al., 2003). Aurora B is not degraded by the same ubiquitin ligase that is responsible of Aurora A proteasome-dependent degradation. Instead, Aurora B undergoes degradation by binding to the human proteasome α subunit C8 (HC8) (Shu et al., 2003).

Aurora C

Very little is known on the function, regulation and the mechanism(s) involved in Aurora C degradation. This kinase appears to be universally expressed among normal human tissues

(Yan et al., 2005). Aurora C expression is maximally elevated during the G_2/M phase but its expression profile in synchronized cells reveals differential temporal regulation through the cell cycle. In diploid human fibroblasts, Aurora C is expressed at approximately an order of magnitude lower than Aurora B. However, Aurora C expression is highest in the reproductive system (sperm and oocytes) in normal tissue (Tseng et al., 1998). Similar to Aurora A and Aurora B, Aurora C levels are elevated in several human cancer cell types. Aurora C, like Aurora B, interacts with the inner centromere protein (INCENP) at the C-terminal end spanning the conserved IN box domain (Li et al., 2004) although compared with Aurora B, Aurora C has a lower binding affinity to INCENP. Aurora C is able to rescue the Aurora B silenced multinucleation phenotype, demonstrating that the function of Aurora C overlaps with and complements that of Aurora B in mitosis (Sasai et al., 2004). The fact that an Aurora C kinase-dead mutant induces multinucleation in a dose-dependent manner and that siRNA mediated silencing of this kinase gives rise to multinucleated cells confirms these features. Taken together, these observations show that Aurora C is a chromosomal passenger protein localizing first to centromeres and then to the midzone of mitotic cells that cooperates with Aurora B to regulate mitotic chromosome segregation and cytokinesis in mammalian cells.

Association with Diseases

All three aurora kinase genes are localized to known poor prognosis chromosomal amplicons that are frequently amplified in human malignancies (reviewed in Matthews et al., 2006). For instance, the *aurora A* gene is located within the chromosome locus 20q13, which is a locus frequently detected in human breast tumours (Lengauer et al., 1998; Bar-Shira et al., 2002). Many studies involving a wide range of techniques, ranging from nude mice, enzymatic assay and immunofluorescent staining in cultured cells to RNAi and comparative genomic hybridization, have shown that Aurora A is over-expressed in a variety of human epithelial carcinomas (pancreatic, ovarian, gastric, breast, colorectal), lymphomas (B- and T-cell non-Hodgkin's lymphoma) and leukemias (acute myeloid leukemia) (Sen et al., 1997; Tanaka et al., 1999; Sakakura et al., 2001; Li et al., 2003; Dove, 2003; Hamada et al., 2003). Fibroblasts transformed with Aurora A from *Xenopus laevis* provided further evidence of the oncogenic role(s) of this kinase as it leads to the development of tumours in mice (Littlepage et al, 2002). Over-expression appears to correlate with severity of the histologic grade and may also be related to the stage of the disease (Goepfert et al., 2002; Keen and Taylor, 2004).

Aurora A overexpression contributes to genetic instability and tumorigenesis by disrupting the proper assembly of the mitotic checkpoint complex at the level of the Cdc20-BubR1 interaction (Zhou et al., 1998; Ke et al., 2003). Its overexpression (either as an active or inactive kinase) causes resistance to apoptosis induced by taxol in human cancer cell lines (Anand et al., 2003), leads to polyploidy and increases p53 degradation, all of which facilitate oncogenic transformation (Katayama et al., 2004). Besides, *aurora A* is a candidate low-penetrance tumour-susceptibility gene due to a T-to-A polymorphism leading to an amino acid change (F31I) that has been associated with aneuploidy in colon and esophageal cancer (Ewart-Toland et al., 2003; Miao et al., 2004). Very importantly, other genetic defects are likely to cooperate with Aurora A to initiate the malignant process because Aurora A over-expression potentiates the oncogenic activity of H-Ras by affecting cell growth (Tatsuka et al., 2005).

Similar to *aurora A*, *aurora B* is a proto-oncogene in the appropriate genetic background; its overexpression is associated with aneuploidy, higher tumour grade and anaplastic features indicative of enhanced tumour proliferation. In cells that over-express Aurora A and B, defects in p53 appear to play a critical role in stabilizing polyploidy. Over-expression of Aurora B in CHE diploid fibroblasts carrying the G245S mutation in both p53 alleles, overrides the p53-dependent G1-checkpoint. These cells when injected into nude mice develop metastases with an associated aneuploid phenotype and increased histone H3 Serine 10 phosphorylation. Further, p53 (G245S) non-transfected cells also form tumours but these do not metastasize (Ota et al., 2002). An extensive systematic analysis of Aurora A, B and C expression levels utilizing mRNA in multiple primary tumour samples of different stages and types (breast, bladder, colon, endometrium, esophagus, liver, lung, ovary, pancreas, prostate, rectum, skin, stomach, and thyroid) demonstrates that Aurora A and B expression levels increase or decrease in parallel and that Aurora C expression is detected in tumour cell lines (Sasai et al., 2004) and patient biopsy material (Takahashi et a., 2000).

Absence or over-expression of Aurora A and B lead to a tetraploid phenotype. However, the mechanism and lethality of tetraploidy are different. Because the absence of Aurora A leads to spindle defects (monopolar and monoastral bipolar spindles) with formation of tetraploid cells arrested in G_2/M by the post-mitotic checkpoint and consequent apoptosis, the lack of Aurora A is not well tolerated by cells. Absence of Aurora B leads to overriding of the mitotic checkpoint with inhibition of cytokinesis. This leads to a tetraploid state with two centrosomes and avoidance of apoptosis. Hence, a lack of Aurora B is better tolerated by cells. On the contrary, over-expression of Aurora B (active and inactive kinase) shows that enzyme inactivation results in disruption of protein-protein interactions.

Structure Features

The size of human Aurora A, Aurora B and Aurora C kinases varies from 309 to 403 amino acid residues. Although the overall sequence identity between human and rodent Aurora A, Aurora B and Aurora C is 82, 84 and 78%, respectively, the three kinases exhibit a relatively high sequence divergence between species. As shown in Fig. 2, Aurora A, Aurora B and Aurora C present a similar domain organisation: a flexible N-terminal domain of 39-129 residues in length, a protein kinase domain and a short C-terminal domain of 15-20 residues. The N-terminal domain of Auroras A, B and C shares low sequence conservation, which determines selectivity for protein-protein interactions (Carmena and Earnshaw, 2003). The alignment of Aurora A and B allows the identification of one distantly conserved 'KEN' motif, spanning 11-18 residues. The 'KEN' motif acts as a Cdh1-dependant anaphase-promoting complex (APC) recognition signal. The N-terminal region of Aurora C also contains the so-called PEST-like motif. The mutation of this motif significantly abrogates Aurora C kinase activity (Chen and Tang, 2002). The C-terminal domain of human Aurora B shares 53% and 73% sequence similarity to human Auroras A and C, respectively. The catalytic domain of the three Aurora kinases is more conserved (Fig. 3). Indeed, the region encoding the catalytic domain of Aurora A, Aurora B and Aurora C exhibits the canonical bilobar protein kinase architecture. The hinge region connects the amino terminal β-sheet domain and the C-terminal domain. The Mg-ATP and substrate binding sites reside between the two domains in an essentially hydrophobic active site pocket. However, the adenosine moiety of ATP binds to the main chain hinge region via hydrogen bond formation that is common to all protein kinases. The C-terminal domain constitutes the docking site for substrates and is implicated in phosphate transfer. The activation loop (T-loop) originating from the C-terminal domain has different conformations depending on the state of the enzyme (i.e., active or inactive).

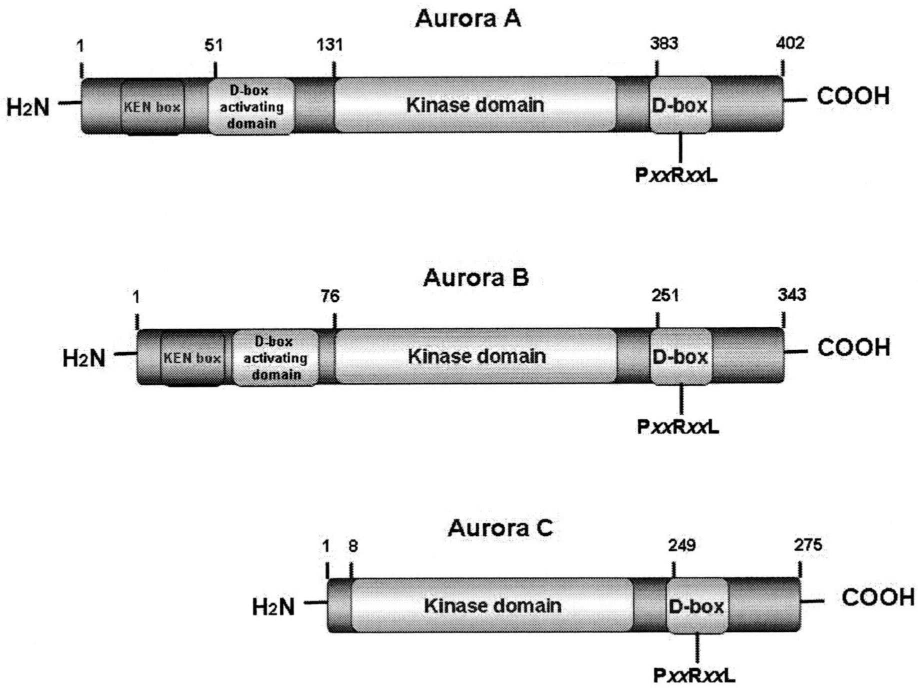

Figure 2. Domain organisation of Aurora kinases A, B and C.

Two different degradation motifs mediate human Aurora A degradation via the anaphase promoting complex/cyclosome (APC/C)-ubiquitin-proteasome pathway in a process that depends on Cdh1 but not on Cdc20 (Taguchi et al., 2002). The first motif corresponds to an N-terminal, D-Box-activating motif (RxLxPS). This motif confers functionality to a second motif, a D-Box, consisting of the sequence RxxLxxG. The D-box is located at the C-terminus of the kinase domain and is the target of Fizzy-related proteins (Castro et al., 2002). Several structures of Aurora kinases solved at atomic resolution have provided insights into how this family of enzymes is regulated. Such detailed structure information has proved to be essential for the development of novel small molecular inhibitors of dual-action and/or specificity, which should also exhibit good resident times and biological half-lives for effective cancer therapy. It can be anticipated that the structure-based approach will help to identify novel compounds that overcome drug resistance if mutations were to arise in the catalytic residues (e.g., gatekeeper residues). Furthermore, the structure-based strategy might be conducted in a similar fashion to that used for the targeting of tyrosine kinases BCR-Abl (chronic myeloid leukaemia) or c-Kit (gastrointestinal stromal tumours) both of which involve the treatment with imatinib.

```
Aurora-A_Hs       YFHDATRVYLILEYAPLGTVYRELQKL--SKFDEQRTATYITELANALSYCHSKRVIHRD 186
Aurora-A_Mm       YFHDATRVYLILEYAPLGTVYRELQKL--SKFDEQRTATYITELANALSYCHSKRVIHRD 177
Aurora-B_Hs       YFYDRRRIYLILEYAPRGELYKELQKS--CTFDEQRTATIMEELADALMYCHGKKVIHRD 130
Aurora-B_Mm       YFYDQQRIYLILEYAPRGELYKELQKS--RTFDEQRTATIMEELSDALTYCHKKKVIHRD 205
Aurora-C_Hs       YFHDARRVYLILEYAPRGELYKELQKS--EKLDEQRTATIIEELADALTYCHDKKVIHRD 96
Aurora-C_Mm       YFYDDTRIYLILEYAPGGELYKELQRH--QKLDQQRTATIIQELSDALTYCHEKKVIHRD 69
Aurora_Dm         YFHDDVRIYLILEYAPQGTLFNALQAQPMKRFDERQSATYIQALCSALLYLHERDIIHRD 289
IPL1_Sc           YFHDEKRVYLLMEYLVNGEMYKLLRLH--GPFNDILASDYIYQIANALDYMHKKNIIHRD 227
Aurora_Dh         FFYDHKNVYLILEYSIHGELYHHLKVQ--RRFNDATASHYIYQVALALDYLHTKHIIHRD 271
Ark1_Sp           HFHDEKRIYLILEFAGRGELYQHLRRA--KRFSEEVASKYIFQMANALSYLHKKHVIHRD 212
Aurora-B_Dr       YFHDDTRVFLILEYAPRGEMYKELQRY--GRFDDQRTATYMEEVSDALQYCHEKKVIHRD 176
Aurora-B_Ss       YFYDRRRIYLILEYAPRXELYKELQKC--RTFDEQRTATIMEELADALIYCHGKKVIHRD 200
Ipl1p-like_Lm     YFFDEHDIYLILEPCSNGMLYSELNRV--KLFPPPTAARYVAQLAEALLYLHQHHILHRD 154
Ark1_Af           HFHDSKRIFLILEFAGRGELYKHLRKE--HRFPEWKAAQYIAQMAAALKYLHKKHVMHRD 243
                  .*.*   ::*::*   #:# :#.*.       :    ::  :  :. ** * * : ::***

Aurora-A_Hs       IKPENLLLGSAGELKIADFGWSVHA-PSS--RRTTLCGTLDYLPPEMIEGRM----HDEK 239
Aurora-A_Mm       IKPENLLLGSNGELKIADFGWSVHA-PSS--RRTTMCGTLDYLPPEMIEGRM----HDEK 230
Aurora-B_Hs       IKPENLLLGLKGELKIADFGWSVHA-PSL--RRKTMCGTLDYLPPEMIEGRM----HNEK 183
Aurora-B_Mm       IKPENLLLGLQGELKIADFGWSVHA-PSL--RRKTMCGTLDYLPPEMIEGRM----HNEM 258
Aurora-C_Hs       IKPENLLLGFRGEVKIADFGWSVHT-PSL--RRKTMCGTLDYLPPEMIEGRT----YDEK 149
Aurora-C_Mm       IKPENLLLGLNGEVKISDFGWSVHT-PSL--RRKTMCGTLDYLPPEMIAQKP----YNEM 122
Aurora_Dm         IKPENLLLGHKGVLKIADFGWSVHE-PNS--MRMTLCGTVDYLPPEMVQGKP----HTKN 342
IPL1_Sc           IKPENILIGFNNVIKLTDFGWSIINPPEN--RRKTVCGTIDYLSPEMVESRE----YDHT 281
Aurora_Dh         IKPENILLSTDNCIKLSDFGWSVKSLPSSSTKRLTICGTLDYLPPEMIESNE----HDYT 327
Ark1_Sp           IKPENILLGIDGEIKLSDFGWSVHA-PSN--RRTTLCGTLDYLPPEMVEGKE----HTEK 265
Aurora-B_Dr       IKPENLLLGYRGELKIADFGWSVHA-PSL--RRRTMCGTLDYLPPEMIEGHS----HDEK 229
Aurora-B_Ss       IKPENLLLGLQGELKIADFGWSVHA-PSL--RRKTMCGTLDYLPPEMIEGRT----HNEK 253
Ipl1p-like_Lm     IKPENILLDHHQNIKLADFGWSVHD-PLN--RRKTSCGTPEYFPPEIVSRQM----YDMS 207
Ark1_Af           IKPENILVGIHGEIKISDFGWSVHA-PNN--RRQTMCGTLDYLPPEMLKPNPQDNYYSEK 300
                  *****:*:.     :*::*****:   *     * * *** :*:.**::   .     :

Aurora-A_Hs       VDLWSLGVLCYEFLVGKPPFEANT-YQETYKRISRVEFTFPDFVTEGARDLISRLLKHNP 298
Aurora-A_Mm       VDLWSLGVLCYEFLVGMPPFEAHT-YQETYRRISRVEFTFPDFVTEGARDLISRLLKHNA 289
Aurora-B_Hs       VDLWCIGVLCYELLVGNPPFESAS-HNETYRRIVKVDLKFPASVPTGAQDLISKLLRHNP 242
Aurora-B_Mm       VDLWCIGVLCYELMVGNPPFESPS-HSETYRRIVKVDLKFPSSVPSGAQDLISKLLKHNP 317
Aurora-C_Hs       VDLWCIGVLCYELLVGYPPFESAS-HSETYRRILKVDVRFPLSMPLGARDLISRLLRYQP 208
Aurora-C_Mm       VDLWCIGVLCYELLVGKPPFESST-SSETYRRIRQVDFKFPSSVPAGAQDLISKLLRYHP 181
Aurora_Dm         VDLWSLGVLCFELLVGHAPFYSKN-YDETYKKILKVDYKLPEHISKAASHLISKLLVLNP 401
IPL1_Sc           IDAWALGVLAFELLTGAPPFEEEM-KDTTYKRIAALDIKMPSNISQDAQDLILKLLKYDP 340
Aurora_Dh         VDIWSLGILCYEFLVGKPPFEEID-KNSTYKRIAKVDLKIPSFLSSEATDLILRLLQKSP 386
Ark1_Sp           VDLWSLGVLTYEFLVGAPPFEDMSGHSATYKRIAKVDLKIPSFVPPDARDLISRLLQHNP 325
Aurora-B_Dr       VDLWSIGVLCYECLVGNPPFETAS-HAETYKRITKVDLQFPKLVSEGARDLISKLLRHSP 288
Aurora-B_Ss       VDLWCIGVLCYELLVGNPPFESAS-HNETYRRIGKVDLKFPPSVPAGAQDLISKLLKHNP 312
Ipl1p-like_Lm     ADLWCLGIFCFELLVGHTPFVSKD-NDQIYKKIHAMQYTIPDSVPPEAKDLISNLLIREG 266
Ark1_Af           VDLWSLGVLTYEFLVGEAPFEDTP--VMTQRRIARADMTVPSFVSPEAKDLIKRLLVLDP 358
                   * *.:*:: :* :.* .**            ::*   :  .*  :.  * .** .**
```

Figure 3. Amino acid sequence conservation of the catalytic domain of Aurora kinases A, B and C. The grey boxes highlight some of the conserved residues that define the active site. The symbol (#) indicates residues that are not conserved among Auroras A, B and C.

Aurora A

Several crystal structures of Aurora A have been solved to date, including the C-terminal catalytic domain of Aurora A bound to ADP, that bound to the small molecule inhibitor PHA-680632 and that of Aurora A bound to a short peptide fragment of TPX2 (Figure 4). For Aurora A bound to adenosine (Figure 4A), the T-loop folds into the active site producing the hydrophobic pocket. In contrast, the crystal structure of Aurora A in complex with the small molecule inhibitor PHA-680632 (Fancelli et al., 2005) (pdb: 2BMC) shows that the T-loop is disordered but the drug occupies the so-called fluorophenyl pocket (Figure 4B), which was originally identified in mitogen-activated protein kinases (MAPK) (Wang et al., 1998).

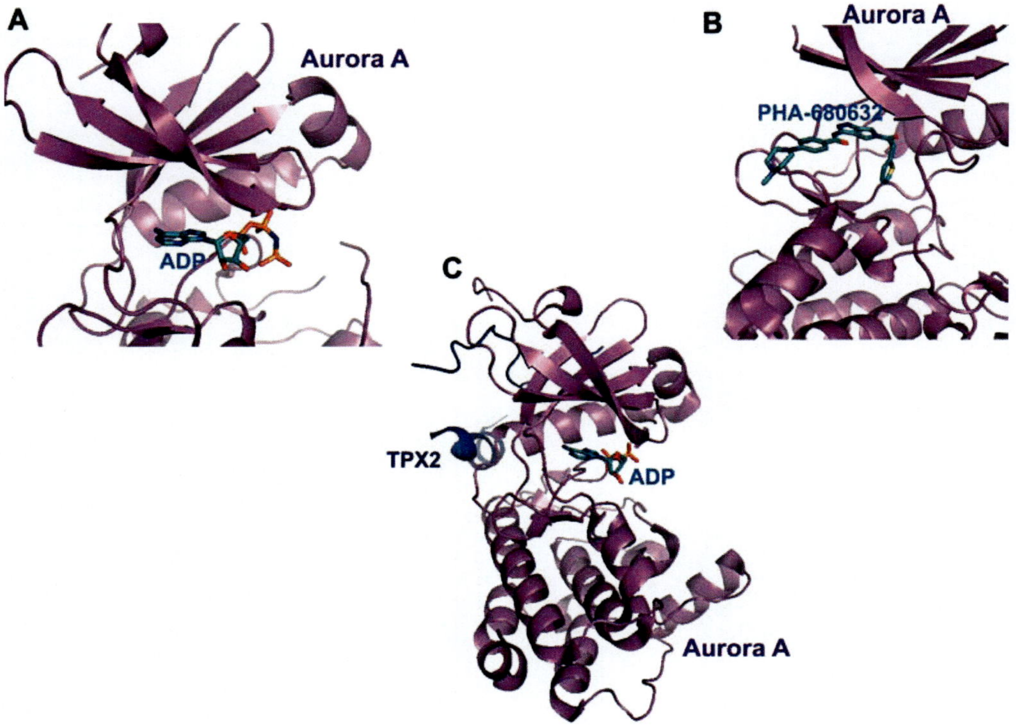

Figure 4. Aurora A kinase is a target of small molecule inhibitors. A) Overall view of the kinase domain in complex with ADP; B) Zoom-in showing the T-loop region and the binding of the inhibitor PHA-680632. C) Crystal structure of the complex formed between Aurora A and TPX2.

The crystal structure of the Aurora A-TPX2 complex (Bayliss et al., 2003) (PDB: 1OL5) shows that the N-terminal segment of TPX2 binds near the amino terminal domain (Figure 4C) at a similar location to where cyclins bind CDKs (Bayliss et al., 2004). However, TPX-2 binding molds an extended active T-loop conformation which requires phosphorylation on Threonine 287 and Threonine 288 in the Aurora consensus target sequence [(R/K)X(T/S)(I/L/V)]. In the absence of the TPX2 peptide, the phospho-T288 is exposed to the solvent while in the TPX2-bound state leads to a buried position which in turn locks the active conformation.

Aurora B

Crystal structure of Aurora B bound to INCENP (Sessa et al., 2005) sheds light into the activation of Aurora B by this protein. Full activation requires phosphorylation of a conserved Thr-Ser-Ser (TSS) motif near the C-terminal of INCENP (Honda et al., 2003). In the Aurora B-IN-box (INCENP) complex the latter forms a molecular 'crown' extending across the N-terminal and the C-terminal domain of Aurora B (PDB: 2BFY) (Figure 5A). This complex represents an intermediate state of activation in comparison to the fully active state show by the Aurora A-TPX2 complex (Bayliss et al., 2003). The binding results in the loss of an ion-pair interaction between residues K122 and E141, which in the active state form a buried ion-pair that orients the α and β phosphates of ATP for phosphotransfer (Huse and Kuriyan, 2002). Also, the catalytic cleft of Aurora B between the two domains opens by approximately 15° with respect to free Aurora A (Figure 5B). The structure of the Aurora B-INCENP-hesperadin ternary complex compared to the binary complex shows very little conformational change except that in this complex the binding of hesperadin results in a much more ordered glycine-rich loop. Hesperadin slots into the open catalytic pocket of the enzyme where the indolinone moiety binds to the main chain carbonyl and amide of E171 and A173 respectively.

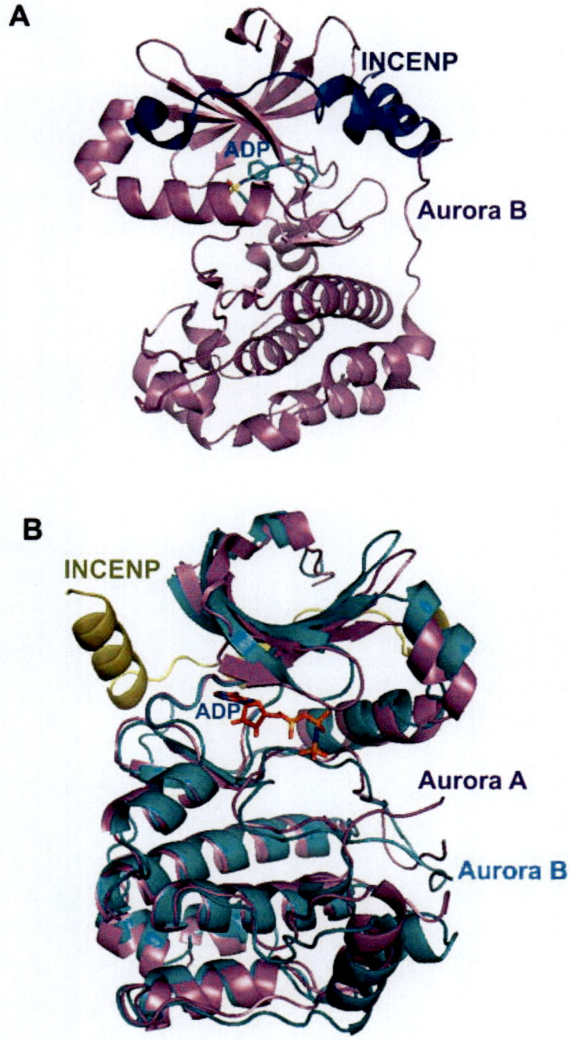

Figure 5. Crystal structure of Aurora B. A) Crystal structure of the ternary complex Aurora B-INCENP-ADP (PDB: 2BFY); B) Superposition of Aurora A and Aurora B reveals an overall little conformation difference between the two structures.

Novel Drugs and Biotherapeutics

Several bio-pharmaceutical groups have developed a series of novel and potent ATP-site competitive small molecule inhibitors of Aurora A and Aurora B that function as anti-proliferative agents (Table 2). The specificity, efficacy and potency of these novel compounds has been carried out in cell-based assays, mouse xenograft models of cancer and *in vitro* kinase assays (Mortlock et al., 2005; Gautschi et al., 2006; Warner et al., 2006a). These series of test have shown that the disruption of the enzymatic activity of Aurora kinases with small molecule inhibitors that selectively target the ATP binding site seems to be effective to proceeds toward the clinic. The first small molecule inhibitors of Aurora kinases described in the literature were hesperadin (Boehringer Ingelheim, Hauf et al., 2003), VX-680 (Vertex, Harrington et al., 2004) and ZM4347439 (AstraZeneca, Ditchfield et al., 2003). These compounds act as inhibitors of cell division and they have no effect on cell cycle progression. The end result is a second round of endoreduplication that leads to tetraploidy and subsequent apoptosis or arrest in pseudo-G_1 dependent on the status of the p53 post mitotic checkpoint (Margolis et al., 2003).

Table 1. Normal sub-cell localisation and diseases associated with the over-expression of Aurora kinases.

Protein	Localisation	Associated disease	References
Aurora A	Mitotic spindle; centrosome	Epithelial carcinoma (pancreatic, ovarian, gastric, colorectal, breast); non-Hodgkin's lymphoma; acute myeloid leukemia	Gritsko et al., 2003; Kurai et al., 2005; Wang et al., 2006a; Chen et al., 2007.
Aurora B	Centrosome; central spindles; chromosome arms	Lung, thyroid, prostate carcinoma cell proliferation; endomitosis	Geddis and Kaushansky, 2004; Sorrentino et al., 2005; Chieffi et al., 2006; Hayama et al., 2007
Aurora C	Central spindles; chromosome arms?	Human thyroid carcinoma cell lines	Ulisse et al., 2006

VX-680 (Vertex Pharmaceuticals) is a 4,6-di-amino pyrimidine that potently inhibits tumour cell proliferation in a panel of tumour cell lines: breast, cervical, colorectal, leukemia, melanoma, pancreatic and prostate. VX-680 is a potent inhibitor of all three Aurora kinases, with apparent inhibition constant ($Ki_{(app)}$) values of 0.6, 18 and 4.6 nM for Aurora A, B and C, respectively (Harrington et al., 2004). VX-680 shows greater selectivity for the Aurora A over many other kinases and has recently entered phase II clinical trials (Table 2). Furthermore VX-680 potently inhibits Abl and the Imatinib resistant mutant T315I. The crystal structure of VX-680 bound to Aurora-A shows that this inhibitor exploits a centrally located hydrophobic pocket in the active site that is only present in an inactive or "closed" kinase conformation. A tight association of VX-680 with this hydrophobic pocket explains its high affinity for the Aurora kinases and also provides an explanation for its selectivity profile (Cheetham et al., 2007).

Table 2. Physicochemical properties of small molecule inhibitors of Aurora kinases currently in pre-clinical and clinical trials. Octanol-water partition coefficient (Log $P_{o/w}$) is used in rational drug design as a measure of molecular hydrophobicity. Hydrophobicity affects drug absorption, bioavailability, hydrophobic drug-receptor interactions, metabolism of molecules, as well as their toxicity. Topological polar surface area (tTPSA) is used to predict drug transport properties as it correlates well with human intestinal absorption. tPSA provides a fast calculation of molecular polar surface area as a sum of fragment based contributions (usually oxygen, nitrogen and attached hydrogen atoms) in the molecule (Ertl et al., 2000).

Compound	Selectivity	MW (Da)	TPSA	Log $P_{o/w}$	Clinical status
Hesperadin	Aurora A, Aurora B	516.6	94.3	3.75	Unknown
MLN8054	Aurora A >> Aurora B	476.8	87.4	5.27	Phase I
VX-680	Aurora A, Aurora B	464.5	102.1	3.33	Phase II
Pyrimido-oxazepine	Aurora A, Aurora B	468.2	92.94	3.91	Unknown
ZM4347439	Aurora A, Aurora B	513.6	97.8	4.27	Unknown
AZD1152	Aurora A, Aurora B	576.6	121.2	4.47	Phase I
PHA-680632	Aurora A > Aurora B	513.7	96.6	2.99	Phase II
MP235	Aurora A > Aurora B	605.6	150.5	3.45	Phase I
JNJ-7706621	Aurora A, Aurora B	394.3	146.0	0.47	Phase I

However, the vast majority of early Aurora kinase inhibitors proved to be non-selective, dual action anti-tumour compounds that inhibited Aurora A but also Aurora B and Aurora C. For instance, the compound ZM4347439 (which was developed by AstraZeneca), inhibits Aurora A and Aurora B kinase activity (Girdler et al., 2006). ZM4347439 induces growth inhibition, causes accumulation of cells with 4N/8N DNA content, and mediates apoptosis of human leukemia cells (Ikezoe et al., 2007). Although VX-680 is relatively specific, it can inhibit other kinases such as Flt-3 (30nM), Fyn (52nM), ITK (22nM), Lck (80nM) and Src (35nM) (Harrington et al., 2004). VX-680 is effective in abolishing colony formation in primary leukemia cells refractory to standard chemotherapy and cells possessing Flt-3 internal tandem duplications. It also inhibits cell cycle progression, induces apoptosis and inhibits tumour growth in xenograft models of leukemia, colon and pancreatic cancers. In all of these cases, VX-680 treatment results in tumour regression. Neutropenia appears to be the main toxic effect (in rats and not mice) but it can be reversed when the drug is discontinued. Aurora B inhibition has no effect on non-dividing cells but in dividing cells exposure to hesperadin or ZM4347439 leads to several checkpoint proteins (BUB1 and BUBR1) not localizing correctly to kinetochores. Interestingly, Aurora kinase small molecule inhibitors do not totally inhibit checkpoint signalling since treated cells expose to microtubule toxins that prevent microtubule assembly (nocodazole) undergo mitotic arrest for several hours.

A second generation of compounds are being developed that target these kinases more specifically (Girdler et al., 2006). Second generation compounds includes ZM3 and MLN8054 (Millennium Pharmaceuticals, Inc., Cambridge, MA 02139) (Girdler et al, 2006; Hoar et al., 2007). ZM3 induces a monopolar phenotype and shows a potency that is more than twenty times higher than that of ZM4347439 (Girdler et al., 2006). MLN8054 possesses a unique scaffold comprised of a benzazepine core with a fused amino pyrimidine ring and an aryl carboxylic acid (Manfredi et al., 2007). At least in tumour cultured cells and xenografts,

MLN8054 acts as an ATP-competitive, reversible inhibitor with approximately 40-fold higher selectivity for Aurora A compared to Aurora B. MLN8054 possesses broad anti-tumour activity and effectively inhibits the growth of tumour cells from diverse tissues (Hoar et al., 2007). Human tumour cells treated with this compound show a high incidence of abnormal mitotic spindles, often with unseparated centrosomes. Consistent with the chromosome segregation defects, cell treatment with MLN8054 results in the development of aneuploidy over time. Anti-tumour activity of MLN8054 in nude mice shows a dose-dependent tumour growth inhibition for colon and prostate cancer after 21 days of treatment. The drug is well tolerated at 30 mg/kg but at this dose there is a significant decrease in body weight of 7.5% with a 73% decline in neutrophil count which can be reversed at termination of therapy (Manfredi et al., 2007). MLN8054 has now entered Phase I trials for advanced solid tumours. In colon cancer cells, MLN8054 in the concentration range 0.25 to 1μM suppresses Aurora A activity while at the higher concentration of 4μM also inhibits Aurora B activity. These findings suggest that Aurora A inhibition kills tumour cells through the development of deleterious aneuploidy (Hoar et al., 2007). Phase I clinical trials to evaluate the potential of MLN8054 as an anticancer drug for advanced solid tumours are underway (Hoar et al., 2007).

AstraZeneca has also produced AZD1152 (Jung et al., 2006) and one of its analogue compounds, ZM2 (Girdler et al., 2006). AZD1152 potently suppresses the phosphorylation of Histone H3 (pHisH3) in tumour cells and in tumour xenografts in mice (Jung et al., 2006) while ZM2 is more than 100 times selective for Aurora B over Aurora A and induces an identical anti-proliferative phenotype. The selective targeting of Aurora B by AZD1152 has been probed in human tumour xenograft models. This inhibitor is currently in phase I trials (Wilkinson et al., 2007).

PHA-680632 (Pfizer-Nerviano Medical Sciences) is an inhibitor that consists of the 1,4,5,6-tetrahydropyrrolo[3,4-c] pyrazole scaffold (Fancelli et al., 2005). PHA-680632 exhibits an IC_{50} of 27, 135 and 120 nM for Aurora A, B and C, respectively (Soncini et al., 2006) and shows minimal cross reactivity to other kinases. The immunohistochemical analysis of tumour cells for histone H3 phosphorylation as a specific biomarker correlates with the anti-tumour response of PHA-680632 (Soncini et al., 2006). The compound showed activity against a wide range of cancer cell lines and had anti-tumour activity in mouse xenograft models of leukemia, ovarian cancer, colon cancer and mammary carcinoma.

PHA-739358 is another Aurora kinase inhibitor currently in clinical development for solid and hematologic malignancies. This compound inhibits in vitro the kinase activity of wild-type Abl and of several mutants, including T315I. The T315I substitution is recognised as a major cause of resistance to these drugs in patients with chronic myelogenous leukemia. The cocrystal structure of PHA-739358-T315I Abl kinase domain shows that the inhibitor associates with an active conformation of the kinase domain in the ATP-binding pocket and lacks the steric hindrance imposed by the substitution of Threonine by Isoleucine (Modugno et al., 2007).

JNJ-7706621 (Johnson & Johnson) is a dual cell cycle inhibitor with activity against Aurora kinases A and B (IC_{50} of 11 and 15nM, respectively) (Emanuel et al., 2005) and cyclin-dependent kinases CDK1, 2, 3 and 6 (IC_{50} of 3-175nM). JNJ-7706621 is selective in inhibiting proliferation of cervical, colon, ovarian, prostate, melanoma, breast and uterine cancer cell lines. The inhibitory activity seems to be irrespective of p53, retinoblastoma and MDR-1/P-glycoprotein status and is approximately 10 times less effective in inhibiting normal cell growth at least in lung fibroblasts, smooth muscle and endothelial cells.

Treatment of a human melanoma mouse tumour xenograft model with JNJ-7706621 indicates that there is a direct correlation between total cumulative dose and anti-tumour efficacy irrespective of dosing schedules.

A pyrimido-oxazepine has been used as a template (ImClone Systems) for the selective inhibition of Aurora and Flt-3 kinases (Pan et al., 2005). A different series of Aurora kinase inhibitors of quinazoline and pyrimidine-based tricyclic scaffolds have been designed utilizing homology modelling, *in silico* fragment search and interactive docking (Warner et al., 2006b). The lead compound MP235 (4-(6,7-dimethoxy-9H-1,3,9-triaza-fluoren-4-yl)-piperazine-1-carbothioic acid [4-9pyrimidin-2-ylsulfamoyl)-phenyl]-amide) (Supergen Pharmaceuticals) demonstrates selectivity for Aurora kinases with a IC_{50} 94nM and led to cell cycle arrest in the G_2/M phase consistent with Aurora B inhibition (Warner et al., 2006b). On the other hand, recombinant enzyme assays on a series of pyrazoloquinazolines have shown greater than 1000-fold selectivity for Aurora B over Aurora A kinase activity. The pyrazoloquinazoline compound AZD1152 is currently in phase 1 clinical trials (Mortlock et al., 2007).

An indirubin analogue, 7-bromoindirubin-3'-oxime (7BIO) inhibits Aurora kinases A, B and C with an IC_{50} of 100, 4.6 and 0.7 μM, respectively. Interestingly, theoretical calculations on 7BIO and various active or inactive indirubin analogues suggest that a single residue in the ATP binding pocket is sufficient to rationalize the activity differences between Auroras A, B, and C (Myrianthopoulos et al., 2007)

The validity for targeting Aurora A in anticancer therapy has been questioned because the major phenotype with pan-Aurora kinase small molecule inhibitors is consistent with Aurora B inhibition (Wang et al., 2006b; Keen and Taylor, 2004). This observation correlates with *S. cerevisiae* Aurora Ipl1, whose function is to resolve syntelic chromosomal orientations in yeast (Tanaka et al., 2002). Taken together, these observations suggest that the development of novel dual-specificity inhibitors of both Aurora kinases might be more appropriate for the effective treatment of human cancer malignancies. Even though the targeting of protein-protein interactions with small molecules remains a challenge (Keen and Taylor, 2004), the alternative strategy of inhibiting Aurora A function through the disruption of the interaction with its interacting partners is worth to explore further.

Conclusion

Aurora kinases A, B and C are frequently overexpressed in many human tumors, including colorectal, breast, pancreatic, and other cancers. Aurora kinases overexpression is not definitely linked with carcinogenesis, and the mechanisms implicating Aurora activity with tumorigenesis remain to be elucidated. Nevertheless, these enzymes constitute promising targets for the development of chemotherapeutic drugs. Indeed, several selective Aurora kinase inhibitors have emerged in the literature, either as drug candidates or as biochemical tools. Examples include AZD1152, Vertex VX-680, Hesperadin, ZM4347439, JNJ-7706621 and MLN8054. Since cancer cells over-express both active Aurora A and B in an abnormal genetic background, it is likely that these kinases collaborate to produce a complex polyploid state that leads to transformation and metastasis. Therefore, the simultaneous inhibition of Aurora A and B would be a feasible and rational approach to cancer therapy.

There are several important aspects that remain to be established, including the functional inter-relationships of human Aurora kinases in initiating and maintaining malignancy, the evaluation of the effect of small molecule inhibitors on different types of tumours most likely to respond to p53 status, kinase selectivity profile and rate of tumour proliferation. It can be envisaged that the identification of genetic defects that contribute to make Aurora A and B oncogenic will shed light on potential mechanisms of resistance and also on how small molecule inhibitors kill tumours.

References

Anand S, Penrhyn-Lowe S, Venkitaraman AR. AURORA-A amplification overrides the mitotic spindle assembly checkpoint, inducing resistance to Taxol. *Cancer Cell* 2003, 3: 51-62.

Bar-Shira A, Pinthus JH, Rozovsky U, Goldstein M, Sellers WR, Yaron Y, Eshhar Z, Orr-Urtreger A. Multiple genes in human 20q13 chromosomal region are involved in an advanced prostate cancer xenograft. *Cancer Res*. 2002, 62: 6803-7.

Bayliss R, Sardon T, Ebert J, Lindner D, Vernos I, Conti E. Determinants for Aurora-A activation and Aurora-B discrimination by TPX2. *Cell Cycle* 2004, 3: 404-7.

Bayliss R, Sardon T, Vernos I, Conti E. Structural basis of Aurora-A activation by TPX2 at the mitotic spindle. *Mol. Cell* 2003, 12: 851-62.

Bischoff JR, Anderson L, Zhu Y, Mossie K, Ng L, Souza B, Schryver B, Flanagan P, Clairvoyant F, Ginther C, Chan CS, Novotny M, Slamon DJ, Plowman GD. A homologue of *Drosophila* aurora kinase is oncogenic and amplified in human colorectal cancers. *EMBO J.* 1998, 17: 3052-65.

Bolanos-Garcia VM. Aurora kinases. *Int. J. Biochem. Cell Biol.* 2005, 37: 1572-7.

Bolton MA, Lan W, Powers SE, McCleland ML, Kuang J, Stukenberg PT. Aurora B kinase exists in a complex with survivin and INCENP and its kinase activity is stimulated by survivin binding and phosphorylation. *Mol. Biol. Cell.* 2002, 13: 3064-77.

Brown JR, Koretke KK, Birkeland ML, Sanseau P, Patrick DR. Evolutionary relationships of Aurora kinases: implications for model organism studies and the development of anti-cancer drugs. *BMC Evol. Biol.* 2004, 4: 39.

Carmena M, Earnshaw WC. The cellular geography of aurora kinases. *Nat. Rev. Mol. Cell Biol.* 2003, 4: 842-54.

Castro A, Arlot-Bonnemains Y, Vigneron S, Labbe JC, Prigent C, Lorca T. APC/Fizzy-related targets Aurora A kinase for proteolysis. *EMBO Reports* 2002, 3: 457-62.

Cheetham GM, Charlton PA, Golec JM, Pollard JR. Structural basis for potent inhibition of the Aurora kinases and a T315I multi-drug resistant mutant form of Abl kinase by VX-680. *Cancer Lett.* 2007, 251: 323-9.

Chen J, Li D, Wei C, Sen S, Killary AM, Amos CI, Evans DB, Abbruzzese JL, Frazier ML. Aurora-A and p16 polymorphisms contribute to an earlier age at diagnosis of pancreatic cancer in Caucasians. *Clin. Cancer Res.* 2007, 13: 3100-4.

Chen SH, Tang TK. Mutational analysis of the phosphorylation sites of the Aie1 (Aurora-C) kinase in vitro. *DNA Cell Biol.* 2002, 21: 41-6.

Chieffi P, Cozzolino L, Kisslinger A, Libertini S, Staibano S, Mansueto G, De Rosa G, Villacci A, Vitale M, Linardopoulos S, Portella G, Tramontano D. Aurora B expression directly correlates with prostate cancer malignancy and influence prostate cell proliferation. *Prostate* 2006, 66: 326-33.

Dewar H, Tanaka K, Nasmyth K, Tanaka TU. Tension between two kinetochores suffices for their bi-orientation on the mitotic spindle. *Nature* 2004, 428: 93-7.

Ditchfield C, Johnson VL, Tighe A, Ellston R, Haworth C, Johnson T, Mortlock A, Keen N, Taylor SS. Aurora B couples chromosome alignment with anaphase by targeting BubR1, Mad2, and Cenp-E to kinetochores. *J. Cell Biol.* 2003, 161: 267-80.

Dove W. Aurora and the hunt for cancer-modifying genes. *Nat. Genet.* 2003, 34: 353-4.

Emanuel S, Rugg CA, Gruninger RH, Lin R, Fuentes-Pesquera A, Connolly PJ, Wetter SK, Hollister B, Kruger WW, Napier C, Jolliffe L, Middleton SA. The in vitro and in vivo effects of JNJ-7706621: a dual inhibitor of cyclin-dependent kinases and aurora kinases. *Cancer Res.* 2005, 65: 9038-46.

Ertl, P, Rohde B, Selzer P. Fast calculation of molecular polar surface area as a sum of fragment based contributions and its application to the prediction of drug transport properties. *J. Med. Chem.* 2000, 43: 3714-7.

Ewart-Toland A, Briassouli P, de Koning JP, Mao JH, Yuan J, Chan F, MacCarthy-Morrogh L, Ponder BA, Nagase H, Burn J, Ball S, Almeida M, Linardopoulos S, Balmain A. Identification of Stk6/STK15 as a candidate low-penetrance tumor-susceptibility gene in mouse and human. *Nat. Genet.* 2003, 34: 403-12.

Fancelli D, Berta D, Bindi S, Cameron A, Cappella P, Carpinelli P, Catana C, Forte B, Giordano P, Giorgini ML, Mantegani S, Marsiglio A, Meroni M, Moll J, Pittala V, Roletto F, Severino D, Soncini C, Storici P, Tonani R, Varasi M, Vulpetti A, Vianello P. Potent and selective Aurora inhibitors identified by the expansion of a novel scaffold for protein kinase inhibition. *J. Med. Chem.* 2005, 48: 3080-4.

Gassmann R, Carvalho A, Henzing AJ, Ruchaud S, Hudson DF, Honda R, Nigg EA, Gerloff DL, Earnshaw WC. Borealin: a novel chromosomal passenger required for stability of the bipolar mitotic spindle. *J. Cell Biol.* 2004, 166: 179-91.

Gautschi O, Mack PC, Davies AM, Lara PN Jr, Gandara DR. Aurora kinase inhibitors: a new class of targeted drugs in cancer. *Clin. Lung Cancer* 2006, 8: 93-8.

Geddis AE, Kaushansky K. Megakaryocytes express functional Aurora-B kinase in endomitosis. *Blood* 2004, 104: 1017-24.

Giet R, Glover DM. Drosophila aurora B kinase is required for histone H3 phosphorylation and condensin recruitment during chromosome condensation and to organize the central spindle during cytokinesis. *J Cell Biol.* 2001, 152: 669-82.

Girdler F, Gascoigne KE, Eyers PA, Hartmuth S, Crafter C, Foote KM, Keen NJ, Taylor SS. Validating Aurora B as an anti-cancer drug target. *J. Cell Sci.* 2006, 119: 3664-75.

Goepfert TM, Adigun YE, Zhong L, Gay J, Medina D, Brinkley WR. Centrosome amplification and overexpression of aurora A are early events in rat mammary carcinogenesis. *Cancer Res.* 2002, 62: 4115-22.

Goto H, Yasui Y, Kawajiri A, Nigg EA, Terada Y, Tatsuka M, Nagata K, Inagaki M. Aurora-B regulates the cleavage furrow-specific vimentin phosphorylation in the cytokinetic process. *J. Biol. Chem.* 2003, 278: 8526-30.

Gritsko TM, Coppola D, Paciga JE, Yang L, Sun M, Shelley SA, Fiorica JV, Nicosia SV, Cheng JQ. Activation and overexpression of centrosome kinase BTAK/Aurora-A in human ovarian cancer. *Clin. Cancer Res.* 2003, 9: 1420-6.

Hamada M, Yakushijin Y, Ohtsuka M, Kakimoto M, Yasukawa M, Fujita S. Aurora2/BTAK/STK15 is involved in cell cycle checkpoint and cell survival of aggressive non-Hodgkin's lymphoma. *Br. J. Haematol.* 2003, 121: 439-47.

Harrington EA, Bebbington D, Moore J, Rasmussen RK, Ajose-Adeogun AO, Nakayama T, Graham JA, Demur C, Hercend T, Diu-Hercend A, Su M, Golec JM, Miller KM.VX-680, a potent and selective small molecule inhibitor of the Aurora kinases, suppresses tumor growth *in vivo*. *Nature Med.* 2004, 10: 262-7.

Hauf S, Cole RW, LaTerra S, Zimmer C, Schnapp G, Walter R, Heckel A, van Meel J, Rieder CL, Peters JM. The small molecule Hesperadin reveals a role for Aurora B in correcting kinetochore microtubule attachment and in maintaining the spindle assembly checkpoint. *J. Cell Biol.* 2003, 161: 281-94.

Hayama S, Daigo Y, Yamabuki T, Hirata D, Kato T, Miyamoto M, Ito T, Tsuchiya E, Kondo S, Nakamura Y. Phosphorylation and Activation of Cell Division Cycle Associated 8 by Aurora Kinase B Plays a Significant Role in Human Lung Carcinogenesis. *Cancer Res.* 2007, 67: 4113-22.

Hirota T, Kunitoku N, Sasayama T, Marumoto T, Zhang D, Nitta M, Hatakeyama K, Saya H. Aurora A and an interacting activator, the LIM protein Ajuba, are required for mitotic commitment in human cells. *Cell* 2003, 114: 585-98.

Hoar K, Chakravarty A, Rabino C, Wysong D, Bowman D, Roy N, Ecsedy JA. MLN8054, a Small-Molecule Inhibitor of Aurora A, Causes Spindle Pole and Chromosome Congression Defects Leading to Aneuploidy. *Mol. Cell Biol.* 2007, 27: 4513-25.

Honda R, Korner R, Nigg EA. Exploring the functional interactions between Aurora B, INCENP, and survivin in mitosis. *Mol. Biol. Cell.* 2003, 14: 3325-41.

Hsu JY, Sun ZW, Li X, Reuben M, Tatchell K, Bishop DK, Grushcow JM, Brame CJ, Caldwell JA, Hunt DF, Lin R, Smith MM, Allis CD. Mitotic phosphorylation of histone H3 is governed by Ipl1/aurora kinase and Glc7/PP1 phosphatase in budding yeast and nematodes. *Cell* 2000, 102: 279-91.

Huse M, Kuriyan J. The conformational plasticity of protein kinases. *Cell* 2002, 109: 275-82.

Ikezoe T, Yang J, Nishioka C, Tasaka T, Taniguchi A, Kuwayama Y, Komatsu N, Bandobashi K, Togitani K, Koeffler HP, Taguchi H. A novel treatment strategy targeting Aurora kinases in acute myelogenous leukemia. *Mol. Cancer Ther.* 2007, 6: 1851-7.

Jung FH, Pasquet G, Lambert-van der Brempt C, Lohmann JJ, Warin N, Renaud F, Germain H, De Savi C, Roberts N, Johnson T, Dousson C, Hill GB, Mortlock AA, Heron N, Wilkinson RW, Wedge SR, Heaton SP, Odedra R, Keen NJ, Green S, Brown E, Thompson K, Brightwell S. Discovery of novel and potent thiazoloquinazolines as selective Aurora A and B kinase inhibitors. *J. Med. Chem.* 2006, 49: 955-70.

Katayama H, Sasai K, Kawai H, Yuan ZM, Bondaruk J, Suzuki F, Fujii S, Arlinghaus RB, Czerniak BA, Sen S. Phosphorylation by aurora kinase A induces Mdm2-mediated destabilization and inhibition of p53. *Nature Genet.* 2004, 36: 55-62.

Ke YW, Dou Z, Zhang J, Yao XB. Function and regulation of Aurora/Ipl1p kinase family in cell division. *Cell Res.* 2003, 13: 69-81.

Keen N, Taylor S. Aurora-kinase inhibitors as anticancer agents. *Nat. Rev. Cancer* 2004, 4: 927-36.

Kufer TA, Nigg EA, Sillje HH. Regulation of Aurora-A kinase on the mitotic spindle. *Chromosoma* 2003, 112: 159-63.

Kurai M, Shiozawa T, Shih HC, Miyamoto T, Feng YZ, Kashima H, Suzuki A, Konishi I. Expression of Aurora kinases A and B in normal, hyperplastic, and malignant human endometrium: Aurora B as a predictor for poor prognosis in endometrial carcinoma. *Hum Pathol*. 2005, 36: 1281-8.

Lengauer C, Kinzler KW, Vogelstein B. Genetic instabilities in human cancers. *Nature* 1998, 396: 643-9.

Li D, Zhu J, Firozi PF, Abbruzzese JL, Evans DB, Cleary K, Friess H, Sen S. Overexpression of oncogenic STK15/BTAK/Aurora A kinase in human pancreatic cancer. *Clin. Cancer Res*. 2003, 9: 991-7.

Li X, Sakashita G, Matsuzaki H, Sugimoto K, Kimura K, Hanaoka F, Taniguchi H, Furukawa K, Urano T. Direct association with inner centromere protein (INCENP) activates the novel chromosomal passenger protein, Aurora-C. *J. Biol. Chem*. 2004, 279: 47201-11.

Littlepage LE, Wu H, Andresson T, Deanehan JK, Amundadottir LT, Ruderman JV. Identification of phosphorylated residues that affect the activity of the mitotic kinase Aurora-A. *Proc. Natl. Acad. Sci. USA* 2002, 99: 15440-5.

Manfredi MG, Ecsedy JA, Meetze KA, Balani SK, Burenkova O, Chen W, Galvin KM, Hoar KM, Huck JJ, LeRoy PJ, Ray ET, Sells TB, Stringer B, Stroud SG, Vos TJ, Weatherhead GS, Wysong DR, Zhang M, Bolen JB, Claiborne CF. Antitumor activity of MLN8054, an orally active small-molecule inhibitor of Aurora A kinase. *Proc. Natl. Acad. Sci. USA* 2007, 104: 4106-11.

Margolis RL, Lohez OD, Andreassen PR. G1 tetraploidy checkpoint and the suppression of tumorigenesis. *J. Cell Biochem*. 2003, 88: 673-83.

Marumoto T, Honda S, Hara T, Nitta M, Hirota T, Kohmura E, Saya H. Aurora-A kinase maintains the fidelity of early and late mitotic events in HeLa Cells. *J. Biol. Chem.* 2003, 278: 51786-95.

Matthews N, Visintin C, Hartzoulakis B, Jarvis A, Selwood DL. Aurora A and B kinases as targets for cancer: will they be selective for tumors? *Expert Rev. Anticancer. Ther.* 2006, 6: 109-20.

Meraldi P, Honda R, Nigg EA. Aurora-A overexpression reveals tetraploidization as a major route to centrosome amplification in p53-/- cells. *EMBO J.* 2002, 21: 483-92.

Miao X, Sun T, Wang Y, Zhang X, Tan W, Lin D. Functional STK15 Phe31Ile polymorphism is associated with the occurrence and advanced disease status of esophageal squamous cell carcinoma. *Cancer Res*. 2004, 64: 2680-3.

Modugno M, Casale E, Soncini C, Rosettani P, Colombo R, Lupi R, Rusconi L, Fancelli D, Carpinelli P, Cameron AD, Isacchi A, Moll J. Crystal Structure of the T315I Abl Mutant in Complex with the Aurora Kinases Inhibitor PHA-739358. *Cancer Res*. 2007, 67: 7987-90.

Morrison C, Henzing AJ, Jensen ON, Osheroff N, Dodson H, Kandels-Lewis SE, Adams RR, Earnshaw WC. Proteomic analysis of human metaphase chromosomes reveals topoisomerase II alpha as an Aurora B substrate. *Nucleic Acids Res*. 2002, 30: 5318-27.

Mortlock AA, Boyle FT, Green S. AZD1152, a selective inhibitor of Aurora B kinase, inhibits human tumor xenograft growth by inducing apoptosis. *Clin. Cancer Res*. 2007, 13: 3682-8.

Mortlock AA, Foote KM, Heron NM, Jung FH, Pasquet G, Lohmann JJ, Warin N, Renaud F, De Savi C, Roberts NJ, Johnson T, Dousson CB, Hill GB, Perkins D, Hatter G, Wilkinson RW, Wedge SR, Heaton SP, Odedra R, Keen NJ, Crafter C, Brown E,

Thompson K, Brightwell S, Khatri L, Brady MC, Kearney S, McKillop D, Rhead S, Parry T, Green S. Discovery, synthesis, and in vivo activity of a new class of pyrazoloquinazolines as selective inhibitors of aurora B kinase. *J. Med. Chem.* 2007, 50: 2213-24.

Mortlock AA, Keen NJ, Jung FH, Heron NM, Foote KM, Wilkinson RW, Green S. Progress in the development of selective inhibitors of Aurora kinases. *Curr. Top. Med. Chem.* 2005, 5: 807-21.

Murata-Hori M, Wang YL. The kinase activity of aurora B is required for kinetochore-microtubule interactions during mitosis. *Curr. Biol.* 2002, 12: 894-9.

Myrianthopoulos V, Magiatis P, Ferandin Y, Skaltsounis AL, Meijer L, Mikros E. An integrated computational approach to the phenomenon of potent and selective inhibition of aurora kinases B and C by a series of 7-substituted indirubins. *J. Med. Chem.* 2007, 50: 4027-37.

Ota T, Suto S, Katayama H, Han ZB, Suzuki F, Maeda M, Tanino M, Terada Y, Tatsuka M. Increased mitotic phosphorylation of histone H3 attributable to AIM-1/Aurora-B overexpression contributes to chromosome number instability. *Cancer Res.* 2002, 62: 5168-77.

Pan W, Liu H, Xu YJ, Chen X, Kim KH, Milligan DL, Columbus J, Hadari YR, Kussie P, Wong WC, Labelle M. Pyrimido-oxazepine as a versatile template for the development of inhibitors of specific kinases. *Bioorg. Med. Chem. Lett.* 2005, 15: 5474-7.

Sakakura C, Hagiwara A, Yasuoka R, Fujita Y, Nakanishi M, Masuda K, Shimomura K, Nakamura Y, Inazawa J, Abe T, Yamagishi H. Tumour-amplified kinase BTAK is amplified and overexpressed in gastric cancers with possible involvement in aneuploid formation. *Br. J. Cancer* 2001, 84: 824-31.

Sasai K, Katayama H, Stenoien DL, Fujii S, Honda R, Kimura M, Okano Y, Tatsuka M, Suzuki F, Nigg EA, Earnshaw WC, Brinkley WR, Sen S. Aurora-C kinase is a novel chromosomal passenger protein that can complement Aurora-B kinase function in mitotic cells. *Cell Motil. Cytoskeleton*. 2004, 59: 249-63.

Sen S, Zhou H, White RA. A putative serine/threonine kinase encoding gene BTAK on chromosome 20q13 is amplified and overexpressed in human breast cancer cell lines. *Oncogene* 1997, 14: 2195-200.

Sessa F, Mapelli M, Ciferri C, Tarricone C, Areces LB, Schneider TR, Stukenberg PT, Musacchio A. Mechanism of Aurora B activation by INCENP and inhibition by hesperadin. *Mol. Cell* 2005, 18: 379-91.

Shu F, Guo S, Dang Y, Qi M, Zhou G, Guo Z, Zhang Y, Wu C, Zhao S, Yu L. Human aurora-B binds to a proteasome alpha-subunit HC8 and undergoes degradation in a proteasome-dependent manner. *Mol. Cell Biochem.* 2003, 254: 157-62.

Soncini C, Carpinelli P, Gianellini L, Fancelli D, Vianello P, Rusconi L, Storici P, Zugnoni P, Pesenti E, Croci V, Ceruti R, Giorgini ML, Cappella P, Ballinari D, Sola F, Varasi M, Bravo R, Moll J. PHA-680632, a novel Aurora kinase inhibitor with potent antitumoral activity. *Clin. Cancer Res.* 2006, 12: 4080-9.

Sorrentino R, Libertini S, Pallante PL, Troncone G, Palombini L, Bavetsias V, Spalletti-Cernia D, Laccetti P, Linardopoulos S, Chieffi P, Fusco A, Portella G. Aurora B overexpression associates with the thyroid carcinoma undifferentiated phenotype and is required for thyroid carcinoma cell proliferation. J. *Clin. Endocrinol. Metab.* 2005, 90: 928-35.

Taguchi S, Honda K, Sugiura K, Yamaguchi A, Furukawa K, Urano T. Degradation of human Aurora A protein kinase is mediated by hCdh1. *FEBS Letters* 2002, 519: 59-65.

Takahashi T, Futamura M, Yoshimi N, Sano J, Katada M, Takagi Y, Kimura M, Yoshioka T, Okano Y, Saji S. Centrosomal kinases, HsAIRK1 and HsAIRK3, are overexpressed in primary colorectal cancers. *Jpn. J. Cancer Res.* 2000, 91: 1007-14.

Tanaka T, Kimura M, Matsunaga K, Fukada D, Mori H, Okano Y. Centrosomal kinase AIK1 is overexpressed in invasive ductal carcinoma of the breast. *Cancer Res.* 1999, 59: 2041-4.

Tanaka TU, Rachidi N, Janke C, Pereira G, Galova M, Schiebel E, Stark MJ, Nasmyth K. Evidence that the Ipl1-Sli15 (Aurora kinase-INCENP) complex promotes chromosome bi-orientation by altering kinetochore-spindle pole connections. *Cell* 2002, 108: 317-29.

Tatsuka M, Sato S, Kitajima S, Suto S, Kawai H, Miyauchi M, Ogawa I, Maeda M, Ota T, Takata T. Over-expression of Aurora A potentiates HRAS-mediated oncogenic transformation and is implicated in oral carcinogenesis. *Oncogene* 2005, 24: 1122-7.

Tsai MY, Wiese C, Cao K, Martin O, Donovan P, Ruderman J, Prigent C, Zheng Y. A Ran signaling pathway mediated by the mitotic kinase Aurora A in spindle assembly. *Nat. Cell Biol.* 2003, 5: 242-8.

Tseng TC, Chen SH, Hsu YP, Tang TK. Protein kinase profile of sperm and eggs: cloning and characterization of two novel testis-specific protein kinases (AIE1, AIE2) related to yeast and fly chromosome segregation regulators. *DNA Cell Biol.* 1998, 17: 823-33.

Ulisse S, Delcros J-G, Baldini E. Toller M, Curcio F, Giacomelli L, Prigent C, Ambesi-Impiombato FS, D'Armiento M, Arlot-Bonnemains Y. Expression of Aurora kinases in human thyroid carcinoma cell lines and tissues. *Int. J. Cancer* 2006, 119: 275-82.

Walter AO, Seghezzi W, Korver W, Sheung J, Lees E. The mitotic serine/threonine kinase Aurora2/AIK is regulated by phosphorylation and degradation. *Oncogene* 2000, 19: 4906-16.

Wang X, Zhou YX, Qiao W, Tominaga Y, Ouchi M, Ouchi T, Deng CX. Overexpression of aurora kinase A in mouse mammary epithelium induces genetic instability preceding mammary tumor formation. *Oncogene* 2006a, 25: 7148-58.

Wang Y, Toppari J, Parvinen M, Kallio MJ. Inhibition of Aurora kinases perturbs chromosome alignment and spindle checkpoint signaling in rat spermatocytes. *Exp. Cell Res.* 2006b, 312: 3459-70.

Wang Z, Canagarajah BJ, Boehm JC, Kassisa S, Cobb MH, Young PR, Abdel-Meguid S, Adams JL, Goldsmith EJ. Structural basis of inhibitor selectivity in MAP kinases. *Structure* 1998, 6: 1117-28.

Warner SL, Bashyam S, Vankayalapati H, Bearss DJ, Han H, Mahadevan D, Von Hoff DD, Hurley LH. Identification of a lead small-molecule inhibitor of the Aurora kinases using a structure-assisted, fragment-based approach. *Mol. Cancer Ther.* 2006a, 5: 1764-73. Erratum in: *Mol. Cancer Ther.* 2006, 5: 3312.

Warner SL, Munoz RM, Stafford P, Koller E, Hurley LH, Von Hoff DD, Han H. Comparing Aurora A and Aurora B as molecular targets for growth inhibition of pancreatic cancer cells. *Mol. Cancer Ther.* 2006b, 5: 2450-8.

Wilkinson RW, Odedra R, Heaton SP, Wedge SR, Keen NJ, Crafter C, Foster JR, Brady MC, Bigley A, Brown E, Byth KF, Barrass NC, Mundt KE, Foote KM, Heron NM, Jung FH, Yan X, Cao L, Li Q, Wu Y, Zhang H, Saiyin H, Liu X, Zhang X, Shi Q, Yu L. Aurora C is directly associated with Survivin and required for cytokinesis. *Genes Cells* 2005, 10: 617-26.

Zhou H, Kuang J, Zhong L, Kuo WL, Gray JW, Sahin A, Brinkley BR, Sen S. Tumour amplified kinase STK15/BTAK induces centrosome amplification, aneuploidy and transformation. *Nat. Genetics* 1998, 20: 189-193.

In: Oncogene Proteins: Structure, Functions and Analyses ISBN: 978-1-60456-259-0
Editors: P.B. Murphy and J.R. Clarke, pp. 269-290

Chapter XI

Diabetes Increases Risk for Oral Carcinogenesis by Induction of Cell Proliferation and Reduction of Cell Adhesion. An Animal Model Study

Christos Yapijakis and Eleftherios Vairaktaris*
Department of Oral and Maxillofacial Surgery, University of Athens Medical School, Attikon Hospital, Athens, Greece

Abstract

Oral cancer is the seventh most common malignancy, worldwide. It constitutes a major health problem because it is still associated with poor prognosis and about 50% five-year survival, despite advances in surgery, radiation and chemotherapy treatment over the past two decades. Oral squamous cell carcinoma (OSCC) accounts for more than 90% of oral cancer cases. In the development of OSCC have been implicated environmental factors such as tobacco smoking and heavy alcohol consumption, infectious agents such as some types of human papilloma virus, genetic alterations involving oncogenes and tumor suppressor genes, as well as epigenetic changes. Recently, some epidemiological studies have also incriminated diabetes mellitus as a risk factor for the development of oral premalignant lesions and OSCC.

In order to investigate the possible association of diabetes with increased risk for oral cancer, as well as the underlying molecular mechanism of this association an experimental model of OSCC induced in diabetic and normal Sprauge–Dawley rats was established. Diabetes type I was induced by a single intraperitoneal injection of streptozotocin in 20 rats and verified by measuring glucose levels in blood samples obtained by tail prick in a strip-operated sensor. In 14 of the diabetic rats, as well as in 12 rats without diabetes, chemical carcinogenesis was induced by a five-month application of the carcinogen 4-nitroquinoline N-oxide on the hard palate of the animals. Six diabetic rats as well as 6 non-diabetic rats were used as controls of normal oral tissue without carcinogenesis. The four groups of animals were used in order to study the putative effect of diabetes on signal transduction pathways involved in the

* E-mail address: lvairakt@med.uoa.gr. Tel: +30 210 6443035, Fax: +30 210 6443803, Correspondence to: Professor Dr. Dr. Eleftherios Vairaktaris, MD, DDS, PhD, PhD, Department of Oral and Maxillofacial Surgery, University of Athens Medical School, Kifissias and Perikleous Stavrou 9, Athens 11524, Greece.

sequential stages of oral carcinogenesis: normal mucosa, hyperplasia, dysplasia, early invasion, well differentiated OSCC and moderately differentiated OSCC.
The diabetic rat model was used to investigate whether the molecular basis for the putative association of diabetes with oral oncogenesis involves the insulin receptor and its substrate-1 (IR and IRS-1, respectively), as well as the influence of diabetes on signal transduction pathways involving oncogene proteins EGFR, erbB2, erbB3, FGFR-2, FGFR-3, FAK, c-myc, N-ras, ets-1, H-ras, c-fos and c-jun, tumor suppressor gene proteins p53 and p16, apoptosis-related markers Bax and Bcl-2, and cell proliferation-related marker Ki-67. The obtained findings suggest that the insulin and insulin-like growth factor-1/IR/IRS-1 signalling pathway is altered by diabetes, leading to reduced cell adhesion and possibly increased risk for oral cancer. In addition, diabetes seems to promote the activation of Ras/Raf/MAPK signal transduction pathway, mainly by induction of receptors erbB2 and erbB3, resulting in increased cell proliferation. Finally, diabetes does not seem to influence apoptotic mechanisms during oral carcinogenesis. An interesting extrapolation of these findings is that the type I diabetic patients who are not insulin-regulated may have a greater risk for developing oral cancer than the general population does. This hypothesis remains to be examined by large epidemiological studies.

Keywords: oral cancer, diabetes, animal model, oncogenes, tumor suppressor genes, cell proliferation, apoptosis, signal transduction pathways, stages of carcinogenesis

1. Diabetes as a Risk Factor for Oral Cancer

As a common chronic and systemic disease, diabetes mellitus has been correlated with a variety of oral conditions as early as the 19th century. These include periodontal disease, which has been reported as the sixth most common complication of diabetes, dental caries, tooth loss, stomatitis, glossitis and sensory changes [1-3]. Oral manifestations of diabetes are so common that dentists are recognized as being among the first health care professionals to whom a previously undiagnosed diabetic patient may present [4,5]. The involvement of diabetes mellitus in the pathogenesis of oral cancer emerged only recently [6,7]. Several independent epidemiological studies have incriminated diabetes as a risk factor for the development of oral premalignant conditions, such as leukoplakia [8-11]. In addition, a Hungarian group has reported epidemiological findings suggesting that diabetes is a risk factor for the development of oral benign tumors and carcinomas [9,12]. Such an association of diabetes with risk for malignancy has already been reported. A number of studies have found an increased incidence of diabetes mellitus in patients with bladder carcinoma [13,14].

Oral cancer is the seventh most common malignancy worldwide and the eighth most common in the European Union [15-17]. It is characterized by severe morbidity and <50% five-year survival, despite advances in treatment with surgery, radiation and chemotherapy in the last two decades [17-19]. More than 90% of oral malignancies represent oral squamous cell carcinoma (OSCC), with 300,000 new cases being diagnosed each year worldwide [15-19]. Environmental factors such as tobacco smoking and heavy alcohol consumption, infectious agents such as some types of human papilloma virus, genetic alterations involving oncogenes and tumor suppressor genes, as well as epigenetic changes have been implicated in the development of OSCC [19,20]. In addition, diabetes recently emerged as a risk factor for OSCC, as already mentioned [21]. Oral oncogenesis is a multi-step process involving progression through several histological stages until the development of carcinomas [19,20]. The deciphering of the molecular mechanisms involved in the initiation and progression to malignancy (including the effect of diabetes) is expected to improve prophylactic measures as well as treatment procedures [19]. In addition, the screening identification of individuals at high risk for oral cancer is expected to offer the opportunity to change their behaviors and optimize the implementation of preventive and therapeutic measures [22].

2. Molecular Mechanisms of Oncogenesis

Unlimited cell proliferation and division may occur due to mutational activation, deregulation or amplification of oncogenes or mutational inactivation, deregulation or loss of tumor suppressor genes [23]. These events may influence key points of signal transduction cascades, which regulate cell proliferation, resulting in cancer progression [23]. The most common pathways leading to cell proliferation act through a cascade activation of cell membrane receptor tyrosine kinase (RTKs), intracellular second messengers and gene expression of transcription factors (Figure 1).

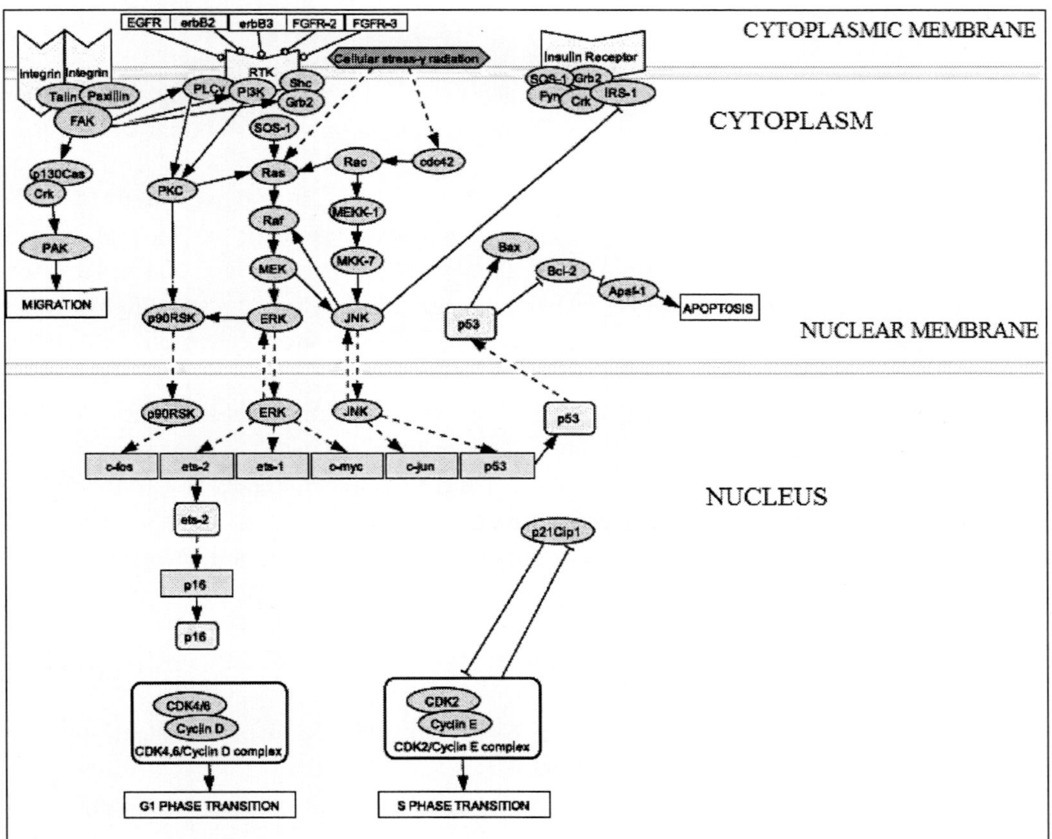

Figure 1. Signal transduction pathways in normal oral mucosa ([21] with permission). Broken or solid lines correspond to activation of specific protein molecules depending whether it occurs with or without passage via a membrane, respectively.

Extracellular mitogenic growth factors bind to RTKs that are activated and transduce a positive growth signal by phosphorylating intracellular second messengers [23,24]. The RTK substrates belong to a protein family with homologous domains, the most important members of which, in regard to cell signalling, are phospholipase-γ (PLCγ), phosphatidylinositol kinase (PI3K) and Grb2 [25-27]. PLC-γ generates the downstream messengers diacylglycerol (DAG) and inositol triphosphate (IP3) by hydrolyzing membrane phospholipids (44). PI3K catalyzes the formation of downstream messenger phosphatidylinositol triphosphate (PIP3) resulting in the ultimate activation of protein kinase. C [28]. Grb2 binds to the RTK and also interact with downstream messenger Son of Sevenless (SOS), which in turn activates Ras [29-31]. The localization of Ras proteins to the inner surface of the cytoplasmic membrane is critical for activation of several downstream effectors, most notably the kinase Raf, the first signalling element in the pathway of the mitogen activated protein kinases (MAPKs)[32]. When activated, Raf in turn activates the MAPKs by phosphorylation, initiating a cascade of activation of kinases that eventually results in the enhanced transcription of genes coding for proteins that regulate cell proliferation and apoptosis [32,33].

3. Animal Models of Oral Cancer

A number of animal models of induced oral oncogenesis have been generated, including hamster, rat and mouse models. These animal models allow reproducible pathological evaluation of all precancerous and cancer stages in comparison with normal oral mucosa [34,35].

The first demonstration of OSCC induction in an animal model was effected half a century ago in hamster cheek pouch with the aid of polycyclic aromatic carcinogens, such as 9,10-dimethyl-1,2-benzanthracene (DMBA), 20 methyl cholanthrene and 3,4-benzopyrene [36]. Thirty-five years ago, the rat model of oral oncogenesis was introduced using the water-soluble carcinogen 4-nitroquinoline-1-oxide (4NQO)[37-39]. A metabolic product of 4NQO binds predominantly at guanine residues of DNA producing a spectrum of precancerous and neoplastic lesions [40]. The rat model of 4NQO-induced carcinogenesis is considered as the preferred model for the study of OSCC, since it parallels oral oncogenesis in humans [41-47].

The use of carcinogens for induction of oral cancer in animal models is warranted, since chemical agents appear to be the dominant etiological factors in oral cavity. Animal models are a very useful tool towards understanding important mechanisms in oral oncogenesis.

4. The Model of Oral Oncogenesis in Diabetic Rats

An experimental animal model was used in order to study the effect of diabetes mellitus type I (also known as insulin-dependent diabetes) on the molecular pathways involved in the sequential stages of oral oncogenesis. Oral cancer was induced in diabetic type I and normal rats. The experimental procedures will be discussed briefly here, since they have been previously described in detail [8].

4.1. Animal Groups

Thirty-seven female Sprague-Dawley rats purchased from the Hellenic Pasteur Institute (Athens) at the age of six weeks and weighing approximately 135 g each were used in this study. The rats were handled in accordance with the Greek Legislation for the Care and Use of Laboratory Animals (P.D. 160/91 and Law 2015/92), conforming to Directive 86/609 of the European Union, as well as the Guide for the Care and Use of Laboratory Animals (U.S. National Institutes of Health Publication No. 85-23, revised 1996). The animals were randomly divided into four groups. A) Group D (n=6): Diabetic rats without carcinogenesis, B) Group Dc (n=13): Diabetic rats used for induced carcinogenesis, C) Group N (n=6): Normal rats without carcinogenesis, D) Group Nc (n=12): Normal rats used for induced carcinogenesis.

4.2. Induction of Diabetes

The induction of diabetes type I was performed in 19 previously overnight fasted animals by a single intraperitoneal injection of streptozotocin (STZ) dissolved in saline buffer at a dose of 70 mg/kg of weight (ZANOSAR, Pharmacia & Upjohn Co., USA) and confirmed by glucose levels in the blood after three weeks, as previously described [8].

4.3. Experimental Carcinogenesis

Oral cancer was induced in Dc and Nc animals by application of the carcinogen 4NQO at a concentration of 5% in propylene glycol 3 times per week for 5 months in the rats' hard palate (Fluca AG Chemische Fabrik, Switzerland), as described elsewhere [8]. Clinical signs of oral lesions putatively tumor-related were observed within 6 months after last application of carcinogen. After sacrifice of animals by ether treatment, the oral regions with cancer (mainly palate and tongue) of Dc and Nc rats and the respective regions of D and N rats were excised for immunohistochemical analysis [8].

4.4. Pathological and Immunohistochemical Evaluation

The histological status of the lesions was defined after examination of the complete section under light microscopy and the tissue profiles were classified into the following categories: normal mucosa, hyperplasia, dysplasia, early invasion, well- and moderately differentiated carcinoma. In every sample, a number of all possible representative histological lesions were analysed. The biopsies from the 37 animals were fixed in 10% neutralized formaldehyde solution and embedded in paraffin. Sections 4 μm-thick were prepared from each specimen and were mounted on Super Frost Plus-coated glass slides (Menzel and Co., Braunschweig, Germany). All sections were used for detection of cell membrane tyrosine kinase receptors (EGFR, erbB2, erbB3, FGFR-2, FGFR-3), a docking cell membrane protein (IRS-1), cytoplasmic proteins (FAK, H-ras, N-ras), apoptosis-related proteins (Bax, Bcl-2), a cell proliferation marker (Ki-67), nuclear transcriptional factors (p53, c-myc, c-fos, c-jun, ets-

1), and a cell cycle protein (p16), using a standard immunohistochemical technique (DAB Immunoperoxidase Secondary Detection System) as described elsewhere [8,48-55].

4.5. Statistical Analysis

In order to evaluate the pattern of antibody expression in relation to the histological status, the various lesions were divided according to tumor progression into a) normal tissue, b) non-cancerous and precancerous conditions (hyperplasia, dysplasia), c) tumor (early invasion, well differentiated OSCC, moderately differentiated OSCC). The mean percentages of positively stained cells were calculated for all the different lesions present in each sample. These values were tabulated for diabetic animals (groups D and Dc) and normal ones (groups N and Nc). Using the two-tailed Student's *t*-test, the mean percentage of positively-stained cells from each non-cancerous and precancerous category was compared with those of the normal tissue, while the percentage of positively stained cells from each tumor category was compared with the average percentage of the non-cancerous and precancerous lesions. In all cases with no normal distribution, the results of both the Wilcoxon test and two-tailed Student's *t*-test provided the same level of significance.

5. Effect of Diabetes on Signal Transduction Pathways

Despite the recent epidemiological studies that associate diabetes and oral oncogenesis, no molecular basis of this association had been proposed in the literature. The model of oral oncogenesis in diabetic rats was employed to studying a possible mechanism linked to both diseases, as well as the expression of several proteins implicated in common signal transduction pathways that regulate cell proliferation and apoptosis. These studies allowed the deciphering of the involvement of 18 signal transduction factors in each sequential stage of oral oncogenesis both in the presence or absence of diabetes type I (Figure 2).

5.1. Insulin Receptor and Its Substrate

A possible mechanism linked both to diabetes type I and cancer involves insulin receptor and its substrate (Figure 1). Binding of insulin to its cell membrane receptor stimulates tyrosine phosphorylation of insulin receptor substrate-1 (IRS-1), which in turn activates PI3K and starts a cascade resulting in cell growth [56]. IRS-1 is a docking cell membrane protein that plays an important role in the insulin signal transduction cascade and its expression is affected both in the insulin deficiency status (diabetes type I) and in the insulin resistance status (diabetes type II)[57,58]. One of the signalling molecules implicated in IRS-1 pathways is focal adhesion kinase pp125 (FAK), also known as cytosolic tyrosine kinase pp125 [59,60]. FAK interacts both with IRS-1 and with the transmembrane integrins, which are involved in cell migration and adhesion, apoptosis, wound repair and angiogenesis [61]. Following insulin stimulation, IRS-1 is detected co-immunoprecipitating with $\alpha_v\beta_3$ integrin [62]. Interaction of both FAK and integrins may induce activation of IRS-1 by tyrosine phosphorylation upon insulin stimulation, but at the same time the FAK/integrin complex may inhibit activation of IRS-1 in an insulin-independent phosphoinositolglycan-mediated fashion [59,63]. Over-expression of FAK and integrins is observed in OSCC and other malignancies, underlying their important role in oncogenesis [63-66]. FAK is a cytoskeletal-associated tyrosine kinase and its over-expression of leads to increased cell migration and survival under anchorage-independent conditions [64]. Interestingly, mutations in the insulin receptor alpha-subunit gene affecting its expression result in altered cell morphology associated with increased cell size, reduced cell adhesiveness and dephosphorylation of FAK in the absence of insulin [67].

In the rat model, expression of IRS-1 was significantly higher in diabetic (D) than normal animals, but it decreased in diabetic animals with tumor, especially in more advanced stages [8]. Furthermore, FAK expression was significantly higher in rats with cancer in comparison to the ones without it, regardless the diabetes status [8]. The expression of IRS-1 and the expression of FAK were not correlated during the sequential stages of oral oncogenesis (Table 1). Since the IRS-1 pathway plays an important role in the post-receptor signaling glucose transport in muscle and adipose tissues as well as in cell growth, the over-expression of IRS-1 in diabetic rats may be attributed to an attempt to cope in an environment of scarce insulin, since the abundance of IRS-1 might facilitate its activation by the relatively few receptors that bind to minimally available insulin, as previously suggested [68]. These data suggest that the role of IRS-1 as a signal-mediating molecule is diminished in an insulin-deprived malignant environment. This study showed that diabetes type I may indeed increase the risk for OSCC through alteration of the IRS-1/FAK pathway, which may possibly end up in cytoskeleton changes and reduced cell adhesion, an initial step towards neoplasia.

5.2. Receptor Tyrosine Kinases

Members of the EGFR family of RTKs include epidermal growth factor receptor (EGFR), erbB2 (also known as HER-2/neu) and erbB3 (also known as HER-3). Significant over-expression of EGFR has been reported in cases of head and neck cancer, correlated with tumor size and advanced stages of oncogenesis [20,69-71]. Over-expression of erbB2 has been significantly associated with shorter survival of patients with OSCC [71-73]. Accordingly, erbB3 over-expression has been correlated to the malignant progression of head and neck SCC [72-75]. Increased expression of both fibroblast growth factor receptors 2 and 3 (FGFR-2, FGFR-3) in oral dysplasia and OSCC has been reported [76].

In the diabetic rat model, there was a trend for reduced EGFR expression in diabetic compared to normal animals throughout oral oncogenesis (Figure 2)[53]. On the contrary, diabetes enhances erbB2 expression during nearly all stages of oral cancer development, as well as erbB3 expression in non-cancerous and precancerous stages (Figure 2)[51]. The expression pattern of FGFR-2 and FGFR-3 was not affected by diabetes during oral oncogenesis (Figure 2)[50].

Table 1. Expression of IRS-1 and of FAK during the sequential stages of oncogenesis in palate and tongue of normal and diabetic rats, as well as examination of their correlation (Spearman's correlation coefficient *r*). All possible histological regions were evaluated for every animal biopsy. Protein expression is given in mean percentage of cells positive for immunochemical staining within all equivalent regions in each biopsy sample. Significant P values are shown in bold italics. A) Normal rats of groups N and Nc. B) Diabetic rats of groups D and Dc.

A

Normal Rats – Tongue										
	Normal Sites		Precancerous Sites				Cancerous Sites			
			Hyperplasia		Dysplasia		Ca in Situ		Ca	
Pairs	**FAK**	**IRS-1**	**FAK**	**IRS-1**	**FAK**	**IRS-1**	**FAK**	**IRS-1**	**FAK**	**IRS-1**
1.	1.18	15.12	1.44	15.27	2.10	19.93	3.15	20.45	1.71	19.88
2.	2.34	21.44	2.91	19.19	3.40	16.82	2.53	21.49	5.06	22.16
3.	2.12	27.05	1.36	26.12	2.69	16.93	3.30	15.73	2.68	22.71
4.	2.10	12.36	2.89	20.19	1.49	15.58				
5.	0.96	18.84	2.00	17.14	1.40	15.26				
6.	1.41	17.07	3.40	18.19	1.25	14.66				
7.	3.41	14.53	1.03	10.13						
8.	0.91	9.15	1.33	16.12						
Mean	1.804	16.945	2.045	17.794	2.055	16.530	2.993	19.223	3.150	21.583
SD	0.854	5.566	0.901	4.559	0.850	1.888	0.408	3.070	1.724	1.501
r	0.205		0.299		0.458		-0.770		0.570	
P	0.626		0.472		0.361		0.441		0.430	

Table 1. Continued

A

Normal Rats – Palate										
	Normal Sites		Precancerous Sites				Cancerous Sites			
			Hyperplasia		Dysplasia		Ca in Situ		Ca	
Pairs	**FAK**	**IRS-1**	**FAK**	**IRS-1**	**FAK**	**IRS-1**	**FAK**	**IRS-1**	**FAK**	**IRS-1**
1.	23.66	26.85	10.10	10.33	20.19	21.16	19.31	18.13	23.61	22.03
2.	25.34	26.71	15.30	6.13	20.34	25.18	19.31	24.19	24.04	19.64
3.	20.52	25.65	14.91	11.37	24.17	19.12	20.90	26.18	21.09	21.10
4.	21.13	19.18	10.10	5.18	21.14	20.18			21.70	25.75
5.			9.76	10.11	21.92	16.70			22.99	22.14
6.			6.79	16.16	24.60	17.74			32.50	20.48
7.			20.14	21.19					23.04	28.87
8.			28.13	27.18					28.43	22.84
9.			20.71	9.16					19.80	20.74
10.			27.62	22.36					26.14	27.97
11.									26.17	22.30
12.									26.72	22.15
Mean	22.663	24.598	16.356	13.917	22.060	20.013	19.840	22.833	24.686	23.001
SD	2.244	3.651	7.567	7.443	1.909	2.999	0.918	4.193	3.522	2.954
r	0.568		0.692		-0.642		0.691		-0.098	
P	0.432		***0.027***		0.169		0.514		0.763	

Table 1. Continued

B

Diabetic Rats – Tongue										
	Normal Sites		Precancerous Sites				Cancerous Sites			
			Hyperplasia		Dysplasia		Ca in Situ		Ca	
Pairs	**FAK**	**IRS-1**	**FAK**	**IRS-1**	**FAK**	**IRS-1**	**FAK**	**IRS-1**	**FAK**	**IRS-1**
1.	4.27	25.30	3.14	29.14	4.24	23.73	5.06	25.69	3.24	26.44
2.	4.14	15.28	0.84	16.19	1.91	17.03	2.12	16.17	1.76	17.42
3.	2.32	21.73	1.41	28.16	2.61	27.15	4.24	18.40	2.25	17.32
4.	4.34	22.01	1.70	16.13	2.31	17.02	4.32	26.25	4.19	19.05
5.	3.29	17.59			4.17	14.17			5.73	18.55
6.	2.40	19.38								
7.	2.34	27.20								
8.	1.75	26.54								
9.	2.15	32.25								
Mean	3.000	23.031	1.773	22.405	3.048	19.820	3.935	21.628	3.434	19.756
SD	1.021	5.310	0.979	7.222	1.085	5.398	1.265	5.101	1.588	3.809
r	-0.480		0.630		0.015		0.789		0.091	
P	0.191		0.370		0.981		0.211		0.885	

Table 1. Continued

B

	Diabetic Rats – Palate									
	Normal Sites		Precancerous Sites				Cancerous Sites			
			Hyperplasia		Dysplasia		Ca in Situ		Ca	
Pairs	**FAK**	**IRS-1**	**FAK**	**IRS-1**	**FAK**	**IRS-1**	**FAK**	**IRS-1**	**FAK**	**IRS-1**
1.	26.49	31.45	9.13	16.16	20.24	17.18	19.21	21.63	24.34	19.71
2.	27.39	21.64	10.33	13.19	18.12	24.38	20.69	20.77	30.17	17.09
3.	22.17	29.01	8.12	15.17	21.51	20.30	24.51	22.19	19.54	25.07
4.	22.79	29.25	10.13	9.86	25.12	21.31	19.14	19.17	24.19	24.17
5.	20.79	37.24	11.37	7.86	24.23	19.76	23.13	28.17	32.68	18.77
6.			17.35	12.13			20.14	24.73	25.32	17.36
7.			7.31	7.35			20.19	11.19	28.30	21.85
8.			12.27	6.18			-	23.28	22.38	24.84
9.			6.17	10.50			24.64	19.38	22.67	23.19
10.			24.14	5.16			20.13	28.66	26.59	20.02
11.			24.36	19.19						
12			19.15	20.18						
13.			24.73	31.18						
Mean	23.926	29.718	14.197	13.393	21.844	20.586	21.309	21.917	25.618	21.207
SD	2.863	5.603	6.863	7.166	2.871	2.612	2.184	4.999	3.924	3.029
r	-0.707		0.472		-0.272		0.129		-0.726	
P	0.182		0.103		0.658		0.742		***0.018***	

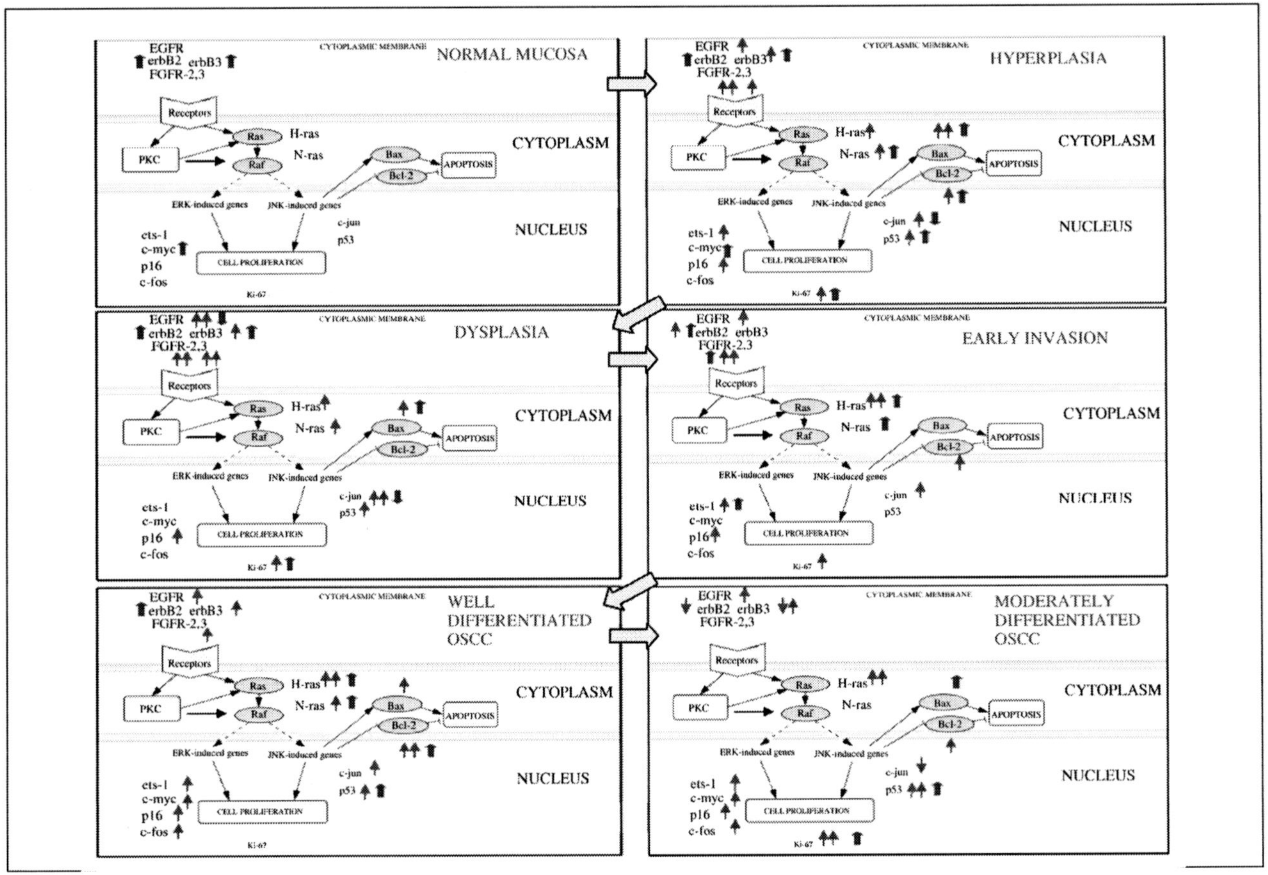

Figure 2. Signal transduction pathways during sequential stages of oral oncogenesis ([21] with permission). Broken or solid lines correspond to activation of specific proteins depending whether it occurs with or without passage via a membrane, respectively. See text (4.5) for statistical comparisons. The arrows by the studied proteins indicate the significant differences in their expression observed in the rat model. Red arrows indicate significant differences in diabetic rats, blue arrows indicate significant differences in normal rats, while green arrows indicate significant differences between normal and diabetic rats in the same stage. An arrow pointing upwards indicates a significant increase of expression in that stage, while one pointing downwards indicates a significant decrease.

5.3. Cytoplasmic Proteins

Mutations in genes coding for Ras proteins seem to be quite rare in OSCC patients from Western countries and more frequent in patients from Southeast Asian countries, such as India and Taiwan [77]. Furthermore, involvement of N-ras in oral oncogenesis is an infrequent occurrence [78,79]. There are conflicting reports in the literature regarding the increased expression of Ras proteins at an early or a later stage in oral carcinogenesis [80,81].

In the rat model, there was a trend for increased H-ras expression in diabetic compared to normal animals during oncogenesis, which reached a statistical significant difference in early invasion and well-differentiated OSCC (Figure 2)[53]. N-ras expression increased with tumor advancement in diabetic rats, while in normal rats N-ras expression was not observed in premalignant stages and increased only in well-differentiated OSCC (Figure 2)[49]. The expression of both H-ras and N-ras seems to be positively affected by diabetes during oral oncogenesis.

5.4. Apoptosis and Cell Proliferation Markers

There are conflicting reports in the literature regarding the levels of the apoptosis-related proteins Bcl-2 and Bax in oral cancer, although in other tumors their levels are so well defined that they are used as prognostic markers [82-84]. Some studies of human OSCC biopsies reported increased levels of Bcl-2 as well as increased levels of Bcl-2/Bax ratio corresponding to reduced apoptosis in comparison to normal oral mucosa [85,86]. Other studies of human biopsies reported reduced Bcl-2 expression and Bcl-2/Bax ratio and increased apoptosis in the initial tumor stages, while Bax was not found to be significantly altered during the different stages of oral oncogenesis [87,88]. Tumor cells are known to manifest elevated levels of certain marker proteins that control cell cycle, such as Ki-67 [89,90]. The cell proliferation marker Ki-67 is highly expressed in initial stages of oral oncogenesis [91].

In the rat model, the Bcl-2/Bax ratio was relatively stable during the oncogenesis process in diabetic animals, contrary to normal rats which appear to have an increased Bcl-2/Bax ratio in the stage of moderately differentiated carcinoma, indicating reduced apoptosis levels (Figure 2)[52]. Furthermore, the expression of Ki-67 was higher in diabetic rats than in normal ones in almost all stages of oral oncogenesis (Figure 2)[52]. Diabetes seems to promote increased cell proliferation during oral oncogenesis, but this is accomplished without affecting the Bax/Bcl-2-mediated apoptotic pathways.

5.5. Nuclear Transcriptional Factors

Gene amplification leading to increased levels of the transcription factor c-myc occurs frequently in poorly differentiated OSCC [92,93]. One of the best studied tumor suppressor genes is *p53*, but its high frequency of alterations in human cancer has not fully elucidated since mutant p53 proteins constitute a complex family of several hundred proteins with heterogeneous properties [94-96]. Mutations that inactivate *p53* gene or impair function of its protein product have been detected in more than 60% of cases of OSCC [97]. The inactivated mutant p53 protein is more stable than the normal one and its accumulation increases gradually during oral oncogenesis [94,97-99]. Together c-fos and c-jun form the composite transcription factor activating protein-1 (AP-1), a mitogen-activated transactivator important for cell proliferation and differentiation [100]. In general, c-jun is considered to be an early marker of early stages of oncogenesis [101]. There are conflicting reports about c-fos, with some studies indicating elevated levels of c-fos in malignant oral lesions, others noting a stable expression at normal levels throughout oral oncogenesis, while still other studies reporting a gradual decrease of c-fos expression from normal oral mucosa towards advanced stages of oral cancer [102-105]. Up-regulation of *ets-1* gene expression has been observed in oral cancer, correlating with the grade of invasiveness and metastasis [106].

Based on the rat model studies, diabetes contributes to increased accumulation of mutant p53 resulting in increased tumor cell proliferation during oral oncogenesis (Figure 2)[48]. Furthemore, c-myc expression is enhanced by diabetes only in the initial stages of oral oncogenesis (Figure 2)[48]. Higher c-jun levels were observed only in non-cancerous and precancerous stages in normal rats compared to diabetic rats, indicating that diabetes does not affect in tumor stages the c-jun *N*-terminal kinase (JNK)/c-jun pathway (Figure 2)[54]. No statistical differences concerning c-fos expression were detected between diabetic and normal animals (Figure 2)[53]. The same pattern of increased ets-1 expression was observed both in diabetic and normal rats, but in cancerous stages this expression was higher in diabetic animals, indicating that diabetes may contribute to enhanced invasion and metastatic potential (Figure 2)[49].

5.6. Cell Cycle Regulators

The negative regulator of cell proliferation p16 is frequently abrogated in head and neck cancer. In OSCC point mutations of the *p16* gene are rare, but alternative mechanisms of p16 function loss have been reported, including homozygous deletions or methylation silencing of the promoter region of the *p16* gene [107-110]. In the experimental rat model of oral oncogenesis, the expression of p16 was not found to be affected by diabetes (Figure 2)[55].

6. Insights from the Diabetic Rat Model Studies

Activation of cell membrane RTKs (such as insulin receptor, EGFR, erbB2, erbB3, FGFR-2 and FGFR-3) leads to the recruitment of signalling molecules (such as IRS-1) to the cell membrane and the initiation of signalling cascades (such as the Ras/Raf/MAPK and PI3K cascades). The animal model studies described here indicate that diabetes mellitus type I seems to influence key points of signal transduction cascades that regulate important cell functions, thereby increasing risk for oral cancer. Specifically, the diabetic rat model provided us with the following insights:

Insulin deficiency seems to result in reduced amounts of insulin receptor and a counter-active increased expression of IRS-1. The latter may lead eventually to cytoskeleton changes and reduced cell adhesion, an initial step towards neoplasia.

Diabetes seems to promote the activation of Ras/Raf/MAPK signal transduction pathway by induction of receptors erbB2 and erbB3 mainly (and not EGFR) leading to increased cell proliferation.

Diabetes does not seem to influence levels of apoptosis related to the Bcl-2/Bax ratio.

An interesting extrapolation of these findings is that the type I diabetic patients who are not insulin-regulated may have a greater risk for developing oral cancer than the general population does. This hypothesis remains to be examined by large epidemiological studies.

Acknowledgements

The authors wish to thank Professor Efstratios Patsouris,MD,PhD for reviewing this manuscript and for valuable suggestions. In addition, the authors thank all colleagues who participated in the studies presented here, especially Ph.D. students Lambros Goutzanis,MD,DDS for the effective development of the animal model, Sofia Spyridonidou,B.S. for excellent laboratory work and data interpretation contributions and Dimitris Avgoustidis,MD for competent assistance in statistical analysis. This work was co-funded by the European Social Fund and National Resources (EPEAEK II “Pythagoras” 70/3/7391) grant.

References

[1] Taylor GW, Manz MC and Borgnakke WS: Diabetes, periodontal diseases, dental caries, and tooth loss: a review of the literature. *Compend Contin Educ Dent* **25**(3):179-190, 2004.

[2] Kim J and Amar S: Periodontal disease and systemic conditions: a bidirectional relationship.*Odontology* **94**(1):10-21, 2006.

[3] Löe H: Periodontal disease. The sixth complication of diabetes mellitus. *Diabetes Care* **16**:329-334, 1993.

[4] Finney LS, Finney MO and Gonzalez-Campoy JM: What the mouth has to say about diabetes. Careful examinations can avert serious complications. *Postgrad Med* **102**(6):117-126, 1997.

[5] Milosevic A: Eating disorders and the dentist. *Br Dent J* **186**(3):109-113, 1999.

[6] Auluck A: Diabetes mellitus: an emerging risk factor for oral cancer? *J Can Dent Assoc* **73**(6):501-503, 2007.

[7] Goutzanis L, Vairaktaris E, Yapijakis C, Kavantzas N, Nkenke E, Derka S, Vassiliou S, Acil Y, Kessler P, Stavrianeas N, Perrea D, Donta I, Skandalakis P and Patsouris E: Diabetes may increase risk for oral cancer through the insulin receptor substrate-1 and focal adhesion kinase pathway. *Oral Oncol* **43**(2): 165-173, 2007.

[8] Albrecht M, Bánóczy J, Dinya E and Tamás G: Occurrence of oral leukoplakia and lichen planus in diabetes mellitus. *J Oral Pathol Med* **21**:364-366, 1992.

[9] Ujpál M, Matos O, Bibok G, Somogyi A, Szabo G and Suba Z: Diabetes and oral tumors in Hungary: epidemiological correlations. *Diabetes Care* **27**(3): 770-774, 2004.

[10] Dietrich T, Peter A. Reichart PA and Scheifele C. Clinical risk factors of oral leukoplakia in a representative sample of the US population. *Oral Oncol* **40**:158-163, 2004.

[11] Dikshit RP, Ramdas K, Hashibe M, Thomas G, Somanathan T and Sankaranarayanan R: Association between diabetes mellitus and premalignant oral diseases: a cross sectional study in Kerala, India. *Int J Cancer* **118**(2):453–457, 2006.

[12] Ujpál M, Barabás J, Kovalszky I, Szabó G, Németh Z, Gábris K and Suba Z: A preliminary comparative study of the prognostic implications of type 2 diabetes mellitus for patients with primary gingival carcinoma treated with surgery and radiation therapy. *J Oral Maxillofac Surg.* **65**(3):452-456, 2007.

[13] Kravchick S, Gal R, Cytron S, Peled R, Weissman Y, Mukamel E and Koren R. Increased incidence of diabetes mellitus in the patients with transitional cell carcinoma of urinary bladder. *Pathol Oncol Res* **7**:56–59, 2001.

[14] Yeung NG, Husain J and Waterfall N. Diabetes mellitus and bladder cancer: an epidemiological relationship? *Pathol Oncol Res* **9**:30–31, 2003.

[15] Parkin DM, Bray F, Ferlay J and Pisani P: Global cancer statistics, 2002. *CA Cancer J Clin* **55**(2): 74-108, 2005.

[16] Boyle P and Ferlay J: Cancer incidence and mortality in Europe, 2004. *Annals of Oncol* **16**: 481-488, 2005.

[17] Silverman S Jr: Demographics and occurrence of oral and pharyngeal cancers. The outcomes, the trends, the challenge. *J Am Dent Assoc* **132**:7S-11S, 2001.

[18] Sudbo J: Novel management of oral cancer: a paradigm of predictive oncology. *Clin Med Res* **2**(4): 233-242, 2004.

[19] Shah JP, Johnson NW and Batsakis JG (editors): *Oral cancer.* Martin Dunitz Publishers, London, UK.

[20] Mehrotra R and Yadav S: Oral squamous cell carcinoma: etiology, pathogenesis and prognostic value of genomic alterations. *Indian J Cancer* **43**(2): 60-66, 2006.

[21] Vairaktaris E, Spyridonidou S, Goutzanis L, Vylliotis A, Lazaris A, Donta I, Perrea D, Yapijakis C and Patsouris E: Diabetes and oral oncogenesis. *Anticancer Res* **27**: 4185-4194, 2007.

[22] Lavelle CL and Scully C: Criteria to rationalize population screening to control oral cancer. *Oral Oncol* **41**(1):11-6, 2005.

[23] Bell HS and Ryan KM: Intracellular signalling and cancer: complex pathways lead to multiple targets. *Eur J Cancer* **41**(2): 206-215, 2005.

[24] Cross M and Dexter TM: Growth factors in development, transformation, and tumorigenesis. *Cell* **64**: 271-280, 1991.

[25] Margolis B: Proteins with SH2 domains: transducers in tyrosine kinase signaling pathway. *Cell Growth Differ* **3**: 73-80, 1992.

[26] Karin M: Signal transduction from cell surface to nucleus in development and disease. *FASEB J* **6**(8): 2581-290, 1992.

[27] Sandy JR: Signal transduction. *Br J Orthod* **25**(4): 269-274, 1998.

[28] Fantl WJ, Escobedo JA, Martin GA, Turck CW, del Rosario M, McCormick F and Williams LT: Distinct phosphotyrosines on a growth factor receptor bind to specific molecules that mediate different signaling pathways. *Cell* **69**: 413-423, 1992.

[29] Lowenstein EJ, Daly RJ, Batzer AG, Li W, Margolis B, Lammers R, Ullrich A, Skolnik EY, Bar-Sagi D and Schlessinger J: The SH2 and SH3 domain containing protein GRB2 links receptor tyrosine kinases to ras signaling. *Cell* **70**: 431-442, 1992.

[30] Li N, Batzer A, Daly R, Yajnik V, Skolnik E, Chardin P, Bar-Sagi D, Margolis B and Schlessinger J: Guanine-nucleotide-releasing factor hsosl binds to Grb2 and links receptor tyrosine kinases to Ras signaling. *Nature* **363**: 85-88, 1993.

[31] Porfiri E and McCormick F: Regulation of epidermal growth factor receptor signaling by phosphorylation of the ras exchange factor hSOS1. *J Biol Chem* **271**: 5871 -5877, 1996.

[32] Beeram M, Patnaik A and Rowinsky EK: Raf: a strategic target for therapeutic development against cancer. *J Clin Oncol* **23**(27): 6771-6790, 2005.

[33] Kamer AR, Krebs L, Hoghooghi SA and Liebow C: Proliferative and apoptotic responses in cancers with special reference to oral cancer. *Crit Rev Oral Biol Med* **10**(1): 58-78, 1999.

[34] Tang XH, Knudsen B, Bemis D, Tickoo S and Gudas LJ: Oral cavity and esophageal carcinogenesis modeled in carcinogen-treated mice. *Clin Cancer Res* **10**(1 Pt 1): 301-313, 2004.

[35] Vairaktaris E, Spyridonidou S, Papakosta V, Vylliotis A, Lazaris A, Perrea D, Yapijakis C and Patsouris E: The hamster model of sequential oral oncogenesis. *Oral Oncol* 2007 PMID 18061531 (in press).

[36] Salley JJ: Experimental carcinogenesis in the cheek pouch of the Syrian hamster. *J Dent Res* **33**(2): 253-262, 1954.

[37] Boyd NM and Reade PC: Temporal alterations in cytokeratin expression during experimental oral mucosal carcinogenesis. *Carcinogenesis* **12**(10): 1767-1771, 1991.

[38] Tanaka T, Kojima T, Okumura A, Yoshimi N and Mori H: Alterations of the nucleolar organizer regions during 4-nitroquinoline 1-oxide-induced tongue carcinogenesis. *Carcinogenesis* **12**: 329-333, 1991.

[39] Vered M, Yarom N and Dayan D: 4NQO oral carcinogenesis: animal models, molecular markers and future expectations. *Oral Oncol* **41**(4): 337-339, 2005.

[40] Kanojia D and Vaidya MM: 4-Nitroquinoline-1-oxide induced experimental oral carcinogenesis. *Oral Oncol* **42**(7): 655-667, 2006.

[41] Itoh T, Yamamoto Y, Saka T, Inoue I and Takahashi H: Estimation of proliferative activity of experimental tongue carcinoma in rats. Immunohistochemical and DNA cytofluorometric analysis. *Acta Otolaryngol* **113**: 568-574, 1993.

[42] Rich AM and Reade PC: Nuclear morphometry in experimental oral mucosal carcinogenesis. *Eur J Cancer B Oral Oncol* **32B**: 169-175, 1996.

[43] Osugi Y: p53 Expression in various stages of 4-nitroquinoline 1-oxide induced carcinoma in rat tongue. *J Osaka Dent Univ* **30**: 29-35, 1996.

[44] Dayan D, Hirshberg A, Kaplan I, Rotem N and Bodner L: Experimental tongue cancer in desalivated rats. *Oral Oncol* **33**: 105-109, 1997.

[45] Nishimura A: Changes in Bcl-2 and Bax expression in rat tongue during 4-nitroquinoline 1-oxide-induced carcinogenesis. *J Dent Res* **78**: 1264-1269, 1999.

[46] Kaplan I, Engelberg J and Dayan D: The effect of desalivation on 4-nitroquinoline-1-oxide-induced tongue carcinogenesis: a morphometric study of nucleolar organizer regions. *J Oral Pathol Med* **30**: 48-52, 2001.

[47] Sakaki T, Tamura I, Katoda H and Kakudo K: Changing expression of E- and P-cadherin during rat tongue carcinogenesis induced by 4-nitroquinoline 1-oxide. *J Oral Pathol Med* **32**: 530-537, 2003.

[48] Vairaktaris E, Kalokerinos G, Goutzanis L, Spyridonidou S, Vassiliou S, Derka S, Nkenke E, Yapijakis C, Vylliotis A, Lazaris A and Patsouris E: Diabetes alters expression of p53 and c-myc in different stages of oral oncogenesis. *Anticancer Res* **27**: 1465-1474, 2007.

[49] Vairaktaris E, Goutzanis L, Kalokerinos G, Yannopoulos A, Yapijakis C, Vassiliou S, Spyridonidou S, Vylliotis A, Nkenke E, Lazaris A, Tesseromatis C and Patsouris E: Diabetes increases both N-ras and ets-1 expression during rat oral oncogenesis resulting in enhanced cell proliferation and metastatic potential. *In Vivo* **21**(4): 615-621, 2007.

[50] Vairaktaris E, Goutzanis L, Nkenke E, Spyridonidou S, Vassiliou S, Derka S, Vylliotis A, Yapijakis C, Lazaris A, Strantzias P, Kalimeras E and Patsouris E: Diabetes does not influence oral oncogenesis through fibroblast growth factor receptors. *In Vivo* **21**(4): 623-628, 2007.

[51] Vairaktaris E, Goutzanis L, Vassiliou S, Spyridonidou S, Nkenke E, Papageorgiou G, Strantzias P, Lazaris A, Yapijakis C and Patsouris E: Enhancement of erbB2 and erbB3 expression during oral oncogenesis in diabetic rats. *J Cancer Res Clin Oncol* 2007 PMID: 17704947 (in press).

[52] Vairaktaris E, Kalokerinos G, Goutzanis L, Yapijakis C, Derka S, Vassiliou S, Spyridonidou S, Vylliotis A, Nkenke E, Lazaris A and Patsouris E: Diabetes enhances cell proliferation but not Bax/Bcl-2 mediated apoptosis during oral oncogenesis. *Int J Oral Maxillof Surg* **37**(1): 60-65, 2008.

[53] Vairaktaris E, Goutzanis L, Yapijakis C, Vassiliou S, Spyridonidou S, Vylliotis A, Nkenke E, Lazaris A, Strantzias P and Patsouris E: Diabetes enhances the expression of H-ras and suppresses the expression of EGFR leading to increased cell proliferation. *In Vivo* 2007 (in press).

[54] Vairaktaris E, Goutzanis L, Kalokerinos G, Vassiliou S, Spyridonidou S, Avgoustidis D, Strantzias P, Lazaris A, Papageorgiou G, Ragos V, Yapijakis C and Patsouris E: Increased risk for oral cancer in diabetic animals is not associated with c-jun activation pathway. *J Craniomaxillofac Surg* **35**(8): 382-387, 2007.

[55] Vairaktaris E, Goutzanis L, Nkenke E, Spyridonidou S, Vassiliou S, Derka S, Vylliotis A, Yapijakis C, Lazaris A, Strantzias P, Kalimeras E and Patsouris E: Expression of cell cycle regulator p16 is not affected by diabetes during oral oncogenesis. *In Vivo* 2007 **21**(5): 745-750, 2007.

[56] White MF and Kahn CR. The insulin signaling system. *J Biol Chem* **269**:1-4, 1994.

[57] Folli F, Saad MJ, Backer JM and Kahn CR. Regulation of phosphatidylinositol 3-kinase activity in liver and muscle of animal models of insulin-resistant and insulin-deficient diabetes mellitus. *Clin Invest* **92**:1787-1794, 1993.

[58] Kondo T and Kahn CR. Altered insulin signaling in retinal tissue in diabetic states. *J Biol Chem* **279**:37997-38006, 2004.

[59] Lebrun P, Mothe-Satney I, Delahaye L, Van Obberghen E and Baron V. Insulin Receptor Substrate-1 as a Signaling Molecule for Focal Adhesion Kinase $pp125^{FAK}$ and $pp60^{src}$. *J Biol Chem* **273**:32244-32253, 1998.

[60] Lebrun P, Baron V, Hauck CR, Schlaepfer DD and Van Obberghen E. Cell Adhesion and Focal Adhesion Kinase Regulate Insulin Receptor Substrate-1 Expression. *J Biol Chem* **275**:38371-38377, 2000.

[61] Hanks SK, Calalb MB, Harper MC and Patel SK. Focal Adhesion Protein-Tyrosine Kinase Phosphorylated in Response to Cell Attachment to Fibronectin. *PNAS USA* **89**:8487-8491, 1992.

[62] Vuori K and Ruoslahti E. Association of insulin receptor substrate-1 with integrins. *Science* **266**:1576-1578, 1994.

[63] Muller G, Wied S and Frick W. Cross talk of pp125(FAK) and pp59(Lyn) non-receptor tyrosine to insulin mimetic signaling in adipocytes. Mol Cell Biol 20:4708-4723, 2000.

[64] Kornberg LJ. Focal adhesion kinase expression in oral cancers. *Head Neck* **20**:634-639, 1998.

[65] Schneider GB Kurago Z, Zaharias R, Gruman LM, Schaller MD and Hendrix MJ. Elevated focal adhesion kinase expression facilitates oral tumor cell invasion. *Cancer* **95**:2508-2515, 2002.

[66] Walker DM, Boey G and McDonald LA. The pathology of oral cancer. *Pathology* **35**(5): 376-383, 2003.

[67] Konstantopoulos N and Clark S. Reduced cell attachment and phosphorylation of focal adhesion kinase associated with expression of a mutant insulin receptor. *J Biol Chem* **27**:28960-28968, 1996.

[68] Friedman JE, Ishizuka T, Liu S, Farrell CJ, Bedol D, Koletsky RJ, Kaung HL and Ernsberger P. Reduced insulin receptor signaling in the obese spontaneously hypertensive Koletsky rat. *Am J Physiol* **273**:1014-1023, 1997.

[69] Issing WJ, Wustrow TPU and Heppt W: Oncogenes related to head and neck cancer. *Anticancer Res* **13**:2541-2552, 1993.

[70] Shin D, Ro JY, Hong WK and Hittelman WN: Dysregulation of epidermal growth factor receptor expression in premalignant lesions during head and neck tumorigenesis. *Cancer Res* **54**: 3153-3159, 1994.

[71] Xia W, Lau YK, Zhang HZ, Liu AR, Li L, Kiyokawa N, Clayman GL, Katz RL and Hung MC: Strong correlation between c-erbB-2 overexpression and overall survival of patients with oral squamous cell carcinoma. *Clin Cancer Res* **33**: 3-9, 1997.

[72] Xia W, Lau YK, Zhang HZ, Xiao FY, Johnston DA, Liu AR, Li L, Katz RL and Hung MC: Combination of EGFR, Her-2/neu, and Her-3 is a stronger predictor for the outcome of oral squamous cell carcinoma than any individual family members. *Clin Cancer Res* **5**: 4164-4174, 1999.

[73] O-charoenrat P, Rhys-Evans PH, Archer DJ and Eccles SA: C-erbB receptors in squamous cell carcinomas of the head and neck: clinical significance and correlation with matrix metalloproteinases and vascular endothelial growth factors. *Oral Oncol* **38**:73-80, 2002.

[74] Funayama T, Nakanishi T, Takahashi K, Taniguchi S, Takigawa M and Matsumura T: Overexpression of c-erbB-3 in various stages of human squamous cell carcinomas. *Oncology* **55**:161-167, 1998.

[75] Shintani S, Funayama T, Yoshihama Y, Alcalde RE and Matsumura T: Prognostic significance of ERBB3 overexpression in oral squamous cell carcinoma. *Cancer Lett* **95**: 79-83, 1995.

[76] Wakulich C, Jackson-Boeters L, Daley TM and Wysocki GP: Immunohistochemical localization of growth factors fibroblast growth factor-1 and fibroblast growth factor-2 and receptors fibroblast growth factor receptor-2 and fibroblast growth factor receptor-3

in normal oral epithelium, epithelial dysplasias, and squamous cell carcinoma. *Oral Surg Oral Med Oral Pathol Oral Radiol Endod* **93**: 573-579, 2002.

[77] Das N, Majumder J and DasGupta UB: ras Gene mutations in oral cancer in eastern India. *Oral Oncol* **36**: 76-80, 2000.

[78] Matsuda H, Konishi N, Hiasa Y, Hayashi I, Tsuzuki T, Tao M, Kitahori Y, Yoshioka N, Kirita T and Sugimura M: Alterations of p16/CDKN2, p53 and ras genes in oral squamous cell carcinomas and premalignant lesions. *J Oral Pathol Med* **25**: 232-238, 1996.

[79] Yeudall WA, Torrance LK, Elsegood KA, Speight P, Scully C and Prime SS: Ras gene point mutation is a rare event in premalignant tissues and malignant cells and tissues from oral mucosal lesions. *Eur J Cancer B Oral Oncol* **29B**(1): 63-67, 1993.

[80] McDonald JS, Jones H, Pavelic ZP, Pavelic LJ, Stambrook PJ and Gluckman JL: Immunohistochemical detection of the H-ras, K-ras, and N-ras oncogenes in squamous cell carcinoma of the head and neck. *J Oral Pathol Med* **23**(8): 342-346, 1994.

[81] Yarbrough WG, Shores C, Witsell DL, Weissler MC, Fidler ME and Gilmer TM: ras mutations and expression in head and neck squamous cell carcinomas. *Laryngoscope* **104**(11): 1337-1347, 1994.

[82] Korkolopoulou P, Lazaris ACh, Konstantinidou AE, Kavantzas N, Patsouris E, Christodoulou P, Thomas-Tsagli E and Davaris P: Differential expression of bcl-2 family proteins in bladder carcinomas. Relationship with apoptotic rate and survival. *Eur Urol* **41**(3):274-283, 2002.

[83] Konstantinidou AE, Korkolopoulou P and Patsouris E: Apoptotic markers for tumor recurrence: a minireview. *Apoptosis*. **7**(5):461-470, 2002.

[84] Konstantinidou AE, Korkolopoulou P and Patsouris E. Apoptotic markers for primary brain tumor prognosis. *J Neurooncol* **72**(2):151-156, 2005.

[85] Chen Y, Kayano T and Takagi M: Dysregulated expression of bcl-2 and bax in oral carcinomas: evidence of post-transcriptional control. *J Oral Pathol Med* **29**: 63-69, 2000.

[86] Loro LL, Vintermyr OK, Liavaag PG, Jonsson R and Johannessen AC: Oral squamous cell carcinoma is associated with decreased bcl-2/bax expression ratio and increased apoptosis. *Hum Pathol* **30**(9): 1097-1105, 1999.

[87] Tang XH, Knudsen B, Bemis D, Tickoo S and Gudas LJ: Oral cavity and esophageal carcinogenesis modelled in carcinogen-treated mice. *Clin Cancer Res* **10**: 301-313, 2004.

[88] Singh BB, Chandler FW Jr, Whitaker SB and Forbes-Nelson AE: Immunohistochemical evaluation of bcl-2 oncoprotein in oral dysplasia and carcinoma. *Oral Surg Oral Med Oral Pathol Oral Radio Endod* **85**: 692-698, 1998.

[89] Konstantinidou AE, Korkolopoulou P, Vassilopoulos I, Tsenga A, Thymara I, Agapitos E, Patsouris E and Davaris P: Reduced retinoblastoma gene protein to Ki-67 ratio is an adverse prognostic indicator for ovarian adenocarcinoma patients. *Gynecol Oncol* **88**(3):369-378, 2003.

[90] Levidou G, Korkolopoulou P, Thymara I, Vassilopoulos I, Saetta AA, Gakiopoulou H, Konstantinidou A, Kairi-Vassilatou E, Pavlakis K and Patsouris E: Expression and prognostic significance of cyclin D3 in ovarian adenocarcinomas. *Int J Gynecol Pathol* **26**(4):410-417, 2007.

[91] Derka S, Vairaktaris E, Papakosta V, Vassiliou S, Acil Y,Vylliotis A, Spyridonidou S, Lazaris AC, Mourouzis C, Kokkori A, Moulavasili P, Perrea D, Donta I, Yapijakis C and Patsouris E: Cell proliferation and apoptosis culminate in early stages of oral oncogenesis. *Oral Oncol* **42**(5): 540-550, 2006.

[92] Spandidos DA, Lamothe A and Field JK: Multiple transcriptional activation of cellular oncogenes in human head and neck solid tumors. *Anticancer Res* **5**: 221-224, 1985.

[93] Schantz SP: Basic science advances in head and neck oncology: the past decade. *Sem Surg Oncol* **11**: 272-279, 1995.

[94] Soussi T: p53 alterations in human cancer: more questions than answers. *Oncogene* **26**(15):2145-2156, 2007.

[95] Goudopoulou A, Saetta A, Korkolopoulou P, Patsouris E, Fanourakis G, Miaouli M, Thomas-Tsagli E and Davaris PS: p53 mutations detection in urinary bladder cancer in the Greek population: application of the NIRCA assay. *J Exp Clin Cancer Res* **22**(1):99-105, 2003.

[96] Saetta AA, Aroni K, Stamatelli A, Lazaris AC and Patsouris E: Expression of mismatch repair enzymes, hMLH1 and hMSH2 is not associated with microsatellite instability and P53 protein accumulation in basal cell carcinoma. *Arch Dermatol Res* **297**(3):99-107, 2005.

[97] Sakai E, Rikimaru K, Ueda M, Matsumoto Y, Ishii N, Enomoto S, Yamamoto H and Tsuchida N: The p53 tumor-suppressor gene and ras oncogene mutations in oral squamous cell carcinoma. *Int J Cancer* **52**: 867-872, 1992.

[98] Tanaka N, Sugihara K, Odajima T, Mimura M, Kimijima Y and Ichinose S: Oral squamous cell carcinoma: electron microscopic and immunohistochemical characteristics. *Med Electron Microse* **35**: 127-138, 2002.

[99] Carlos de Vicente J, Junquera Gutierrez LM, Zapatero AH, Fresno Forcelledo MF, Hernandez-Vallejo G and Lopez Arranz JS: Prognostic significance of p53 expression in oral squamous cell carcinoma without neck node metastasis. *Head Neck* **26**: 22-30, 2004.

[100] Shaulian E and Karin M: AP-1 in cell proliferation and survival. *Oncogene* **20**: 2390-2400, 2001.

[101] Szabo E, Riffe ME, Steinberg SM, Birrer MJ and Linnoila RI: Altered c-JUN expression: an early event in human lung carcinogenesis. *Cancer Res* **56**: 305-315, 1996.

[102] De Sousa SO, Mesquita RA, Pinto DS JR and Gutkind S: Immunolocalization of c-Fos and c-Jun in human oral mucosa and in oral squamous cell carcinoma. *J Oral Pathol Med* **31**: 78-81, 2002.

[103] Mangone FR, Brentani MM, Nonogaki S, Begnami MD, Campos AH, Walder F, Carvalho MB, Soares FA, Torloni H, Kowalski LP and Federico MH: Overexpression of Fos-related antigen-1 in head and neck squamous cell carcinoma. *Int J Exp Pathol* **86**: 205-212, 2005.

[104] Ohyama M, Hirayama Y, Tanuma JI, Hirano M, Semba I, Shisa H, Hiai H, Sugihara K and Kitano M: Expressions of junB and c-fos are enhanced in 4-nitroquinoline 1-oxide-induced rat tongue cancers. *Pathol Int* **54**: 35-40, 2004.

[105] Turatti E, Da Costa Neves A, De Magalhaes MH and De Sousa SO: Assessment of c-Jun, c-Fos and cyclin D1 in premalignant and malignant oral lesions. *J Oral Sci* **47**: 71-76, 2005.

[106] Oikawa T and Yamada T: Molecular biology of the ETS family of transcription factors. *Gene* **303**: 11-34, 2003.

[107] Cairns P, Mao L, Merlo A, Lee DJ, Schwab D, Eby Y, Tokino K, van der Riet P, Blaugrund JE and Sidransky D: Rates of p16 (MTS1) mutations in primary tumors with 9p loss. *Science* **265**(5170): 415-417, 1994.

[108] Merlo A, Herman JG, Mao L, Lee DJ, Gabrielson E, Burger PC, Baylin SB and Sidransky D: 5' CpG island methylation is associated with transcriptional silencing of the tumor suppressor p16/CDKN2/MTS1 in human cancers. *Nat Med* **1**(7): 686-692, 1995.

[109] Reed AL, Califano J, Cairns P, Westra WH, Jones RM, Koch W, Ahrendt S, Eby Y, Sewell D, Nawroz H, Bartek J and Sidransky D: High frequency of p16 (CDKN2/MTS-1/INK4A) inactivation in head and neck squamous cell carcinoma. *Cancer Res* **56**(16): 3630-3633, 1996.

[110] Prime SS, Eveson JW, Guest PG, Parkinson EK and Paterson IC: Early genetic and functional events in the pathogenesis of oral cancer. *Radiat Oncol Investig* **5**(3): 93-96, 1997.

In: Oncogene Proteins: Structure, Functions and Analyses ISBN: 978-1-60456-259-0
Editors: P.B. Murphy and J.R. Clarke, pp. 291-307

Chapter XII

Apoptotic Cell Death Inhibitors: Implications of Clusterin and BCL-2 in Carcinogenesis and Treatment

Rafael Fúnez[1] and Maximino Redondo[2]
[1]Department of Pathology
[2]Department of Biochemistry
CIBER Epidemiologia y Salud Publica (CIBERESP),
Hospital Costa del Sol, Marbella, Universidad de Malaga, Spain.

Abstract

Apoptotic cell death plays a key role in the pathogenesis, aggressiveness and therapy response of cancer. Besides pathways that accelerate cell division, a disruption of the cell death pathway may contribute to selective expansion of malignant clones. The proteins clusterin and bcl-2 have been reported to present a powerful antiapoptotic activity in some systems. Their implications in carcinogenesis and the progression of some carcinomas make clusterin and bcl-2 interesting genes worthy of investigation. Resistance to anticancer chemotherapeutic drugs remains a major obstacle in cancer chemotherapy and novel therapeutic strategies that target the molecular basis of chemo-resistance are required. The response of cytotoxic drugs is modulated by pro and anti-apoptotic proteins, and defects in apoptosis pathways or the activation of antiapoptotic mechanisms may confer resistance to cytotoxic drug treatment. Combined treatment with clusterin or bcl-2 inhibitors and chemotherapeutic drugs has been shown to be a novel and attractive strategy to inhibit the progression of some human carcinomas.

Introduction

Tissue homeostasis requires a carefully-orchestrated balance between cell proliferation, cellular senescence and cell death. Cells proliferate through a cell cycle that is tightly regulated by cyclin-dependent kinase activities. Cellular senescence is a safeguard program limiting the proliferative competence of cells in living organisms. Apoptosis eliminates unwanted cells by the coordinated activity of gene products that regulate and effect cell death. The intimate link between the cell cycle, cellular senescence, apoptosis regulation, cancer development and tumor responses to cancer treatment has become eminently apparent [1].

The Bcl-2 family of proteins stands among the most crucial regulators of apoptosis and performs vital functions in deciding whether a cell will live or die after cancer chemotherapy and irradiation. In addition, several studies have now revealed that members of the Bcl-2 family also interface with the cell cycle, DNA repair/recombination and cellular senescence, effects that are generally distinct from their function in apoptosis [2].

Indeed both proliferation and apoptotic cell death are very complex process that involves the participation of many genes. Thus, another gene as clusterin has been implicated in anti-apoptotic fuctions, althought it is known that the glycoprotein clusterin has two known isoforms generated in human cells. A nuclear form of CLU protein (nCLU) is pro-apoptotic, while a secretory form (sCLU) is pro-survival. Both forms are implicated in various cell functions, including DNA repair, cell cycle regulation, and apoptotic cell death. CLU expression has been associated with tumorigenesis and the progression of various malignancies [3].

One of the most effective ways to combat different types of cancer is through early diagnosis and administration of effective treatment, followed by efficient monitoring that will allow physicians to detect relapsing disease and treat it at the earliest possible time. Dysregulation of programmed cell death mechanisms plays an important role in the pathogenesis and progression of cancer as well as in the responses of tumours to therapeutic interventions. Clusterin and many members of the bcl-2 family of apoptosis-related genes have been found to be differentially expressed in various malignancies. Results have made it clear that a number of coordinating alterations in apoptosis-related genes must occur to inhibit apoptosis and provoke carcinogenesis in a wide variety of cancers. However, more research is required to increase our understanding of the extent to which and the mechanisms by which they are involved in cancer development, providing the basis for earlier and more accurate cancer diagnosis, prognosis and therapeutic intervention that targets the apoptosis pathways [4].

Clusterin

Originally named for its activity in aggregating spermatozoa [5]· On the basis of the high level of expression in apoptotic tissues, it was originally proposed that the protein might be causally involved in apoptosis [6]. Thus, clusterin mRNA has been widely used as a genetic marker of apoptosis cell death [7]. More recently, however, evidence has accumulated suggesting that clusterin expression is not enhanced, but rather down-regulated in the cells undergoing apoptosis, and that its expression in the apoptotic tissue is restricted to the surviving bystander cells [8,9,10]. Moreover, the expression of the gene is induced in processes other than apoptosis, including mitosis and necrosis, as has been shown in the rat liver [11]. These processes involve substantial membrane remodeling, suggesting that clusterin expression may be involved in maintaining membrane integrity and prevents complement attack during membrane remodeling [12]. If clusterin is involved in preventing immune attack during membrane remodeling, aberrant expression of clusterin may be involved in the pathogenesis of inflammatory and neoplastic diseases. In addition, several experimental works have suggested a role for clusterin in morphologic transformation [13,14,15]. As clusterin may provide a means of eluding immune surveillance and apoptosis, a better understanding of the biology of its expression might lead to new insights into cancer progression. Clusterin has been implicated in a number of biological processes, including sperm maturation, lipid transport, programmed cell death, regulation of the complement cascade, membrane recycling, cell-adhesion, and *src*-induced transformation [16]. Because it has been studied independently by a number of laboratories and in different systems, it has been given various names, such as SGP-2, clusterin, apolypoprotein J, SP 40-40, complement lysis inhibitor, gp80, glycoprotein III and T64 [16]. The human homologue of clusterin is comprised of 449 amino acids with two 40kDa subunits (alpha and beta) joined by a unique five disulfide bond motif [17]. The protein precursor is encoded on a single 2 Kb mRNA that is transcribed from a single copy gene located on chromosome 8 (8p21) [18]. The high degree of sequence conservation across species, the widespread tissue distribution and the high circulating concentration suggest that clusterin plays an important biologic role. Whether clusterin is truly a multifunctional protein or whether a common mechanism underlies its various functions is unclear.

Some functions proposed for clusterin may be relevant in the setting of tumorigenesis: these include complement defense, the initiation of apoptosis, membrane protection and the maintenance of cell-cell or cell-substratum contacts. There is no definitive biochemical evidence to support a specific function for clusterin, except for its role in the modulation of the immune system [6]. Clusterin is part of the fluid phase membrane attack complex (SC5b-9) and functions as a complement inhibitor [6,12,19]. One of the physiological roles of the complement system is the lysis of foreign cells, including virus-infected and tumor cells [20]. Therefore, tumor cells that express surface components that inactivate the complement system would be able to evade destruction by the complement system and thus continue to grow in size as well as progress into more malignant tumors.

In prior studies have been described increased expression of clusterin in several types of carcinomas. It increases following malignant transformation of retinal cells by a mutant Rous sarcoma virus [13]. High levels of clusterin expression occur in experimental prostate carcinomas after initiation with N-nitroso-N-methylurea and promotion with testosterone propionate [14]. Up-regulation of clusterin is associated with premalignant prostatic lesions [15] and clusterin expression is higher in prostate carcinomas when compared with normal prostate epithelium or benign prostate lesions [21]. Transcriptional activity of clusterin is three to four fold higher in clear cell carcinomas of the kidney when compared to its mRNA levels in the normal kidney tissue of the same patients [22]. We examined the expression of this molecule in a large cohort of breast tumors, showing that clusterin is expressed in the malignant epithelium from early to late stages of carcinogenesis in the breast [23].

Our findings demonstrate that the up-regulation of this gene is closely associated with the different steps of tumor progression from normal tissue to premalignant and malignant breast lesions, with a greater expression in lymph node metastasis. Overexpression of clusterin (fig. 1) may thus represent an acquired phenotypic feature which facilitates the local invasion and dissemination of breast tumor cells, in fact, our results also show a positive correlation between clusterin expression and tumor size [23]. Therefore, clusterin could be an important factor in determining the aggressive nature of a given breast tumor. We also correlated clusterin over-expression with negative estrogen and progesterone receptor status. In addition, nuclear clusterin expression was detected in three invasive carcinomas which were also negative for the expression of estrogen and progesterone receptors. In a previous study, Akakura et al [24] reported nuclear clusterin staining in recurrent androgen-independent tumor cells.

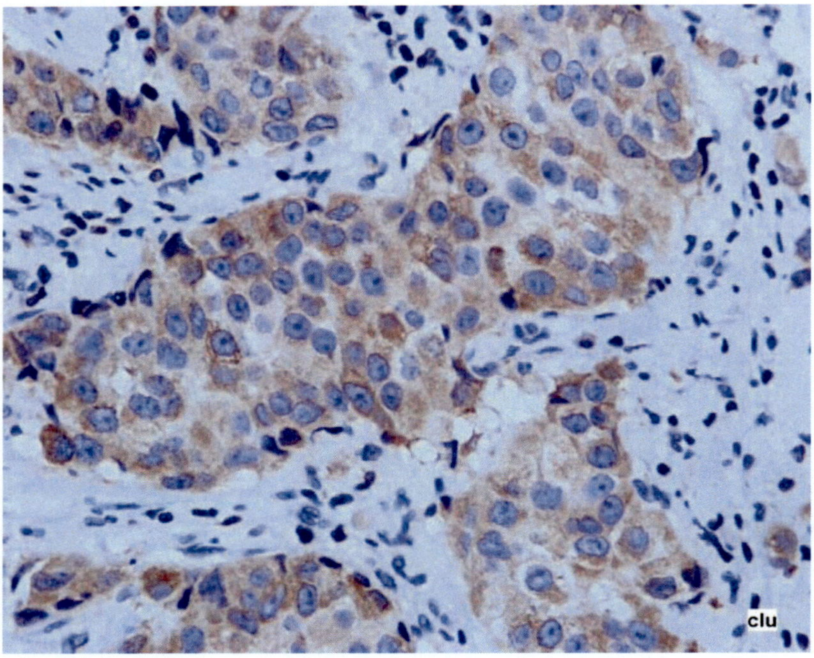

Figure 1. Clusterin. Cytoplasmic expression of clusterin in a moderately-differentiated breast carcinoma (x400).

We also found that clusterin gene expression is not directly associated with programmed cell death in breast tumors, in accordance with other reports [8,11,25]. In fact, we found a lower apoptotic index in clusterin positive tumors than in negative ones. This inverse correlation may support the hypothesis that protection from apoptotic death may account, in part, for biologically aggressive tumor behavior. However, the precise biological function of clusterin in mechanisms of cell death/survival remain to be defined. In this sense, a study reported different isoforms of clusterin in apoptotic and surviving cells in regressing rat ventral prostate. Apoptosis-associated isoforms were predominantly localized to perinuclear regions while surviving associated isoforms were localized to the cytoplasm as is found in most of the primary and metastatic breast carcinomas included in our study. Further studies are needed to determine the exact function of the different isoforms of clusterin and how changes in their expression may modify properties of tumor cells [26].

Clusterin mediated inhibition of complement-induced cytolysis probably protects carcinoma cells from complement-mediated lysis and contributes to the highly metastatic phenotype of those cells. Its presence at cell surfaces and its ubiquitous presence in virtually all types of biological fluids may protect cell membranes exposed to deleterious components. This protective effect may favor the metastatic phenotype of tumor cells which may migrate through tissues and fluids without excessive damage [16]. Thus, an attractive hypothesis is that clusterin induction is a reactive response to environmental changes rather than a causative factor in cell death. In this manner, clusterin could be thought of as an extracellular version of a heat shock protein [27].

We also studied a large series of laryngeal carcinomas. However, clusterin was up-regulated in a low proportion of laryngeal tumors compared with tumors of glandular origin as prostate or breast [28,29]. In these tumors its up-regulation is associated with carcinoma progression. Although the clusterin gene is involved in the carcinogenic pathways in some types of tumors, in laryngeal carcinomas it seem not exert a significant role in carcinogenesis because its mRNA and proteins are detected in few specimens. However, the low proportion of clusterin positive tumors seems to present an aggressive tumor behavior since their expression were associated with local invasiveness.

Interestingly, we found an statistical significant association between the expression of clusterin and bcl-2 proteins. In a recently report [30] clusterin knockdown by siRNA induced down-regulation of bcl-2 in two sarcoma cell lines. The exact mechanism of this association should be elucidated.

We also tried to correlate the expression of clusterin mRNA by in situ hybridization with its protein expression by immunohistochemistry. One study reported differences between both methods in astrocytes and neurons [31]. In our series of laryngeal carcinomas only one tumor presented RNA expression in cytoplasm without clusterin protein expression, which is in accordance with previous studies in other localization [23]. Differences in antibody affinity according to cell types or cell state may explain this opposite results.

The clusterin protein is an inhibitor of apoptosis with a cytoprotective function [32] and thus represents a promising target for molecular intervention strategies such as antisense therapy designed to inhibit its expression, [33]. The over-expression of exogenous clusterin has been shown to result in resistance to paclitaxel [34], doxorubicin [35], cisplatin [36] and radiation therapy [37]. In contrast, decreased clusterin expression by antisense or siRNA expression enhances the chemosensitivities of various cell lines [38,39,40,41], suggesting that clusterin expression is a prominent resistance factor in cancer cells.

On the other hand, glucocorticoids (such as dexamethasone) are routinely used in the clinical application of chemotherapy to prevent adverse effects. A previous study [42] reported the inhibitory action of glucocorticoids on chemotherapy-induced apoptosis, which also raises a clinically relevant question as to whether the pretreatment with glucocorticoids might interfere with the therapeutic efficacy of chemotherapy.

Glucocorticoids play a major role in attenuation of the inflammatory response. These steroid hormones are able to induce apoptosis in cells of the hematopoietic system such as the monocytes, macrophages and T-lymphocytes that are involved in the inflammation reaction. In contrast, it has recently been discovered that in glandular cells such as the mammary gland epithelia, hepatocytes and ovarian follicular cells, and in fibroblasts, glucocorticoids protect against the apoptotic signals evoked by cytokines, cAMP, tumor suppressors and death genes. It is well known that the anti-apoptotic effect of glucocorticoids is exerted by the modulation of survival genes such as Bcl-2, Bcl-x(L) and NFkappaB, in a cell type-specific manner [43]. We hypothesize that clusterin may be one of these genes responsible for the anti-apoptotic effect of glucocorticoids.

Clusterin overexpression has been shown to be associated with the anti-HER 2 antibody trastuzumab (Herceptin) treatment resistance through the inhibition of apoptosis [44]. To explore the potential of the clusterin inhibition approach in breast cancer therapy, the cytotoxic interaction between antisense clusterin oligonucleotide or anticlusterin antibody and the drugs commonly used in breast cancer treatment such as dexamethasone, doxorubicin, paclitaxel and tamoxifen were analyzed in vitro using the breast carcinoma cell lines MCF-7 and MDA-MB-231 [45]. In this study we confirms in the ER positive breast cancer cell line MCF-7 that clusterin levels increase after treatment with tamoxifen and dexamethasone. Previous studies have reported increased clusterin expression after treatment with toremifene [46] and tamoxifen [47]. We also evaluated the effect of cytotoxic treatment on the level of clusterin expression in breast tumor cell lines, because this has been shown to be highly up-regulated in various tissues undergoing apoptotic cell death [48,49]. As expected, clusterin expression in MCF-7 cells was found to increase considerably after cytotoxic treatment, suggesting that clusterin up-regulation is likely to be an adaptative response that mediates chemoresistance. We also investigated the capacity of anti-clusterin treatment to sensitize breast carcinoma cells to chemotherapy, and searched for drug combinations that produce additive cytotoxicity. Antisense clusterin oligonucleotides or anti-clusterin antibodies efficiently inhibited clusterin expression in MCF-7 and MDA-MB-231 cell lines, and this activity was associated with a decrease in cell viability. These findings confirm the cytoprotective function of clusterin in breast carcinoma cells, and suggest there is a role for anticlusterin therapy in the treatment of breast carcinomas which mainly express clusterin protein [41,23].

In addition, we found that in both cell lines, combinations of anticlusterin treatment with chemotherapeutic agents or tamoxifen performed better than the respective single-agent treatments alone. A sequence control oligonucleotide and non-immune IgG fraction did not increase the effect of chemotherapy, which suggests that the sensitization of cells to apoptosis was due to the specific down-regulation of clusterin. A previous report confirmed the chemosensitivity to paclitaxel in breast cancer cell lines [41]. In addition, in other tumors, anticlusterin therapy with antisense oligonucleotides [50,51,52,53] or SiRNA [39] have been evaluated as potential therapeutic agents.

As mention, Dexamethasone, a synthetic steroid, is routinely given to women just before they receive chemotherapy with either paclitaxel or doxorubicin, two drugs commonly used to treat breast cancer. Recent findings show that corticoids could protect against chemotherapy-induced apoptosis [53]. Our results showed that clusterin expression is increased by dexamethasone treatment. A similar response by clusterin after treatment with glucocorticoids was obtained on haemangioma *in vitro* [49]. These results, together with the facts that dexamethasone inhibits chemotherapy-induced cytotoxicity and that antisense oligonucleotides to clusterin increase chemotherapy-induced cytotoxicity clearly show that dexamethasone modulates the expression of clusterin. This effect is in agreement with the known anti-inflammatory properties of dexamethasone in up-regulating the expression of other complement inhibitors [54].

Bcl-2

The bcl-2 gene has also been implicated in oncogenesis by its ability to prolong cell survival through the inhibition of apoptosis, without increasing cell proliferation. The expression of bcl-2 has been measured in a variety of human neoplastic tissues [55,56,57,58].

However, the prognostic significance of bcl-2 protein expression in tumor pathology is contradictory. Some authors have suggested a favorable prognostic role of bcl-2 positivity in breast and non-small cell carcinomas [59,60,61] while others have found a direct relationship between bcl-2 protein expression and worse prognosis in prostate carcinomas and meningiomas [62,63].

We studied the expression of bcl-2 protein and clusterin in a well-documented serie of laryngeal carcinomas with regard to the clinico-pathologic parameters and patients′ disease-free survival[29]. Bcl-2 expression was found to correlate with tumor grade with a significant higher incidence of bcl-2 expression in poorly-differentiated tumors compared with well-differentiated tumors. The same finding has been reported by others [64,65,66,67]. Nodal metastasis was also found to be associated with the expression of bcl-2 which has also been previously reported [64,66], although in one series the opposite result was found [65]. Our results also showed a higher incidence of bcl-2 expression in supraglotic tumors compared to other location. Similar findings has been described previously [66,68]. This association may suggests that tumours originating from different sites in larynx may have different tumor biology.

bcl-2 was shown to be an independent predictor of good prognosis in these tumors Our data are in agreement with several reports that found an association between high expression of bcl-2 and increased survival in larynx cancer [69,70,71]. However, other reports did not find any relationship between bcl-2 and prognosis [68,72]. A smaller number of patients included in these latter studies may have influence this negative result. It cannot be discerned whether bcl-2 is involved directly in contributing to this more indolent phenotype or is simply an epiphenomenon that is a marker for another molecular or biologic process. As bcl-2 does not promote cell proliferation, in the absence of additional genetic alterations bcl-2 positive tumors tend to be relatively non-aggressive. Therefore, bcl-2 expression can be a useful prognostic marker in laryngeal carcinomas and might be helpful in distinguish which tumors with pathological aggressive characteristic (poorly differentiated or with node metastasis) might present a better outcome.

We can conclude that the antiapoptotic proteins bcl-2 and clusterin are up-regulated in a group of larynx carcinomas influencing the progression of these tumors.

In addition, we studied the degree of apoptosis and the apoptosis-influencing protein *bcl-2* during breast carcinoma progression [73].

Apoptosis was rarely present in normal and benign breast lesions. However, in carcinomas the mean apoptotic index was 0.76 (range 0.01-2.8) (fig.2) which is in agreement with previous reports in breast carcinomas [74]. We detected an increase in the extent of apoptosis in metastatic tumours compared to primitive carcinomas, although the association did not reach statistical significance. There are no other reports evaluating apoptosis in metastatic breast carcinomas, although Vakkala et al. [74] reported a significant increase in the extent of apoptosis in recurrent tumours compared to the primary lesions.

Our results also indicate that high AI is related to a number of clinico-pathologic parameters known to predict a higher risk of recurrence, such as grade 3 tumours, large tumour size, and high Ki67 expression and mitotic count. These results are in line with two previous reports [75,76].

The strong association between proliferation and apoptosis indicates that proliferating tumours have an increased chance of apoptosis, which suggest that interactive mechanisms may be involved in the regulation of these pathways. One example of this is the possible role played by *c-myc* in both processes [55].

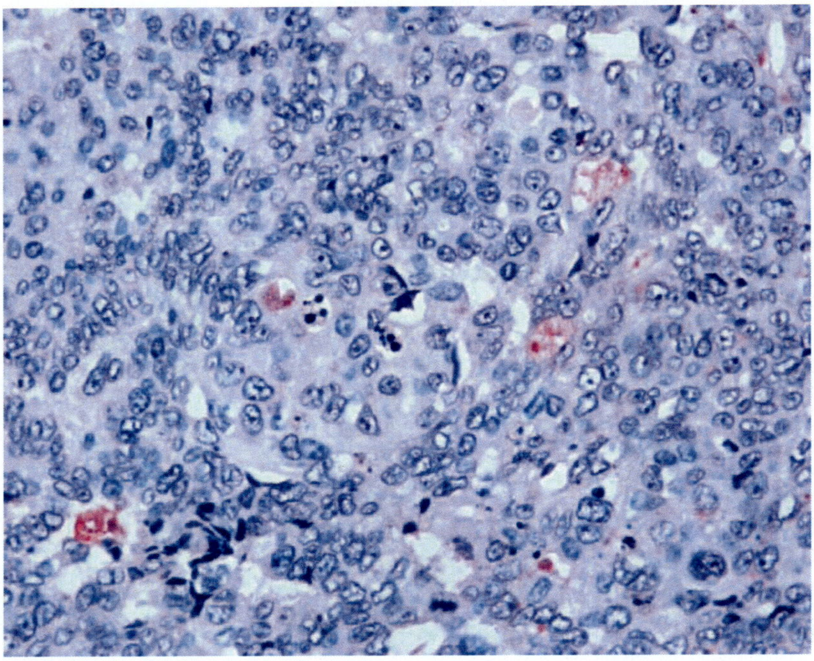

Figure 2. In situ detection of apoptotic cells by TUNEL-staining in breast carcinoma (X 400).

In our study, apoptosis in primary breast carcinomas was independently associated with shorter survival of the patients. A similar association has been observed in two previous reports [66]. Furthermore, increased apoptosis is associated with shortened survival in other epithelial tumours, such as non-small-cell lung carcinomas. Therefore, a high apoptotic index is associated with a number of poor prognostic factors and is independently associated with a worse clinical outcome.

In agreement with previous works on breast [77,78], we found that normal breast tissues have significantly higher levels of *bcl-2* than do neoplastic breast tissues. These data suggest the possibility that the role of this gene in apoptosis may differ in carcinoma and normal tissue. The frequency of *bcl-2* expression in our series (59.5%) is in accordance with several previous studies [58,59,60,79,80]. Moreover, we confirm that *bcl-2* expression is strongly correlated with oestrogen and progesterone receptor status, which is in agreement with a number of reports which indicate that *bcl-2* is an estrogen-receptor regulated gene [60,79,81,82]. As *bcl-2* is present in the normal breast, it may have a role in normal cyclical breast development and be under the control of an oestrogen-dependent transcriptional pathway.

Although the role of the *bcl-2* gene in the development or progression of breast carcinoma has not yet been established, our findings of a significant loss of *bcl-2* expression in high stage breast carcinomas and its more frequent expression in normal epithelium, benign lesions and smaller-diameter tumours suggests that *bcl-2* expression may be an early event in breast tumorigenesis. Furthermore, our results show that overexpression of *bcl-2* increases the risk of lymph node metastases in T_1 (2cm or less) breast carcinomas, which is in agreement with the results previously reported by other authors [83,84]. Therefore, different pathways of progression may be involved in early or late stages of progression. A previous study described the frequent upregulation of *bcl-2* in premalignant lesions in carcinomas in the glandular tissues of the colon, stomach, endocervix and endometrium [85]. However, *bcl-2* might be involved in the carcinogenetic pathway in a different manner in the breast from that of the carcinomas developing from premalignant lesions. Moreover, *bcl-2* expression has been correlated with the tumorigenesis of malignancies of the gastrointestinal tract [86], the prostate [62] and endometrial carcinomas [87]. However, the functional relationship between *bcl-2* expression and malignancy requires further investigation. Furthermore, it has beensuggested that the different roles of *bcl-2* in various types of neoplasia might be partly explained by the fact that the functional activity of *bcl-2* in different tissues is determined by the tissue specific expression of *bcl-2* binding proteins [88,89].

Although several papers have reported *bcl-2* expression in invasive breast cancer, there is considerably less information available on *bcl-2* expression in metastatic breast carcinoma. Our results show that the distribution of *bcl-2* levels did not show any variation between primary and metastatic tumours, suggesting that modification of this gene may influence the acquisition of a malignant phenotype but seems to have no further impact on tumour progression [73].

In our series *bcl-2* expression by univariate analysis was statistically associated with a better clinical outcome and with a number of favourable prognostic features, which is in agreement with previous work on breast carcinomas [58,59,60,77,78] and bronchogenic carcinomas [61]. However, in this series, *bcl-2* was not an independent prognostic factor in the entire cohort by multivariate analysis. Similar results were reported in other series [58,60]. By logistic regression analysis, *bcl-2* was only directly related to estrogen and progesterone receptors, which may indicate that the associations with favorable prognostic features are largely due to its association with hormone receptors. However, this statement is not applicable to patients with lymph node involvement since *bcl-2* represents an independent prognostic factor. It has been hypothesized that the favourable clinical outcome of patients with highly *bcl-2* expressing tumours could be better explained by a relationship between *bcl-2* expression and differentiation than by the role played by the proto-oncogene in the apoptosis process. In this sense, the presence of *bcl-2* immunostaining in normal tissue and its persistence in tumour tissue could indicate a predisposition to differentiation. However in our series, *bcl-2* expression showed no statistical relationship with differentiation (fig. 3), as had been reported in other series [78]. On the other hand, the prognostic information provided by *bcl-2* could be explained by its strict interrelationship with p53 [58]. *bcl-2* and p-53 genes may participate in a common pathway for apoptosis regulation. *bcl-2* has been shown to inhibit apoptosis triggered by wild-type p53 while mutant p53 inhibits apoptosis [90]. In fact, in breast cancer an inverse relationshiphas been observed between *bcl-2* and p53 expression [58]. An important observation in our study was the lack of relationship between *bcl-2* expression and Ki67 immunoreactivity, supporting the hypothesis that *bcl-2* is not a major regulator of proliferation, and therefore, in the absence of additional genetic alterations *bcl-2* positive tumours tend to be relatively non-aggressive. Finally, the increase of associated antiapoptotic proteins may govern cell death regulation in this tumour. In fact, despite of in vitro studies, there is not yet direct evidence supporting the role of *bcl-2* in protecting malignant solid tumors from apoptotic cell death. *bcl-2* reactivity did not correlate with the apoptotic index in our study, suggesting that *bcl-2* expression may not be important in the apoptotic pathway in breast carcinoma. This fact is in agreement with the data from other studies showing that AI is independent of *bcl-2* expression [91,92], although the opposite has also been reported [60,72,74]. It is well recognized that apoptosis regulation is very complex and that *bcl-2* is only one member of a family of genes with different roles in the regulation of cell death. The ratio between *bcl-2* and *bax* ultimately decides whether a cell will undergo apoptosis [92]. The mechanisms by which the interactions between the various members of the *bcl-2* family finally lead to apoptosis are still unknown. It is likely that the control of apoptosis in breast cancer involves genes other than *bcl-2*. Furthermore, more *bcl-2* family members are still being discovered and it has yet to be defined how they are regulated and to what extent their interaction is modulated in a tissue-specific manner. Finally, a study has demonstrated that caspase 3 is able to cleave the loop domain of *bcl-2* at Asp 34 and that the carboxyl-terminal cleavage product triggers and accelerates apoptosis [93]. It remains to be seen whether in tumours showing a positive association between *bcl-2* expression and the extent of apoptosis, *bcl-2* is present in a cleaved, and thus apoptosis-promoting form.

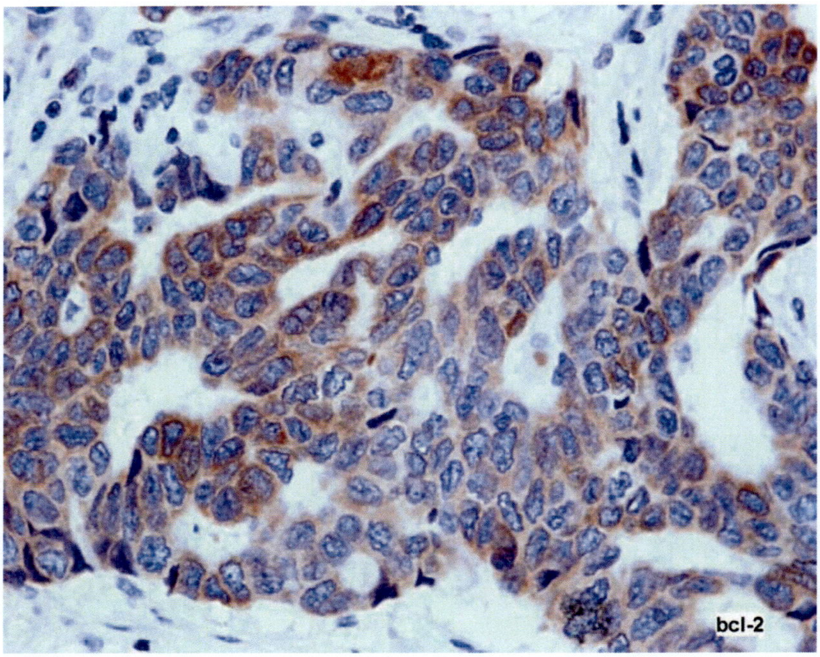

Figure 3. Poorly-differentiated breast carcinoma with strong expression for bcl-2 immunostaining (X 400).

As our study only included a small number of patients who received single hormonal agents as adjuvant therapy, we could not analyze a response to specific treatments. However, some authors have reported significantly longer survival irrespective of the type of adjuvant therapy [59]. Treatment resistance in breast carcinoma may be mediated to a large extent by apoptotic inhibitors other than *bcl-2*. In this context, the tumours in our study with high *bcl-2* did not have low levels of apoptosis. Assessment of the expression of these other regulatory proteins in the study population will help to answer this question.

In summary, *bcl-2* and apoptosis, independently of their functional meaning, are potentially useful prognostic factors in breast carcinoma patients.

Acknowledgements

This research was supported by grants from Fondo de Investigaciones Sanitarias (04/1731 and 06/1062) and Junta de Andalucía (06/377), Spain. M. Redondo belongs to the program <Intensificación de la Investigación> del Instituto de Salud Carlos III

References

[1] Villard JF, Lacombe F, Belloc F, Pellegrin JL, Reiffers J. Molecular mechanisms controlling the cell cycle: fundamental aspects and implications for oncology. *Cancer Radiother*. 2001, 5:109-129.

[2] Zhan Q, Kontny U, Iglesias M, Alamo I JR, Yu K, Hollander MC, Woodworth CD, Fornace AJ, Inhibitory effect of Bcl-2 on p53-mediated transactivation following genotoxic stress. *Oncogene*. 1999 ,18:297-304.

[3] Shannan B, Seifert M, Boothman DA, Tilgen W, Reichrath J. Clusterin and DNA repair: a new function in cancer for a key player in apoptosis and cell cycle control. *J Mol Histol* 2006, 37:183-188.

[4] Thomadaki H, Scorilas A. BCL2 family of apoptosis-related genes: functions and clinical implications in cancer. *Crit Rev Clin Lab Sci*. 2006, 43:1-67.

[5] Fritz IB, Burdzy C, Setchell B and Blaschuck O: Ram rete testes fluid contains a protein (clusterin) which influence cell-cell interactions in vitro. *Biol Reprod* 1993, 28:1173-1188

[6] Jenne DE and Tschopp J. Clusterin: The intriguing guises of a widely expressed glycoprotein. *Trends Biochem Sci* 1992, 17:154-159

[7] Tenniswood MP, Guenette RS, Lakins Y, Mooibroek M, Wong P, Welsh JE: Active cell death in hormone-dependent tissue. *Cancer Metastasis Rev* 1992, 11:197-200

[8] French LE, Wohlwend A, Sappino A-P, Tschopp J, Schifferli JA: Human clusterin gene expression is confined to surviving cells during in vitro programmed cell death. *J Clin Invest* 1994, 93:877-884

[9] French LE, Sappino A-P, Tschopp J, Schifferli JA. Clusterin gene expression in the rat thymus is not modulated by dexamethasone treatment. *Immunology* 1994, 82:328-331

[10] Sensibar J.A, Sutkowski DM, Raffo A, Buttyan R, Griswold MD, Sylvester SR, Kozlowski JM, Lee C: Prevention of cell-death induced by tumor necrosis factor alpha in LNCaP cells by overexpression of sulfated glycoprotein-2 (clusterin). *Cancer Res* 1995, 55:2431-2437

[11] Bursch W, Gleeson T, Kleine L, Teniswood M: Expression of clusterin (testosterone-repressed prostate message-2) mRNA during growth and regeneration of rat liver. *Arch Toxicol* 1995, 69:253-258

[12] Kirsbaum L, Sharpe JA, Murphy B, d´Apice AJF, Classon B, Hudson P, Walkeer ID: Molecular clonning and characterization of the novel, human complement-associated protein, SP40,40: a link between the complement and reproductive system. *EMBO J* 1989, 8:711-718

[13] Michel D, Gillet G, Volovitch M, Pessac B, Calothy G, Brun G: Expression of a novel gene encoding a 51.5 kD precursor protein is induced by different retroviral oncogenes in quail neuroretinal cells. *Oncogene Res* 1989, 4:127-136

[14] Kadomatsu K, Anzano MA, Slayter MV, Winokur TS, Smith JM, Sporn MB: Expression of sulfated Glycoprotein 2 is associated with carcinogenesis induced by Nnitroso-N-methylurea in rat prostate and seminal vesicle. *Cancer Res* 1993, 53:1480-1483

[15] Ho SM, Leav I, Ghatak S, Merk F, Jagannathan VS, Mallery K: Lack of association between enhanced TRPM-2/clusterin expression and increased apoptotic activity in sex-hormone-induced prostatic dysplasia of the noble rat. *Am J Pathol* 1998; 153:131-139

[16] Jordan-Stark TC, Witte DP, Aronow B and Harmony JAK : Apolipoprotein J: a membrane policeman? Curr Op Lipidol 1992, 3:75-85

[17] Kirszbaum L, Bozas SE, Walker ID: SP-40-40 a protein involved in the control of the complement pathway possesses a unique array of disulphide bridges. *FEBS lett* 1992, 297:70-76

[18] Fink TM, Zimmer M, Tschopp J, Etienne J, Jenne DE, and Lichter P: Human clusterin (CLI) maps to 8p21 in proximity to the lipoprotein lipase (LPL) gene. *Genomics* 1993, 16:526-528

[19] Choi NH, Mazda T, Tomita M: A serum protein SP40,40 modulates the formation of membrane attack complex of complement on erythrocytes. *Mol Immunol* 1989, 26:835-840

[20] Hakulinen J and Meri S: Complement-mediated killing of microtumors in vitro. *Am J Pathol* 1998, 153:845-855

[21] Steinberg J, Oyasu R, Lang S, Sintich S, Rademaker A, Lee C, Kozlowski JM, Sensibar JA: Intracellular levels of SGP-2 (clusterin) correlate with tumor grade in prostate cancer. *Clin Cancer Res* 1997, 3:1707-1711

[22] Parczyk K, Pilarsky C, Rachel U, Koch-Brandt C: Gp80 (Clusterin; TRPM2) mRNA level is enhanced in human renal clear cell carcinomas. *J Cancer Res Clin Oncol* 1994, 120:186-188

[23] Redondo M, Villar E, Torres JE, Téllez T, Petito CK, Morell M. Overexpression of clusterin in human breast carcinoma. *Am J Pathol.* 2000, 157:393-399.

[24] Akakura K, Bruchousky N, Rennie PS, Coldman AJ, Goldenberg SL, Tenniswood M, Fox K: Effects of intermittent androgen suppression on the stem cell composition and the expression of the TRPM-2 (clusterin) gene in the Shionogi carcinoma. *J Steroid Biochem Molec Biol* 1996, 59:501-511

[25] Fratelli M, Galli G, Minto M, Pasinetti GM: Role of clusterin in cell adhesion during early phases of programmed cell death in P19 embryonic carcinoma cells. *Biochim Biophys Acta* 1996,1311:71-76

[26] Lakins J, Bennett SAL, Chen JH, Arnold JM, Morrissey C, Wong P, O'Sullivan J, Tenniswood M. Clusterin Biogenesis Is Altered during Apoptosis in the Regressing Rat Ventral Prostate. *J Biol Chem*, **273**: 27887-27895, 1998.

[27] Humphreys DT, Carver JA, Easterbroock-Smith SB, Wilson MR: Clusterin has chaperone-like activity similar to that of small heat shock protein. *J Biol Chem* 1999, 274:6875-6881

[28] Steinberg J, Oyasu R, Lang S, Sintich S, Rademaker A, Lee C, Kozlowski JM, Sensibar JA. Intracellular levels of SGP-2 (clusterin) correlate with tumor grade in prostate cancer. *Clin Cancer Res* 1997, 3:1707-1711.

[29] Redondo M, Esteban F, Gonzalez-Moles MA, Delgado-Rodriguez M, Nevado M, Torres-Muñoz JE, Tellez T, Villar E, Morell M, Petito CK. Expression of the antiapoptotic proteins clusterin and bcl-2 in laryngeal squamous cell carcinomas. *Tumour Biol* 2006, 27:195-200

[30] Trougakos IP, So A, Jansen B, Gleave ME, Gonos ES. Silencing expression of the clusterin/apolipoprotein j gene in human cancer cells using small interfering RNA induces spontaneous apoptosis, reduced growth ability, and cell sensitization to genotoxic and oxidative stress. *Cancer R*es. 2004; 64 :1834-1842.

[31] Pasinetti GM, Johnson S.A, ODA T, Rozovsky I, Finch CE. Clusterin (SGP-2): A multifunctional glycoprotein with regional expression in astrocytes and neurons of the adult rat brain. *J.Comp. Neurol.* 1994; 339:387-400.

[32] 32.Zhang H , Kim JK, Edwards CA, Xu Z, Taichman R, Wang CY: Clusterin inhibits apoptosis by interacting with activated Bax. *Nat Cell Biol* 2005, 7: 909-915.

[33] Olie RA, Zangemeister-Wittke U: Targeting tumor cell resistance to apoptosis induction with antisense oligonucleotides: progress and therapeutic potential. *Drug Resist Update* 2001, 4: 9-15

[34] Miyake H, Nelson C, Rennie PS, Gleave ME: Acquisition of Chemoresistant Phenotype by Overexpression of the Antiapoptotic Gene Testosterone-repressed Prostate Message-2 in Prostate Cancer Xenograft Models. *Cancer Research* 2000, 60:2547-2554

[35] Cervellera M, Raschella G, Santilli G, Tanno B, Ventura A, Mancini C, Sevignani C, Calabretta B, Sala A: Direct transactivation of the anti-apoptotic gene apolipoprotein J (clusterin) by B-MYB. *J Biol Chem* 2000, 275: 21055-21060.

[36] Chung J, Kwak C, Jin RJ, Lee CH, Lee KH, Lee SE: Enhanced chemosensitivity of bladder cancer cells to cisplatin by suppression of clusterin in vitro. *Cancer Lett* 2004, 203:155-161.

[37] Zellweger T, Miyake H, July LV, Akbari M, Kiyama S, Gleave ME: Chemosensitization of human renal cell cancer using antisense oligonucleotides targeting the antiapoptotic gene clusterin. *Neoplasia* 2001, 3:360-367.

[38] Gleave M, Miyake H. Use of antisense oligonucleotides targeting the cytoprotective gene, clusterin, to enhance androgen- and chemo-sensitivity in prostate cancer. *World J Urol* 2005, 23: 38-46

[39] Trougakos IP, So A, Jansen B, Gleave ME, Gonos ES: Silencing expression of the clusterin/apolipoprotein j gene in human cancer cells using small interfering RNA induces spontaneous apoptosis, reduced growth ability, and cell sensitization to genotoxic and oxidative stress. *Cancer Res* 2004, 64:1834-1842.

[40] Klokov D, Criswell T, Leskov KS, Araki S, Mayo L, Boothman DA: IR-inducible clusterin gene expression: a protein with potential roles in ionizing radiation-induced adaptive responses, genomic instability, and bystander effects. *Mutat Res* 2004, 568:97-110.

[41] So A, Sinnemann S, Huntsmann D, Fazli L, Gleave M: Knockdown of the cytoprotective chaperone, clusterin, chemosensitizes human breast cancer cells both in vitro and in vivo. *Mol Cancer Ther* 2005, 4:1837-1849.

[42] Wu W, Chaudhuri S, Brickley DR, Pang D, Karrison T, Conzen SD: Microarray analysis reveals glucocorticoid-regulated survival genes that are associated with inhibition of apoptosis in breast epithelial cells. *Cancer Res* 2004, 64:1757-64.

[43] Amsterdam A , Sasson R: The anti-inflammatory action of glucocorticoids is mediated by cell type specific regulation of apoptosis. *Mol Cell Endocrinol* 2002, 189:1-9.

[44] Biroccio A, D'Angelo C, Jansen B, Gleave ME, Zupi G: Antisense clusterin oligodeoxynucleotides increase the response of HER-2 gene amplified breast cancer cells to Trastuzumab. *J Cell Physiol* 2005, 204: 463-469.

[45] Redondo M, Téllez T, Roldan MJ, Serrano A, García-Aranda M, Gleave ME , Hortas ML, Morell M. *Anticlusterin treatment of breast cancer cells increases chemotherapy- and tamoxifen-sensitivity and counteracts the inhibitory action of dexamethasone on chemotherapy-induced cytotoxicity*. Submitted

[46] Warri AM, Huovinen RL, Laine AM, Martikainen PM, Harkonen PL: Apoptosis in toremifene-induced growth inhibition of human breast cancer cells in vivo and in vitro. *J Natl Cancer Inst* 1993, 85:1412-1418.

[47] Chen H, Tritton TR, Kenny N, Absher M, Chiu JF: Tamoxifen induces TGF-beta 1 activity and apoptosis of human MCF-7 breast cancer cells in vitro. *J Cell Biochem* 1996, 61:9-17.

[48] Sensibar JA, Griswold MD, Sylvester SR, Buttyan R, Bardin CW, Cheng CY, Dudek S, Lee C: Prostatic ductal system in rats: regional variation in localization of an androgen-repressed gene product, sulfated glycoprotein-2. *Endocrinology* 1991, 128:2091-2102.

[49] Kyprianou N, English HF, Davidson NE, Isaacs JT: Programmed cell death during regression of the MCF-7 human breast cancer following estrogen ablation. *Cancer Res* 1991, 51:162-166

[50] Miyake H, Hara I, Kamidono S, Gleave ME: Synergistic chemosensitization and inhibition of tumor growth and metastasis by the antisense oligodeoxynucleotide targeting clusterin gene in a human bladder cancer model. *Clin Cancer Res* 2001, 7:4245-4252

[51] Zellweger T, Miyake H, July LV, Akbari M, Kiyama S, Gleave ME: Chemosensitization of human renal cell cancer using antisense oligonucleotides targeting the antiapoptotic gene clusterin. *Neoplasia* 2001, 3:360-367.

[52] Orlandi A, Pucci S, Ciucci A, Pichiorri F, Ferlosio A, Spagnoli LG: Modulation of clusterin isoforms is associated with all-trans retinoic acid-induced proliferative arrest and apoptosis of intimal smooth muscle cells. *Arterioscler Thromb Vasc Biol* 2005, 25:348-353.

[53] Hasan Q, Tan ST, Xu B, Davis PF: Effects of five commonly used glucocorticoids on haemangioma in vitro. *Clin Exp Pharmacol Physiol* 2003, 30:140-144.

[54] Imai M, Hwang HY, Norris JS, Tomlinson S: The effect of dexamethasone on human mucin 1 expression and antibody-dependent complement sensitivity in a prostate cancer cell line in vitro and in vivo. *Immunology* 2004, 111:291-297.

[55] Fanidi A, Harrington EA, Evan GL, 1992, Cooperative interaction between c-myc and Bcl-2 protooncogene. *Nature (Lond),* **59**, 554-556.

[56] Saegusa M, Takano I, Hashimura M, Shoji Y, Okayasu I, 1995a, The possible role of bcl-2 expression in the progression of tumours of the uterine cervix. Cancer, 76, 2297-2303.

[57] Saegusa M, Takano Y, Okayasu I, 1995b, Bcl-2 expression and its association with cell kinetics in human gastric carcinomas and intestinal metaplasia. *J Cancer Res Clin Oncol*, **121**, 357-363.

[58] Silvestrini R, Veneroni S, Daidone MG, Benini E, Boracchi P, Mezzetti M, Di Fronzo G, Rilke F, Veronesi U, 1994, The bcl-2 protein, a prognostic indicator strongly related to p53 protein in lymph-node-negative breast cancer patients. *J Natl Cancer Inst,* **86**, 499-504.

[59] Gasparini G, Barbareschi M, Doglioni C, Palma PD, Mauri FA, Boracchi P, Bevilaacqua P, Caffo O, Morelli L, Verderio P, 1995, Expression of bcl-2 protein predicts efficacy of adjuvant treatments in operable node-positive breast cancer. *Clin Cancer Res,* **1**,189-198.

[60] Joensuu H, Pylkkanen L, Toikkanen S, 1994, bcl-2 protein expression and long term survival in breast cancer. *Am. J. Pathol,* **145**, 1191-1198.

[61] Pezzella F, Turley H, Kuzn I, Tungekar MF, Dunnill MS, Pierce CB, Harris A, Gatter KC, Mason DY, 1993, bcl-2 protein expression in non-small cell lung carcinoma. *N Engl J Med,* **329**: 690-69.

[62] Krajewska M, Krajewski S, Epstein J.I, Shabaik A, Sauvageot J, Song K, Kitada S, Reed J.C, 1996, Immunohistochemical analysis of bcl-2, bax, bcl-x, and mcl-1 expression in prostate cancers. *Am J Pathol,* **148**: 1567-1576.

[63] Karamitopoulou E, Perentes E, Tolnay M, Probst A. Prognostic significance of MIB-1, p53, and bcl-2 immunoreactivity in meningiomas. *Hum Pathol* 1998, 29: 140-145.

[64] Yuen AP, Lam KY, Choy JT Ho WK, Wei WI. The clinicopathological significance of bcl-2 expression in the surgical treatment of laryngeal carcinoma. *Clin Otolaryngol* 2001; 26:129-133.

[65] Fracchiolla NS, Capaccio P, Carboni N Pagliari AV, Neri A, Ronchett D, Pruner G, Silvotti MG, Pignataro L, Buffa R, Broich G. Immunohistochemical and molecular analysis of bax, bcl-2 and agnor counts as prognostic factors in laryngeal squamous cell carcinomas. *Anticancer Res* 1999; 19:1043-1052.

[66] Jackel MC, Dorudian MA, Marx D, Brinck U, Schauer A, Steiner W. Spontaneous apoptosis in laryngeal squamous cell carcinoma is independent of bcl-2 and bax protein expression. *Cancer* 1999; 85:591-599.

[67] Whisler LC, Wood NB, Caldarelli DD, Hutchinson JC, Panje WR, Friedman M, Preisler HD, Leurgans S, Nowak J, Coon JS. Regulators of proliferation and apoptosis in carcinoma of the larynx. *Laryngoscope* 1998; 108: 630-638.

[68] Hirvikoski P, Kumpulainen E, Virtaniemi J, Pirinen R, Salmi L, Halonen P, Johansson R, Kosma VM. Enhanced apoptosis correlates with poor survival in patients with laryngeal cancer but not with cell proliferation, bcl-2 or p53 expression. *Eur J Cancer* 1999; 35:231-237.

[69] Xie X, Clausen O, De Angelis P, Boysen M. The prognostic value of spontaneous apoptosis, bax, bcl-2, and p53 in oral squamous cell carcinoma of the tongue. *Cancer* 1999; 86:913-920.

[70] Wilson GD, Grover R, Richman PI, Daley FM, Saunders MI, Dische S. Bcl-2 expression correlates with favourable outcome in head and neck cancer treated by accelerated radiotherapy. *Anticancer Res* 1996; 16: 2403-2408.

[71] Homma, A., Furuta, Y., Oridate, N., Nakano Y, Yagi K, Nagahashi T, Fukuda S, Inuyama Y. Correlation of clinicopathological parameters and biological markers related to apoptosis and proliferative activity with a clinical outcome in squamous cell carcinoma of the larynx treated with concurrent chemoradiotherapy. *Auris Nasus Larynx* 2001; 28 Suppl: S87-94.

[72] 72 . Friedman, M., Lim, J.W., Manders, E. Schaffner AD, Kirshenbaum GL, Tanyeri HM, Caldarelli DD, Coon JS. Prognostic significance of Bcl-2 and p53 expression in advanced laryngeal squamous cell carcinoma. *Head Neck* 2001; 23: 280-285.

[73] Villar E, Redondo M, Rodrigo I, Garcia J, Avila E, Matilla A.. .bcl-2 Expression and apoptosis in primary and metastatic breast carcinomas. *Tumour Biol* 2001, 22:137-145.

[74] Vakkala M, Lahteenmaki K, Raunio H, Paakko P, Soini Y, 1999, Apoptosis during breast carcinoma progression. *Clin Cancer Res*, **5**,:319-324.

[75] Zhang GJ, Kimijima I, Abe R, Watanabe T, Kanno M, Hara K, Tsuchiya A, 1998, Apoptotic index correlates to bcl-2 and p53 protein expression, histological grade and prognosis in invasive breast cancers. *Anticancer Res*, **18**: 1989-1998.

[76] Pillai MR, Kesari AL, Chellam VG, Madhavan J, Nair P, Nair MK, 1998, Spontaneous programmed cell death in infiltrating duct carcinoma: association with p53, BCL-2, hormone receptors and tumor proliferation. *Pathol Res Pract* 1998, 194: 549-557.

[77] Hockenbery DM, Zutter M, Hickey W, Nahm M, Korsmeyer SJ, 1991, Bcl-2 protein is topographically restricted in tissues characterized by apoptotic cell death. *Proc Natl Acad Sci.* 1991, 88: 6961-6965.

[78] Kobayashi S, Iwase H, Ito Y, Yamashita H, Iwata H, Yamashita T, Ito K, Toyama T, Nakamura T, Masaoka A,. Clinical significance of bcl-2 gene expression in human breast cancer tissues. *Breast Cancer Res Treat* 1997, 42: 173-181.

[79] Bhargava V, Kell DL, van de Rijn M, Warnke RA,. Bcl2 immunoreactivity in breast carcinoma correlates with hormone receptor positivity. *Am J Pathol* 1994, 145: 535-540.

[80] Leek R, Kaklamanis L, Pezzella F, Gatter KC, Harris AL, 1994, bcl-2 in normal human breast and carcinoma, association with oestrogen receptor positive , epidermal growth factor receptor-negative tumors and in situ cancer. *Br. J. Cancer* 1994, 69: 135-139.

[81] Lipponen P, Pietiläinen T, Kosma VM, Aaltomaa S, Eskelinen M, Syrjänen K.. Apoptosis suppressing protein bcl-2 is expressed in well-differentiated breast carcinomas with favourable prognosis. *J. Pathol* 1995, 177, 49-55.

[82] Teixeira C, Redd JC, Prat MA,. Estrogen promotes chemotherapeutic drug resistance by a mechanism involving bcl-2 proto-oncogene expression in human breast cancer cells. *Cancer Res* 1995, 55: 3902-3907.

[83] Sierra A, Castellsague X, Tortola S, Escobedo A, LLoveras B, Peinado MA, Moreno A, Fabra A, 1996, Apoptosis loss and bcl-2 expression :key determinants of lymph node metastases in T_1 breast cancer. *Clin Cancer Res* 1996, 2: 1887-1894.

[84] Singh Y, Nambu H, Yoshizawa K, Hatano T, Hioki K, Tsubura A. Factors related to axillary lymph node metastasis in T1 breast carcinoma. *Oncol Rep* 1998, 5: 459-462.

[85] Nakamura T, Nomura S, Sakai T, Nariya S. Expression of bcl-2 oncoprotein in gastrointestinal and uterine carcinomas and their premalignant lesions. *Hum Pathol* 1997, 28: 309-315.

[86] Bronner MP, Culin C, Reed JC, Futh EE, The bcl-2 protooncogen and the gastrointestinal epithelial tumor progression model. *Am J Pathol* 1995, 146: 20-26.

[87] Chan WK, Mole MM, Levison DA, Ball RY, Lu QL, Patel K, Hanby AM. Nuclear and cytoplasmic bcl-2 expression in endometrial hyperplasia and adenocarcinoma. *J Pathol* 1995, 177, 241-246.

[88] Boise LH, Gónzalez-García M, Postema CE, Ding L, Lindstein T, Turka LA, Mao X, Nuñez G, Thompson CB. bcl-x, a bcl-2 related gene that functions as a dominant regulator of apoptotic cell death. *Cell* 1993, 74: 597-608.

[89] Oltvai ZN, Millimann CL, Korsmeyer SJ. bcl-2 heterodimerizes in vivo with a conserved homolog, bax, that accelerates programmed cell death. *Cell* 1993*,* 74: 609-619.

[90] Wang Y, Szekely L, Okam I, Klein G, Wiman KG,1993, Wild-type p53 triggered apoptosis is inhibited by bcl-2 in a v-myc-induced T-cell lymphoma line. *Oncogene* **8:** 3427-3431.

[91] Berardo MD, Elledge RM, de Moor C, Clark GM, Osborne CK, Allred DC. bcl-2 and apoptosis in lymph node positive breast carcinoma. *Cancer* 1998, 82: 1296-1302.

[92] Lu Q, Abel P, Foster CS, Lalani E,. bcl-2: role in epithelial differentiation and oncogenesis. *Hum. Pathol* 1996, 27: 102-110.

[93] Cheng EH-Y, Kirsch DG, Clem RJ, Ravi R, Kastan MB, Bedi A, Veno K, Hardwick JM, Conversion of bcl-2 to a bax-like death effector by caspases. *Science* 1997, 278: 1966-1968.

In: Oncogene Proteins: Structure, Functions and Analyses ISBN: 978-1-60456-259-0
Editors: P.B. Murphy and J.R. Clarke, pp. 309-321

Chapter XIII

Proteomics, Bioinformatics and Advanced Biotechnology for Oncogene Protein Research

Viroj Wiwanitkit
Department of Laboratory Medicine, Faculty of Medicine, Chulalongkorn University, Bangkok Thailand 10330

Abstract

Oncogene protein is an important group of protein. It strongly related to the oncogenesis mechanism. Application of proteomics techniques to oncogene protein research can help finding new knowledge. Basic biocheminformatics applications for oncogene protein research included genomics, proteomics, transcriptomics, immunomics, metabolomics, cytomics and systomics. Expression analysis of the oncogene protein can also be performed by the in vivo, in vitro and in silico techniques. In addition, the application of other new biotechnology including nanomedicine technology can help driven the difficult oncogene protein research to the new easier generation.

What's Oncogene Protein

Basically, cancer is caused by the malfunction of genes that regulate cell proliferation [1]. Two kinds of regulatory genes have been discovered in the search for cancer genes: those that promote growth, called oncogenes, and those that suppress growth, called anti-oncogenes or cancer suppressor genes [1]. Oncogene protein or oncoprotein is an important group of protein. It strongly related to the oncogenesis mechanism. Oncogene protein is the product of oncogene. The oncogene encodes the mRNAs then sends them out of the nucleus for protein production in the cytosol and be secreted out to and trans cell membrane into the external environment. The main function of an oncogene is to control the growth of cells to make excessive proteins which further make the cells grow out of control. Abnormal expression and function of these genes and proteins are responsible for transformation to the malignant phenotype. Several oncogene proteins are mentioned in oncology. Several models can be proposed for such a mechanism, including the insertion nearby of a viral promoter, alteration of the physiological promoter by a mutagenic agent, gene amplification, relocation in a transcriptionally active region of the genome as a consequence of chromosomal rearrangements and point mutations induced by external factors [2]. Examples of these different mechanisms of oncogene activation can be found in animal and human tumors [2].

Application of proteomics techniques to oncogene protein research can help finding new knowledge. Basic biocheminformatics applications for oncogene protein research included genomics, proteomics, transcriptomics, immunomics, metabolomics, cytomics and systomics.

Genomics for Oncogene Proteins Research

Genomics can be applied for study of the oncogenes. Genomic approaches to cancer are beginning to have an important impact in unraveling the complex etiologies of this disease, as well as allowing us to rationally treat afflicted patients. Many new applied gene expression profiling, high-density single-nucleotide polymorphism arrays, and systematic resequencing techniques can be used for identification novel oncogenes and signaling pathways in cancer that might represent therapeutic targets [3]. Comparative genomics has emerged as a valuable tool for locating genes, transcription factor motifs and other putative control regions [4]. This technique can be applied for oncogene proteins research. Based on comparative genomics, many researches focus largely on approaches to cancer susceptibility. In a tumor cell, dozens of different genes may be aberrant in structure or copy number, and hundreds or thousands of genes may be differentially expressed [5]. A number of familial cancer genes with high-penetrance mutations have been identified, but the contribution of low-penetrance genetic variants or polymorphisms to the risk of sporadic cancer development remains unclear [5]. Studies of the complex somatic genetic events that take place in the emerging cancer cell may aid the search for the more elusive variants that confer increased susceptibility. With new comparative genomics tool, the mutation prone point within the gene can also be easily detected. The knowledge from the comparative genomics research on oncogene can be useful for prediction of the possible oncogene protein presentation. For example, Lillington et al study primary retinoblastoma tumors by the use of comparative genomic hybridization and clinical/histological correlations [6]. According to this work, it is possible that tumors displaying loss of 13q and 5q indicate those patients who may suffer an adverse outcome and who would require alternative or more intensive therapy [6]. Herzog et al used comparative genomic hybridization to search for DNA copy number changes in isolated unilateral retinoblastomas [7]. They suggested that the progression of retinoblastomas from older patients followed mutational pathways different from those of younger patients [7]. Zhou et al investigated gastric MALT lymphomas with and without t(11;18) by comparative genomic hybridization and the results showed the genetic difference between MALT lymphomas with and without t(11;18) [8]. Zhou et al suggested that genomic gain of genes that modulate NFkappaB activation, such as MALT1, TRAF2 and CARD9, might play a role in the pathogenesis of the translocation-negative MALT lymphoma [8]. In additional to comparative genomics, functional genomics can be applied for study of oncogene. The effect of polymorphism of gene can be easily tested by gene ontology technique. The gene ontology tool can be used for clarified the effect of different genetic polymorphisms.

Proteomics for Oncogene Proteins Research

In recent years the discovery of cancer biomarkers has become a major focus of cancer research. With the advent of proteomic technologies such as DNA and tissue microarray, two-dimensional gel eletrophoresis, mass spectrometry and protein assays coupled with advanced bioinformatic tools, it is possible to develop biomarkers that are able to reliably and accurately predict outcomes during cancer management and treatment [9]. In addition, proteomic profiling of altered proteins by anticancer drugs on cancer cells may contribute to providing the fundamental resources for investigation of disease-specific target proteins, elucidation of the novel mechanisms of action and development of new drugs [10]. The advent of proteomics has provided the hope of discovering novel biological markers for use in the screening, early diagnosis and prediction of response to therapy [10]. It is accepted that functional proteomics is a promising technique for the rational identification of biomarkers and novel therapeutic targets for cancers.

For diagnosis, proteomics profiling has recently been investigated using blood or other body fluid for the detection purpose. A key challenge in the clinical proteomics of cancer is the identification of biomarkers that would enable early detection, diagnosis and monitoring of disease progression to improve long-term survival of patients [11 – 12]. Cancer proteomics is a main part of the proteomics, by which we can understand the protein profiles during the different stages of the oncogenesis and it can help discover of the tumor-specific biomarkers. Recent advances in proteomic instrumentation and computational methodologies offer a unique chance to rapidly identify these new candidate markers or pattern of markers [11 – 12]. Proteomics has distinct advantages over genomic and ribonucleic acid expression studies because it is the proteins that are ultimately responsible for the malignant phenotype [11]. The goal of traditional proteomic studies is to identify disease-specific biomarkers [11]. Two-dimensional (2-D) gel electrophoresis (polyacrylamide gel electrophoresis; PAGE) coupled with mass spectrometry is the most widely used experimental strategy and has yielded several potentially relevant cancer biomarkers [11]. Novel mass spectrometry-(MS) based technologies in particular, such as surface-enhanced laser desorption/ionisation time of flight (SELDI-ToF-MS), have shown promising results in the recent literature [12]. The combination of retentate affinity chromatography and MS is one of the most interesting new approaches for cancer diagnostics using proteomic profiling [13]. Despite promising results, the use of the proteomic profiling as a diagnostic tool brought some controversies and technical problems and still requires some efforts to be standardised and validated [14]. For example, Jones et al performed 2-D PAGE on lysates from microdissected ovarian tumors (three invasive ovarian cancers and two noninvasive, low malignant potential (LMP) ovarian tumors) and then compared silver stained 2-D gels created from microdissected lysates with SYPRO-Ruby stained 2-D PAGE profiles of the patient-matched undissected bulk tumor lysates from all patients [15]. They proposed that the direct comparison of laser capture microdissection generated proteomic profiles of invasive versus LMP ovarian cancer might more directly generate important markers for early detection and/or therapeutic targets unique to the invasive phenotype [15]. Ornstein et al performed another study by proteomic analysis of laser capture microdissected human prostate cancer and in vitro prostate cell lines [16]. According to this study, they demonstrated that 2-D PAGE analysis of laser capture microdissected -derived cells could reliably detect alterations in protein expression associated with prostate cancer, and that these differentially expressed proteins were produced in high enough levels which could allow for their clinical utility as new targets for therapeutic intervention and serum markers [16]. Skekouh et al performed another similar proteomics analysis study on pancreas cancer [17]. According to this work, they performed laser capture microdissection to enrich for both normal and malignant pancreatic ductal epithelial cells [17]. They indicated that the combination of laser capture microdissection with 2-DE provided an effective strategy to discover proteins that are differentially expressed in pancreatic ductal adenocarcinoma [17]. In addition, the proteomics analysis can also bring the new knowledge in the pathogenesis of cancer. For example, Vandermoere et al studied proteins co-immunoprecipitated with Akt in MCF-7 breast cancer cells [18]. According to this work, cortical remodeling of actin associated with cell migration was reversed by small interfering RNA directed against Akt, indicating the involvement of Akt in the dynamic reorganization of actin cytoskeleton germane to breast cancer cell migration. Together these data identify actin as a new functional target of Akt signaling [18]. Jin et al implemented a proteomics approach to the systematical analysis of the alterations in the proteome of NIH3T3 cells transformed by oncogenic H-RasV12 [19]. A main finding was the up-regulation of gankyrin, which was the first evidence to show that gankyrin pathway was implicated in Ras-activated transformation [19].

Proteomics, the global analysis of expressed cellular proteins, provides powerful tools for the detailed comparison of proteins from normal and neoplastic tissue [20 – 22]. In additional to diagnostic purpose, there are also many researches on proteomics analysis for therapeutic purpose. In the preproteomic studies, one protein had generally been held responsible for a given chemoresistance [20]. Basically, chemotherapy forms a significant component of the treatment of many human cancers. The failure of malignant cells to respond to the action of cytotoxic drugs is referred to as drug resistance. Different types of cancer are naturally resistant to many anticancer drugs [21]. Additionally, these tumours develop acquired drug resistance, which includes the classical multidrug resistance accompanied by the synthesis of P-glycoprotein, a member of the superfamily of ATP-binding cassette transporters [21]. The challenge of elucidating the link between drug response and resistance, and then integrating the information for the purpose of therapeutic intervention, is a major task that awaits scientists in the post-genomic era [23]. To overcome chemoresistance problems, antineoplastic drugs are often combined with other modes of therapy where good response has been reported in several experimental tumour models and in advanced cancer patients [21]. Several laboratory techniques, such as polymerase chain reaction, immunocytochemistry, flow cytometry, blotting, and fluorescent microscopy have been used for the identification of drug resistance markers and mechanisms [24]. As previously mentioned, proteomic analysis can help solve this problem. In particular, cancer cell culture models are suited for applying proteomics techniques, such as 2D-PAGE and MS, to identify specific protein expression profiles and/or proteins that may be associated with a defined phenotype of the cancer cells [22]. For example, Sinha et al used proteomic tools to study drug resistance in melanoma [25]. Sinha et al impressed that it was clear that this proteomic approach for studying chemoresistance phenomena was a prerequisite before further investigation could yield insight into the biology and development of drug resistance in malignant melanoma [25]. Castagna performed a similar study to study the chemoresistance in cervical cancer [26]. According to this work, the basal levels of HSC71 and HSP60 were increased in A431/Pt cells as compared to A431 cells, and cisplatin exposure resulted in up-regulation of HSP60 and HSP90 only in A431 cells [26]. Liu et al compared the global protein profiles between MCF7 and MCF7/AdVp3000 cells in breast cancer [27]. They reported that 14-3-3sigma was a contributor to drug resistance in human breast cancer cells [27]. The rapid accumulation of genomic and proteomic databases for complex biological systems, such as cancer, together with advances in technology platforms, have paved the way to an increased molecular understanding and prediction of antitumor drug response [28]. Several microarray studies of genetic patterns from untreated and pre-treated cancers have provided "fingerprints" that can predict response to therapeutics [28]. However, analysis via proteomic tools may reveal entire sets of proteins that are up- or downregulated in chemoresistant tumor cell lines compared with parental tumor lines. Therefore, it appears more realistic to expect that exposure of cells to drugs results in the activation of different mechanisms of resistance [20].

Oncoproteomics

Oncoproteomics is the application of proteomics technologies in oncology [29 - 30]. Recent progress in proteomics has opened new avenues for tumor-associated biomarker discovery [29]. With the advent of new and improved proteomics technologies, such as the development of quantitative proteomic methods, high-resolution, -speed and -sensitivity mass spectrometry and protein arrays, as well as advanced bioinformatics for data handling and interpretation, it is now possible to discover biomarkers that can reliably and accurately predict outcomes during cancer management and treatment [29]. Oncoproteomics has the potential to revolutionize clinical practice, including cancer diagnosis and screening based on proteomic platforms as a complement to histopathology, individualized selection of therapeutic combinations that target the entire cancer-specific protein network, real-time assessment of therapeutic efficacy and toxicity, and rational modulation of therapy based on changes in the cancer protein network associated with prognosis and drug resistance [31]. Furthermore, oncoproteomics is also applied to the discovery of new therapeutic targets and to the study of drug effects [31]. Molecular diagnostics, laser capture microdissection and protein biochips are among the technologies that are having an important impact on oncoproteomics [32]. The discovery of new protein patterns help distinguish cancer and disease-free states with high sensitivity and specificity and can facilitate the development of personalized therapy of cancer [32].

Diagnostic oncoproteomics is the application of proteomic techniques for the diagnosis of malignancies. One probable solution to finding cancer-associated protein signatures is the emerging technology of affinity proteomics [33]. This approach addresses some of the shortcomings of traditional proteomics and combines it with the power of microarrays [33]. A new mass spectrometric technology involves surface enhanced laser desorption ionization combined with time-of flight mass analysis (SELDI-TOF-MS), using special protein chips [34]. Based on SELDI-TOF MS, Ciphergen's proteinchip system offers a single, unified, high-throughput platform for a multitude of proteomic research applications [35]. The remarkable findings in disease biomarkers have shed light to the early diagnosis, monitoring and predicting prognosis of cancer [35]. Its high-throughput nature and simplicity in its experimental procedures have allowed this technology to become a popular research tool for biomarker discovery in the past recent years [36]. However, Henderson and Steele said that the reproducibility, standardization and feasibility of this technology was required to be addressed before these proteomic approaches to the discovery of novel, highly sensitive diagnostic tools could become routine clinical care [37]. Jain concluded that continued refinement of techniques and methods to determine the abundance and status of proteins in vivo brought great promise for future study of cancer and development of personalized cancer treatment [38].

Pharmacogenomics and Toxicogenomics for Oncogene Proteins Research

Pharmacogenomics is the advent in pharmacology. With advanced techniques, pharmacogenetics can help develop screening mechanisms to optimise drug therapy. DNA variations in metabolism, transport and drug target genes may contribute to chemotherapy efficacy and toxicities [39]. Molecular signature of cancer guiding drug choice toward personalized therapy is the new modality in cancer treatment [40 - 41].

There has been a change from an emphasis on non-specific, cytotoxic agents to specific, molecular-based cancer therapy. Mechanism-based therapy is designed to hit on cellular and molecular targets that are causally involved in the formation, growth, and progression of human malignancies [42]. New targets are identified and validated by state-of-the-art approaches, including high-throughput screening, combinatorial chemistry, and gene expression arrays, which increase the speed and efficiency of drug discovery and development [42]. Various studies have also demonstrated the potential for using gene expression profiles for the analysis of oncogenic pathways [43]. The data from those study are useful for designing of new oncogene-based molecular therapeutic options Examples of oncogene-based, molecular therapeutics that show promising clinical activity include trastuzumab (Herceptin), imatinib (Gleevec), and gefitinib (Iressa) [42]. However, the full potential of oncogenes as novel targets for cancer therapy has not been realized and many challenges remain, from the validation of novel targets, to the design of specific agents, to the evaluation of these agents in both preclinical and clinical settings [42]. It can be seen that pharmacogenomics research can bring a lot of advantage to cancer therapy. For example, Xu et al studied the pharmacogenomic profiling of the PI3K/PTEN-AKT-mTOR pathway in common human tumors [44]. This is the first report on the PI3K/PTEN-AKT-mTOR pathway in common human tumors and evaluation of the coordinated expression of different signaling proteins [44]. This study also provide a useful tool for selecting future targeted phase II and III clinical trials in the development of this exciting class of agents [44].

In additional to pharmacogenomics, toxicogenomics can also be applied to study the oncogene. Basically, toxicogenomics is the study of the structure and output of the genome as it responds to adverse xenobiotic exposure [45]. Large-scale transcriptional analysis, made possible through microarray technologies, enables us to study and understand the complexity of the biological effects of drugs and chemicals, with the ultimate goal of separating wanted effects from adverse effects [45]. The need for a thorough assessment of pharmacogenomic effects of polymer therapeutics to maximize the clinical outcomes and understand the pharmacological and toxicological effects of polymer-based drugs and delivery systems is mentioned and this bring us to the new omics science of polymer genomics [46 - 47]. Synthetic polymers and nanomaterials display selective phenotypic effects in cells and in the body signal transduction mechanisms involved in inflammation, differentiation, proliferation, and apoptosis [46 -47]. When physically mixed or covalently conjugated with cytotoxic agents, bacterial DNA or antigens, polymers can drastically alter specific genetically controlled responses to these agents [46 – 47]. These are the new facets to be concerned in cancer treatment.

Transcriptomics for Oncogene Proteins Research

Expression genomics can be defined as the study of the dynamic transciptome and its regulatory elements [48]. Technologies are available that can assess transcripts on a genome-wide scale over time and across many samples [48]. This comprehensive and dynamic database is being used to decipher signaling pathways and to identify new biomarkers and targets [48]. In silico transcriptomics can be used for searching for tumor-associated antigens for immunotherapy [49]. A complementary approach to experimental tumor antigen discovery is to screen the human genome in silico, particularly the databases of "Expressed Sequence Tags" (ESTs), in search of tumor-specific and tumor-associated antigens [49]. The public databases currently provide a massive amount of ESTs from several hundreds of cDNA tissue libraries, including tumoral tissues from various types [49]. Integrated application of transcriptomics and metabonomics can also yield new insight into the toxicity of anticancer drug.

Immunomics for Oncogene Proteins Research

Immumomics for oncogene is the new omics science. The main application of immunomics is to find the vaccine candidate. For an example of application of immunomics in cancer, searching for autoantibodies in malignancies by immunomics is presently widely used. Several proteomics approaches for the identification of antigens recognized by these autoantibodies can be used. Serological proteome analysis combines separation of tumor cell proteins on 2D gel electrophoresis gels, western blotting with sera of patients and healthy subjects, and identification of the detected antigens by MS [50]. Alternatively multiple affinity protein profiling combines isolation of the antigens recognized by patient antibodies by two-dimensional immunoaffinity chromatography and identification by MS/MS [50]. Bradford et al used the described technique to study prostate cancer [51]. Using a panel of 22 peptides, Bradford et al were able to detect prostate cancer with high specificity and sensitivity [51]. Glycoimmunomics is another important application. Since future strategies for the treatment of human cancer require a full appreciation of the intracellular and extracellular changes that accompany neoplastic transformation, elucidation of the profiles and properties of all molecules associated with tumor progression is required to develop a successful strategy to treat human cancer [52]. Apart from general peptides, glycoantigens and lipoantigens are usually neglected and the cancer glycoimmunomics has been launched to fulfill this gap for a few recent years. However, the most focus for the present oncogene protein research is development of new vaccine. The development of DNA microarray technology a decade ago led to the establishment of functional genomics as one of the most active and successful scientific disciplines today [53]. With the ongoing development of immunomic microarray technology-a spatially addressable, large-scale technology for measurement of specific immunological response-the new challenge of functional immunomics is emerging, which bears similarities to but is also significantly different from functional genomics [53]. Based on functional immunomics, several new in silico techniques can be used for vaccine candidate finding. For example, Wiwanitkit recently find a T-cell epitope for a melanoma vaccine by an immunomics technique [54]. The data from this study is useful for further vaccine development because these promiscuous peptide binders allows to minimize the total number of predicted epitopes without compromising the population coverage required in the design of multi-epitope vaccines [54].

Metabolomics for Oncogene Proteins Research

Metabolomics is one of the newest "omics" sciences. It is a dynamic portrait of the metabolic status of living systems [55]. Metabolomics has brought new insights on metabolic fluxes and a more comprehensive and holistic understanding of a cell's environment [55]. This omic science can be well applied for oncology. This technology permits simultaneous monitoring of many hundreds, or thousands, of macro- and small molecules, as well as functional monitoring of multiple pivotal cellular pathways [55]. In addition, elucidation of cellular responses to molecular damage, including evolutionarily conserved inducible molecular defense systems, could be derived with metabolomics and could help find new biomarkers of molecular responses to functional perturbations [55]. For example, inherent and specific metabolic profiles of major brain tumour cell types, as determined by proton nuclear magnetic resonance spectroscopy ((1)H MRS), have also been used to define metabolite phenotypes in tumours in vivo [56]. However, the application of metabolomics is rather from the main focus on oncogene protein research.

Cytomics and Systomics for Oncogene Proteins Research

Cytomics and systomics can be applied in oncology [57 – 58]. The quantitative analysis of networking molecules within the cells might be applied to understand native-state tissue signalling biology, complex drug actions and dysfunctional signalling in transformed cells, that is, in cancer cells [57]. High-content and high-throughput single-cell analysis can lead to systems biology and cytomics [57]. The application of cytomics in cancer research and diagnostics is very broad, ranging from the better understanding of the tumour cell biology to the identification of residual tumour cells after treatment, to drug discovery [57]. Similar to metabolomics, cytomics and systomics are rather from the main focus on oncogene protein research.

Cancer Phenomics

Finally, cancer phenomics, the systematic acquisition and objective documentation of host and/or somatic cancer phenotypic data at many levels, is a new field compared with other molecular-based 'omics' [59]. Basically, connecting genotype to phenotype is fundamental in biomedical research and in our understanding of disease [60]. Phenomics, the large scale quantitative phenotypic analysis of genotypes on a genome-wide scale, connects automated data generation with the development of novel tools for phenotype data integration, mining and visualization [60]. This can be the new trend in oncogene protein research.

Application of Nanomedicine Technology for Oncogene Protein Research

Nanomedicine can be applied for oncogene protein research. Nanodiagnosis is already applied for biomarker diagnosis. Quantitative protein analysis from formalin-fixed tissues is an example of new nanoscale molecular diagnosis for oncogene protein research [61]. This technique is a robust, fast, standardized, and easy to use technique for the solubilization of non-degraded, full length, and immunoreactive proteins from formalin-fixed tissues for western blot and protein microarray analysis [61]. Other new applications of nanomedicine is expected to be launched within the next few years.

References

[1] Yarbro JW. Oncogenes and cancer suppressor genes. *Semin Oncol Nurs*. 1992 Feb;8(1):30-9.

[2] Wong-Staal F. The oncogene and its potential role in carcinogenesis. *Arch Toxicol Suppl*. 1985;8:61-72.

[3] Thomas RK, Weir B, Meyerson M. Genomic approaches to lung cancer. *Clin Cancer Res*. 2006 Jul 15;12(14 Pt 2):4384s-4391s.

[4] Ptacek T, Sell SM. A tiered approach to comparative genomics. *Brief Funct Genomic* Proteomic. 2005 Jul;4(2):178-85.

[5] Balmain A, Gray J, Ponder B. The genetics and genomics of cancer. *Nat Genet*. 2003 Mar;33 Suppl:238-44.

[6] Lillington DM, Kingston JE, Coen PG, Price E, Hungerford J, Domizio P, Young BD, Onadim Z. Comparative genomic hybridization of 49 primary retinoblastoma tumors identifies chromosomal regions associated with histopathology, progression, and patient outcome. *Genes Chromosomes Cancer*. 2003 Feb;36(2):121-8..

[7] Herzog S, Lohmann DR, Buiting K, Schüler A, Horsthemke B, Rehder H, Rieder H. Marked differences in unilateral isolated retinoblastomas from young and older children studied by comparative genomic hybridization. *Hum Genet*. 2001 Feb;108(2):98-104.

[8] Zhou Y, Ye H, Martin-Subero JI, Hamoudi R, Lu YJ, Wang R, Siebert R, Shipley J, Isaacson PG, Dogan A, Du MQ. Distinct comparative genomic hybridisation profiles in

gastric mucosa-associated lymphoid tissue lymphomas with and without t(11;18)(q21;q21). *Br J Haematol.* 2006 Apr;133(1):35-42.

[9] Chatterjee SK, Zetter BR Cancer biomarkers: knowing the present and predicting the future. *Future Oncol.* 2005 Feb;1(1):37-50.

[10] Righetti PG, Castagna A, Antonioli P, Cecconi D, Campostrini N, Righetti SC. Proteomic approaches for studying chemoresistance in cancer. *Expert Rev Proteomics.* 2005 Apr;2(2):215-28.

[11] Ornstein DK, Tyson DR. Proteomics for the identification of new prostate cancer biomarkers. *Urol Oncol.* 2006 May-Jun;24(3):231-6.

[12] Seibert V, Ebert MP, Buschmann T. Advances in clinical cancer proteomics: SELDI-ToF-mass spectrometry and biomarker discovery. *Brief Funct Genomic Proteomic.* 2005 May;4(1):16-26.

[13] Solassol J, Jacot W, Lhermitte L, Boulle N, Maudelonde T, Mangé A. Clinical proteomics and mass spectrometry profiling for cancer detection. *Expert Rev Proteomics.* 2006 Jun;3(3):311-20.

[14] Solassol J, Boulle N, Maudelonde T, Mangé A. Clinical proteomics: towards early detection of cancers. *Med Sci (Paris).* 2005 Aug-Sep;21(8-9):722-9.

[15] Jones MB, Krutzsch H, Shu H, Zhao Y, Liotta LA, Kohn EC, Petricoin EF. Proteomic analysis and identification of new biomarkers and therapeutic targets for invasive ovarian cancer. *Proteomics.* 2002 Jan;2(1):76-84.

[16] Ornstein DK, Gillespie JW, Paweletz CP, Duray PH, Herring J, Vocke CD, Topalian SL, Bostwick DG, Linehan WM, Petricoin EF, Emmert-Buck MR. Proteomic analysis of laser capture microdissected human prostate cancer and in vitro prostate cell lines. *Electrophoresis.* 2000 Jun;21(11):2235-42.

[17] Shekouh AR, Thompson CC, Prime W, Campbell F, Hamlett J, Herrington CS, Lemoine NR, Crnogorac-Jurcevic T, Buechler MW, Friess H, Neoptolemos JP, Pennington SR, Costello E. Application of laser capture microdissection combined with two-dimensional electrophoresis for the discovery of differentially regulated proteins in pancreatic ductal adenocarcinoma. *Proteomics.* 2003 Oct;3(10):1988-2001.

[18] Vandermoere F, El Yazidi-Belkoura I, Demont Y, Slomianny C, Antol J, Lemoine J, Hondermarck H. Proteomics exploration reveals that actin is a signaling target of the kinase *Akt Mol Cell Proteomics.* 2007 Jan;6(1):114-24.

[19] Jin BF, He K, Wang HX, Bai B, Zhou T, Li HY, Man JH, Liu BY, Gong WL, Wang J, Li AL, Zhang XM. Proteomics analysis reveals insight into the mechanism of H-Ras-mediated transformation. *J Proteome Res.* 2006 Oct;5(10):2815-23.

[20] Righetti PG, Castagna A, Antonioli P, Cecconi D, Campostrini N, Righetti SC. Proteomic approaches for studying chemoresistance in cancer. *Expert Rev Proteomics.* 2005 Apr;2(2):215-28

[21] Poland J, Schadendorf D, Lage H, Schnolzer M, Celis JE, Sinha P. Study of therapy resistance in cancer cells with functional proteome analysis. *Clin Chem Lab Med.* 2002 Mar;40(3):221-34.

[22] Lage H. Proteomics in cancer cell research: an analysis of therapy resistance. *Pathol Res Pract.* 2004;200(2):105-17.

[23] Verrills NM, Kavallaris M. Drug resistance mechanisms in cancer cells: a proteomics perspective. *Curr Opin Mol Ther.* 2003 Jun;5(3):258-65.

[24] Hutter G, Sinha P. Proteomics for studying cancer cells and the development of chemoresistance. *Proteomics*. 2001 Oct;1(10):1233-48.

[25] Sinha P, Poland J, Kohl S, Schnolzer M, Helmbach H, Hutter G, Lage H, Schadendorf D. Study of the development of chemoresistance in melanoma cell lines using proteome analysis. *Electrophoresis*. 2003 Jul;24(14):2386-404.

[26] Castagna A, Antonioli P, Astner H, Hamdan M, Righetti SC, Perego P, Zunino F, Righetti PG. A proteomic approach to cisplatin resistance in the cervix squamous cell carcinoma cell line A431. *Proteomics*. 2004 Oct;4(10):3246-67.

[27] Liu Y, Liu H, Han B, Zhang JT. Identification of 14-3-3sigma as a contributor to drug resistance in human breast cancer cells using functional proteomic analysis. *Cancer Res.* 2006 Mar 15;66(6):3248-55.

[28] Alaoui-Jamali MA, Dupre I, Qiang H. Prediction of drug sensitivity and drug resistance in cancer by transcriptional and proteomic profiling. *Drug Resist Updat*. 2004 Aug-Oct;7(4-5):245-55.

[29] Cho WC, Cheng CH. Oncoproteomics: current trends and future perspectives. Expert Rev Proteomics. 2007 Jun;4(3):401-10.

[30] Wingren C.First world congress meeting on OncoProteomics April 27-28, 2006, CA, USA. *Expert Rev Proteomics*. 2006 Jun;3(3):267-70.

[31] Cho WC. Contribution of oncoproteomics to cancer biomarker discovery. Mol Cancer. 2007 Apr 2;6:25. Contribution of oncoproteomics to cancer biomarker discovery. *Mol Cancer*. 2007 Apr 2;6:25.

[32] Jain KK. Role of oncoproteomics in the personalized management of cancer. *Expert Rev Proteomics*. 2004 Jun;1(1):49-55.

[33] Borrebaeck CA. Antibody microarray-based oncoproteomics. *Expert Opin Biol Ther.* 2006 Aug;6(8):833-8.

[34] Roboz J. Mass spectrometry in diagnostic oncoproteomics. *Cancer Invest.* 2005;23(5):465-78.

[35] Cho WC. Research progress in SELDI-TOF MS and its clinical applications. Sheng Wu Gong Cheng Xue Bao. 2006 Nov;22(6):871-6.

[36] Poon TC. Opportunities and limitations of SELDI-TOF-MS in biomedical research: practical advices. Expert Rev Proteomics. 2007 Feb;4(1):51-65.

[37] Henderson NA, Steele RJ. SELDI-TOF proteomic analysis and cancer detection. *Surgeon*. 2005 Dec;3(6):383-90, 422.

[38] Jain KK. Recent advances in oncoproteomics. *Curr Opin Mol Ther.* 2002 Jun;4(3): 203-9.

[39] Marsh S, McLeod HL. Pharmacogenetics and oncology treatment for breast cancer. *Expert Opin Pharmacother*. 2007 Feb;8(2):119-27.

[40] Downward J. Cancer biology: signatures guide drug choice. *Nature*. 2006 Jan 19;439(7074):274-5.

[41] Herbst RS, Lippman SM. Molecular signatures of lung cancer--toward personalized therapy. *N Engl J Med*. 2007 Jan 4;356(1):76-8.

[42] Zhang Z, Li M, Rayburn ER, Hill DL, Zhang R, Wang H. Oncogenes as novel targets for cancer therapy (part I): growth factors and protein tyrosine kinases. *Am J Pharmacogenomics*. 2005;5(3):173-90.

[43] Bild AH, Yao G, Chang JT, Wang Q, Potti A, Chasse D, Joshi MB, Harpole D, Lancaster JM, Berchuck A, Olson JA Jr, Marks JR, Dressman HK, West M, Nevins JR.

Oncogenic pathway signatures in human cancers as a guide to targeted therapies. *Nature.* 2006 Jan 19;439(7074):353-7.

[44] Xu G, Zhang W, Bertram P, Zheng XF, McLeod H. Pharmacogenomic profiling of the PI3K/PTEN-AKT-mTOR pathway in common human tumors. *Int J Oncol.* 2004 Apr;24(4):893-900.

[45] Ulrich RG. The toxicogenomics of nuclear receptor agonists. *Curr Opin Chem Biol.* 2003 Aug;7(4):505-10.

[46] Kabanov AV. Polymer genomics: an insight into pharmacology and toxicology of nanomedicines. *Adv Drug Deliv Rev.* 2006 Dec 30;58(15):1597-621.

[47] Kabanov AV, Batrakova EV, Sriadibhatla S, Yang Z, Kelly DL, Alakov VY. Polymer genomics: shifting the gene and drug delivery paradigms. *J Control Release.* 2005 Jan 3;101(1-3):259-71.

[48] Liu ET. Expression genomics and drug development: towards predictive pharmacology. Brief Funct Genomic Proteomic. 2005 Feb;3(4):303-21.

[49] Vinals C, Gaulis S, Coche T. Using in silico transcriptomics to search for tumor-associated antigens for immunotherapy. *Vaccine.* 2001 Mar 21;19(17-19):2607-14.

[50] Caron M, Choquet-Kastylevsky G, Joubert-Caron R. Cancer immunomics using autoantibody signatures for biomarker discovery. *Mol Cell Proteomics.* 2007 Jul;6(7):1115-22.

[51] Bradford TJ, Wang X, Chinnaiyan AM. Cancer immunomics: using autoantibody signatures in the early detection of prostate cancer. *Urol Oncol.* 2006 May-Jun;24(3):237-42.

[52] Ravindranath MH, Yesowitch P, Sumobay C, Morton DL. Glycoimmunomics of human cancer: current concepts and future perspectives. *Future Oncol.* 2007 Apr;3(2):201-14.

[53] Braga-Neto UM, Marques ET Jr. From functional genomics to functional immunomics: new challenges, old problems, big rewards. *PLoS Comput Biol.* 2006 Jul 28;2(7): e81.

[54] Wiwanitkit V. Finding a T-cell epitope for a melanoma vaccine by an immunomics technique. *Asian Pac J Cancer Prev.* 2006 Oct-Dec;7(4):659-60.

[55] Claudino WM, Quattrone A, Biganzoli L, Pestrin M, Bertini I, Di Leo A. Metabolomics: available results, current research projects in breast cancer, and future applications. *J Clin Oncol.* 2007 Jul 1;25(19):2840-6.

[56] Griffin JL, Kauppinen RA. A metabolomics perspective of human brain tumours. FEBS J. 2007 Mar;274(5):1132-9.

[57] Tarnok A, Bocsi J, Brockhoff G. Cytomics - importance of multimodal analysis of cell function and proliferation in oncology. *Cell Prolif.* 2006 Dec;39(6):495-505.

[58] Tarnok A, Valet GK, Emmrich F. Systems biology and clinical cytomics: The 10th Leipziger Workshop and the 3rd International Workshop on Slide-Based Cytometry, Leipzig, Germany, April 2005. *Cytometry A.* 2006 Jan;69(1):36-40.

[59] Zbuk KM, Eng C. Cancer phenomics: RET and PTEN as illustrative models. *Nat Rev Cancer.* 2007 Jan;7(1):35-45.

[60] Fernandez-Ricaud L, Warringer J, Ericson E, Glaab K, Davidsson P, Nilsson F, Kemp GJ, Nerman O, Blomberg A. PROPHECY--a yeast phenome database, update 2006. *Nucleic Acids Res.* 2007 Jan;35(Database issue):D463-7.

[61] Becker KF, Schott C, Hipp S, Metzger V, Porschewski P, Beck R, Nährig J, Becker I, Höfler H. Quantitative protein analysis from formalin-fixed tissues: implications for translational clinical research and nanoscale molecular diagnosis. *J Pathol.* 2007 Feb;211(3):370-8.

Index

A

C

D

E

F

G

H

I

M

N

O

P

R

S

T

U

V

W

X

Y

Z